Lecture Notes in Computer Science

Lecture Notes in Artificial Intelligence 15412
Founding Editor

Jörg Siekmann

Series Editors

Randy Goebel, *University of Alberta, Edmonton, Canada*
Wolfgang Wahlster, *DFKI, Berlin, Germany*
Zhi-Hua Zhou, *Nanjing University, Nanjing, China*

The series Lecture Notes in Artificial Intelligence (LNAI) was established in 1988 as a topical subseries of LNCS devoted to artificial intelligence.

The series publishes state-of-the-art research results at a high level. As with the LNCS mother series, the mission of the series is to serve the international R & D community by providing an invaluable service, mainly focused on the publication of conference and workshop proceedings and postproceedings.

Aline Paes · Filipe A. N. Verri
Editors

Intelligent Systems

34th Brazilian Conference, BRACIS 2024
Belém do Pará, Brazil, November 17–21, 2024
Proceedings, Part I

Editors
Aline Paes ⓘ
Universidade Federal Fluminense
Niterói, Brazil

Filipe A. N. Verri ⓘ
Instituto Tecnológico de Aeronáutica
São José dos Campos, Brazil

ISSN 0302-9743　　　　　　　ISSN 1611-3349　(electronic)
Lecture Notes in Artificial Intelligence
ISBN 978-3-031-79028-7　　　ISBN 978-3-031-79029-4　(eBook)
https://doi.org/10.1007/978-3-031-79029-4

LNCS Sublibrary: SL7 – Artificial Intelligence

© The Editor(s) (if applicable) and The Author(s), under exclusive license to Springer Nature Switzerland AG 2025

This work is subject to copyright. All rights are solely and exclusively licensed by the Publisher, whether the whole or part of the material is concerned, specifically the rights of translation, reprinting, reuse of illustrations, recitation, broadcasting, reproduction on microfilms or in any other physical way, and transmission or information storage and retrieval, electronic adaptation, computer software, or by similar or dissimilar methodology now known or hereafter developed.
The use of general descriptive names, registered names, trademarks, service marks, etc. in this publication does not imply, even in the absence of a specific statement, that such names are exempt from the relevant protective laws and regulations and therefore free for general use.
The publisher, the authors and the editors are safe to assume that the advice and information in this book are believed to be true and accurate at the date of publication. Neither the publisher nor the authors or the editors give a warranty, expressed or implied, with respect to the material contained herein or for any errors or omissions that may have been made. The publisher remains neutral with regard to jurisdictional claims in published maps and institutional affiliations.

This Springer imprint is published by the registered company Springer Nature Switzerland AG
The registered company address is: Gewerbestrasse 11, 6330 Cham, Switzerland

If disposing of this product, please recycle the paper.

Preface

The 34th Brazilian Conference on Intelligent Systems (BRACIS 2024) stands as one of the most significant events in Brazil in 2024 for researchers focused on publishing groundbreaking work in Artificial and Computational Intelligence. BRACIS was formed from the merger of Brazil's two leading scientific events in these fields: the Brazilian Symposium on Artificial Intelligence (SBIA, 21 editions) and the Brazilian Symposium on Neural Networks (SBRN, 12 editions). This year, the community decided to continue the numbering from the first SBIA edition to honor its history.

Supported by the Brazilian Computer Society (SBC), together with the Special Committees on Artificial Intelligence (CEIA) and Computational Intelligence (CEIC), BRACIS 2024 served as a platform to promote both theoretical advancements and practical applications in these areas. The conference promoted the exchange of innovative ideas among researchers, practitioners, and industry leaders.

In 2024, BRACIS was held in Belém do Pará, from November 17th to 21st, 2024, on the *Universidade Federal do Pará* Campus. The conference took place alongside three other events: the National Meeting on Artificial and Computational Intelligence (ENIAC), the Symposium on Information and Human Language Technology (STIL), and the Symposium on Knowledge Discovery, Mining and Learning (KDMiLe).

BRACIS 2024 received 307 paper submissions, each undergoing a rigorous review process conducted by a Program Committee of 198 experienced AI researchers, with the support of 71 additional reviewers, all within a double-anonymous format. This year, we accepted 116 (38%) of the submissions in three key tracks: 70 articles in the main track, showcasing cutting-edge AI methods and solid results; 10 articles in the AI for Social Good track, featuring innovative applications of AI for societal benefit using established methodologies; and 36 articles in other AI applications, presenting novel applications using established AI methods, naturally considering the ethical aspects of the application. In addition, seven top-tier papers published in leading AI conferences and journals were presented during the event.

The topics of interest included, but were not limited to, the following:

- Agent-Based and Multi-agent Systems
- AI for Social Good
- Bioinformatics and Biomedical Engineering using AI
- Cognitive Modeling and Human Interaction
- Combinatorial and Numerical Optimization
- Commonsense Reasoning
- Computer Vision
- Constraints and Search
- Foundations of AI
- Deep Learning and Neural Networks
- Distributed AI
- Education for AI and AI for Education

- Ethics in AI
- Evolutionary Computation and Metaheuristics
- Foundation Models
- Fuzzy Systems
- Game Playing and Intelligent Interactive Entertainment
- Generative AI
- Human-centric AI
- Information Retrieval, Integration, and Extraction
- Intelligent Robotics
- Knowledge Representation and Reasoning
- Large Language Models
- Logic-Based Knowledge Representation and Reasoning
- Machine Learning and Data Mining
- Meta-learning
- Model-Based Reasoning
- Multidisciplinary AI and CI
- Natural Language Processing
- Ontologies and the Semantic Web
- Pattern Recognition and Cluster Analysis
- Planning and Scheduling
- Probabilistic Reasoning and Approximate Reasoning
- Reinforcement Learning

We extend our heartfelt thanks to everyone who contributed to the success of BRACIS 2024. We are especially grateful to the Program Committee members and reviewers for their generous and voluntary efforts in the review process. Special thanks also go to all the authors who submitted their papers and worked tirelessly to refine them into the best possible versions. We also acknowledge the General Chairs, the Local Organization Committee for their unwavering support, the Brazilian Computing Society (SBC), and all our sponsors and supporters. We are confident that these proceedings showcase the exceptional work being done in the artificial and computational intelligence fields.

November 2024

Aline Paes
Filipe A. N. Verri

Organization

General Chairs

André Carlos Ponce de L. F. de Carvalho	Universidade de São Paulo, Brazil
Carlos Renato Lisboa Francês	Universidade Federal do Pará, Brazil
Evelin Helena Silva Cardoso	Universidade Federal do Pará, Brazil

Program Chairs

Aline Paes	Universidade Federal Fluminense, Brazil
Filipe A. N. Verri	Instituto Tecnológico da Aeronáutica, Brazil

Steering Committee (CEIA and CEIC)

Anna Helena Reali Costa	Universidade de São Paulo, Brazil
Anne Magaly de Paula Canuto	Universidade Federal do Rio Grande do Norte, Brazil
Denis Deratani Mauá	Universidade de São Paulo, Brazil
Felipe Rech Meneguzzi	University of Aberdeen, UK
Filipe De Oliveira Saraiva	Universiade Federal do Pará, Brazil
Gisele Lobo Pappa	Universidade Federal de Minas Gerais, Brazil
Leliane Nunes de Barros	Universidade de São Paulo, Brazil
Maria Viviane de Menezes	Universidade Federal do Ceará, Brazil
Murilo Coelho Naldi	Universidade Federal de São Carlos, Brazil
Renato Tinós	Universidade de São Paulo, Brazil
Ricardo M. Marcacin	Universidade de São Paulo, Brazil
Tatiane Nogueira Rios	Universidade Federal da Bahia, Brazil

Local Organizers

Jose Jailton Henrique Ferreira Junior	Universidade Federal do Pará, Brazil
Hugo Pereira Kuribayashi	Universidade do Sul e Sudeste do Pará, Brazil
Jorge Antonio Moraes de Souza	Universidade Federal Rural da Amazônia, Brazil

Program Committee

Adenilton da Silva	Universidade Federal de Pernambuco, Brazil
Adrião Duarte Dória Neto	Universidade Federal do Rio Grande do Norte, Brazil
Alex Fernandes de Souza	Universidade Federal de Itajubá, Brazil
Alexandre Ferreira	Universidade Estadual de Campinas, Brazil
Aline Neves	Universidade Federal do ABC, Brazil
Aline Paes	Universidade Federal Fluminense, Brazil
Alison R. Panisson	Universidade Federal de Santa Catarina, Brazil
Altigran Soares da Silva	Universidade Federal do Amazonas, Brazil
Alvaro Moreira	Universidade Federal do Rio Grande do Sul, Brazil
Amedeo Napoli	Centre Inria de l'Université de Lorraine, France
Ana Bazzan	Universidade Federal do Rio Grande do Sul, Brazil
Ana Carolina Lorena	Instituto Tecnológico de Aeronáutica, Brazil
Ana Cristina B. K. Vendramin	Universidade Tecnológica Federal do Paraná, Brazil
Ana Cristina Garcia	Universidade Federal do Estado do Rio de Janeiro, Brazil
André Britto	Universidade Federal do Sergipe, Brazil
André Carlos P. de L. F. de Carvalho	Universidade de São Paulo, Brazil
André Luis Debiaso Rossi	Universidade Estadual Paulista, Brazil
Andrés Eduardo Coca Salazar	Universidade Tecnológica Federal do Paraná, Brazil
Angelo Loula	Universidade Estadual de Feira de Santana, Brazil
Anna Helena Reali Costa	Universidade de São Paulo, Brazil
Anne Canuto	Universidade Federal do Rio Grande do Norte, Brazil
Antonio Parmezan	Universidade de São Paulo, Brazil
Araken Santos	Universidade Federal Rural do Semi-Árido, Brazil
Ariane Machado-Lima	Universidade de São Paulo, Brazil
Artur Jordão	Universidade de São Paulo, Brazil
Aurora Pozo	Universidade Federal do Paraná, Brazil
Bernardo Gonçalves	Universidade de São Paulo, Brazil
Bianca Zadrozny	IBM Research, Brazil
Bruno Nogueira	Universidade Federal de Mato Grosso do Sul, Brazil
Bruno Souza	Universidade Federal do Maranhão, Brazil
Carla Amor Divino M. Delgado	Universidade Federal do Rio de Janeiro, Brazil
Carlos Ferreira	Universidade Federal de Ouro Preto, Brazil

Carlos Silla	Pontifícia Universidade Católica do Paraná, Brazil
Carlos Thomaz	Centro Universitário FEI, Brazil
Carlos Henrique Ribeiro	Instituto Tecnológico de Aeronáutica, Brazil
Carolina Paula de Almeida	Universidade Estadual do Centro-Oeste, Brazil
Celia Ralha	Universidade de Brasília, Brazil
Cleber Zanchettin	Universidade Federal de Pernambuco, Brazil
Cristiano Torezzan	Universidade Estadual de Campinas, Brazil
Daniel Araújo	Universidade Federal do Rio Grande do Norte, Brazil
Daniel Oliveira Dantas	Universidade Federal de Sergipe, Brazil
Daniela Barreiro Claro	Universidade Federal da Bahia, Brazil
Danilo Sanches	Universidade Tecnológica Federal do Paraná, Brazil
Debora Medeiros	Universidade Federal do ABC, Brazil
Denis Fantinato	Universidade Estadual de Campinas, Brazil
Denis Mauá	Universidade de São Paulo, Brazil
Diana Adamatti	Universidade Federal do Rio Grande, Brazil
Diego Furtado Silva	Universidade de São Paulo, Brazil
Diego Mesquita	Fundação Getúlio Vargas, Brazil
Diego Pinheiro	Universidade de Pernambuco, Brazil
Eder Mateus Gonçalves	Universidade Federal do Rio Grande, Brazil
Edson Gomi	Universidade de São Paulo, Brazil
Edson Matsubara	Fundação Universidade Federal de Mato Grosso do Sul, Brazil
Eduardo Borges	Universidade Federal do Rio Grande, Brazil
Eduardo Goncalves	Escola Nacional de Ciências Estatísticas, Brazil
Eduardo Luz	Universidade Federal de Ouro Preto, Brazil
Eduardo Spinosa	Universidade Federal do Paraná, Brazil
Emerson Paraiso	Pontificia Universidade Católica do Paraná, Brazil
Evandro Costa	Universidade Federal de Alagoas, Brazil
Fabiano Silva	Universidade Federal do Paraná, Brazil
Fabio Faria	Universitdade Federal de São Paulo, Brazil
Fabiola Souza Fernandes Pereira	Universidade Federal de Uberlândia, Brazil
Fabrício Enembreck	Pontificia Universidade Católica do Paraná, Brazil
Fabrício A. Silva	Universidade Federal de Viçosa, Brazil
Fabricio Olivetti de França	Universidade Federal do ABC, Brazil
Fábio Cozman	Universidade de São Paulo, Brazil
Felipe Meneguzzi	University of Aberdeen, UK
Felix Antreich	Instituto Tecnológico de Aeronáutica, Brazil
Fernanda Maria da Cunha Santos	Universidade Federal de Uberlândia, Brazil
Fernando Maciano de Paula Neto	Universidade Federal de Pernambuco, Brazil
Fernando Osório	Universidade de São Paulo, Brazil

Filipe Saraiva	Universidade Federal do Pará, Brazil
Filipe Verri	Instituto Tecnológico da Aeronáutica, Brazil
Flavia Bernardini	Universidade Federal Fluminense, Brazil
Flavio Tonidandel	Centro Universitario da FEI, Brazil
Flavius Gorgônio	Universidade Federal do Rio Grande do Norte, Brazil
Flávio Soares Corrêa da Silva	Universidade de São Paulo, Brazil
Francisco De Carvalho	Universidade Federal de Pernambuco, Brazil
George D. da Cunha Cavalcanti	Universidade Federal de Pernambuco, Brazil
Geraldo Pereira Rocha Filho	Universidade Estadual do Sudoeste da Bahia, Brazil
Gerson Zaverucha	Universidade Federal do Rio de Janeiro, Brazil
Giancarlo Lucca	Universidade Católica de Pelotas, Brazil
Gisele Pappa	Universidade Federal de Minas Gerais, Brazil
Gleifer Vaz Alves	Universidade Tecnologica Federal do Paraná, Brazil
Guilherme Barreto	Universidade Federal do Ceará, Brazil
Guilherme Palermo Coelho	Universidade Estadual de Campinas, Brazil
Guilherme Derenievicz	Universidade Federal do Paraná, Brazil
Guilherme Dean Pelegrina	Universidade Presbiteriana Mackenzie, Brazil
Guillermo Simari	Universidad Nacional del Sur in Bahia Blanca, Argentina
Gustavo Giménez-Lugo	Universidade Tecnológica Federal do Paraná, Brazil
Heder Bernardino	Universidade Federal de Juiz de Fora, Brazil
Helen Senefonte	Universidade Estadual de Londrina, Brazil
Helena Caseli	Universidade Federal de São Carlos, Brazil
Helena Maia	Universidade Estadual de Campinas, Brazil
Helida Santos	Universidade Federal do Rio Grande, Brazil
Heloisa Camargo	Universidade Federal de São Carlos, Brazil
Islame Felipe da Costa Fernandes	Universidade Federal da Bahia, Brazil
Ivan Reis Filho	Universidade do Estado de Minas Gerais, Brazil
Ivette Luna	Universidade Estadual de Campinas, Brazil
Jadson Gertrudes	Universidade Federal de Ouro Preto, Brazil
Jaime Sichman	Universidade de São Paulo, Brazil
Jean Paul Barddal	Pontificia Universidade Católica do Paraná, Brazil
João Bertini	Universidade Estadual de Campinas, Brazil
João Papa	Universidade Estadual Paulista, Brazil
João C. Xavier-Júnior	Universidade Federal do Rio Grande do Norte, Brazil
João Paulo Canário	Stone Co., Brazil
Joniel Barreto	Instituto Tecnológico de Aeronáutica, Brazil

José Antonio Sanz	Universidad Pública de Navarra, Spain
José Augusto Baranauskas	Universidade de São Paulo, Brazil
Jose Eduardo Ochoa Luna	Universidad Católica San Pablo, Peru
Júlio Cesar Nievola	Pontifícia Universidade Católica do Paraná, Brazil
Julio C. S. Reis	Universidade Federal de Viçosa, Brazil
Karla Roberta Lima	Universidade de São Paulo, Brazil
Leonardo Emmendörfer	Universidade Federal de Santa Maria, Brazil
Leonardo Matos	Universidade Federal de Sergipe, Brazil
Leonardo Tomazeli Duarte	Universidade Estadual de Campinas, Brazil
Li Weigang	Universidade de Brasília, Brazil
Livy Real	B2W Digital/GLiC, Brazil
Lucelene Lopes	Universidade de São Paulo, Brazil
Luciano Digiampietri	Universidade de São Paulo, Brazil
Luiz Henrique de C. Merschmann	Universidade Federal de Lavras, Brazil
Luiza de Macedo Mourelle	Universidade do Estado do Rio de Janeiro, Brazil
Marcela Ribeiro	Universidade Federal de São Carlos, Brazil
Marcelo Bruno	Instituto Tecnológico de Aeronáutica, Brazil
Marcelo Finger	Universidade de São Paulo, Brazil
Marcilio de Souto	Université d'Orléans, France
Marco A. G. de Carvalho	Universidade Estadual de Campinas, Brazil
Marcos Domingues	Universidade Estadual de Maringá, Brazil
Marcos Quiles	Universidade Federal de São Paulo, Brazil
Marcos R. O. A. Maximo	Instituto Tecnológico de Aeronáutica, Brazil
Mariela Morveli-Espinoza	Universidade Tecnologica Federal do Paraná, Brazil
Marilton Aguiar	Universidade Federal de Pelotas, Brazil
Mariza Ferro	Universidade Federal Fluminense, Brazil
Marley M. B. R. Vellasco	Pontifícia Universidade Católica do Rio de Janeiro, Brazil
Marlo Souza	Universidade Federal da Bahia, Brazil
Marlon Mathias	Universidade de São Paulo, Brazil
Matheus Giovanni Pires	Universidade Estadual de Feira de Santana, Brazil
Mauri Ferrandin	Universidade Federal de Santa Catarina, Brazil
Murillo Carneiro	Universidade Federal de Uberlândia, Brazil
Murilo Loiola	Universidade Federal do ABC, Brazil
Murilo Naldi	Universidade Federal de São Carlos, Brazil
Myriam Delgado	Universidade Tecnológica Federal do Paraná, Brazil
Nádia Felix	Universidade Federal de Goiás, Brazil
Omar Andres Carmona Cortes	Instituto Federal do Maranhão, Brazil
Paula Costa	Universidade Estadual de Campinas, Brazil
Paulo Gabriel	Universidade Federal de Uberlândia, Brazil

Paulo Mann	Universidade do Estado do Rio de Janeiro, Brazil
Paulo Pirozelli	Universidade de São Paulo, Brazil
Paulo Quaresma	Universidade de Évora, Portugal
Paulo Henrique Pisani	Universidade Federal do ABC, Brazil
Paulo T. Guerra	Universidade Federal do Ceará, Brazil
Pedro Silva	Universidade Federal de Ouro Preto, Brazil
Petrucio Viana	Universidade Federal Fluminense, Brazil
Priscila Lima	Universidade Federal do Rio de Janeiro, Brazil
Rafael H. Bordini	Pontifícia Universidade Católica do Rio Grande do Sul, Brazil
Rafael Giusti	Universidade Federal do Amazonas, Brazil
Reinaldo Bianchi	Centro Universitario FEI, Brazil
Renato Tinos	Universidade de São Paulo, Brazil
Ricardo Cerri	Universidade de São Paulo, Brazil
Ricardo Dutra da Silva	Universidade Tecnológica Federal do Paraná, Brazil
Ricardo Marcacini	Universidade de São Paulo, Brazil
Ricardo Prudêncio	Universidade Federal de Pernambuco, Brazil
Ricardo Rios	Universidade Federal da Bahia, Brazil
Ricardo Suyama	Universidade Federal do ABC, Brazil
Ricardo Augusto Souza Fernandes	Universidade Federal de São Carlos, Brazil
Roberto Fray da Silva	Universidade de São Paulo, Brazil
Roberto Douglas G. de Aquino	Instituto Tecnológico de Aeronáutica, Brazil
Robespierre Pita	Universidade Federal da Bahia, Brazil
Rodrigo Lira	Instituto Federal de Pernambuco, Brazil
Rodrigo Silva	Universidade Federal de Ouro Preto, Brazil
Rodrigo Veras	Universidade Federal do Piauí, Brazil
Romis Attux	Universidade Estadual de Campinas, Brazil
Ronaldo Prati	Universidade Federal do ABC, Brazil
Roseli Ap. Francelin Romero	Universidade de São Paulo, Brazil
Rosiane de Freitas Rodrigues	Universidade Federal do Amazonas, Brazil
Rui Camacho	Universidade do Porto, Portugal
Sandro Fiorini	IBM Research, Brazil
Sarajane Marques Peres	Universidade de São Paulo, Brazil
Sílvia Maia	Universidade Federal do Rio Grande do Norte, Brazil
Sílvio César Cazella	Universidade Federal de Ciências da Saúde de Porto Alegre, Brazil
Silvia Botelho	Universidade Federal do Rio Grande, Brazil
Solange Rezende	Universidade de São Paulo, Brazil
Talles Medeiros	Universidade Federal de Ouro Preto, Brazil
Tatiane Nogueira	Universidade Federal da Bahia, Brazil

Thiago Covoes — Thomson Reuters Labs, USA
Thiago Pardo — Universidade de São Paulo, Brazil
Tiago Almeida — Universidade Federal de São Carlos, Brazil
Tiago Tavares — Instituto de Ensino e Pesquisa Insper, Brazil
Valéria Santos — Universidade Federal de Ouro Preto, Brazil
Valmir Macario — Universidade Federal Rural de Pernambuco, Brazil
Vander Freitas — Universidade Federal de Ouro Preto, Brazil
Vasco Furtado — Universidade de Fortaleza, Brazil
Vinicius de Carvalho — Universidade de São Paulo, Brazil
Vinicius Souza — Pontifícia Universidade Católica do Paraná, Brazil
Vitor Curtis — Instituto Tecnológico de Aeronáutica, Brazil
Viviane Moreira — Universidade Federal do Rio Grande do Sul, Brazil
Viviane Torres da Silva — IBM Research, Brazil
Wagner Meira Jr. — Universidade Federal de Minas Gerais, Brazil

Additional Reviewers

Alessandra Aparecida Paulino
Alexandre Alcoforado
Aline Del Valle
Amanda Perez
Anderson Moraes
André De Oliveira
Annie Amorim
Arthur Torres
Arthur Valencio
Astrid Wiens
Bernardo Scapini Consoli
Caroline Pires Alavez Moraes
Débora Engelmann
Diedre Santos
Diego Minatel
Eduardo Silva
Elaine Gatto
Eliezer de Souza da Silva
Elineide Silva dos Santos
Eliton Perin
Estela Ribeiro
Eulanda Miranda dos Santos
Fabian Cardoso
Fabiano Silva
Felipe Serras
Filipe Cordeiro
Gabriel de Castro Michelassi
Gustavo Henrique do Nascimento
Hermon Faria Araújo
Israel Fama
Ivandro Sanches
João Barbon
Johny Moreira
Jose Jailton Henrique Ferreira Junior
Juan Colonna
Justino Santos
Karliane Vale
Keylly Santos
Leonardo da Silva Costa
Leonardo Sousa
Letícia Freire
Lucas Alegre
Lucas dos Santos
Lucas Rodrigues
Lucas Santos
Luciana Bencke
Luis Vogado
Luiz Felipe Vercosa

Miguel de Mello Carpi
Napoleao Galegale
Natanael Batista
Patrick Ferreira
Patrick Terrematte
Paulo Girardi
Rafael Berri
Renata Wassermann
Rodrigo Maia
Sarah Negreiros
Selma Carloto
Silvio Romero de Araujo Junior
Tamara Pereira
Thais Luca
Thiago Carvalho
Thiago Freire de Oliveira
Thiago Miranda
Thomas Palmeira Ferraz
Tiago Botelho
Tiago Pinho da Silva
Vítor Lourenço
Wandry Faria
Wesley Seidel

Contents – Part I

Main Track

A Contrastive Objective for Training Continuous Generative Flow Networks 3
Tiago da Silva and Diego Mesquita

A Data Distribution-Based Ensemble Generation Applied to Wind Speed Forecasting 17
Diogo M. Almeida, Paulo S. G. de Mattos Neto, and Daniel C. Cunha

A Large Dataset of Spontaneous Speech with the Accent Spoken in São Paulo for Automatic Speech Recognition Evaluation 33
Rodrigo Lima, Sidney E. Leal, Arnaldo Candido Junior, and Sandra M. Aluísio

A Multi-level Semantics Formalism for Multi-Agent Microservices 48
Alison R. Panisson and Giovani P. Farias

A Novel Genetic Algorithm Approach for Discriminative Subspace Optimization 64
Bernardo B. Gatto, Marco A. F. Mollinetti, Eulanda M. dos Santos, Alessandro L. Koerich, and Waldir S. da Silva Junior

A Performance Increment Strategy for Semantic Segmentation of Low-Resolution Images from Damaged Roads 80
Rafael S. Toledo, Cristiano S. Oliveira, Vitor H. T. Oliveira, Eric A. Antonelo, and Aldo von Wangenheim

A Unified Framework for Average Reward Criterion and Risk 96
Willy Arthur Silva Reis, Karina Valdivia Delgado, and Valdinei Freire

Adaptive Client-Dropping in Federated Learning: Preserving Data Integrity in Medical Domains 111
Arthur Negrão, Guilherme Silva, Rodrigo Pedrosa, Eduardo Luz, and Pedro Silva

An Ensemble of LLMs Finetuned with LoRA for NER in Portuguese
Legal Documents 127
  *Rafael Oleques Nunes, Letícia Maria Puttlitz, Antonio Oss Boll,
  Andre Spritzer, Carla Maria Dal Sasso Freitas,
  Dennis Giovani Balreira, and Anderson Rocha Tavares*

An Instance Level Analysis of Classification Difficulty for Unlabeled Data 141
  Patricia S. M. Ueda, Adriano Rivolli, and Ana Carolina Lorena

Analyzing the Impact of Coarsening on k-Partite Network Classification 156
  *Thiago de Paulo Faleiros, Paulo Eduardo Althoff,
  and Alan Demétrius Baria Valejo*

Applying Transformers for Anomaly Detection in Bus Trajectories 169
  Michael Cruz and Luciano Barbosa

Aroeira: A Curated Corpus for the Portuguese Language with a Large
Number of Tokens 185
  *Thiago Lira, Flávio Cação, Cinthia Souza, João Valentini,
  Edson Bollis, Otavio Oliveira, Renato Almeida, Marcio Magalhães,
  Katia Poloni, Andre Oliveira, and Lucas Pellicer*

Assessing Adversarial Effects of Noise in Missing Data Imputation 200
  *Arthur Dantas Mangussi, Ricardo Cardoso Pereira,
  Pedro Henriques Abreu, and Ana Carolina Lorena*

Assessing European and Brazilian Portuguese LLMs for NER
in Specialised Domains 215
  *Rafael Oleques Nunes, Joaquim Santos, Andre Spritzer,
  Dennis Giovani Balreira, Carla Maria Dal Sasso Freitas,
  Fernanda Olival, Helena Freire Cameron, and Renata Vieira*

BASWE: Balanced Accuracy-Based Sliding Window Ensemble
for Classification in Imbalanced Data Streams with Concept Drift 231
  *Douglas Amorim de Oliveira, Karina Valdivia Delgado,
  and Marcelo de Souza Lauretto*

Beyond Audio Signals: Generative Model-Based Speaker Diarization
in Portuguese 247
  *Antônio Oss Boll, Letícia Maria Puttlitz, Heloísa Oss Boll,
  and Rodrigo Mor Malossi*

Classification of Non-alcoholic Fatty Liver Disease in Thermal Images
of the Liver Using a Siamese Neural Network 260
    *Maxwell Pires Silva, Aristófanes Corrêa Silva,
    and Anselmo Cardoso de Paiva*

Classifying Graphs of Elementary Mathematical Functions Using
Convolutional Neural Networks 270
    Joaquim Viana, Helder Matos, Marcelle Mota, and Reginaldo Santos

Comparing Neural Network Encodings for Logic-Based Explainability 281
    Levi Cordeiro Carvalho, Saulo A. F. Oliveira, and Thiago Alves Rocha

Deep Learning Approach to Temporal Dimensionality Reduction
of Volumetric Computed Tomography 296
    Lucas Almeida da Silva, Eulanda Miranda dos Santos, and Rafael Giusti

Deployment of IBM Federated Learning Platform and Aggregation
Algorithm Comparison: A Case Study Using the MNIST Dataset 310
    Hans Herbert Schulz and Benjamin Grando Moreira

Detection of Pathological Regions of the Gastrointestinal Tract in Capsule
Images Using EfficientNetV2 and YOLOv8 324
    *Anderson Lopes Silva, Hellen Guterres França,
    Carlos Mendes dos Santos Neto, Alexandre César Pinto Pessoa,
    Darlan Bruno Pontes Quintanilha, Aristófanes Corrêa Silva,
    and Anselmo Cardoso de Paiva*

Dual-Bandwidth Spectrogram Analysis for Speaker Verification 340
    *Rafaello Virgilli, Arnaldo Candido Junior, Augusto Seben da Rosa,
    Frederico S. Oliveira, and Anderson da Silva Soares*

Dynamicity Analysis in the Selection of Classifier Ensembles Parameters 352
    *Jesaías Carvalho Pereira Silva, Anne Magaly de Paula Canuto,
    and Araken de Medeiros Santos*

Embedding Representations for AutoML Pipelines 368
    Camila Santana Braz, Matheus Cândido Teixeira, and Gisele Lobo Pappa

Enhancing Graph Data Quality by Leveraging Heterogeneous Node
Features and Embeddings 383
    Silvio Fernando Angonese and Renata Galante

Ensemble of CNNs for Enhanced Leukocyte Classification in Acute
Myeloid Leukemia Diagnosis 399
    *Leonardo P. Sousa, Romuere R. V. Silva, Maíla L. Claro,
Flávio H. D. Araújo, Rodrigo N. Borges, Vinicius P. Machado,
and Rodrigo M. S. Veras*

ERASMO: Leveraging Large Language Models for Enhanced
Clustering Segmentation 414
    *Fillipe dos Santos Silva, Gabriel Kenzo Kakimoto,
Julio Cesar dos Reis, and Marcelo S. Reis*

Euclidean Alignment for Transfer Learning in Multi-band Common
Spatial Pattern 430
    *Marcelo M. Amorim, Leonardo Prata, João Stephan Maurício,
Alex Borges, Heder Bernardino, and Gabriel de Souza*

Evaluating CNN-Based Classification Models Combined
with the Smoothed Pseudo Wigner-Ville Distribution to Identify
Low Probability of Interception Radar Signals 444
    Edgard B. Alves, Jorge A. Alves, and Ronaldo R. Goldschmidt

Evaluating Large Language Models for Tax Law Reasoning 460
    *João Paulo Cavalcante Presa, Celso Gonçalves Camilo Junior,
and Sávio Salvarino Teles de Oliveira*

Explaining Biomarker Response to Anticoagulant Therapy in Atrial
Fibrillation: A Study of Warfarin and Rivaroxaban with Machine Learning
Models 475
    Adriano Veloso, Gianlucca Zuin, and Luan Sena

Author Index 489

Main Track

A Contrastive Objective for Training Continuous Generative Flow Networks

Tiago da Silva and Diego Mesquita[✉]

Getulio Vargas Foundation, Rio de Janeiro, Brazil
{tiago.henrique,diego.mesquita}@fgv.br

Abstract. Generative Flow Networks (GFlowNets) are a novel class of flexible amortized samplers for distributions supported on complex objects (e.g., graphs and sequences), achieving significant success in problems such as combinatorial optimization, drug discovery and natural language processing. Nonetheless, training of GFlowNets is challenging—partly because it relies on estimating high-dimensional integrals, including the log-partition function, via stochastic gradient descent (SGD). In particular, for distributions supported on non-discrete spaces, which have received far less attention from the recent literature, every previously proposed learning objective either depends on estimating a log-partition function or is restricted to on-policy training, which is susceptible to mode-collapse. In this context, inspired by the success of contrastive learning for variational inference, we propose the *continuous contrastive loss* (CCL) as the first objective function natively enabling off-policy training of continuous GFlowNets without reliance on the approximation of high-dimensional integrals via SGD, extending previous work based on discrete distributions. Additionally, we show that minimizing the CCL objective is empirically effective and often leads to faster training convergence than alternatives.

Keywords: GFlowNets · Diffusion models · Bayesian inference

1 Introduction

The evaluation of high-dimensional expectations under unnormalized distributions is one of the main challenges of statistics and machine learning [4]. In Bayesian decision theory, for instance, one seeks to minimize a risk function that is defined as an expectation of a functional under a often intractable and unnormalized posterior distribution provisioned by Bayes' rule [21]. Similarly, many model-free reinforcement learning algorithms for searching an optimal policy in a contextual Markov decision process rely on integrating out a high-dimensional context variable of the quality function. Under these circumstances, a common approach consists of substituting the unmanageable integration problem by a manageable optimization problem, which is known as *variational inference*. For this, one introduces a tractable and large set of parametric distributions, known as the *variational family*, and finds the family's member that most closely matches the unnormalized target by solving a stochastic optimization problem. Then, one uses the solution to this problem as a surrogate to the intractable target and evaluate the desired expectations using Monte Carlo estimators.

Recently, Generative Flow Networks (GFlowNets) [2,3,13] were proposed as a flexible variational family parameterized by neural networks to approximate unnormalized distributions with compositional supports [15], with remarkable success in relevant applications including causal inference [5,24], drug discovery [2], refinement of large language models [9], and combinatorial optimization [28]. To implement a GFlowNet, we first define a flow network over an extension of the target distribution's support, which thereafter corresponds to the sink nodes, and then search for a consistent flow assignment for which the flow going through each sink node matches its unnormalized probability. During inference, we travel across the defined network by starting at the origin and choosing each transition according to the estimated flow therein. If the flow is correctly estimated, this procedure is guaranteed to yield independent and identically distributed (i.i.d.) samples from the intractable target distribution [2,13].

To successfully find a consistent and balanced flow assignment, we conventionally parameterize the network's transition policies with versatile models as, e.g., multilayer perceptrons or graph neural networks. Then, to estimate the model's parameters, we minimize a loss function consisting of the expectation of log-squared local balance violations according to a prescribed *exploratory policy* via a variant of SGD. Importantly, this exploratory policy may be different of the learned policy, in which case the training is said to be carried in an *off-policy* fashion. We remark that, when dealing with high-dimensional and highly sparse target distributions, an off-policy sampling scheme is paramount for avoiding the collapse of the variational distribution [14,15]. Nonetheless, as we note in Sect. 2, such a possibility of off-policy learning comes at the cost of introducing additional parameters that often entail the estimation of high-dimensional integrals as a subproblem to the successful training of GFlowNets, potentially hindering the sample-efficiency of these models.

Seeking to bypass this problem, recent work [23,26] developed a contrastive learning objective for GFlowNets. In the spirit of Hinton's contrastive divergence learning [8], these approaches sidestep the estimation of difficult-to-approximate quantities by, instead of directly optimizing an optimality criteria, minimizing the difference between them. This ensures that the intractable terms cancel out and that the resulting problem is computationally easier to solve. Remarkably, however, these works are constrained to sampling from unnormalized distributions with finite support; a formal extension and an empirical validation of these objectives to the non-discrete setting is lacking in the literature. In this context, we extend both the contrastive balance condition and the contrastive loss function [23] to train GFlowNets defined on measurable topological spaces [13]. In addition, we empirically show that the resulting *continuous contrastive loss* (CCL) frequently leads to faster training convergence with respect to competing learning objectives for the standard tasks of approximating a sparse mixture of multivariate Gaussian distributions and a banana-shaped distribution.

In summary, our contributions are:

1. We derive a contrastive balance condition for continuous GFlowNets and rigorously show that it is a sufficient for ensuring sampling correctness;
2. We propose the *continuous contrastive loss* (CCL) as a theoretically sound learning objective for GFlowNets that does not rely on the estimation of high-dimensional integrals via SGD during training;
3. We empirically compare the performance of CCL against alternative loss functions and show that CCL leads to faster convergence in most cases.

The paper is organized as follows. In Sect. 2, we recall the notions of a measurable pointed directed acyclic graph (DAG), transition kernels and flow networks, laying out the theoretical framework upon which GFlowNets are built, and also review popular learning objectives for continuous GFlowNets. Then, in Sect. 3, we rigorously derive the contrastive balance condition and propose CCL as a sound learning objective for GFlowNets, the performance of which is assessed in Sect. 4 for the approximation of a mixture of multivariate Gaussians and of a banana-shaped distributions [16]. Finally, we provide a summary of the relevant literature in Sect. 5 and a discussion of future directions, along with some concluding remarks, in Sect. 6.

2 Background

This section reviews the main components of and the commonly used learning objectives for GFlowNets. To build intuition, we firstly define a GFlowNet for finitely supported distributions in Sect. 2.1. Then, we review the formal framework of Lahlou et al. [13] for extending the notion of a flow networks to arbitrary topological spaces by substituting an adjacency matrix by a *transition kernel* that describes the continuous network's connectivity.

2.1 GFlowNets in Discrete Spaces

Discrete GFlowNets. Let R be a (potentially unnormalized) probability distribution on a finite space $\mathcal{X}$ and $S \supseteq \mathcal{X}$ be an extension of $\mathcal{X}$ with two distinguished elements, s_o and s_f. Our objective is to randomly sample elements $x \in \mathcal{X}$ proportionally to $R(x)$. For this, we let $G = (S, E)$ be a DAG with edges E such that (i) the only nodes connected to s_f are those in $\mathcal{X}$, (ii) there is a path from s_o to every $x \in \mathcal{X}$, (iii) there are no outgoing edges from s_f and there are no incoming edges to s_o. We call $\mathcal{G}$ the *state graph*. In this setting, let $p_F \colon S \times S \to \mathbb{R}_+$ be a *forward policy* on $\mathcal{G}$, i.e., a function such that $p_F(s, \cdot)$ is a probability measure on S supported on the children of s in $\mathcal{G}$ and $p_F(\tau) = \prod_{1 \le i \le n} p_F(s_i | s_{i-1})$ be the corresponding distribution over trajectories $\tau = (s_o, \ldots, s_{n-1}, s_n = s_f)$. To accomplish our objective, we must find a p_F for which

$$\sum_{\tau \colon \tau \rightsquigarrow x} p_F(\tau) \propto R(x). \tag{1}$$

In practice, the left-hand-side of Eq. (1) is computationally intractable and is satisfied by a possibly infinite number of p_F's. To ensure tractability and uniqueness, we also introduce a *backward policy* $p_B \colon S \times S \to \mathbb{R}_+$, which is a forward policy over the transpose of $\mathcal{G}$, and rewrite Eq. (1) as $p_F(\tau) \propto p_B(\tau|x) R(x)$. Malkin et al. [14] showed that such technique induces a well-posed problem with a unique solution.

Training GFlowNets. To find a forward policy abiding by Eq. (1), we parameterize p_F with a neural network with parameters θ and fix $p_B(s, \cdot)$ as an uniform policy for each $s \in S$. Then, by introducing an additional learnable function $F \colon S \to \mathbb{R}_+$, which is also parameterized by a (different) neural network γ, we concomitantly estimate θ and γ by minimizing one of the following objectives. Below, p_e is an *exploratory policy*, i.e., a forward policy that does not depend on (θ, γ) and has full support over the space of trajectories. Also, we denote by x (resp. x') the state immediately preceding s_f in a trajectory τ (resp. τ').

1. The *detailed balance* (DB) loss [3] corresponds to $\mathcal{L}_{DB}(\theta, \gamma) =$

$$\mathbb{E}_{\tau \sim p_e} \left[\frac{1}{|\tau|} \sum_{(s_{i-1}, s_i) \in \tau} \left(\log \frac{F_\gamma(s_{i-1}) p_F(s_i|s_{i-1})}{p_B(s_i|s_{i-1}) F_\gamma(s_{i-1})} \right)^2 + \left(\log \frac{F_\gamma(x)}{R(x)} \right)^2 \right]$$

which enforces the DB condition, $F(s) p_F(s|s') = F(s') p_B(s|s')$ for $(s, s') \in E$ and $F(x) = R(x)$ for the states $x \in S$.

2. The trajectory balance (TB) loss [14] introduces $\log Z(\gamma) = \log F_\gamma(s_o)$ as the log-partition function of R and is then defined by

$$\mathcal{L}_{TB}(\theta, \gamma) = \mathbb{E}_{\tau \sim p_e} \left[(\log Z(\gamma) + \log p_F(\tau; \theta) - \log p_B(\tau|x) - \log R(x))^2 \right],$$

enforcing the TB condition $F(s_o) p_F(\tau) = p_B(\tau|x) R(x)$.

3. The *contrastive balance* (CB) loss [23] avoids parameterizing F by sampling a pair (τ, τ') of trajectories from p_e, being defined by

$$\mathcal{L}_{CB}(\theta) = \mathbb{E}_{(\tau, \tau') \sim p_e} \left[\left(\log \frac{p_F(\tau)}{p_B(\tau|x) R(x)} - \log \frac{p_F(\tau')}{p_B(\tau'|x') R(x')} \right)^2 \right], \quad (2)$$

and ensuring the CB condition, $p_F(\tau) p_B(\tau'|x') R(x') = R(x) p_F(\tau') p_B(\tau|x)$.

Notably, the learning of discrete GFlowNets is widely studied, and many strategies for enhancing training convergence were proposed [3,11,17,18,22,25,28,30]. In this work, we focus on extending the CB objective to the continuous setting.

2.2 GFlowNets in Continuous Spaces

Notations. We let $(\mathcal{X}, T_\mathcal{X}, \Sigma_\mathcal{X}, \mu)$ be a topological measurable space with reference measure μ, topology $T_\mathcal{X}$, σ-algebra $\Sigma_\mathcal{X}$, and underlying space $\mathcal{X}$, and consider a non-negative function $r \colon \mathcal{X} \to \mathbb{R}_+$ over $\mathcal{X}$ such that $\int_\mathcal{X} r(x) \mu(dx) < \infty$.

In what follows, we will assume that each measure, denoted by a capital letter, has a density relatively to the reference measure μ, which will be denoted by the corresponding lowercase letter. Under these conditions, our objective is to learn a generative process over $\mathcal{X}$ that samples each measurable subset A proportionally to $\int_A r(x)\mu(dx)$, and we define the target measure $R\colon \Sigma_{\mathcal{X}} \to \mathbb{R}_+$ as $R(A) = \int_A r(x)\mu(dx)$, which is absolutely continuous with respect to μ. Additionally, for a given set S and a corresponding σ-algebra Σ_S (which will be specified later), we define a *Markov kernel* on S as a function $k\colon S \times \Sigma_S \to \mathbb{R}_+$ such that $k(s,\cdot)$ is a measure in (S, Σ_S). Also, we denote by $S^{\otimes n}$ the nth-order Cartesian product of S and by $\Sigma_S^{\otimes n}$ the product σ-algebra of Σ_S. Naturally, a transition kernel k on S is naturally extended to a transition kernel $\kappa^{\otimes n}\colon S \times \Sigma_S^{\otimes n} \to \mathbb{R}_+$ on $S^{\otimes n}$ recursively via $\kappa^{\otimes n}(s, (A, A_n)) = \kappa^{\otimes (n-1)}(s, A)\kappa(s, A_n)$, with $A \in \Sigma_S^{\otimes(n-1)}$, $A_n \in \Sigma_S$, $s \in S$, which exists uniquely due to Carathéodory's extension theorem. Finally, we denote by $S^{\otimes \leq N} = \bigcup_{1 \leq i \leq N} S^{\otimes i}$ and $\Sigma_S^{\otimes \leq N} = \sigma\left(S^{\otimes \leq N}\right)$ the σ-algebra generated by the space of up-to-Nth order product σ-algebras.

Measurable Pointed DAGs (MP-DAG). We first present the definition of an MP-DAG, originally proposed by Lahlou et al. [13], which formally extends the concept of a flow network to a potentially non-countable state space.

Definition 1 (MP-DAG). *Let $(\bar{S}, \mathcal{T})$ be a topological space and Σ be the Borel σ-algebra on $\mathcal{T}$. Define the* source $s_o \in \bar{S}$ *and* sink $s_f \in \bar{S}$ *states. Also, let κ, the* reference kernel, *and κ_b, the* backward kernel, *be σ-finite and continuous Markov kernels in $(\bar{S}, \mathcal{T})$; and let μ be a reference measure on Σ. Also, let $S = \bar{S} \setminus \{s_f\}$. A measurable pointed DAG is a tuple $\mathcal{G} = (\bar{S}, \mathcal{T}, \Sigma, \kappa, \kappa_b, \mu)$ such that*

1. *(**finality**) $\kappa(s_f, \cdot) = \delta_{s_f}$ and if $\kappa(s, \{s_f\}) > 0$ for a $s \in \bar{S}$, $\kappa(s, \{s_f\}) = 1$;*
2. *(**reachability**) $\forall B \in \mathcal{T} \setminus \{\emptyset\}$, $\exists n \geq 0$ with $\kappa^{\otimes n}(s_o, B) > 0$;*
3. *(**initialness**) $\forall B \in \Sigma$, $\kappa_b(s_o, B) = 0$; and*
4. *(**consistency**) $\forall B \in \Sigma \times \Sigma$, $\nu \otimes \kappa(B) = \nu \otimes \kappa_b(B)$.*

Also, a MP-DAG is said to be *finitely absorbing* if there is a $N \in \mathbb{N}$ for which $\kappa^{\otimes N}(s_o, \{s_f\}) > 0$ and, by the *finality* property, $\kappa^{\otimes N}(s_o, \{s_f\}) = 1$. In other words, any iterative generative process starting at s_o and sampling novel states according to κ will reach s_f in at most N steps. We call N the *maximum trajectory length* of the MP-DAG, and define $\mathcal{X} = \{x \in S\colon \kappa(x, \{s_f\}) > 0\}$ as the set of *terminal states*. Under these circumstances, the set $\{s_o\} \times S^{\otimes \leq N-1} \times \{s_f\}$ fully characterizes the trajectories starting at s_o, and we will denote it by $\mathbb{T}$; also, we denote by $\mathbb{T}' = \{s_o\} \times (S \setminus \mathcal{X})^{\otimes \leq N-1}$ the space of trajectories without terminal states. We highlight that, when S is finite and μ is set as the counting measure in S, then the definition above recovers the traditional flow network outlined in Sect. 2.1 for the discrete setting, with $\{\kappa(s, \{s'\})\}_{(s,s') \in S \times S}$ characterizing the network's adjacency matrix and N denoting its (directed) diameter.

Continuous GFlowNets. Similarly to its discrete counterpart, we characterize a continuous GFlowNet by a MP-DAG $\mathcal{G} = (\bar{S}, \mathcal{T}, \Sigma, \kappa, \kappa_b, \mu)$, a measure R on $\mathcal{X}$, a *forward policy* $P_F\colon \bar{S} \times \Sigma_{\bar{S}} \to \mathbb{R}_+$, defined as a Markov kernel that is absolutely continuous relatively to κ, and a *backward policy* $P_B \colon \bar{S} \times \Sigma_{\bar{S}} \to \mathbb{R}_+$, defined as a Markov kernel absolutely continuous relatively to κ_b. Our objective is to find a P_F such that the marginal distribution over $\mathcal{X}$ of the κ-Markov chain deterministically starting at s_o matches R, that is, informally,

$$\int_{\mathbb{T}} \mathbb{1}_{\{\tau \text{ finishes at } A\}} P_F^{\otimes N}(s_o, \mathrm{d}\tau) \propto R(A) \tag{3}$$

for every measurable subset A of $\mathcal{X}$.

Training GFlowNets. To achieve this objective, we let p_F and p_B be the densities of P_F and P_B relatively to κ and κ_b, respectively, and parameterize p_F as a neural network θ. To avoid notational overload, we also denote by p_F and p_B the densities of $P_F^{\otimes i}$ and $P_B^{\otimes i}$ with respect to the corresponding product kernels. Analogously to the discrete GFlowNets, we define an auxiliary function $u \colon \bar{S} \to \mathbb{R}_+$ parameterized by a neural network γ. Additionally, we introduce an exploratory policy $P_E \colon \bar{S} \times \Sigma_{\bar{S}} \to \mathbb{R}_+$ absolutely continuous with respect to the forward kernel κ. Under these conditions, Lahlou et al. [13] showed that the trajectory and detailed balance losses of Sect. 2.1 are sound objectives for learning continuous GFlowNets when we substitute the transition kernels by the corresponding densities. So, for instance, we let $Z = u_\gamma(s_o)$ and then

$$\mathcal{L}_{TB}(\theta, \gamma) = \mathbb{E}_{\tau \sim P_E(s_o, \cdot)} \left[\left(\log Z(\gamma) + \log p_F(\tau) - \log p_B(\tau|x) - \log r(x) \right)^2 \right] \tag{4}$$

becomes the continuous equivalent of the TB loss. Remarkably, both the TB and DB losses rely on the estimation of the auxiliary function u, which often corresponds to a high-dimensional integral (such as the partition function) that is difficult to compute and has no inferential purpose.

Illustration. To clarify the above definitions and emphasize their versatility, we show how to instantiate a continuous GFlowNet to approximate an arbitrary distribution in an Euclidean space $\mathbb{R}^n$, an example that will be thoroughly explored in our experimental campaign in Sect. 4. For this setting, we may define $S = \mathbb{R}^n$ and $\mathcal{X} = \{(x, \top) \colon x \in \mathbb{R}^n\}$, with $\top$ artificially distinguishing the elements of $\mathcal{X}$ from those of S. Then, $s_o = \mathbf{0}$ and a transition corresponds to $x^{t+1} = x^t + \alpha e^t$, with $\alpha \sim Q_\theta$ sampled from an appropriate distribution Q_θ with parameters estimated by a neural network that receives x^t as input and e^t denoting the tth line of the identity matrix. Critically, this iterative generative process induces a finitely absorbing MP-DAG satisfying all the desired properties when we define κ to satisfy $\kappa(x, \cdot) = \delta_{s_f}(\cdot)$ for every $x \in \mathcal{X}$ and $\kappa_b(x^t, \cdot) = \delta_{x^t - x_t^t e_t}(\cdot)$.

3 A Contrastive Objective

We now present our main theoretical results regarding the soundness of the contrastive balance objective for training GFlowNets in a non-countable state space. Firstly, Sect. 3.1 defines the *CB condition* as a provably sufficient requirement for ensuring the sampling correctness of GFlowNets. Secondly, Sect. 3.2 derives the *contrastive continuous loss* (CCL) as a realization of the CB condition that, when minimized, ensures that the resulting model is correct almost everywhere. Finally, we analyze the connection between the CCL and the popular variance loss [20] for carrying out generic VI.

3.1 Contrastive Balance

Contrastive Balance. To get a finer understanding of the contrastive balance condition, we first remark that the TB condition may be equivalently stated as

$$\text{Var}\left[\frac{p_F(\tau)}{p_B(\tau|x)r(x)}\right] = 0, \tag{5}$$

i.e., the quotient $\beta(\tau) = p_F(\tau)/p_B(\tau|x)r(x)$ is constant as a function of τ. Thus, if $\beta(\tau) = \beta(\tau')$ for any pair of trajectories (τ, τ'), the TB condition must be satisfied and the GFlowNet should sample correctly from the target. Importantly, however, the β does not depend on the estimation of intractable quantities and enforcing the equality $\beta(\tau) = \beta(\tau')$, which is the essence of Definition 2 below, avoids the estimation of high-dimensional integrals such as $\log Z(\gamma)$ in Eq. (4). Nextly, we formally define the contrastive balance objective. Notably, we slightly abuse notation by denoting $P_B(x, d\tau)$ for the product kernel $P_B(x, ds_n) \otimes P_B(s_n, ds_{n-1}) \otimes \cdots \otimes P_B(s_1, ds_o)$ associated to $\tau = (s_o, s_1, \ldots, s_n, x)$.

Definition 2 (Contrastive balance). *Let $(\mathcal{G}, P_F, P_B, R, \mu)$ be a GFlowNet. The* contrastive balance *(CB) condition is defined by*

$$\int_{\mathcal{X}} \int_{\mathcal{X}} \int_{\mathbb{T}'} \int_{\mathbb{T}'} f(\tau, x, \tau', x') P_F(s_o, d\tau dx) P_B(x', d\tau') R(dx') \\ = \int_{\mathcal{X}} \int_{\mathcal{X}} \int_{\mathbb{T}'} \int_{\mathbb{T}'} f(\tau, x, \tau', x') R(dx) P_B(x, d\tau) P_F(s_o, d\tau' dx') \tag{6}$$

for every bounded measurable function $f \colon \mathbb{T}' \times \mathcal{X} \times \mathbb{T}' \times \mathcal{X} \to \mathbb{R}$.

Illustratively, when S is finite, the reference measure is the counting measure, and we let f represent the indicator functions of elements in the space $\mathbb{T}' \times \mathcal{X} \times \mathbb{T}' \times \mathcal{X}$, the condition above reduces to the discrete contrastive balance condition, namely, $p_F(\tau, x) p_B(x', \tau') r(x') = r(x) p_B(x, \tau) p_F(\tau', x')$ for every pair (τ, τ') of trajectories and (x, x') of terminal states. In the light of this, Definition 2 generalizes the discrete CB condition to a significantly broader setting.

Sufficiency of CB. We rigorously show in the theorem below that the continuous variant of the CB condition outlined in Definition 2 is sufficient for ensuring that the corresponding GFlowNet generates samples distributed according to the target measure R. Note, for this, that we may rewrite the Eq. (6) in term of the densities of the corresponding measures.

Theorem 1. *Let $(\mathcal{G}, P_F, P_B, R, \mu)$ be a GFlowNet abiding by the contrastive balance condition. Then, the marginal distribution over $\mathcal{X}$ induced by the P_F-Markov chain starting at s_o matches R.*

Proof. To start with, we define $t \colon \mathbb{T} \to \mathcal{X}$ as the function taking a complete trajectory and returning its terminal state. By the definition of $\mathcal{X}$, t is well-defined almost everywhere with respect to the measure $\kappa(s_o, \cdot)$ over trajectories. Then, we show that Eq. (6) is enough to ensure, for all measurable $A \subseteq \mathcal{X}$, that

$$\int_{\mathbb{T}} \mathbb{1}_{\{t(\tau) \in A\}} P_F(s_o, \mathrm{d}\tau) \propto R(A), \tag{7}$$

i.e., the marginal distribution of P_F on $\mathcal{X}$ matches R. Under these conditions, we first note that, when the function f does not depend on the last two parameters, Eq. (6) becomes

$$\int_{\mathcal{X}} \int_{\mathbb{T}'} f(\tau, x) P_F(s_o, \mathrm{d}\tau \mathrm{d}x) \underbrace{\left(\int_{\mathcal{X}} \int_{\mathbb{T}'} R(\mathrm{d}x') P_B(x', \mathrm{d}\tau') \right)}_{=Z}$$
$$= \underbrace{\left(\int_{\mathcal{X}} \int_{\mathbb{T}'} P_F(s_o, \mathrm{d}\tau' \mathrm{d}x') \right)}_{=1} \int_{\mathcal{X}} \int_{\mathbb{T}'} f(\tau, x) R(\mathrm{d}x) P_B(x, \mathrm{d}\tau), \tag{8}$$

which recovers Lahlou et al.'s TB condition [13, Definition 6]. Consequently, the result follows from [13, Theorem 2, (4)]. For completeness, however, it is easy to see that Eq. (8) ensures that $P_F(s_o, \cdot)$ and $R(\mathrm{d}x) P_B(x, \cdot)$, when interpreted as distributions over $S^{\otimes \leq N-1} \times \{s_f\}$, are equivalent in terms of expectations of bounded measurable functions and, thus, must be indistinguishable up to a multiplicative constant. Hence, by integrating out the elements of τ that are not members of $\mathcal{X}$, we obtain Eq. (7).

Finally, before we delve into the algorithmic aspects of contrasitve learning of GFlowNets, we again emphasize the result above is general enough to contemplate distributions supported on continuous, discrete, and mixed spaces.

3.2 Contrastive Loss

Contrastive Loss. Realistically, finding a policy P_F satisfying the CB condition is an analytically intractable problem. As a consequence, we parameterize P_F with a neural network with weights θ and estimate the optimal θ^* by minimizing a loss function that provably enforces Eq. (6). The corollary below, which is an immediate consequence of Theorem 1, provides such a loss function.

Corollary 1 (Continuous Contrastive Loss). *Under the notations of Theorem 1, let p_F, p_B and r be the densities of P_F, P_B, and R relatively to κ, κ_b and μ, respectively. Then, we define the* continuous contrastive loss *(CCL) as*

$$\mathcal{L}_{CB}(\theta) = \mathbb{E}_{\tau,\tau' \sim P_E(s_o,\cdot) \otimes P_E(s_o,\cdot)} \left[\left(\frac{p_F(\tau)}{p_B(\tau|x)r(x)} - \frac{p_F(\tau')}{p_B(\tau'|x')r(x')} \right)^2 \right], \quad (9)$$

in which $P_E(s_o, \cdot)$ is an exploratory policy with full support over the trajectories on $\Sigma_S^{\otimes \leq N}$ starting at s_o and we use x (resp. x') to denote the unique terminal state within τ (resp. τ'). Then, if $\mathcal{L}_{CB}(\theta^\star) = 0$, the GFlowNet parameterized by $\theta^\star$ abides by Eq. (8) and samples correctly from the target.

Intuitively, the condition $\mathcal{L}_{CB}(\theta) = 0$ ensures that the quotients $\beta(\tau)$ defined in Sect. 3.1 are constant $\kappa(s_o, \cdot)$-almost everywhere, which then implies Theorem 1 and ultimately guarantees GFlowNet's distributional correctness.

Algorithm 1. Contrastive learning of GFlowNets

Require: $\mathcal{G}$, p_F parameterized by θ, p_B, r, exploratory policy P_E, batch size B
Ensure: marginal of p_F on $\mathcal{X}$ matches r
   repeat
      $\{(\tau_i, \tau_i') \sim P_E(s_o, \cdot) \otimes P_E(s_o, \cdot) : 1 \leq i \leq B\}$
      $L(\theta) \leftarrow 0$
      for $i \in \{1, \ldots, B\}$ **do**
         $f_i \leftarrow \log p_F(\tau_i) - \log p_F(\tau_i')$,
         $b_i \leftarrow \log p_B(\tau_i|x_i) + \log r(x_i) - \log p_B(\tau_i'|x_i') - \log r(x_i)$
         $L(\theta) \leftarrow f_i - b_i$
      end for
      $L(\theta) \leftarrow L(\theta)^2$
      Gradient step on θ with $\nabla L(\theta)$.
   until some convergence criterion is satisfied.

Algorithm. Algorithm 1 illustrates the training of GFlowNets through the minimization of the CCL. In practice, both the sampling and neural network forward passes can be massively parallelized. Also, for the optimization step, we use Adam [12] to estimate the stochastic gradients. Importantly, the architecture of the parameterizing neural network with weights θ and the nature of exploratory policy P_E are very problem-dependent and are abstracted away from Algorithm 1; we provide some examples in the next section.

Connection to Other Losses. To conclude, we remark that the CB loss is tightly connected to the variance loss for variational inference [20,26]; indeed, it is well known that the expectation of the squared-difference between two independently sampled variables equal their variance, i.e., $\mathbb{E}_{\tau,\tau'}[(\beta(\tau) - \beta(\tau'))^2] = \text{Var}_\tau[\beta(\tau)]$.

4 Experiments

This section shows that the CCL is a sound and effective learning objective for continuous GFlowNets in varied experimental setups, which are described below.

Tasks. We underline the effectiveness of the CB loss in the tasks of sampling from a mixture of Gaussian distributions (GM) [28] and of a banana-shaped distribution (BS) [16], which are commonly implemented benchmark models respectively for continuous GFlowNets and variational inference algorithms in general. For the GMs, we evaluate GFlowNet's performance for a sparse 2-dimensional model and a 15-dimensional model, both with isotropic variances. For the BS distribution, we consider the target distribution

$$\mathbb{R}^2 \ni \mathbf{x} \sim \mathcal{N}\left(\begin{bmatrix} x_1 \\ x_2 + x_1^2 + 1 \end{bmatrix} \middle| \begin{bmatrix} 0 \\ 0 \end{bmatrix}, \begin{bmatrix} 1 & 0.9 \\ 0.9 & 1 \end{bmatrix}\right). \tag{10}$$

Experimental Setup. For each experiment, we consider the iterative generative process described at the end of Sect. 2.2. To parameterize Q_θ, which is here defined as a Gaussian mixture distribution, we employ an MLP with 2 64-dimensional layers that receives the current state as an input and return the log-weights, averages, and log-variances of the mixture. In Algorithm 1, we fix a batch size of 128 and use Adam with a linearly decaying learning rate of 10^{-3} for optimization. However, when minimizing the TB loss, we adopt Malkin et al.'s [14] suggestion to implement a fixed learning rate of 10^{-1} for $\log Z(\gamma)$. Code for reproducing the results will be released upon acceptance.

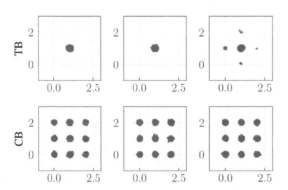

Fig. 1. Minimizing the CB loss (bottom row) drastically improves the training convergence when compared to TB minimization (top row) when the target is a 2-dimensional sparse Gaussian Mixture (see Fig. 3). This corroborates our hypothesis that avoiding parameterizing $\log Z(\gamma)$ facilitates the learning of GFlowNets. Columns represent results of three randomly initialized runs.

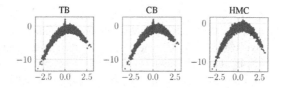

Fig. 2. CB and TB minimization perform similarly for the (relatively easy-to-approximate) banana-shaped target. We consider HMC (right plot) as a gold-standard reference for sample generation due to its accuracy and reliability.

Results. Figure 1 highlights that a GFlowNet trained by minimizing $\mathcal{L}_{CB}$ leads to a substantial improvement in terms of approximation quality when compared to the minimization of $\mathcal{L}_{TB}$ given a fixed computational budget and a sparse target, which is illustrated in Fig. 3. This validates our hypothesis that the training of $\mathcal{L}_{TB}$ is significantly constrained by the obligatory estimation of the intractable $\log Z(\gamma)$. For the experiments based on the high-dimensional Gaussian mixture, which cannot be simply visualized, we show in Fig. 4 the Jensen-Shannon divergence between the learned distribution q and the target p,

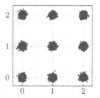

Fig. 3. A sparse Gaussian mixture.

$$\mathcal{D}_{JS}[p||q] = \frac{1}{2}\left(\mathcal{D}_{KL}[p||m] + \mathcal{D}_{KL}[q||m]\right), \tag{11}$$

with $m := {p+q}/{2}$ and $\mathcal{D}_{KL}[p||m] = \mathbb{E}_{x \sim p}[\log {p}/{m}]$ representing the Kullback-Leibler divergence between p and m. Importantly, $\mathcal{D}_{JS}$ is a frequently implemented metric for assessing GFlowNets [6,13]. Finally, for the low-dimensional and non-sparse targets, the training of GFlowNets based on both the minimization of CB and TB yield similar approximations given an equivalent computation resources, as Fig. 2 attests, which is potentially a consequence of the relative simplicity of the target distribution.

5 Related Works

Discrete GFlowNets. GFlowNets were originally conceived as a reinforcement learning algorithm tailored to the search of a diversity of high-valued states [2,3]. Since then, this family of models was successfully applied to Bayesian structure learning and causal discovery [1,5,6], natural language processing [9], robust scheduling [26], phylogenetic inference [30], and probabilistic modelling in general [10,15,27,29]. Correspondingly, a large effort was devoted to improving the training convergence of discrete GFlowNets [11,14,17,19,25].

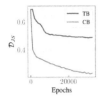

Fig. 4. Learning curve for CB/TB.

Continuous GFlowNets. In contrast, the literature on continuous GFlowNet, introduced by Lahlou et al. [13], is relatively thin. To the best of our knowledge, the only fruitful application of these models occurred recently in the context of Bayesian structure learning [6]. The reason for this, we believe, is the difficulty of efficiently training these models. Indeed, Zhou et al. [30] discretized the continuous component of the state space to avoid training a model to sample from a continuous distribution. Under these conditions, our work is an important step towards the improvement the usability of continuous GFlowNets.

6 Conclusions

We rigorously introduced a continuous variant of the *contrastive balance condition* and an accompanying *continuous contrastive loss* for efficiently training GFlowNets over distributions supported on discrete, continuous, and mixed spaces. All in all, our theoretical analysis assured the reliability of the proposed learning objective and our empirical results suggested that minimizing the CCL often significantly speeds up the model's training convergence.

Yet, the training of continuous GFlowNet remains challenging, and extending the recent advancements in the training of discrete GFlowNets to the continuous setting, along the lines of our work, is a promising endeavor that may greatly diminish these difficulties. For instance, one may build upon the framework of Pan et al.'s [19] Generative Augmented Flow Networks to incorporate intermediate reward-related signals within a trajectory for enhanced credit assignment [22]. Similarly, the theoretical developments of Tiapkin et al. [7,25] showing the relationship between RL and GFlowNets paves the road for the design of learning objectives motivated by techniques in the theory of control for continuous states.

Acknowledgements. This work was supported by Fundação Carlos Chagas Filho de Amparo à Pesquisa do Estado do Rio de Janeiro FAPERJ (SEI-260003/000709/2023), São Paulo Research Foundation FAPESP (2023/00815-6), and Conselho Nacional de Desenvolvimento Científico e Tecnológico CNPq (404336/2023-0).

Disclosure of Interests. The authors have no competing interests to declare that are relevant to the content of this article.

References

1. Atanackovic, L., Tong, A., Hartford, J., Lee, L.J., Wang, B., Bengio, Y.: DynGFN: towards Bayesian inference of gene regulatory networks with GFlowNets. In: Advances in Neural Processing Systems (NeurIPS) (2023)
2. Bengio, E., Jain, M., Korablyov, M., Precup, D., Bengio, Y.: Flow network based generative models for non-iterative diverse candidate generation. In: NeurIPS (2021)

3. Bengio, Y., Lahlou, S., Deleu, T., Hu, E.J., Tiwari, M., Bengio, E.: GFlowNet foundations. J. Mach. Learn. Res. (JMLR) **24**, 1–55 (2023)
4. Blei, D.M., et al.: Variational inference: a review for statisticians. J. Am. Stat. Assoc. **112**, 859–877 (2017)
5. Deleu, T., et al.: Bayesian structure learning with generative flow networks. In: UAI (2022)
6. Deleu, T., Nishikawa-Toomey, M., Subramanian, J., Malkin, N., Charlin, L., Bengio, Y.: Joint Bayesian inference of graphical structure and parameters with a single generative flow network. In: Advances in Neural Processing Systems (NeurIPS) (2023)
7. Deleu, T., Nouri, P., Malkin, N., Precup, D., Bengio, Y.: Discrete probabilistic inference as control in multi-path environments (2024)
8. Hinton, G.E., Osindero, S., Teh, Y.W.: A fast learning algorithm for deep belief nets. Neural Comput. **18**, 1527–1554 (2006)
9. Hu, E.J., Jain, M., Elmoznino, E., Kaddar, Y., et al.: Amortizing intractable inference in large language models (2023)
10. Hu, E.J., Malkin, N., Jain, M., Everett, K.E., Graikos, A., Bengio, Y.: GFlowNet-EM for learning compositional latent variable models. In: International Conference on Machine Learning (ICLR) (2023)
11. Jang, H., Kim, M., Ahn, S.: Learning energy decompositions for partial inference in GFlownets. In: The Twelfth International Conference on Learning Representations (2024)
12. Kingma, D.P., Ba, J.: Adam: a method for stochastic optimization. arXiv preprint arXiv:1412.6980 (2014)
13. Lahlou, S., et al.: A theory of continuous generative flow networks. In: ICML Proceedings of Machine Learning Research, vol. 202, pp. 18269–18300. PMLR (2023)
14. Malkin, N., Jain, M., Bengio, E., Sun, C., Bengio, Y.: Trajectory balance: improved credit assignment in GFlowNets. In: NeurIPS (2022)
15. Malkin, N., et al.: GFlowNets and variational inference. In: International Conference on Learning Representations (ICLR) (2023)
16. Mesquita, D., Blomstedt, P., Kaski, S.: Embarrassingly parallel MCMC using deep invertible transformations. In: UAI (2019)
17. Pan, L., Malkin, N., Zhang, D., Bengio, Y.: Better training of GFlowNets with local credit and incomplete trajectories. In: International Conference on Machine Learning (ICML) (2023)
18. Pan, L., Malkin, N., Zhang, D., Bengio, Y.: Better training of GFlowNets with local credit and incomplete trajectories. arXiv preprint arXiv:2302.01687 (2023)
19. Pan, L., Zhang, D., Courville, A., Huang, L., Bengio, Y.: Generative augmented flow networks. In: International Conference on Learning Representations (ICLR) (2023)
20. Richter, L., Boustati, A., Nüsken, N., Ruiz, F.J.R., Akyildiz, Ö.D.: VarGrad: a low-variance gradient estimator for variational inference (2020)
21. Robert, C.P., et al.: The Bayesian Choice: From Decision-Theoretic Foundations to Computational Implementation, vol. 2. Springer, Heidelberg (2007). https://doi.org/10.1007/0-387-71599-1
22. Shen, M.W., Bengio, E., Hajiramezanali, E., Loukas, A., Cho, K., Biancalani, T.: Towards understanding and improving GFlowNet training. In: International Conference on Machine Learning (2023)
23. da Silva, T., Carvalho, L.M., Souza, A., Kaski, S., Mesquita, D.: Embarrassingly parallel GFlowNets (2024)

24. da Silva, T., et al.: Human-in-the-loop causal discovery under latent confounding using ancestral GFlowNets. arXiv preprint arXiv:2309.12032 (2023)
25. Tiapkin, D., Morozov, N., Naumov, A., Vetrov, D.: Generative flow networks as entropy-regularized RL (2024)
26. Zhang, D.W., Rainone, C., Peschl, M., Bondesan, R.: Robust scheduling with GFlowNets. In: International Conference on Learning Representations (ICLR) (2023)
27. Zhang, D., Chen, R.T., Malkin, N., Bengio, Y.: Unifying generative models with GFlowNets and beyond. In: ICML Beyond Bayes Workshop (2022)
28. Zhang, D., Dai, H., Malkin, N., Courville, A., Bengio, Y., Pan, L.: Let the flows tell: solving graph combinatorial optimization problems with GFlowNets. In: NeurIPS (2023)
29. Zhang, D., Malkin, N., Liu, Z., Volokhova, A., Courville, A., Bengio, Y.: Generative flow networks for discrete probabilistic modeling. In: International Conference on Machine Learning (ICML) (2022)
30. Zhou, M.Y., et al.: PhyloGFN: phylogenetic inference with generative flow networks. In: The Twelfth International Conference on Learning Representations (2024)

A Data Distribution-Based Ensemble Generation Applied to Wind Speed Forecasting

Diogo M. Almeida[✉], Paulo S. G. de Mattos Neto, and Daniel C. Cunha

Centro de Informática (CIn), Universidade Federal de Pernambuco (UFPE), Recife, PE, Brazil
{dma4,psgmn,dcunha}@cin.ufpe.br

Abstract. Variability in wind intensity and direction causes unstable electricity supply to the power system, making integrating this energy into the electrical system a significant challenge for operations and planning practices. Ensembles can be used as an alternative to address the complex patterns over time in wind speed time series. The appropriate size of the training partition for ensemble models depends on dataset characteristics. And, to enhance model training efficiency, it is important to minimize repetitive data. By maintaining a concise training set, it becomes easier to meet the requirements of software and hardware constraints. Mainly because, there is a growing interest in deploying machine learning models on edge devices. For these reasons, a new method called Local Distribution (LocDist) has been introduced in this paper to predict wind speed, utilizing local pattern recognition based on data distribution. In testing with three wind speed time series, LocDist created a compact training subset with less than 20% of the training data. The Diebold-Mariano hypothesis test was employed to assess the significance of the forecast errors of the proposal compared to individual and bagging methods that use the entire training set. The LocDist method with long short-term memory (LSTM) and gated recurrent unit (GRU) won in 100% of the cases. Additionally, the LocDist with extreme learning machines (ELM) and autoregressive integrated moving average (ARIMA) won or tied in 83% and 66% of the cases, respectively.

Keywords: Time series · Forecasting · Ensembles · Data distribution · Wind speed

1 Introduction

The use of wind energy is increasing worldwide each year. Over the past 20 years, wind energy has become the primary clean and cost-competitive energy globally [1]. In 2023, wind energy reached the historic milestone of 1 TW of installed capacity, and the coming years will mark a crucial transition period for the global wind industry. It took 40 years to reach this milestone, but the next 1 TW will take less than a decade [2]. The Brazilian Northeast is a region with some of the best winds in the world for wind energy production, as they are more constant, have stable speeds, and do not frequently change direction. Therefore, 90% of Brazilian wind farms are located in the Northeast Region [3]. Although wind energy is considered an attractive option due to its abundance and ecological benefits, several challenges are still faced in its exploitation, such as the variability in wind intensity and direction, causing unstable electricity supply to the power system [4]. Integrating this energy into the electrical system represents a significant challenge for operations and planning practices [5]. Hence, wind energy systems' planning, scheduling, maintenance, and control depend on wind speed forecasting, making it essential to obtain wind speeds in advance [4]. For these reasons, the analysis and evaluation of this type of energy have attracted the attention of researchers worldwide [6].

Various machine learning methods have already been applied to wind speed forecasting, such as support vector regression (SVR) [7–9], extreme learning machines (ELM) [10], long short-term memory (LSTM) [11,12], and gated recurrent unit (GRU) [12]. Each model can recognize nonlinear patterns in a series and has its particularities. However, wind speed time series display complex patterns [6]. Factors such as regions' surface roughness, vegetation variability, and land use influence wind behavior [13]. Furthermore, research shows that wind speed data can exhibit a wide range of distributions [14], and these time series may also reveal hidden patterns with chaotic characteristics [5]. As a result, accurately forecasting wind speed poses a significant challenge [6].

An alternative to modeling complex patterns over time in wind speed time series is using ensembles [6]. For example, applying traditional procedures to a time series with a mixed distribution can lead to misleading results [15]. Therefore, we understand that ensembles offer superior prediction accuracy and reduced uncertainty compared to individual models, as they combine the different characteristics of individual prediction models [16–20], especially in the field of wind speed forecasting [6,13,21–23]. Ensemble is a technique used to combine prediction models [6]. It involves at least two steps: generation and integration. As stated in [6], individual models (also known as base models) use time series as input to produce predictions in generation. In integration, the predicted outputs of the base models are combined to obtain the final prediction. Diversity among predictions is crucial to recognizing different patterns in the time series, and it is one of the primary objectives of using an ensemble [20]. The authors in [17] have reported that the appropriate size of the training partition for ensemble models varies depending on dataset characteristics [17]. It is highlighted in [24] that to enhance model training efficiency, it is important to minimize repetitive data. By maintaining a concise training set, it becomes easier to meet the requirements of software and hardware constraints. Additionally, efficient artificial intelligence (AI) everywhere is a goal pursued by humans. As a result, there is a growing interest in deploying machine learning models on edge devices (computation and data storage closer to the sources of data) that have stringent constraints on resources and energy [25].

For these reasons, studies on the ensemble generation stage that appropriately define partitions for the base models are desirable in wind speed prediction.

Especially because in [26], it has been shown that it is achievable to obtain a subset of wind speed data partitions that adequately represent the data dynamics throughout the entire analysis period. This fact suggests that wind speed time series exhibit redundancy in their patterns, rendering it unnecessary to utilize the entire training set for model training. So, the proposal of this paper introduces a new generation method for homogeneous ensembles, composed of a single type of base model, to predict wind speed, called Local Distribution (LocDist), which utilizes local pattern recognition based on data distribution. This method generates partitions based on local data distribution using a divide-and-conquer strategy. Instead of using a single prediction model to identify patterns in different distributions in the time series, base models of an ensemble are applied to separate partitions selected from the time series to identify distribution patterns. The predictions of the base models are then combined to aggregate the mapping of different distribution patterns of the time series into an ensemble. Our hypothesis suggests that if expert models are generated in partitions containing data with different distributions among the partitions, it will result in models with diversity in their predictions. Moreover, combining these models will create an ensemble of specialists, each well-versed in different distributions of the time series data, containing concise information about the data patterns. Thus, the primary contribution of this proposal is to create a compact training subset that minimally represents the data distribution of the time series. The training phase can be executed more efficiently by reducing the amount of training data.

The outline of the paper is organized as follows. Section 2 describes the related works. Section 3 introduces the proposal in detail. Section 4 is the description of the processes through which results are generated. In Sect. 5, results are presented by performing a comparison among individual models, ensembles and the proposal. Finally, in Sect. 6, conclusions are drawn.

2 Related Works

The use of individual models for wind speed time series forecasting has been widespread [8,10,12,27]. In [27], it was found that autoregressive moving average (ARMA) models are more effective than the persistence model in predicting future observations. The persistence model simply repeats the most recent past observation. Furthermore, in [8], it was shown that support vector regression (SVR) outperforms multilayer perceptron (MLP). In [10], the extreme learning machines (ELM) demonstrated competitive prediction errors compared to MLP and SVR while requiring lower computational costs. Reference [12] introduced a new seasonal autoregressive integrated moving average (ARIMA) model for predicting wind speed time series and compared it with long short-term memory (LSTM) and gated recurrent unit (GRU). Despite the more advanced machine learning algorithms of LSTM and GRU, the seasonal ARIMA model exhibited superior predictive performance and faster training. These individual models employ a global mapping of a time series, using the entire training set for model training. However, this approach is not the optimal choice due to the weaknesses, disadvantages, and limitations of modeling complex energy systems [6].

Some methods for the generation stage in ensembles have been proposed to identify complex patterns in time series data. In [21–23], wind speed time series are decomposed into a finite and small number of intrinsic mode functions (IMFs) using empirical mode decomposition (EMD). These IMFs contain information about local trends and fluctuations at different scales of the original signal, which is valuable for understanding the actual physical significance of the signal [23]. A homogeneous ensemble utilizing a resampling technique is developed in a different application in [19] and used on the M3-Competition dataset. The time series is bootstrapped using the moving block bootstrap (MBB). This process creates new time series from the original, intending that the predictors of these time series will make the final forecast resilient to the uncertainty observed in the data [18]. This method has a significant computational cost, involving training 100 base models on the entire training set.

However, some generation methods have been proposed focusing on identifying different local patterns in a time series, thus providing specialized local predictors [17]. Therefore, in [17], the authors propose a homogeneous ensemble for various time series applications. They create training partitions of equal size, allowing for an overlap between two adjacent partitions, which enables the model to adapt to progressive pattern changes. The authors also investigate four different partition sizes. In [28], a generation strategy is used to obtain partitions of the training set with the most informative data through clustering. This approach can accelerate the learning phase without compromising accuracy or improving it. It is particularly beneficial when the training dataset is large, as it can significantly reduce computation time by avoiding the processing of redundant data [28]. However, in the proposed method of [28], it is initially necessary to train 200 prediction models on training partitions before applying clustering. The authors argue that the execution time of their method decreases because these 200 models can be run in parallel.

In most of the previously cited works, excluding [28], no analysis is conducted on the partition data to identify different data patterns or to remove redundancy. There is no evidence that the constructed partitions have different data patterns to meet the requirement of generating diverse models, nor any analysis on the most appropriate size or region for the training partitions. In [25], it is reported that developing an automatic solution for model compression is challenging. The aim is to achieve a higher compression ratio while minimizing accuracy loss, posing a challenging trade-off. Therefore, further research is desirable to investigate the ensemble generation step by analyzing the data patterns in the partitions. Mainly because in [26], it was demonstrated that it is possible to obtain a subset of wind speed data partitions that reasonably reflect the data dynamics over the entire analysis period. This fact indicates that wind speed time series may exhibit redundancy in their patterns, making it unnecessary to use the entire training set for training a model.

3 Proposed New Ensemble

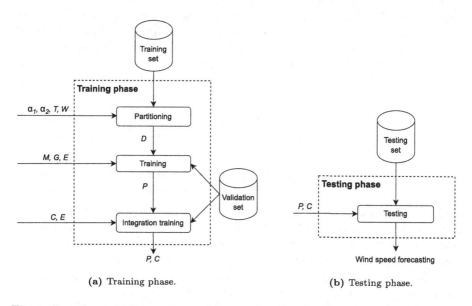

Fig. 1. Training and testing phases of the proposed local pattern recognition based on data distribution.

The proposal is a homogeneous ensemble comprising two stages: generation and integration. We define partitions with different data distributions and eliminate other ones with redundant data distributions within the training set during the generation stage. Therefore, a concise training set that reflects the data distribution patterns is obtained. This process enables the creation of a pool of trained models $P = \{m_1, m_2, \ldots, m_N\}$, that is, a group of N local expert predictors, which have been trained on partitions generated with different data distributions validated by the Kolmogorov-Smirnov (KS) hypothesis test. The null hypothesis of KS test states that the two samples share the same distribution [29]. Subsequently, the predictors are combined in the integration stage to achieve the final forecasting of the ensemble.

Figure 1(a) shows the training phase of the LocDist method. The inputs consist of the training and validation sets of the time series, the significance levels of the KS hypothesis test, α_1 and α_2, the minimum partition size T, the sliding window size W, the base model M with a technique for hyperparameter selection G, the evaluation measure E for the evaluation of M and, lastly, a combination method C for integration. The output is the pool of trained models $P = \{m_1, m_2, \ldots, m_N\}$ of N local expert models and combination method C.

First, the ensemble generation step is employed through partitioning, which is constructed as follows: (i) Windows w_t are created in the training set with a fixed size W. The difference between two adjacent windows $(w_t - w_{t-1})$ is only one observation. The first window w_1 is defined as the initial reference window Ref_1; (ii) Subsequently, a significance level α_1 is defined for applying the KS hypothesis test between the reference window and each subsequent window sequentially; (iii) The method constructs the partition by joining adjacent windows that have a p-value $> \alpha_1$, otherwise a new partition is initiated with a new reference window. In the end, we will have the defined partitions and their respective reference windows; (iv) Next, partitions with a size smaller than T are eliminated; (v) Return to step (iii) applying a grid search with different significance levels for α_1, aiming to maximize the expression

$$\text{Maximize } \frac{\sum_{i<j}^{u} KS(Ref_i, Ref_j)}{B_{u,2}} \qquad (1)$$

$$\text{Subject to: } KS(Ref_i, Ref_j) = \begin{cases} 1, & \text{if } p\text{-value} \leq 0.05; \\ 0, & \text{otherwise.} \end{cases} \qquad (2)$$

where B is a combination in the field of combinatorial analysis and u is the number of reference windows at the end of step (iv). Therefore, $B_{u,2}$ represents the total number of possible two-by-two combinations between the reference windows of each partition. KS refers to the Kolmogorov-Smirnov (KS) hypothesis test. Thus, $KS(Ref_i, Ref_j)$ is equal to 1 if the reference windows i and j of partitions have different distributions (p-value ≤ 0.05), and equal to 0, otherwise. Maximizing the expression (1) is a key approach for identifying the most significant number of partitions with different distributions in the training set; (v) After determining the best significance level for α_1 in the grid search to maximize the expression (1), the KS test is applied once again to eliminate partitions with redundant distributions at a significance level of $\alpha_2 = 0.05$. At the end of step (v), the disjoint partitions D with solid evidence of different data distributions are obtained, and the partitioning procedure is concluded. Succeeding generating disjoint partitions D with different distributions, time series prediction models M are trained on these disjoint partitions D using technique G and evaluation metric E for hyperparameter selection. Linear models are chosen based on the lowest error E in the training partition, while nonlinear models are selected based on the lowest error E in the validation partition to avoid overfitting. This process creates a pool $P = \{m_1, m_2, \ldots, m_N\}$ consisting of N local expert predictors specialized in different distribution patterns of the time series. This pool P is combined using a combination method C, which can be a statistical measure, such as the average, or a nonlinear model M trained with technique G and evaluation metric E for hyperparameter selection. The best combination method C is selected based on the lowest error E in the validation set.

Finally, Fig. 1(b) shows the testing phase of the LocDist method. The input involves the testing set of the time series, the pool P of N local expert predictors, and the combination method C for integration. The output is the wind speed forecasting. This phase consists of testing the predictive capacity of the proposal with the models P and the combination method C on an out-of-sample set.

Table 1. Statistical summary of wind speed in the three time series databases.

Time series	Maximum	Mean	Minimum	Standard deviation	Coefficient of variation
1	10.52 m/s	5.95 m/s	1.59 m/s	1.40 m/s	0,234
2	16.76 m/s	9.35 m/s	3.17 m/s	2.68 m/s	0.287
3	28.00 m/s	18.04 m/s	8.51 m/s	2.70 m/s	0.149

4 Experimental Protocol

The databases considered in this work are from the Brazilian Institute of Space Research (INPE, in Portuguese), obtained through sensors at a height of 50 m [30]. The Northeastern Region of Brazil has some of the world's best winds for wind energy production, which is why 90% of Brazilian wind parks are located in this region [3]. Database 1 refers to Petrolina-PE, while Databases 2 and 3 were obtained in Triunfo-PE. The first database covers the period from July to September 2010. Databases 2 and 3 cover the periods from October to December 2006 and March to May 2006, respectively. There are 13,248 instances at ten-minute intervals between consecutive instances for each database, and no missing values. Every six instances were averaged, resulting in 2,208 instances for each time series with one-hour time intervals to reduce the size of the database. Subsequently, each time series was fully normalized to the interval [0, 1]. Thus, the input data for the neural networks are appropriately mapped according to the codomain of the sigmoid activation function. Then, each time series was divided into three subsets: the first 50% of the data for training, then 25% for validation, and the last 25% for testing. Table 1 presents the statistical summary of the three pre-processed and non-normalized databases.

The LocDist method is a homogeneous ensemble and was separately analyzed with the following base models: ARIMA, ELM, SVR, LSTM, and GRU. The value 24 was chosen for the sliding window size W due to the seasonality of the time series, which is 24 observations (24 h). The grid search for the values of α_1 was conducted in the range [0.05, 0.1, 0.15, ..., 0.3]. Values of $\alpha_1 > 0.3$ tend to construct minimal partitions, so the search ends at $\alpha_1 = 0.3$. Furthermore, the behavior of LocDist was evaluated for different values of the variable T, which is the minimum allowed size for the generated partitions. In [31], mentions that 66 observations are considered a minimal amount for adjusting the parameters of a machine learning model. This manner, the variable T was evaluated with values 45, 60, and 75. For values above 60, such as 75, the generation method could not find partitions with this minimum size for all time series considered in this work. Thus, the evaluation was limited to the values $T = 45$ and $T = 60$. The elimination of partitions with redundant data distributions was done so that the less recent distributions were discarded. It was found that maintaining the more recent distributions resulted in better LocDist predictive performance compared to the approach that discards the more recent redundant partitions. The ARIMA base model used the training partition for training and parameter adjustment. The ELM, SVR, LSTM and GRU models used the training partition for training and the validation set for parameter adjustment to avoid overfitting. At last, in the integration stage, the LocDist used the simple average or the ELM nonlinear model as a combination method. The simple average is a statistical measure known in the ensemble literature for its robustness [20]. The combination with a nonlinear model ELM is a more sophisticated approach that can surpass the simple average combination in the presence of predictors with significant differences in accuracy [32]. The validation set was used to select the best combination approach according to the lowest RMSE.

Firstly, we compared LocDist with the respective individual models: ARIMA, SVR, ELM, LSTM, or GRU. Each model was globally trained on the complete training set. Subsequently, the LocDist was compared to the respective homogeneous ensemble using a resampling technique through bagging [19]. The resampling technique was applied to ARIMA, ELM, and SVR base models. Thus, three homogeneous ensembles were obtained, and the base models were globally trained one hundred times on resampled series [19]. This approach has a high computational cost, and because of that, LSTM and GRU were not used in this resampling technique. In the integration stage, the combination method of the ensemble with bagging was the simple average.

The same parameter selection technique was applied to the individual and base models of the ensembles. For the ARIMA model, the integer parameter d was chosen according to the augmented Dickey-Fuller test to identify stationarity. Meanwhile, p and q were obtained in the range $[0, 1, ..., 24]$, as there is a seasonality of 24 lags in the hourly time series. The conditional sum of squares method was applied for model fitting, and the model with the lowest corrected Akaike information criterion was selected. The grid search technique was applied to the remaining models. The SVR type was ϵ-regression with kernel radial basis function (RBF). The input used the sliding window method whose width values were taken from $[1, 6, 12, 18, 24]$. The regularization parameter was adjusted in the range $[2^0, 2^1, ..., 2^{10}]$, while the tolerance region ϵ and the parameter λ from kernel RBF were obtained from the values $[2^{-10}, 2^{-9}, ..., 2^0]$. For ELM, the parameters were selected in the range $[1, 2, ..., 24]$ for the input and hidden layer nodes. The activation functions for hidden and output layers were the sigmoid and the identity, respectively. However, during the integration stage with the ELM model in the LocDist, the number of neurons in the hidden layer was selected from the range $[1, 2, ..., 42]$ to enhance the combination. Concerning LSTM and GRU, a stateful network with parameter values taken from the set $[1, 6, 12, 18, 24]$ was assumed for the input nodes. The network parameters were obtained from the set $[6, 12, 18, 24]$ for the hidden layer nodes. Still, for the hidden layer, the activation function was the sigmoid. The LSTM and GRU were trained through 100 epochs and batch size equal to one. Finally, the optimization solver adopted was Adam.

Table 2 shows all individual models and ensembles implemented and their respective acronyms. All simulations were performed for one-step-ahead forecasting using the R language with the forecast [33] library for ARIMA, E1071 [34] for

Table 2. Individual models and ensemble forecasting approaches used in this work with corresponding methods and respective acronyms.

Approach	Method	Acronym
Individual modeling	Autoregressive integrated moving average [27]	ARIMA
	Support vector regression [8]	SVR
	Extreme learning machines [10]	ELM
	Long short-term memory [12]	LSTM
	Gated recurrent units [12]	GRU
Homog. ens. with bagging based on [19]	Mean(ARIMA$_1$,ARIMA$_2$,...,ARIMA$_{100}$)	$BAGG_{ARIMA}$
	Mean(ELM$_1$,ELM$_2$,...,ELM$_{100}$)	$BAGG_{ELM}$
	Mean(SVR$_1$,SVR$_2$,...,SVR$_{100}$)	$BAGG_{SVR}$
Proposal	C(ARIMA$_1$,ARIMA$_2$,...,ARIMA$_N$)	$LocDist_{ARIMA}$
	C(ELM$_1$,ELM$_2$,...,ELM$_N$)	$LocDist_{ELM}$
	C(SVR$_1$,SVR$_2$,...,SVR$_N$)	$LocDist_{SVR}$
	C(LSTM$_1$,LSTM$_2$,...,LSTM$_N$)	$LocDist_{LSTM}$
	C(GRU$_1$,GRU$_2$,...,GRU$_N$)	$LocDist_{GRU}$

SVR, ElmNNRcpp [35] for ELM, and Keras [36] for LSTM and GRU. The grid search technique selected the nonlinear models parameters based on the smallest RMSE in the validation set. The RMSE is defined by Equation (3), such that

$$RMSE = \sqrt{\left(\frac{\sum_{t=1}^{L}(y_t - \widehat{y}_t)^2}{L}\right)}, \qquad (3)$$

where L is the size of the series, y_t is the actual value of the time series in period t, and $\widehat{y}_t$ is the predicted value of the time series in period t. The RMSE was also employed to assess predictive performance on the test set, along with three other metrics: mean absolute error (MAE), mean absolute percentage error (MAPE), and predicted change in direction (POCID). For MAE and MAPE metrics, the lower their values, the better the model's accuracy. In the case of POCID, the higher the value, the better the model's performance. The MAE is defined by

$$MAE = \frac{\sum_{t=1}^{L}|y_t - \widehat{y}_t|}{L}, \qquad (4)$$

where $|\cdot|$ is the absolute value operator. The MAPE metric is given by

$$MAPE = \frac{100}{L}\sum_{t=1}^{L}\frac{|y_t - \widehat{y}_t|}{y_t}, \qquad (5)$$

while the POCID is designated by

$$POCID = 100\frac{\sum_{t=1}^{L}Trend_t}{L}, \qquad (6)$$

where

$$Trend_t = \begin{cases} 1, \text{ if } (y_t - y_{t-1})(\widehat{y}_t - \widehat{y}_{t-1}) > 0; \\ 0, \quad \text{otherwise.} \end{cases} \quad (7)$$

Finally, the Diebold-Mariano (DM) hypothesis test [37] was applied between LocDist and the global mapping methods of the respective base model. So, the RMSE loss function is considered for the errors of both competing predictors. The null hypothesis H_0 assumes equality in forecasting accuracy for both predictors. For H_0 to be rejected, the p-value must be less than the statistical significance level $\alpha = 0.05$. Some signals were used to interpret the results in Tables 3, 4 and 5 of Sect. 5. The sign "+" indicates that H_0 was rejected and the proposed method outperforms the method used for comparison. For instance, if $LocDist_{ARIMA}$ outperformed ARIMA with statistical significance, the sign "+" will appear in the row of ARIMA. The sign "−" indicates that the proposed method underperforms the method used for comparison. The sign "=" indicates that the null hypothesis was not rejected, i.e., equality in forecasting accuracy for both predictors.

Table 3. Evaluation measurements and DM hypothesis test for the testing set obtained from the wind speed time-series Database 1.

Model	Method	RMSE (m/s)	MAE (m/s)	MAPE (%)	POCID (%)	DM test	p-value
ARIMA	ARIMA [27]	0.8142	**0.6141**	10.04	48.81	=	0.4079
	$BAGG_{ARIMA}$ based on [19]	0.8273	0.6292	10.30	48.21	=	0.0890
	$LocDist_{ARIMA}$	**0.8058**	0.6172	10,29	**49.01**		
ELM	ELM [10]	0.8126	0.6227	10.15	**48.71**	=	0.4907
	$BAGG_{ELM}$ based on [19]	**0.8041**	**0.6154**	**10.05**	47.82	=	0.6623
	$LocDist_{ELM}$	0.8071	0.6184	10.25	46.83		
SVR	SVR [8]	0.7804	0.6038	**9.84**	**48.51**	−	0.0120
	$BAGG_{SVR}$ based on [19]	**0.7799**	**0.6003**	9.88	48.21	−	0.0055
	$LocDist_{SVR}$	0.8107	0.6248	10.36	47.82		
LSTM	LSTM [12]	0.9105	0.7014	10.85	48.61	+	$2.49 \cdot 10^{-14}$
	$LocDist_{LSTM}$	**0.7966**	**0.6101**	**9.96**	**49.21**		
GRU	GRU [12]	0.8812	0.6796	10.65	**49.01**	+	$5.45 \cdot 10^{-11}$
	$LocDist_{GRU}$	**0.8116**	**0.6228**	**10.15**	48.61		

5 Results

The best results were obtained for all three databases with the value $T = 60$, which is the minimum allowed size for the generated partitions. In Database 1, LocDist generated three partitions with the following observations: 69, 62, and 96. This corresponds to 20.6% of the entire training set, which consists of 1104 observations. Table 3 shows the evaluation measurements and DM hypothesis test for the testing set obtained from the wind speed Database 1. The first column indicates the five model types, ARIMA, ELM, SVR, LSTM, and GRU, which were used in the methods of this research. The second column lists the

Table 4. Evaluation measurements and DM hypothesis test for the testing set obtained from the wind speed time-series Database 2.

Model	Method	RMSE (m/s)	MAE (m/s)	MAPE (%)	POCID (%)	DM test	p-value
ARIMA	ARIMA [27]	1.4085	1.0249	12.22	55.16	+	0.0463
	$BAGG_{ARIMA}$ based on [19]	1.4032	1.0253	**12.14**	**56.94**	=	0.1480
	$LocDist_{ARIMA}$	**1.3659**	**1.0087**	12.43	55.36		
ELM	ELM [10]	1.3765	1.0227	12.78	54.47	=	0.2132
	$BAGG_{ELM}$ based on [19]	1.3641	1.0208	12.69	54.96	=	0.3117
	$LocDist_{ELM}$	**1.3436**	**0.9965**	**12.18**	**56.55**		
SVR	SVR [8]	1.2806	**0.9449**	11.73	56.86	–	0,0044
	$BAGG_{SVR}$ based on [19]	**1.2730**	0.9450	**11.71**	**57.14**	–	0.0009
	$LocDist_{SVR}$	1.3403	0.9966	12.27	56.55		
LSTM	LSTM [12]	1.3771	1.0247	12.78	**55.95**	+	0.0302
	$LocDist_{LSTM}$	**1.3370**	**0.9918**	**12.16**	**55.95**		
GRU	GRU [12]	1.3929	1.0426	13.11	55.36	+	0.0017
	$LocDist_{GRU}$	**1.3411**	**1.0014**	**12.44**	**56.55**		

constructed methods, such as individual, ensemble with bagging, and the proposed LocDist, according to the respective model type. Then, the four evaluation metrics, RMSE, MAE, MAPE, and POCID, are presented for each forecasting method. Finally, the DM hypothesis test result and the respective p-value are shown. The best measure among the forecasting methods for each model type is highlighted in bold. In the case of ARIMA, the $LocDist_{ARIMA}$ achieved the best results in RMSE and POCID, though with no statistical significance compared to individual ARIMA and $BAGG_{ARIMA}$. About ELM, $LocDist_{ELM}$ did not stand out in any metric, but the DM hypothesis test demonstrated no statistical significance in the errors compared to individual ELM and $BAGG_{ELM}$. In the case of SVR, $LocDist_{SVR}$ loses to individual SVM and $BAGG_{SVR}$ with statistical significance. Concerning LSTM, the $LocDist_{LSTM}$ outperformed individual LSTM in all metrics with statistical significance. The last type of model is GRU, and the $LocDist_{GRU}$ achieved the best results in RMSE, MAE, and MAPE with statistical significance compared to individual GRU. The integration step selected the combination via ELM for the ARIMA, LSTM, and GRU approaches, and the combination by averaging for the ELM and SVM.

Regarding Database 2, LocDist generated two partitions with the following observations: 122 and 89. This corresponds to 19.1% of the entire training set. Table 4 illustrates the evaluation measures and the DM hypothesis test for the wind speed Database 2 testing set. About ARIMA, $LocDist_{ARIMA}$ achieved the best results in RMSE and MAE, with statistical significance compared to individual ARIMA. In the case of ELM, $LocDist_{ELM}$ achieved the best results in all metrics but with no statistical significance compared to the individual ELM and $BAGG_{ELM}$. About SVR, $LocDist_{SVR}$ loses to individual SVM and $BAGG_{SVR}$ with statistical significance. Concerning LSTM, the $LocDist_{LSTM}$ achieved the best results in RMSE, MAE, and MAPE with statistical significance compared to individual LSTM. Finally, the $LocDist_{GRU}$ method outperforms the individual GRU in all metrics and with statistical significance. The integration step selected the combination via ELM for all models except for the SVM, which used the average.

In Database 3, LocDist generated three partitions with the following observations: 66, 65, and 71. This corresponds to 18.3% of the entire training set. Table 5 exhibits the results for the testing set acquired from the wind speed time series Database 3. $LocDist_{ARIMA}$ reached the best result in POCID, though it had lost to individual ARIMA and $BAGG_{ARIMA}$ with statistical significance. Regarding ELM, the $LocDist_{ELM}$ did not exhibit superior performance in any metric but showed statistical significance in the DM hypothesis test, indicating lower forecasting errors compared to the individual ELM but higher errors compared to $BAGG_{ELM}$. In the case of SVR, the $LocDist_{SVR}$ achieved the best result in RMSE, MAE, and MAPE, and with statistical significance. Concerning LSTM, the $LocDist_{LSTM}$ achieved the best results in RMSE, MAE, and MAPE with statistical significance compared to individual LSTM. At last, the $LocDist_{GRU}$ method one more time outperformed the individual GRU in all four metrics with statistical significance. The integration step selected the combination via ELM for all models.

Concerning the results of the DM hypothesis test in Tables 3, 4 and 5, we can see that the methods $LocDist_{LSTM}$ and $LocDist_{GRU}$ won 100% of the cases. In addition, the $LocDist_{ELM}$ and $LocDist_{ARIMA}$ won or tied in 83% and 66% of the cases, respectively. These results corroborate our hypothesis that the LocDist method can generate a concise and representative training set compared to the complete training dataset, potentially enabling more appropriate recognition of local patterns. $LocDist_{SVR}$ has lost in 66% of the cases. This reveals some limitations, indicating that it is not always the best approach for all models.

We then chose some models, such as ARIMA, ELM, LSTM, and GRU, for graphical evaluation of the methods in each dataset. The SVR was excluded from

Table 5. Evaluation measurements for the test set obtained from the wind speed time-series Database 3.

Model	Method	RMSE (m/s)	MAE (m/s)	MAPE (%)	POCID (%)	DM test	p-value
ARIMA	ARIMA [27]	**1.0055**	**0.7939**	4.08	52.38	–	$1.00 \cdot 10^{-5}$
	$BAGG_{ARIMA}$ based on [19]	1.0165	0.8026	4.12	52.18	–	0.0001
	$LocDist_{ARIMA}$	1,1009	0,8725	4,41	**52,78**		
ELM	ELM [10]	1.1147	0.8838	4.45	51.39	+	0,0004
	$BAGG_{ELM}$ based on [19]	**1.0592**	**0.8394**	**4.26**	52.18	–	$3.71 \cdot 10^{-7}$
	$LocDist_{ELM}$	1,1000	0,8704	4,39	51,98		
SVR	SVR [8]	1.1616	0.9014	4.51	51.59	+	0,0074
	$BAGG_{SVR}$ based on [19]	1.2513	0.9547	4.73	**53.17**	+	$4.79 \cdot 10^{-6}$
	$LocDist_{SVR}$	**1,0792**	**0,8545**	**4,33**	52,18		
LSTM	LSTM [12]	1.1706	0.9328	4.72	**53.17**	+	$2.25 \cdot 10^{-7}$
	$LocDist_{LSTM}$	**1.1208**	**0.8874**	**4.46**	51.98		
GRU	GRU [12]	1.1564	0.9231	4.70	51.79	+	0.0001
	$LocDist_{GRU}$	**1.1155**	**0.8830**	**4.44**	**52.18**		

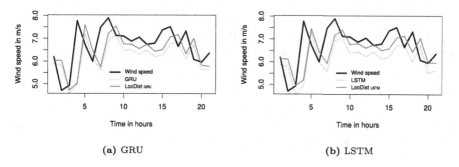

Fig. 2. One-step ahead prediction results for LocDist and global mapping methods on dataset 1 considering 20 sequential observations from the test set.

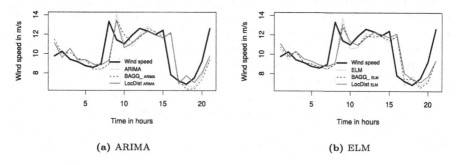

Fig. 3. One-step ahead prediction results for LocDist and global mapping methods on dataset 2 considering 20 sequential observations from the test set.

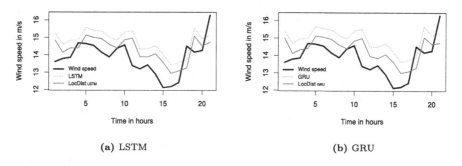

Fig. 4. One-step ahead prediction results for LocDist and global mapping methods on dataset 3 considering 20 sequential observations from the test set.

the analysis based on the results of the DM hypothesis test. Figures 2(a) and 2(b) show, respectively, the one-step-ahead prediction using the method LocDist built from GRU and LSTM models. The superiority of LocDist is noticeable compared to the individual methods. In Fig. 3, we have dataset 2, and the chosen models were ARIMA and ELM. We can observe that the local mapping of LocDist exhibits similar or even smaller prediction errors than the global mapping of individual and bagging methods. Finally, Fig. 4 illustrates the results on dataset 3 for the LSTM and GRU models. It is also observed that LocDist exhibits lower prediction errors compared to methods trained on the entire training set.

6 Conclusions

The LocDist introduces a new generation method for ensembles to predict wind speed, which utilizes local pattern recognition based on data distribution. It creates a compact training subset that minimally represents the data distribution of the time series. The results show promise for the LocDist applied to three wind speed time series. Creating a training subset for the LocDist ensemble was possible using less than 20% of the training data. The predictive performance of LocDist was competitive with or even better than that of the individual and bagging models trained on the entire training set. Concerning the Diebold-Mariano hypothesis test applied in the RMSE, the LocDist method with LSTM and GRU won in 100% of the cases. In addition, the LocDist with ELM and ARIMA won or tied in 83% and 66% of the cases, respectively, while the proposed ensemble method with SVR lost in 66% of the cases, revealing some limitations. For future work, it would be interesting to investigate the behavior of this new generation method with more sophisticated ensemble approaches, such as dynamic selection.

References

1. GWE Council: GWEC| global wind report 2021. Global Wind Energy Council, Brussels, Belgium (2019)
2. GWE Council: GWEC| global wind report 2023. Global Wind Energy Council, Brussels, Belgium (2023)
3. A. B. d. E. E. ABEEólica, "Abeeólica | infovento," INFOVENTO 31, 15 de junho de 2023 (2023)
4. Qu, Z., Mao, W., Zhang, K., Zhang, W., Li, Z.: Multi-step wind speed forecasting based on a hybrid decomposition technique and an improved back-propagation neural network. Renew. Energy **133**, 919–929 (2019)
5. Jiang, P., Wang, B., Li, H., Lu, H.: Modeling for chaotic time series based on linear and nonlinear framework: application to wind speed forecasting. Energy **173**, 468–482 (2019)
6. Ahmadi, M., Khashei, M.: Current status of hybrid structures in wind forecasting. Eng. Appl. Artif. Intell. **99**, 104133 (2021)
7. Mohandes, M.A., Halawani, T.O., Rehman, S., Hussain, A.A.: Support vector machines for wind speed prediction. Renew. Energy **29**(6), 939–947 (2004)
8. Salcedo-Sanz, S., Ortiz-Garcı, E.G., Pérez-Bellido, Á.M., Portilla-Figueras, A., Prieto, L., et al.: Short term wind speed prediction based on evolutionary support vector regression algorithms. Exp. Syst. Appl. **38**(4), 4052–4057 (2011)
9. Kong, X., Liu, X., Shi, R., Lee, K.Y.: Wind speed prediction using reduced support vector machines with feature selection. Neurocomputing **169**, 449–456 (2015)

10. Saavedra-Moreno, B., Salcedo-Sanz, S., Carro-Calvo, L., Gascón-Moreno, J., Jiménez-Fernández, S., Prieto, L.: Very fast training neural-computation techniques for real measure-correlate-predict wind operations in wind farms. J. Wind Eng. Ind. Aerodyn. **116**, 49–60 (2013)
11. Memarzadeh, G., Keynia, F.: A new short-term wind speed forecasting method based on fine-tuned LSTM neural network and optimal input sets. Energy Convers. Manage. **213**, 112824 (2020)
12. Liu, X., Lin, Z., Feng, Z.: Short-term offshore wind speed forecast by seasonal ARIMA - a comparison against GRU and LSTM. Energy **227**, 120492 (2021)
13. Ferreira, M., Santos, A., Lucio, P.: Short-term forecast of wind speed through mathematical models. Energy Rep. **5**, 1172–1184 (2019)
14. Wu, J., Li, N.: Impact of components number selection in truncated gaussian mixture model and interval partition on wind speed probability distribution estimation. Sci. Total Environ. **883**, 163709 (2023)
15. Robinson, P.: Analysis of time series from mixed distributions. Ann. Stat., 915–925 (1982)
16. Santos Júnior, D.S.O., de Mattos Neto, P.S.G., de Oliveira, J.F.L., Cavalcanti, G.D.C.: A hybrid system based on ensemble learning to model residuals for time series forecasting. Inf. Sci. **649**, 119614 (2023)
17. de Mattos Neto, P.S., Cavalcanti, G.D., Firmino, P.R., Silva, E.G., Nova Filho, S.R.V.: A temporal-window framework for modelling and forecasting time series. Knowl. Based Syst. **193**, 105476 (2020)
18. Petropoulos, F., Hyndman, R.J., Bergmeir, C.: Exploring the sources of uncertainty: why does bagging for time series forecasting work? Eur. J. Oper. Res. **268**(2), 545–554 (2018)
19. Bergmeir, C., Hyndman, R.J., Benítez, J.M.: Bagging exponential smoothing methods using STL decomposition and Box-Cox transformation. Int. J. Forecast. **32**(2), 303–312 (2016)
20. Sergio, A.T., de Lima, T.P., Ludermir, T.B.: Dynamic selection of forecast combiners. Neurocomputing **218**, 37–50 (2016)
21. Hu, J., Wang, J., Zeng, G.: A hybrid forecasting approach applied to wind speed time series. Renew. Energy **60**, 185–194 (2013)
22. Ruiz-Aguilar, J.J., Turias, I., González-Enrique, J., Urda, D., Elizondo, D.: A permutation entropy-based EMD-ANN forecasting ensemble approach for wind speed prediction. Neural Comput. Appl. **33**(7), 2369–2391 (2021)
23. Jiang, Z., Che, J., Wang, L.: Ultra-short-term wind speed forecasting based on EMD-VAR model and spatial correlation. Energy Convers. Manage. **250**, 114919 (2021)
24. Bowden, G.J., Maier, H.R., Dandy, G.C.: Optimal division of data for neural network models in water resources applications. Water Resour. Res. **38**(2), 2–1 (2002)
25. Deng, L., Li, G., Han, S., Shi, L., Xie, Y.: Model compression and hardware acceleration for neural networks: a comprehensive survey. Proc. IEEE **108**(4), 485–532 (2020)
26. Ribeiro, R., Fanzeres, B.: Identifying representative days of solar irradiance and wind speed in Brazil using machine learning techniques. Energy AI **15**, 100320 (2024)
27. Torres, J.L., Garcia, A., De Blas, M., De Francisco, A.: Forecast of hourly average wind speed with ARMA models in Navarre (Spain). Sol. Energy **79**(1), 65–77 (2005)
28. Tetko, I.V., Villa, A.E.: Efficient partition of learning data sets for neural network training. Neural Netw. **10**(8), 1361–1374 (1997)

29. Hodges, J., Jr.: The significance probability of the Smirnov two-sample test. Ark. Mat. **3**(5), 469–486 (1958)
30. INPE: Rede do sistema de organização nacional de dados ambientais (2020). http://sonda.ccst.inpe.br/index.html. Accessed 27 Jul 2023
31. Cerqueira, V., Torgo, L., Soares, C.: A case study comparing machine learning with statistical methods for time series forecasting: size matters. J. Intell. Inf. Syst. **59**(2), 415–433 (2022)
32. Andrawis, R.R., Atiya, A.F., El-Shishiny, H.: Forecast combinations of computational intelligence and linear models for the NN5 time series forecasting competition. Int. J. Forecast. **27**(3), 672–688 (2011)
33. Hyndman, R.J., et al.: "Package 'forecast'," Online (2024). https://cran.r-project.org/web/packages/forecast/forecast.pdf
34. Meyer, D., et al.: e1071: Misc functions of the department of statistics, probability theory group (formerly: E1071), TU Wien. R package version (2023)
35. Mouselimis, L., Gosso, A., Jonge, E.: elmNNRcpp: the extreme learning machine algorithm. R package (2023)
36. Chollet, F.: et al. "Keras" (2015). https://keras.io
37. Diebold, F.X., Mariano, R.S.: Comparing predictive accuracy. J. Bus. Econ. Stat. **13**(3) (1995)

A Large Dataset of Spontaneous Speech with the Accent Spoken in São Paulo for Automatic Speech Recognition Evaluation

Rodrigo Lima[1], Sidney E. Leal[1(✉)], Arnaldo Candido Junior[2], and Sandra M. Aluísio[1]

[1] University of São Paulo, São Carlos, SP 13566-590, Brazil
guico21@usp.br, sidleal@gmail.com, sandra@icmc.usp.br
[2] Universidade Estadual Paulista, São José do Rio Preto, SP 15054-000, Brazil
arnaldo.candido@unesp.br

Abstract. We present a freely available spontaneous speech corpus for the Brazilian Portuguese language and report preliminary automatic speech recognition (ASR) results, using both the Wav2Vec2-XLSR-53 and Distil-Whisper models fine-tuned and trained on our corpus. The NURC-SP Audio Corpus comprises 401 different speakers (204 females, 197 males) with a total of 239.30 h of transcribed audio recordings. To the best of our knowledge, this is the first large Paulistano accented spontaneous speech corpus dedicated to the ASR task in Portuguese. We first present the design and development procedures of the NURC-SP Audio Corpus, and then describe four ASR experiments in detail. The experiments demonstrated promising results for the applicability of the corpus for ASR. Specifically, we fine-tuned two versions of Wav2Vec2-XLSR-53 model, trained a Distil-Whisper model using our dataset with labels determined by Whisper Large-V3 model, and fine-tuned this Distil-Whisper model with our corpus. Our best results were the Distil-Whisper fine-tuned over NURC-SP Audio Corpus with a WER of 24.22% followed by a fine-tuned versions of Wav2Vec2-XLSR-53 model with a WER of 33.73%, that is almost 10% point worse than Distil-Whisper's. To enable experiment reproducibility, we share the NURC-SP Audio Corpus dataset, pre-trained models, and training recipes in Hugging-Face and Github repositories.

Keywords: Automatic speech recognition evaluation · Spontaneous speech · Brazilian Portuguese · Public speech corpora

1 Introduction

Public or open datasets for training and evaluating automatic speech recognizers (ASR) in Brazilian Portuguese (BP) have increased in number and hours since mid-2020, when there were approximately 60 h available, divided into four small datasets of read speech. In 2024, more than 8 thousand hours are available for training ASR models, either from datasets automatically labelled or manually revised (see Table 1 for a list of datasets). These BP resources allow the training of state-of-the-art ASR models, such as Wav2Vec2-XLSR-53 [8] and Distil-Whisper [11]. Specifically, both models are suitable for low- or medium-resource languages, providing consistent results without the need for training sets having tens of thousand of hours. Wav2Vec2-XLSR-53 is pre-trained in a large dataset of unlabeled speech encompassing 53 languages, including BP, through auto-supervised learning. The pre-training process allows for efficient fine-tuning on available BP datasets, generating competitive ASR results. Distil-Whisper is a distilled version of Whisper [19]. The latter, trained on 680,000 h of multilingual data in a multitask approach, has shown superior performance across multiple datasets and domains when compared to speech recognizers using the Wav2Vec2 model. A disadvantage of the various large Whisper models is their requirement for a large dataset for increased performance and the inability to use them in computationally limited environments. Distil-Whisper, on the other hand, is a small variant based on knowledge distillation focused on efficiency which is 6 times faster, 49% smaller, and performs within 1% word error rate (WER) on out-of-distribution evaluation sets. To the best of our knowledge, Distil-Whisper has never been evaluated for BP datasets.

Regarding the data used for training speech recognizers, [10] characterizes them in two dimensions: (i) **production style**, which describes a continuum of spontaneity, coming from one side, which is planned speech, to the other, which would be unplanned speech; (ii) **mode**, which characterizes the elicitation process of the training pairs, in which on the one hand the text is used as a stimulus for speech (leading to a read speech) and on the other hand it is the speech that is transcribed into text. Thus, read speech is planned speech that comes from a recited text and, at the other extreme, spontaneous/conversational speech is characterized as speech transcribed in an unplanned style. As a compromise, we have an speech prepared to be spoken later (e.g. the Ted Talks[1], in which exponents from all over the world present a talk in 18 min or less, resulting in the need to use a planned text to transmit the message in the short time).

Spontaneous speech has phenomena that make its recognition more complex than that of read or prepared speech. As a result, datasets with read/prepared speech were first and widely disseminated for training speech recognizers or speech synthesizers, mainly for the English language (see [17,24]), since the words in the audios correspond directly to the words of the transcript, as they were read. However, more naturally occurring conversational and spontaneous speech contains many prosodic phenomena that are not present in read speech, such

---

[1] https://www.ted.com/talks.

as higher degrees of segmental reduction[2] and more complex forms of variation (fillers, self-corrections and repetitions) affected by speech rate.

Thus, the various types of spontaneous speech that are common in everyday life, such as classes, conversations between two or more people, and interviews, are important materials for the datasets used to train ASRs, as they bring natural intonation to questions, statements, and expressions of emotions such as surprise, admiration, indignation, anger, astonishment, fright, exaltation, enthusiasm, among others [4,22]. Furthermore, they bring linguistic phenomena such as filled pauses, generally written as "eh", "ah", "ahh", "mm", "uhn" and editing disfluencies (repetitions of words or parts of words, revisions of what is intended to be said, with speech restarts) [16]. Consequently, speech recognizers are expected to perform worse with spontaneous speech than when applied to read speech. [19] shows the application of ASR Whisper to 14 datasets of read and spontaneous speech in the English language. The highest WER (Word Error Rate) values occur for the spontaneous speech datasets, for example CHiME6[3] has a WER of 25.5 and AMI SDM1[4] has a WER of 36.4 while Common Voice[5] has a WER of 9.0 and Tedlium[6] has a WER of 4.0, using the large-v2 model. For Brazilian Portuguese, [15] reports a WER of 14.50% for the Whisper large-v2 model applied to a Portuguese dataset of approximately 17 h of spontaneous speech in interviews about life histories.

This paper introduces a new corpus of spontaneous BP speech, particularly of paulistano accent (São Paulo city), suitable for training and evaluating ASR systems. The NURC-SP Audio Corpus is part of São Paulo division of the NURC project (*Norma Urbana Linguística Culta*), a project designed to document and study Portuguese spoken language by people with a high degree of formal education in five Brazilian capitals (see details of this corpus in Sect. 3). It contains 239.30 h of audios sampled at 16 kHz and their respective transcriptions, totalling 170k segmented audios. The audios were automatically transcribed for the first time and manually revised aiming at the ASR task. Therefore, this new corpus adds 239 h to the amount available for BP (see Table 1), totaling 8,598 h for training and evaluating speech recognition systems.

In order to compare the quality of our corpus, four ASR models are made available in this work: (i) a fine-tuned version of Wav2Vec2.0 XLSR-53 with the train and validation subsets of NURC-SP Audio Corpus, (ii) the same as the first, but using as start point the model that [6] trained for CORAA-ASR v1.1[7], (iii) a distilled version of Whisper Large-v3 model [19], a model that has support for the Portuguese language, trained using our dataset with labels determined

---

[2] For example, [5] comments on vowel elision between words that, in the São Paulo dialect, affects the final posttonic vowels /a/, /o/ and /u/. For instance, in the example (a) *me'ren[da es]co'lar* (school lunch) –> *me'ren[des]co'lar*, the vowel /a/ is deleted and a new syllable is created ([des]).

[3] https://openslr.org/150/.

[4] https://groups.inf.ed.ac.uk/ami/corpus/overview.shtml.

[5] https://commonvoice.mozilla.org/en/datasets.

[6] https://www.openslr.org/51/.

[7] https://github.com/nilc-nlp/CORAA.

by Whisper Large-v3, and (iv) a fine-tuned version of the third model with the train and validation subsets of NURC-SP Audio Corpus.

The corpus and trained models are publicly available in our Github repository[8] under the CC BY-NC-ND 4.0 license. The main contributions made in this paper are summarised as follows:

1. A large BP corpus of human validated audio-transcription pairs containing 239.30 h of spontaneous speech.
2. The first corpus, according to our knowledge, tackling a large amount of accented paulistano speech for ASR in BP (CORAA ASR brings 31.14 h of speech from São Paulo capital)
3. Four ASR models, publicly available, based on the presented corpus.

2 Related Work on Datasets of Brazilian Portuguese for ASR

Table 1 presents large corpora for building ASR systems focused on the Portuguese language. Some resources are multilingual, however Table 1 specifically details the statistics for the Portuguese language. Among the resources presented, there is a slightly greater preponderance of the Brazilian variant in the existing resources, although the European Portuguese variant is also included in some of them.

Table 1. Statistics of the main datasets available for ASR in Portuguese. Some datasets are multilingual, however, the numbers shown are for the Portuguese language.

Corpora (Launching Date)	Speaking Style	Hours	Number of Audios	Number of Speakers	License
MultiLingual LibriSpeech (MLS) (2020)	read	130.1	-	54	CC BY
Multilingual TEDx (2021)	prepared	164	93,000	-	CC BY-NC-ND 4.0
Spotify Podcast Dataset (2022)	spontaneous	7,600	123,000	-	proprietary dataset
CORAA ASR 1.1 (2022)	spontaneous	290	402,466	1,689	CC BY-NC-ND 4.0
Common Voice 17.0 (2024)	read	175	-	3,453	CC-0
NURC-SP Audio Corpus (2024)	spontaneous	239.30	177,224	401	CC BY-NC-ND 4.0

One of the best-known projects dealing with read/prepared speech in English is Librivox[9], which distributes public domain books in audio format. These audios were used in several projects to create resources for processing speech in English, such as LibriSpeech ASR Corpus[10] and LibriTTS[11], both hosted in the Open Speech and Language Resources repository.

The MultiLingual LibriSpeech (MLS) corpus [18] is a large multilingual corpus suitable for speech research, derived from read audiobooks from LibriVox

---
[8] github.com/nilc-nlp/nurc-sp-audio-corpus.
[9] https://librivox.org/pages/about-librivox/.
[10] https://www.openslr.org/12.
[11] https://www.openslr.org/60/.

and consists of 8 languages, including about 32K hours of English and a total of 4.5K hours for other languages. For Portuguese, there are 131 h and 54 speakers. Specifically for the recognition task, it can be combined with other resources, as it has relatively few speakers (audiobook speakers). These resources consist of cleaner audio, generally in studio quality. Because of this, models built solely on this type of audio are only suitable for speech recognition in low-noise scenarios. To overcome this feature, one solution would be to inject noise into the audio or combine it with audio from other projects at different quality levels.

The MultiLingual TeDx Corpus [21] was proposed to enable research in the areas of automatic speech recognition and speech-to-text translation. For Portuguese, there are 164 h available in 93k audios. The corpus is made up of talks on a wide range of subjects, being managed within the scope of the TEDx project, linked to the TED group (Technology, Entertainment and Design). In the case of Portuguese, there are also translations of the transcriptions into English and Spanish. In addition, audios in Spanish and French also have translations into Portuguese.

The Spotify corpus [7] was first released in the English language. In 2022, the company launched a new version[12] incorporating Portuguese [12], offering several audios for the Portuguese language coming mainly from podcasts available on the platform. In total, 76k hours of audio were made available from 123k episodes of shows on the platform. Transcripts were automatically generated and are subject to transcription errors. It has a free license for academic use, but researchers interested in accessing the audios must submit a request for access on the organizers' website.

CORAA ASR is the combination of five independent projects dealing with speech in the interior of São Paulo (with 35.96 hours – ALIP Project), Minas Gerais (with 9.64 hours — C-ORAL Brasil Project), Recife (141.31 hours – NURC-Recife Project) and São Paulo capital (31.14 hours – SP2010 Project), in addition to the prepared speech of TeDx Talks in Brazilian and European Portuguese (72.74 hours), totaling 290 hours and 402k audios.

The Common Voice corpus [1] is an open-use project created by the Mozilla Foundation. The project is a response to the lack of resources for several languages, including Portuguese. In the project, users can simultaneously contribute to the growth of the base and access other people's audio. To collaborate with the project, users can donate audio in their own voices and review donations from other users. The project has tools for collection, validation and internationalization (adaptation to different languages). The permissive use license of this project allows the exploration of the corpus including for commercial purposes. In version 17, the subcorpus for the Portuguese language has 211 h of audio and transcriptions, of which 175 were validated.

The NURC-SP Audio Corpus, the focus of this paper, is described in detail in Sect. 3.

---

[12] htttps://arxiv.org/abs/2209.11871.

3 The NURC-SP Corpus

NURC-SP was the São Paulo division of the NURC (*Norma Urbana Linguística Culta*) project. NURC-SP collected more than 300 h of São Paulo speakers throughout the 1970s. The NURC-SP corpus is made up of 375 audio recordings of three genre types:

1. (DID) Documenter and Participant - Dialogue carried out between a documenter and a participant directly;
2. (D2) Participant and Participant - Dialogue between two participants mediated by a documenter; and
3. (EF) Participant - Lectures, seminars, classes, speeches in general, given by a participant in a formal context.

which are divided into three subcorpora: the Minimum Corpus (21 audio recordings), the Corpus of Non-Aligned Audios and Transcriptions (26 audio recordings), and the Audio Corpus.

NURC-SP had its original analog audios digitized by the Alexandre Eulalio Cultural Documentation Center (CEDAE/UNICAMP) and in December 2020 it was made available to the Tarsila Project[13] as a resource to: (i) build training data sets for spontaneous speech recognition systems, and (ii) facilitate future linguistic studies. The three subcorpora that integrate the NURC-SP Digital repository were made available in the Portal NURC-SP Digital [20] for researchers from fields of Linguistics as well for the general public, due to easy access and filtering tools to use the material.

Here, in this paper, we focus on the NURC-SP Audio Corpus. It was originally composed of 328 audio recordings without transcriptions, which have been automatically transcribed by WhisperX [3]. The revision of automated WhisperX transcription segments were performed from June 2023 to December 2023 by 14 native speakers of BP. The revision process was based on an annotation guideline designed to: (i) help making the revision uniform and (ii) remove segments with high amount of noise and overlapping voices. The guideline contains 11 rules, dealing with (i) orality marks, (ii) how to transcribe filled pauses, (iii) repetitive hesitations, (iv) numbers, (v) individual letters, (vi) acronyms, (vii) foreign terms, (viii) punctuation and capitalization, (iv) emotion sounds (ex. laughter) which were annotated in parentheses, (x) misunderstanding of words or passages and (xi) how to deal with automatic segmentation failures.

3.1 Pre-processing and Filtering of NURC-SP Audio Corpus for Evaluating ASR Models

Audio quality was a very important characteristic when preparing the dataset for evaluating ASR models. In this section, we present the processing steps of the NURC-SP Audio Corpus to generate the version used to evaluate the four ASR models in this paper:

---

[13] https://sites.google.com/view/tarsila-c4ai/home.

1. Removal of those audios (and all of its segments) with poor/fair quality (with high number of problems with voice distortion, hissing, background noise and interruptions);
2. Annotation of the segments using "high/low" quality labels for all remaining audios, based on what was reported by human annotators during the automatic transcription revision;
3. Removal of part of the low quality segments, as detailed further in this section, in order to generate the final dataset for training ASR models; and
4. Generation of both statistics of NURC-SP Audio Corpus and the train, validation and test sets.

Of the 328 audios, four of them had poor quality, 92 had fair quality, 52 had good quality and 180 had excellent quality. Therefore, of these 328 audios, five of them were not used for training the ASR models in this paper: SP_EF_395, SP_DID_283, SP_DID_194, SP_D2_397, SP_D2_337, as they had problems with voice distortion, hissing, background noise and interruptions, as they were responsible for generating transcriptions with a high rate of errors.

The transcriptions were processed to remove the labels used in the automatic transcription revision: (i) "###" was used to indicate segments with high background noise, very low audio, overlapping voices and music; (ii) paralinguistic sounds such as laughter or cough were annotated in parentheses—(laughter), (cough); (iii) misunderstanding of words or passages were also marked with parentheses "( )"; and (iv) words truncated at the end or the beginning of the audio due to automatic segmentation failure were partially transcribed with the help of ">" and "<" (for example: if the word "casa" – house) where truncated at the end of the segment, it was annotated as "ca>" and if it was at the beginning, it was annotated as "<sa".

Segments marked with "###" are removed completely, as well as segments having only paralinguistic sounds (e.g., laughter) or containing only truncated words. Segments having truncated and non-truncated words (marked with ">" or "<") had the truncated word removed and were marked as low quality. Pairs that have words or passages marked as misunderstood, using the "( )" tags, had the tags removed and the pairs were marked as "low quality". Audio-transcription pairs with paralinguistic sounds tags only had the tags removed and were marked as "high quality". The dataset available in this work has 323 audio-transcription pairs, which were divided into 177,224 segments transcribed by WhisperX, and revised by native BP speakers (see more details in Table 2).

3.2 Statistics of NURC-SP Audio Corpus

Table 2 presents statistics of NURC-SP Audio Corpus Dataset. Overall, almost 240 h of speech were generated in the final export, distributed in around 177 thousand segments with an average duration of approximately 5 s. In total, more than 2 million tokens were transcribed, resulting in approximately 12 tokens per segment. It should be noted that the sum of audios with females and males is higher than the total number of audios, because a given audio can have simultaneously both types of voices.

Table 2. NURC-SP Audio Corpus Statistics

	Training	Validation	Test	Total
Number of Audios (unique)	303	6	14	323
Female voices (number)*	191	4	9	204
Male voices (number)*	186	3	8	197
Male/Female Ratio	0.97	0.75	0.89	0.97
Duration (hours)	224.47	4.60	10.23	239.30
Quantity of Audios (segmented)	166,971	3,142	7,111	177,224
Segment Duration (avg seconds)	4.83	5.27	5.18	4.86
Segment Duration (max seconds)	29.87	29.70	28.98	29.87
Avg Tokens	11.81	12.67	12.32	11.85
Avg Types	10.61	11.73	12.29	10.69
Total Tokens	1,971,993	39,715	87,598	2,099,306
Total Types	84,767	8,005	12,218	88,004
Type/Token Ratio	0.043	0.202	0.139	0.042

*There are audios with two speakers, in several combinations: Male and Male, Male and Female, Female and Female.

Figure 1 shows the duration of audio segments. It can be noted that audios smaller than 5 are predominant, with a peak in approximately 3 s, while the interval from 5 to 20 s is also well represented. There are also audios longer than 20 s, but their occurrence is less common.

Figure 2 shows the number of audios by age and gender. The number of audios of females and males are similar, resulting reasonable balancing regarding gender, which can also be seem in male/female ratio in Table 2. Regarding age, there are more speakers in age group II (from 36 to 55 years old) than the others groups (I = 25–35 and III = 36–55). There also a small number of audios with speakers of unknown age.

Figure 3 shows the number of audios by speech gender (Lectures and Talks—(EF), Interviews (DID) and Dialogs (D2), respectively). Interviews and dialogues are much more common than lectures, totalling 90% of the audios. Thus, NURC-SP is a corpus predominantly made of spontaneous speech.

4 Experiments on NURC-SP Audio Corpus

We performed experiments over NURC-SP Audio Corpus in order to assess the dataset quality and limitations. We based our experiments in two main architectures: Wav2Vec2 [8] and Distil-Whisper [11]. Wav2Vec2 has the advantage of presenting a good performance in low and middle resource languages. Wa2Vec2 is an auto-supervised pretraining in different languages, which speeds up the training process for ASR, allowing robustness against noises that are common in spontaneous speech. Distil-Whisper [11] is a distilled version of Whisper, the

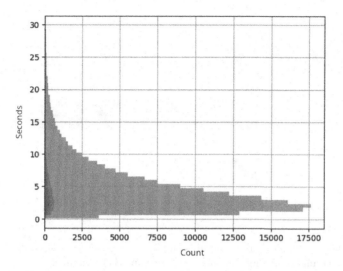

Fig. 1. Audio Segments Duration

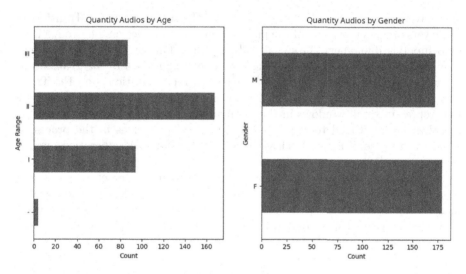

Fig. 2. (a) Audios by Speaker's Age Range and (b) Gender (M/F). Speakers are distributed into three age groups (I = 25–35, II = 36–55, and III = 56 onwards).

state-of-the-art in ASR in several languages. We opted for the distilled version because original Whisper is costly, being originally trained over thousands hours of labelled speech, while Distil-Whisper can perform efficiently with less training data. Distil-Whisper is smaller and faster, while preserving many of original Whisper's strengths.

Wav2vec2 architecture operates directly on the wave signal, without the need of intermediary representations such as spectrograms. Furthermore, there are two

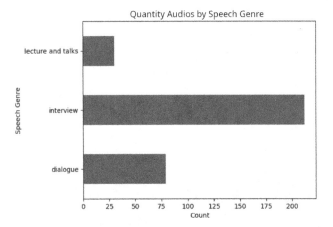

Fig. 3. Audios by Speech Genre. Lectures and Talks receive the acronym EF, Interviews are called DID and Dialogues between two participants are named D2.

main layer types: convolutional layers process the raw wave signal; Transformer [23] layers process the result of the previous step. The model can be trained in an auto-supervised manner. The signal is also quantized in the after passing through convolutional layers and masked language modelling is used in the Transformer layers. A rich loss takes into account quantized representations and the Transformer layer predictions in order to produce diversified quantized representations that represent small windows in the original audio. The learned representations can then be fine-tuned to represent, for example, phonemes in the process of ASR. In this case, it is used a classical seq2seq loss for ASR generation, namely the CTC loss [13].

Whisper proposes an off-the-shelf approach to its architecture, consisting on a traditional transformer, with encoding and decoding blocks, and focusing more on the training data than on architectural adjustments. The input must be converted to log-mel spectrogram format. Distil-Whisper is the distilled version of Whisper. Model distillation is a technique that allows for a smaller model (the student) to be trained on both data labels as well as predictions from a bigger model (the professor) in a given prediction task. The main idea behind distillation is that a model's predictions have richer information about the data distribution than original labels. Therefore, a bigger model can capture data distribution and transmit it to the student. Both Whisper and Distil-Whisper are available in different versions. Comparing Whisper Large-v3 with Distil-Whisper Large-v3 (used in this work), it can be observed an improvement of 5.8 times in inference time while using 51% fewer parameters in the distilled version.

Table 3 presents statistics of training, validation and test partitions of each NURC-SP Audio Corpus subset.

Table 3. Statistics of training, validation and test partitions of each NURC-SP Audio Corpus subset: the total duration in hours and number of audio recordings (left) and number of speakers in each gender (right).

Subset	Hours/Number			Speakers (M|F)		
	Training	Validation	Test	Training	Validation	Test
D2	75.93/75	1.45/1	3.12/3	66|82	1|1	3|3
DID	132.39/205	1.46/2	3.83/6	103|103	1|1	3|3
EF	16.15/23	1.69/3	3.87/5	17|6	1|2	2|3
Total	224.47/303	4.60/6	10.23/14	186|191	3|4	8|9

4.1 Baseline Models Development

To assess the dataset quality, we trained four new ASR models and compared the results with previous related works. Two of them are fine-tuned versions of Wav2Vec2 and the third and fourth models are approaches with Distil-Whisper. All four models are described below.

Wav2Vec2-NURC-SP-1. This model is a fine-tuned version of Wav2Vec 2.0 XLSR-53 [2] [9], pre-trained over 53 languages (Portuguese included). The pre-trained model was fine-tuned with our train and validation subsets of (NURC-SP Audio Corpus) NURC-SP-AC in one GPU Nvidia DGX A100 80GB for 16 epochs, with early stop of 10. The other settings were the same from [6], training code is available at github[14].

Wav2Vec2-NURC-SP-2. The second model is almost the same as the first one, but using as start point the model that [6] trained for CORAA-V1 and made publicly available at HuggingFace[15]. It also was trained for 16 epochs and finished with a better WER that the first model.

Distil-Whisper-NURC-SP. The third model is a distilled version of Whisper Large-v3 [19], a model that has support for the Portuguese language, trained using our dataset with labels determined by Whisper Large-v3, with the reason being that more knowledge can be passed from the teacher model to the student model this way [11]. The model was trained with our train and validation subsets of NURC-SP-AC in one GPU Nvidia DGX A100 80GB for 48 epochs, following to the steps given by the Distil-Whisper github[16].

Distil-Whisper-NURC-SP-Fine-Tuned. This model is a fine-tuning of the third one with our train and validation subsets of NURC-SP-AC, following the steps recommended by the Distil-Whisper developers[17]. It was also trained with

---

[14] https://github.com/Edresson/Wav2Vec-Wrapper.
[15] https://huggingface.co/Edresson/wav2vec2-large-xlsr-coraa-portuguese.
[16] https://github.com/huggingface/distil-whisper/tree/main/training.
[17] https://huggingface.co/blog/fine-tune-whisper.

48 epochs in one GPU Nvidia DGX A100 80GB and achieved a better WER and CER (Character Error Rate)[18] than the third model.

4.2 Normalization

We performed the following steps to normalize the train, validation and test subsets into a standardized form in order to penalize only when a word error is caused by failure in transcribing the word, and not by formatting, punctuation or spontaneous speech differences. The revised transcriptions of NURC-SP Audio Corpus are of different text genders (EF, D2 and DID) and use upper and lower case letters and punctuation, as well as filled pause markers such as *eh, hum, ãh,* etc. To simplify the training and calculation of the CER and WER metrics, the following normalization was performed:

1. The texts were transformed into lowercase;
2. All punctuation marks generated by Whisper were removed (ellipsis, exclamation mark, fullstop, question mark, and comma);
3. The filled pauses were standardized to: *eh, uh, ah,* as follows: *eh = eh, éh, ehm, ehn; uh = uh, hm, uhm, hmm, mm, mhm; ah = ah, huh, ãh, ã*;
4. Any successive blank spaces have been replaced with one space.

4.3 Experiments Results

For comparison purposes, we kept the tests on the datasets used in previous works and added the new dataset made available by this work. The [14] and [6] models were rerun on all datasets, as were the four new trained models. WER and CER metrics were evaluated for each run. The new dataset proved to be quite challenging for the ASR task, as can be seen in Table 4.

Table 4. NURC-SP Audio Corpus Baseline Models results compared with previous works. The best values for CER and WER appear in bold.

Datasets	CommonVoice		CORAA v1		NURC-SP Audio Corpus		Mean	
Metrics	CER	WER	CER	WER	CER	WER	CER	WER
Gris et al. (2022) [14]	**4.50**	**16.32**	22.32	43.70	26.52	47.74	17.78	35.92
Candido Jr et al. (2023) [6]	6.99	24.44	**11.02**	**24.18**	22.87	40.29	13.63	29.64
Wav2Vec2-NURC-SP-1	10.41	35.74	24.24	49.13	23.69	43.44	19.45	42.77
Wav2Vec2-NURC-SP-2	8.07	26.74	14.59	31.19	19.30	33.73	13.99	30.55
Distil-Whisper-NURC-SP	7.14	18.66	23.53	36.17	25.03	36.25	18.57	30.36
Distil-Whisper-NURC-SP-Fine-Tuned	5.70	17.76	14.89	26.91	**15.77**	**24.22**	**12.12**	**22.96**

Below, we show four examples of result instances from our best model (Distil-Whisper NURC-SP-Fine-Tuned)—the first is the Original Normalized (ON) and

---

[18] Here, we also focus our analysis on the metric CER, because for smaller audios, with just a few words, this metric tends to be more reliable.

the second is the Model's Prediction Normalized (MPN). In the first two, there are problems with named entities as the "Martinelli" building was transcribed as "martini" and the name "Paulo Emilio Salles Gomes" had a wrong transcription as "paulo e milho fales gomes". The model used more frequent common names than the proper names "Martinelli" and "Emilio Salles". Rare words related to certain domains (e.g. food) or terminologies of research areas (e.g. linguistics) also bring some difficulties to the model (see third and fourth examples):

1. (ON) *o martinelli ficou célebre em todo o exterior do estado no interior do estado de são paulo e mesmo pelo brasil afora como um arranha-céu notável para a época*; (MPN) *o martini ficou célebre em todo o exterior do estado do interior de estado de são paulo e mesmo pelo brasil afora como arranha-céu notável para a época*;
2. (ON) *você me falou em cinema eu lembrei de paulo emilio salles gomes que foi meu colega na faculdade e é um entendidíssimo de cinema né*; (MPN) *você me falou em cinema eu me lembrei de paulo e milho fales gomes que foi minha colega na faculdade e é um entendidíssimo de cinema né*;
3. (ON) *cuscuz paulista bobó de camarão essas coisas assim*; (MPN) *cuscos paulista babota de camarão essas coisas*;
4. (ON) *de um lado objeto direto do outro adjunto*; (MPN) *de um lado é o chefe do e o outro é de junho*.

5 Discussion and Conclusions

In this paper, we present a freely available spontaneous speech corpus for the Brazilian Portuguese language, totaling 239.30 h, and report preliminary ASR results, using both the Wav2Vec2-XLSR-53 and Distil-Whisper models fine-tuned and trained on our corpus. To the best of our knowledge, Distil-Whisper has never been evaluated for BP datasets.

For comparison purposes, we also bring the tests on the datasets used in previous works. The Wav2vec 2.0 based model from [14] remains the best for the Common Voice dataset with a WER of 16.32%, even in the most recent version (CV version 17.0), as it is trained in prepared/read speech. In the CORAA v1 dataset, [6]'s models, which is also trained an Wav2Vec 2.0 model, but mostly based on spontaneous speech, continues presenting the best values for this dataset with 24.18% WER, but our Wav2Vec2-NURC-SP-2 model was in second position (since it is also trained for spontaneous speech). For the new dataset, focus of this paper, the best results were the Distil-Whisper fine-tuned over NURC-SP Audio Corpus with a WER of 24.22% followed by a fine-tuned versions of Wav2Vec2-XLSR-53 model with a WER of 33.73%, that is almost 10% point worse than Distil-Whisper's. These results indicates that Distil-Whisper is promising for low and medium resource languages and should be evaluated with more BP datasets in the future.

Acknowledgements. This work was carried out at the Center for Artificial Intelligence (C4AI-USP), with support by the São Paulo Research Foundation (FAPESP grant #2019/07665-4) and by the IBM Corporation. This project was also supported by the Ministry of Science, Technology and Innovation, with resources of Law No. 8.248, of October 23, 1991, within the scope of PPI-SOFTEX, coordinated by Softex and published Residence in TIC 13, DOU 01245.010222/2022-44.

References

1. Ardila, R., et al.: Common voice: a massively-multilingual speech corpus. In: Calzolari, N., et al. (eds.) Proceedings of the Twelfth Language Resources and Evaluation Conference, May 2020, Marseille, France, pp. 4218–4222. European Language Resources Association (2020). https://aclanthology.org/2020.lrec-1.520
2. Baevski, A., Zhou, H., Mohamed, A., Auli, M.: wav2vec 2.0: a framework for self-supervised learning of speech representations. Adv. Neural Inf. Process. Syst. **33**, 12449–12460 (2020). https://arxiv.org/abs/2006.11477
3. Bain, M., Huh, J., Han, T., Zisserman, A.: WhisperX: time-accurate speech transcription of long-form audio. In: INTERSPEECH 2023, pp. 4489–4493 (2023). https://doi.org/10.21437/Interspeech.2023-78
4. Beckman, M.E.: A typology of spontaneous speech. In: Computing Prosody: Computational Models for Processing Spontaneous Speech, pp. 7–26. Springer, New York (1997). https://doi.org/10.1007/978-1-4612-2258-3_2
5. Bohn, G.P.: Processos e representações lexicais: o caso das vogais posteriores do dialeto paulista. DELTA: Documentação e Estudos em Linguística Teórica e Aplicada **33**(2), September 2017. https://revistas.pucsp.br/index.php/delta/article/view/34370
6. Candido_Junior, A., et al.: CORAA ASR: a large corpus of spontaneous and prepared speech manually validated for speech recognition in Brazilian Portuguese. Lang. Resour. Eval. **57**, 1139–1171 (2023). https://doi.org/10.1007/s10579-022-09621-4. https://link.springer.com/article/10.1007/s10579-022-09621-4
7. Clifton, A., et al.: 100,000 podcasts: a spoken English document corpus. In: Scott, D., Bel, N., Zong, C. (eds.) Proceedings of the 28th International Conference on Computational Linguistics, Barcelona, Spain (Online), December 2020, pp. 5903–5917. International Committee on Computational Linguistics (2020). https://aclanthology.org/2020.coling-main.519
8. Conneau, A., Baevski, A., Collobert, R., Mohamed, A., Auli, M.: Unsupervised cross-lingual representation learning for speech recognition. In: Proceedings of the INTERSPEECH 2021, pp. 2426–2430 (2021). https://doi.org/10.21437/Interspeech.2021-329
9. Conneau, A., et al.: Unsupervised cross-lingual representation learning at scale. In: Proceedings of the 58th Annual Meeting of the Association for Computational Linguistics, pp. 8440–8451 (2020)
10. Gabler, P., Geiger, B.C., Schuppler, B., Kern, R.: Reconsidering read and spontaneous speech: causal perspectives on the generation of training data for automatic speech recognition. Information **14**(2) (2023)
11. Gandhi, S., von Platen, P., Rush, A.M.: Distil-whisper: robust knowledge distillation via large-scale pseudo labelling (2023)
12. Garmash, E., et al.: Cem mil podcasts: a spoken Portuguese document corpus for multi-modal, multi-lingual and multi-dialect information access research. In: Arampatzis, A., et al. (eds.) Experimental IR Meets Multilinguality. Multimodality, and Interaction: 14th International Conference of the CLEF Association, CLEF 2023,

Thessaloniki, Greece, 18–21 September 2023, Proceedings, pp. 48–59. Springer, Heidelberg (2023). https://doi.org/10.1007/978-3-031-42448-9_5
13. Graves, A., Fernández, S., Gomez, F., Schmidhuber, J.: Connectionist temporal classification: labelling unsegmented sequence data with recurrent neural networks. In: Proceedings of the 23rd International Conference on Machine Learning, pp. 369–376 (2006)
14. Gris, L.R.S., Casanova, E., de Oliveira, F.S., da Silva Soares, A., Candido Junior, A.: Brazilian Portuguese speech recognition using Wav2vec 2.0. In: Pinheiro, V., et al. (eds.) Computational Processing of the Portuguese Language, pp. 333–343. Springer, Cham (2022). https://doi.org/10.1007/978-3-030-98305-5_31
15. Gris, L.R.S., Marcacini, R., Junior, A.C., Casanova, E., Soares, A., Aluísio, S.M.: Evaluating OpenAI's whisper ASR for punctuation prediction and topic modeling of life histories of the museum of the person (2023)
16. Liu, Y., Shriberg, E., Stolcke, A., Hillard, D., Ostendorf, M., Harper, M.: Enriching speech recognition with automatic detection of sentence boundaries and disfluencies. IEEE Trans. Audio Speech Lang. Process., 1526–1540 (2006)
17. Panayotov, V., Chen, G., Povey, D., Khudanpur, S.: Librispeech: an ASR corpus based on public domain audio books. In: 2015 IEEE International Conference on Acoustics, Speech and Signal Processing (ICASSP), pp. 5206–5210 (2015)
18. Pratap, V., Xu, Q., Sriram, A., Synnaeve, G., Collobert, R.: MLS: a large-scale multilingual dataset for speech research. In: Proceedings of the INTERSPEECH 2020, pp. 2757–2761 (2020). https://doi.org/10.21437/Interspeech.2020-2826
19. Radford, A., Kim, J.W., Xu, T., Brockman, G., Mcleavey, C., Sutskever, I.: Robust speech recognition via large-scale weak supervision. In: Krause, A., Brunskill, E., Cho, K., Engelhardt, B., Sabato, S., Scarlett, J. (eds.) Proceedings of the 40th International Conference on Machine Learning, Proceedings of Machine Learning Research, 23–29 July 2023, vol. 202, pp. 28492–28518. PMLR (2023)
20. Rodrigues, A.C., et al: Portal NURC-SP: design, development, and speech processing corpora resources to support the public dissemination of Portuguese spoken language. In: Gamallo, P., et al (eds.) Proceedings of the 16th International Conference on Computational Processing of Portuguese, pp. 187–195. Association for Computational Lingustics (2024)
21. Salesky, E., et al.: The Multilingual TEDx corpus for speech recognition and translation. In: Proceedings of the INTERSPEECH 2021, pp. 3655–3659 (2021)
22. Éva Székely, Henter, G.E., Beskow, J., Gustafson, J.: Spontaneous conversational speech synthesis from found data. In: Proceedings of the INTERSPEECH 2019, pp. 4435–4439 (2019). https://doi.org/10.21437/Interspeech.2019-2836
23. Vaswani, A., et al.: Attention is all you need. In: Advances in Neural Information P0rocessing Systems, vol. 30 (2017)
24. Zen, H., et al.: LibriTTS: a corpus derived from LibriSpeech for text-to-speech (2019)

A Multi-level Semantics Formalism for Multi-Agent Microservices

Alison R. Panisson[✉] and Giovani P. Farias

Department of Computing (DEC), Federal University of Santa Catarina (UFSC), Araranguá, Brazil
alison.panisson@ufsc.br

Abstract. Operational semantics is a fundamental approach to the formalisation of programming languages and almost a standard when it comes to agent-oriented programming languages. It helps ensure the correctness of interpreters, facilitates their implementation, and supports proofs of important properties. However, in the literature, there are few attempts of formalising multi-level operational semantics for multi-agent systems which allows the formalisation of all levels of abstraction in those systems. Also, recently, authors have started to discuss a new approach to build multi-agent systems known as multi-agent microservices, arguing that microservices represent a potential point of convergence between modern software engineering and multi-agent systems. We agree that microservices could be a useful approach to think of multi-agent systems and, in this paper, we propose a multi-level semantics to formalise this new approach to build multi-agent systems based on the abstraction of microservices. We not only introduce a template for those formalisation but we also show how our operational semantics style promotes desired characteristics of modularity, reusability, and a clear specification of shared communication artifacts in multi-agent microservices.

Keywords: Multi-agent Systems · Microservices · Multi-level Semantics

1 Introduction

Multi-agent systems involve core concepts such as loose-coupling, distribution, reactivity and isolated (local) state. Thus, given also its characteristics, the concept of agents has been widely studied and a large range of tools, programming languages and methodologies developed [5]. However, the adoption of these tools, programming languages and methodologies is not largely seen within industry [10]. In the same time that multi-agent systems have been largely studied, the industry has evolved through a range of enterprise paradigms and models, that have shifted from centralised systems towards highly-decentralised systems, showing properties that were exclusively found in multi-agent systems [9]. Nowadays, the industry has focused on the so-called microservice paradigm, in which systems are built from small, loosely-coupled services that maintain their own state [16]. With microservices, the industry has found a manner to develop systems towards an ecosystem of tools and components that are easy to extend, strong on fault tolerance and scaling [34].

Studies on the convergence of software engineering for microservices and multi-agent systems brings many benefits for both areas [9]. On the one side, from the multi-agent systems perspective, the main benefits are: (i) the embracing of Web models for agents; (ii) the decomposition of agent systems into self-contained subsystems, which can be independently tested, improving reliability and the creation of libraries of agents; and (iii) the alignment of multi-agent systems deployment with the best practice of software engineering into the industry. On the other side, from the microservices perspective, the main benefit is the seamless integration of agent technologies into microservices-based systems [9]. Although the convergence between modern software engineering for microservices and multi-agent systems has been studied, for example, in [9], still it is missing a formalisation for such systems. In this paper, we propose a formal operational semantics to multi-agent microservices. Thus, a formalisation for multi-agent microservices using our semantics can be directly implemented in any agent-oriented programming language that supports such components. The operational semantics provide means to specify programs which are independent of any particular programming language, taking in consideration the main components of multi-agent microservices.

The main contributions of this work are: (i) a template to define the formal semantics of multi-agent microservices, which allows us to specify the evident multiple levels of abstraction present in multi-agent microservices; (ii) abstract instances of the multi-level semantic rules, showing how the operational semantics can be used to define formally the communication processes not only between microservices and agents but also for the treatment of external communication; (iii) a discussion on how the multi-level semantics provide a specification of multi-agent microservices that promote desired characteristics like modularity and reusability. Also, we discuss how the multi-level components provide a clear specification for shared communication artifacts, which is a fundamental characteristic of microservices, given they reduce chattiness [28].

2 Background

2.1 Multi-Agent Microservices (MAMS)

In [9], the authors introduce Multi-Agent MicroServices (MAMS), which is a class of systems comprised of Agent-Oriented MicroServices (AOMS) and Plain-Old MicroServices (POMS). AOMS are microservices built using MAS methodology and they are exposed through well-defined interfaces modelled as a set of Representation State Transfer (REST). REST promotes a view in which all resources are isolated and treated as distinct services. In order to avoid an abundance of network calls, chattiness [28], a microservice may often play the role of a host to multiple resources. Thus, microservices share some characteristics with agents. Agents are not just resources; they are complex rational and decision-making entities. However, an agent can be seen itself as a resource as well as it can be seen as a host for others resources it has access to perceive and act. Also, agents rarely work separately, and multi-agent systems are normally modelled by means of a society of agents that will collaborate, through message-passing, to reach an organisational goal [33]. Another related concept present in microservices and multi-agent systems is visibility. As human organisations, some individuals are supposed to be more communicative than other. For example, in a company, sales and customers support staff are supposed to interact with the external public, while other roles within that organisation will implement internal functions necessary to keep the company running. Concepts of *interface* agents for organisations are already discussed in the multi-agent system literature [12,22].

With these ideas of convergent concepts between microservices and multi-agent systems technologies, the authors in [9] have proposed that an agent (or group of agents) can implement a microservice and an organisation of agents (or multi-agent system) could be seen as a service, an approach they named as *Organisation as a Service (OaaS)*. Figure 1 shows an example of two multi-agent systems (services), each system configured through group/set of agents representing microservices within the multi-agent system. In MAMS, agents communicate within their groups, microservices interact through interfaces usually handled by an agent, and multi-agent systems (services) interact through interfaces usually handled by a service.

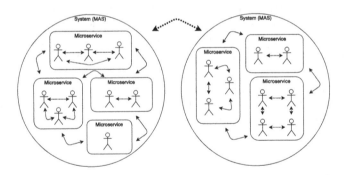

Fig. 1. Multi-Agent Microservices

2.2 Multiple levels in Multi-agent Systems

There are various approaches to developing multi-agent systems that consider different levels of abstraction. In this section, we discuss some of such approaches in order to clarify the existence of such multiple levels in MAS, which supports the manner we are going to define the operational semantics for multi-agents microservices. One well-known practical multi-agent programming framework is JaCaMo [3,4]. JaCaMo is built upon three existing platforms: Jason [7] for programming autonomous agents, $\mathcal{M}$OISE [19] for programming agent organisations, and CArtAgO [27] for programming shared environments. JaCaMo emphasises the existence of three levels, in that work called "dimensions", in a single framework so that all levels can be programmed using the languages provided by the combined platforms mentioned above. Also, the development of MAS based on Electronic Institutions [21,32] considers the existence of multiple levels in MAS, emphasising that, while electronic institutions focus on the so-called macro-level (societal) aspects, the so-called micro-level has its focus on the agents [30]. Further, the PopOrg (Populational-Organisational) Model [12] also aims to structure the MAS in various levels of abstraction. The model focuses on the dynamics of the different internal organisational levels of the organisation of MAS. In addition, the work in [11] introduces so-called *social sub-systems*, which account for the structuring of organisational entities into sub-systems within the MAS. Other aspects that emphasises the existence of multiple levels in MAS can be found in the survey [18].

Furthermore, multi-agent microservices inherit such characteristics of multiple levels of abstraction from multi-agent systems. In multi-agent microservices, the multiple levels of abstraction not only promote the system organisation but also promote the modularity and reusability of the system's components, i.e., microservices.

3 Operational Semantics for Multi-Agent Microservices

The main idea behind structured operational semantics proposed by [26] is to define the behaviour of a program in terms of the behaviour of its parts, providing a structural, i.e., syntax-oriented and inductive, view on operational semantics. Thus, specifying a structured operational semantics, it is possible to define the behaviour of a program in terms of a (set of) transition relation(s). The transition relation(s) take the form of a set of inference rules that define the valid transitions of composite tuples representing the transitions of its components. Defining the operational semantics allows us not only a formal analysis of the behaviour of systems but also studies of relations between systems. Furthermore, operational semantics has an intuitive look and easy-to-follow structure.

In this section, we propose a style of formal semantics for multi-agent microservices, in which their components and interfaces are formalised through an operational semantics. Updates at the system, microservices (group/set of agents), and agents internal state are given by an elegant transition system proposed by [26] extended to embrace multiple levels of abstraction. As will be discussed in this work, the operational semantics provide a formal specification of the system, guaranteeing the modularity necessary in those systems, encouraging the reuse of those formalisations.

We argue that our style of operational semantics promotes the following desired characteristics of multi-agent microservices: (i) The modularity of the

CONDITIONS
(a)   System $\rightarrow_s$ Updated System
(b) Set of Microservices $\rightarrow_{ms}$ Updated Set of Microservices
(c)   Set of Agent $\rightarrow_{ag}$ Updated Set of Agents

where:
(a) Updates at Components of the System
(b) Updates at Components of the Microservices
(c) Updates at Components of the Agents

(SEMANTICRULENAME)

formalisation, which not only fits into methodologies of software engineering to systems based on services but also simplifies the specification of those systems through a multi-level composition, and clear interaction between components from those multiple levels. It also promotes the scalability, considering that the operational semantics provides a modular approach to formalise the modern forefront on the development of distributed computing towards containerisation. It allows us to specify containers individually; (ii) The reusability of the formalisation, in which components (or even entire levels of the formalisation) could be reused in the development of similar systems; (iii) The clear specification of shared communication artifacts, which are essential in systems built based on the paradigm of microservices, when it is necessary that agents, microservices and system share information, thus avoiding chattiness [28].

3.1 The Basis for MAMS Operational Semantics

In our approach, a system will be represented as a tuple $\langle s_{id}, \mathcal{MS}_s, \mathcal{D}_s, \mathcal{M}_s \rangle$ with s_{id} the service identifier, $\mathcal{MS}_s$ a set of microservices (the microservices populating the main service), $\mathcal{D}_s$ a set of data that service stores at the system level, and $\mathcal{M}_s$ is the message pooling component for that system. The microservices are represented by the tuple $\langle ms_{id}, Ag_{ms}, \mathcal{D}_{ms}, \mathcal{M}_{ms} \rangle$ with ms_{id} the microservice identifier, Ag_{ms} a set of agents (which are populating that microservice), $\mathcal{D}_{ms}$ a set of data that microservice stores at the microservice level, and $\mathcal{M}_{ms}$ is the message pooling component for that microservice. Each agent in the multi-agent system (a member of the microservices) is represented by a tuple $\langle ag_{id}, \mathcal{G}_{ag}, \mathcal{B}_{ag}, \mathcal{A} \rangle$ with ag_{id} the agent identifier, $\mathcal{G}_{ag}$ the agent goals, $\mathcal{B}_{ag}$ the agent beliefs, and $\mathcal{A}$ the set of actions that the agent is able to execute.

In order to define the semantic rules, we are going to use a particular style of operational semantics introduced in [23,24], in which multiple levels of a multi-agent system are considered. Thus, we have the template to the operational semantic rules represented in the semantic rule below, named SemanticRuleName.

In the semantic rule SemanticRuleName, we present a template to define the operational semantics of multi-agent microservices considering their multiple levels of abstraction. At the top of the semantic rule, we have CONDITIONS representing the preconditions necessary to apply such semantic rule, which also represents the context of the system at that moment. Below the conditions, there are three transitions systems: $\rightarrow_s$, $\rightarrow_{ms}$, $\rightarrow_{ag}$, representing the transitions occurring in the system, set of microservices, and set of agents, respectively. Note that the set of agents being considered at that particular semantic rule represents the set of agents populating the microservice being considered from the set of microservices populating that system. We use the notation "$\{\langle ag_{id}, \mathcal{G}_{ag}, \mathcal{B}_{ag}, \mathcal{A}\rangle \ldots\}$" as a simplified representation for the set of agents, highlighting the particular agent being considered, i.e., ag_{id} is the agent currently being considered in that semantic rule. The same notation is used to highlight the microservice, from the set of microservices, being considered at that particular semantic rule. Thus, using transitions in three levels, i.e., the service, the microservices, and the agents, we are able to define the operational semantics for multi-agent microservices which is modular and reusable.

3.2 Abstract Examples

In this section, we show abstract examples of common operations regarding multi-agent microservices, for example, the creation of a new microservice in the system, and the treatment of external messages[1].

Both, microservices approaches and multi-agent systems communicate through message passing and it should not be different in multi-agent microservices. In order to exemplify some operations in multi-agent microservices, we use the following format for messages: $\langle$SENDER, RECEIVER, PERFORMATIVE, CONTENT$\rangle$, in which SENDER represents the sender of the message, RECEIVER represents the receiver of the message (a system, a particular microservice, or a particular agent), PERFORMATIVE represents the speech-act used, representing the intention of the sender, and CONTENT represents the content of the message, which together with the performative will provide the meaning of that message. For example, $\langle$bob, alice, achieve, closed(door)$\rangle$ has a different meaning than $\langle$bob, alice, tell, closed(door)$\rangle$, in the first message bob's desires that alice closes the door, while in the second message bob is telling alice that the door is closed. This representation for messages is commonly used in the multi-agent literature, for example [25, 31]. In the semantic rules, we use a select message function, S_M, which selects one of the messages from the message pooling being considered at that semantic rule, according to some priority for treating those messages.

$$S_M(\mathcal{M}_s) = \langle sender, s_{id}, \texttt{create\_microservice}, \langle ms_{id}, Ag_{ms}, \mathcal{D}_{ms}, \mathcal{M}_{ms}\rangle\rangle$$

$$\frac{\begin{array}{ll}(a) \ \langle s_{id}, \mathcal{MS}_s, \mathcal{D}_s, \mathcal{M}_s\rangle \rightarrow_s & \langle s_{id}, \mathcal{MS}'_s, \mathcal{D}_s, \mathcal{M}'_s\rangle \\ (b) \qquad \{\ldots\} & \rightarrow_{ms} \{\langle ms_{id}, Ag_{ms}, \mathcal{D}_{ms}, \mathcal{M}_{ms}\rangle \ldots\}\end{array}}{}$$

where:
(a) $\mathcal{MS}'_s = \mathcal{MS}_s \cup \{\langle ms_{id}, Ag_{ms}, \mathcal{D}_{ms}, \mathcal{M}_{ms}\rangle\}$
$\mathcal{M}'_s = \mathcal{M}_s \setminus \{\langle sender, s_{id}, \texttt{create\_microservice}, \langle ms_{id}, Ag_{ms}, \mathcal{D}_{ms}, \mathcal{M}_{ms}\rangle\rangle\}$
(CREATEMICROSERVICE)

---

[1] For all semantics rules presented in this section, when there is no update at some of the three levels, that transition is omitted from the rule.

In the first semantic rule, CreateMicroService, we show how the system treats a request for a new microservice. In this semantics rule, the select message function, $S_M()$, selects the message $\langle sender, s_{id}, \texttt{create\_microservice}, \langle ms_{id}, Ag_{ms}, \mathcal{D}_{ms}, \mathcal{M}_{ms} \rangle \rangle$, executing the creation of such microservice. We assume that the sender of the message has permission to create a new microservice. In this case, both levels, the system and the microservices levels suffer a transition, represented in (a) and (b), respectively. Basically, the system updates the message pooling removing that message, considering that it already has been treated by the system, and creates a new microservice $\langle ms_{id}, Ag_{ms}, \mathcal{D}_{ms}, \mathcal{M}_{ms} \rangle$ that will populate the system, represented by "$\{...\} \rightarrow_{ms} \{\langle ms_{id}, Ag_{ms}, \mathcal{D}_{ms}, \mathcal{M}_{ms}\rangle ...\}$", i.e., the set of microservices is updated with that particular new microservice. Such request should include all information necessary to create the microservice, including the microservice identifier ms_{id}, the set of agents, Ag_{ms}, that will populate that microservice, the relevant data to keep at the microservice level, $\mathcal{D}_{ms}$, and a message pooling for that microservice, $\mathcal{M}_{ms}$.

$$\frac{S_M(\mathcal{M}_s) = \langle sender, ms_{id}, \texttt{create\_agent}, \langle ag_{id}, \mathcal{G}_{ag}, \mathcal{B}_{ag}, \mathcal{A}\rangle\rangle}{\begin{array}{l}(a)\quad \langle s_{id}, \mathcal{MS}_s, \mathcal{D}_s, \mathcal{M}_s\rangle \quad\rightarrow_s\quad \langle s_{id}, \mathcal{MS}'_s, \mathcal{D}_s, \mathcal{M}'_s\rangle\\(b)\ \{\langle ms_{id}, Ag_{ms}, \mathcal{D}_{ms}, \mathcal{M}_{ms}\rangle ...\} \rightarrow_{ms} \{\langle ms_{id}, Ag_{ms}, \mathcal{D}_{ms}, \mathcal{M}'_{ms}\rangle...\}\end{array}} \text{(CreateAgentS)}$$

where:
(a) $\mathcal{M}'_s = \mathcal{M}_s \setminus \{\langle sender, ms_{id}, \texttt{create\_agent}, \langle ag_{id}, \mathcal{G}_{ag}, \mathcal{B}_{ag}, \mathcal{A}\rangle\rangle\}$
(b) $\mathcal{M}'_{ms} = \mathcal{M}_{ms} \cup \{\langle sender, ms_{id}, \texttt{create\_agent}, \langle ag_{id}, \mathcal{G}_{ag}, \mathcal{B}_{ag}, \mathcal{A}\rangle\rangle\}$

In the semantic rule CreateAgentS, we show how the system treats a request for the creation of a new agent inside a particular microservice ms_{id}. Considering that this request comes from a source that is external to the system, this request first will be treated by the system message pooling, as we show in the semantics rule CreateAgentS, which forwards the message to the corresponding microservice ms_{id}. Note that, at (a), both components $\mathcal{MS}_s$ and $\mathcal{M}_s$ are updated, give that updates at (b) always will cause an update to the $\mathcal{MS}_s$ component, given it represents the set of microservices populating that system.

When the microservice treats a request for creating an agent, represented in the semantic rule CreateAgentMS, by selecting the message $\langle sender, ms_{id}, \texttt{create\_agent}, \langle ag_{id}, \mathcal{G}_{ag}, \mathcal{B}_{ag}, \mathcal{A}\rangle\rangle$ from its message pooling $\mathcal{M}_{ms}$, it updates the message pooling component, $\mathcal{M}_{ms}$, and the set of agents populating that microservice, Ag_{ms}, including the agent $\langle ag_{id}, \mathcal{G}_{ag}, \mathcal{B}_{ag}, \mathcal{A}\rangle$ to the set of agents populating that microservice, represented in (c) by "$\{...\} \rightarrow_{ag} \{\langle ag_{id}, \mathcal{G}_{ag}, \mathcal{B}_{ag}, \mathcal{A}\rangle ...\}$". Note that, similar to the previous semantic rule, both components at (b) are updated: the set of agents populating the microservice ms_{id} represented by Ag_{ms} and the messsage pooling $\mathcal{M}_{ms}$.

We assume that external communication is treated by the system, in which an interface for authentication could be implemented through the message pooling component, $\mathcal{M}_s$. However, microservices communicate directly through shared artifacts, which reduces the chattiness [28]. In the next semantic rules, we present an example of communication between microservices, following the message format introduced above.

In the semantic rule SendRequestInfoMS, an agent ag_{id} selects a goal[2] to be executed using the goal selection function $S_G()$, which returns the goal request(ψ, ms_2). Consequently, ag_{id} posts a message $\langle ms_1, ms_2, \text{Request}, \psi \rangle$ in the message pooling of the microservice ms_2, requesting the information ψ from the microservice ms_2. Then, the agent ag_{id} updates its goals, considering that it reached that goal. Note that ag_{id}'s goals are updated at (c), which causes an update to the set of agents, Ag_{ms_1}, populating the microservice ms_1 at (b). Also, the message pooling of the microservice ms_2 is updated with that new message at (b). Finally, given that both microservices ms_1 and ms_2 have been updated, consequently the set of microservices of the system s_{id}, $\mathcal{MS}_s$, also is updated at (a). Note that, the operational semantics allows us to specify multiple elements from the sets of microservices and agents. In order to make it clear to which set of agents we refer at the transition (c), those sets are labelled with their respective microservices. In the semantic rule SendRequestInfoMS, $\{\langle ag_{id}, \mathcal{G}_{ag}, \mathcal{B}_{ag}, \mathcal{A}\rangle \ldots\}_{ms_1}$ represents the set of agents $\{\langle ag_{id}, \mathcal{G}_{ag}, \mathcal{B}_{ag}, \mathcal{A}\rangle \ldots\}$ populating the microservice ms_1.

$$\frac{S_M(\mathcal{M}_{ms}) = \langle sender, ms_{id}, \texttt{create\_agent}, \langle ag_{id}, \mathcal{G}_{ag}, \mathcal{B}_{ag}, \mathcal{A}\rangle\rangle}{\begin{array}{lll}(a) & \langle s_{id}, \mathcal{MS}_s, \mathcal{D}_s, \mathcal{M}_s \rangle & \rightarrow_s \quad \langle s_{id}, \mathcal{MS}'_s, \mathcal{D}_s, \mathcal{M}_s\rangle \\ (b) & \{\langle ms_{id}, Ag_{ms}, \mathcal{D}_{ms}, \mathcal{M}_{ms}\rangle \ldots\} & \rightarrow_{ms} \{\langle ms_{id}, Ag'_{ms}, \mathcal{D}_{ms}, \mathcal{M}'_{ms}\rangle \ldots\} \\ (c) & \{\ldots\} & \rightarrow_{ag} \quad \{\langle ag_{id}, \mathcal{G}_{ag}, \mathcal{B}_{ag}, \mathcal{A}\rangle \ldots\}\end{array}} \quad (\textsc{CreateAgentMS})$$

where:
(b) $\mathcal{M}'_{ms} = \mathcal{M}_{ms} \setminus \{\langle sender, ms_{id}, \texttt{create\_agent}, \langle ag_{id}, \mathcal{G}_{ag}, \mathcal{B}_{ag}, \mathcal{A}\rangle\rangle\}$
$Ag'_{ms} = Ag_{ms} \cup \{\langle ag_{id}, \mathcal{G}_{ag}, \mathcal{B}_{ag}, \mathcal{A}\rangle\}$

In the semantic rule ReceiveRequestInfoMS, the microservice ms_2 selects the message $\langle ms_1, ms_2, \text{Request}, \psi \rangle$ from the microservice message pooling $\mathcal{M}_{ms_2}$ through the select message function $S_M()$. Also, considering that ms_2 owns the requested information, i.e., $\psi \in \mathcal{D}_{ms_2}$, it answers the request message posting a message $\langle ms_1, ms_2, \text{Answer}, \psi \rangle$ in the message pooling of the microservice ms_1. Both messages pooling are updated at (b), there is no update at (c), and, given that two microservices of the system have been updated, then the set of microservices of that system, $\mathcal{MS}_s$, also is updated at (a).

Finally, in the semantic rule ReceiveInfoMS, the microservice ms_1 selects the message $\langle ms_2, ms_1, \text{Answer}, \psi \rangle$ from its message pooling $\mathcal{M}_{ms_1}$ through the select message function $S_M()$. Then, the message is removed from the message pooling $\mathcal{M}_{ms_1}$, and the microservice data $\mathcal{D}_{ms_1}$ is updated with that information. In that semantic rule, no agent is updated, thus there is no update at (c). At (b), both, the message pooling $\mathcal{M}_{ms_1}$ and data $\mathcal{D}_{ms_1}$ components of the microservice

---

[2] Note that, considering the practical reasoning to reach that goal is out of the scope of this paper, thus we assume that such goal is reached by the agent executing that particular action. For details about the mental attitudes we use in this paper from the BDI architecture we refer to Bratman's paper [8].

$$S_G(\mathcal{G}_{ag}) = \text{request}(\psi, ms_2)$$

$$\frac{(a) \quad \langle s_{id}, \mathcal{MS}_s, \mathcal{D}_s, \mathcal{M}_s \rangle \qquad \qquad \langle s_{id}, \mathcal{MS}'_s, \mathcal{D}_s, \mathcal{M}_s \rangle}{(b) \quad \{\langle ms_1, Ag_{ms_1}, \mathcal{D}_{ms_1}, \mathcal{M}_{ms_1} \rangle, \langle ms_2, Ag_{ms_2}, \mathcal{D}_{ms_2}, \mathcal{M}_{ms_2} \rangle, \ldots\} \rightarrow_s \rightarrow_{ms} \{\langle ms_1, Ag'_{ms_1}, \mathcal{D}_{ms_1}, \mathcal{M}_{ms_1} \rangle, \langle ms_2, Ag_{ms_2}, \mathcal{D}_{ms_2}, \mathcal{M}'_{ms_2} \rangle, \ldots\}} \quad (\text{SendRequestInfoMS})$$
$$(c) \quad \{\langle ag_{id}, \mathcal{G}_{ag}, \mathcal{B}_{ag}, \mathcal{A} \rangle \ldots\}_{ms_1} \quad \rightarrow_{ag} \quad \{\langle ag_{id}, \mathcal{G}'_{ag}, \mathcal{B}_{ag}, \mathcal{A} \rangle \ldots\}_{ms_1}$$

where:
(b) $\mathcal{M}'_{ms_2} = \mathcal{M}_{ms_2} \cup \langle ms_1, ms_2, \text{Request}, \psi \rangle$
(c) $\mathcal{G}'_{ag} = \mathcal{G}_{ag} \setminus \{\text{request}(\psi, ms_2)\}$

$$S_M(\mathcal{M}_{ms_2}) = \langle ms_1, ms_2, \text{Request}, \psi \rangle \qquad \psi \in \mathcal{D}_{ms_2}$$

$$\frac{(a) \quad \langle s_{id}, \mathcal{MS}_s, \mathcal{D}_s, \mathcal{M}_s \rangle \qquad \qquad \langle s_{id}, \mathcal{MS}'_s, \mathcal{D}_s, \mathcal{M}_s \rangle}{(b) \quad \{\langle ms_1, Ag_{ms_1}, \mathcal{D}_{ms_1}, \mathcal{M}_{ms_1} \rangle, \langle ms_2, Ag_{ms_2}, \mathcal{D}_{ms_2}, \mathcal{M}_{ms_2} \rangle, \ldots\} \rightarrow_s \rightarrow_{ms} \{\langle ms_1, Ag_{ms_1}, \mathcal{D}_{ms_1}, \mathcal{M}'_{ms_1} \rangle, \langle ms_2, Ag_{ms_2}, \mathcal{D}_{ms_2}, \mathcal{M}'_{ms_2} \rangle, \ldots\}} \quad (\text{ReceiveRequestInfoMS})$$

where:
(b) $\mathcal{M}'_{ms_2} = \mathcal{M}_{ms_2} \setminus \langle ms_1, ms_2, \text{Request}, \psi \rangle$
$\mathcal{M}'_{ms_1} = \mathcal{M}_{ms_1} \cup \langle ms_2, ms_1, \text{Answer}, \psi \rangle$

$$S_M(\mathcal{M}_{ms_1}) = \langle ms_2, ms_1, \text{Answer}, \psi \rangle$$

$$\frac{(a) \quad \langle s_{id}, \mathcal{MS}_s, \mathcal{D}_s, \mathcal{M}_s \rangle \qquad \qquad \langle s_{id}, \mathcal{MS}'_s, \mathcal{D}_s, \mathcal{M}_s \rangle}{(b) \quad \{\langle ms_1, Ag_{ms_1}, \mathcal{D}_{ms_1}, \mathcal{M}_{ms_1} \rangle, \ldots\} \rightarrow_s \rightarrow_{ms} \{\langle ms_1, Ag_{ms_1}, \mathcal{D}'_{ms_1}, \mathcal{M}'_{ms_1} \rangle, \ldots\}} \quad (\text{ReceiveInfoMS})$$

where:
(b) $\mathcal{M}'_{ms_1} = \mathcal{M}_{ms_1} \setminus \langle ms_2, ms_1, \text{Answer}, \psi \rangle$
$\mathcal{D}'_{ms_1} = \mathcal{D}_{ms_1} \cup \{\psi\}$

ms_1 are updated, and; therefore, the set of microservices of that system, $\mathcal{MS}_s$, also is updated at (a).

4 Example - Taxi Service Application

In this section, we discuss how our approach for multi-level semantics for multi-agent microservices can be used to specify systems that already have been implemented according to the MAMS paradigm. In particular, we discuss how some processes from a taxi service application, described in [2], could be formalised using our approach. In [2], the authors propose a multi-agent system that is responsible for providing a taxi allocation service, in which taxis drivers dynamically enter the system, becoming available for allocation. When a client requests the system for transportation, the microservice responsible for treating requests in that particular area that the client departs from allocates one of the available taxis drivers to that client. In the system, different areas are covered by different microservices according to an architecture based on MAMS [9], and the microservices are dynamically created according to the system's demand.

Figure 2 shows the dynamism of the microservices according to the areas covered by the system. Each area is covered by a microservice responsible for registering, receiving requests, and allocating taxis drivers.

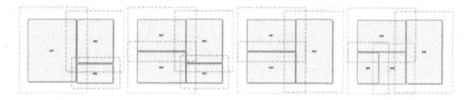

Fig. 2. Dynamic reorganisation for geometric treatment. It shows an evolution from the top-left configuration to the bottom-right configuration [2].

In the system proposed by [2], one of the main operations is to create a new microservice always when a particular region (controlled by a microservice) has too many registered taxis drivers and requests, splitting such region into two small regions, one of them controlled by the new microservice. This situation can be observed in Fig. 2, in which region A01 is divided in two regions, the old microservice remains responsible for A01, and a new microservice becomes responsible for that new region A04.

First, the agent called **controller01**, which populates and is responsible for the region A01, requests to the taxi service system **TX01** to create a new microservice A04, populated by **controller04**, which will be responsible for that region and its registered taxis, i.e., $\Delta_{A01'}$ contains all registered taxis in the region of A04. The rule CreateMicroServiceA04 shows how TX01 processes this message creating the new microservice A04.

In the application proposed in [2], each user's application (clients and taxis drivers) runs an agent that interacts with the user. While clients' application can request for taxi drivers, taxi drivers' applications allow them to register in the systems, being notified always that a client is allocated to them.

Continuing our example, imagine that a client in region A01 requests a taxi driver to go from a particular address at region A01 to another address at region A04. Also, imagine that the microservice responsible for A01 has no taxi drivers available, thus it requests to microservice A04 the closest taxi driver in the region A04. This request can be observed in the semantic rule SendRequestInfoMSA04.

In the semantic rule SendRequestInfoMSA04, the *controller* agent, in the microservice A01, responsible for allocating taxi drivers to clients, i.e., *controller01*, selects the goal request(closest(TD,A01),A04) from its goals, which can be achieved by sending a message to A04 requesting the information closest(TD,A01). Thus, a message of the type $\langle A01, A04, \texttt{Request}, closest(TD, A01), A04 \rangle$ is posted in the message poll $\mathcal{M}_{A04}$ and its goals are updated accordingly.

In the semantic rule ReceiveRequestInfoMSA01, when the microservice A04 selects that message from its message pool $\mathcal{M}_{A04}$, and the microservice A04 has that particular information in their data base $\mathcal{D}_{A04}$, then it answer that particular request posting a message $\langle A04, A01, \texttt{Answer}, closest(tx245, A01), A04 \rangle$ into A01's message poll $\mathcal{M}_{A_{\infty}}$, informing that the taxi tx245 is the closest taxi from A01.

Finally, in the semantic rule ReceiveInfoMSA04, when the microservice A01 receives the requested information, selecting that message from the message poll $\mathcal{M}_{A01}$, it updates its database $\mathcal{D}_{A01}$ with that particular information, also removing the message $\langle A04, A01, \texttt{Answer}, closest(tx245, A01), A04 \rangle$ from its message poll. Now, A01 microservice is able to allocate that taxi driver tx245 to that particular client.

5 Related Work

In [15], the authors propose an *Agent Infrastructure Layer* (AIL) for BDI-style programming languages, aiming to (i) provide a common semantic foundation for various BDI languages, and (ii) support formal verification through an optimized *model-checker*. They design AIL based on comprehensive operational semantics from [14], claiming it encompasses all key features of common BDI languages.

In [13], the authors introduce operational semantics for modularisation, which they argue simplifies the implementation of agents, roles, and profiles—key aspects in structured programming. They provide operational semantics for creating, executing, testing, updating, and realizing module instances, essential for agent programming.

In [17], operational semantics are provided for managing goals—dropping, aborting, suspending, and resuming them. The authors argue that these semantics clarify how an agent handles its goals and support the verification of correct

A Multi-level Semantics Formalism for Multi-Agent Microservices   59

$$\frac{}{S_M(\mathcal{M}_{TX01}) = \langle controller01, TX01, \texttt{create\_microservice}, \langle A04, \{controller04\}, \Delta_{A01'}, \{\}\rangle\rangle} \text{ (CREATEMICROSERVICEA04)}$$

$$(a) \ \langle TX01, \mathcal{MS}_{TX01}, \mathcal{D}_{TX01}, \mathcal{M}_{TX01}\rangle \rightarrow_s \frac{\langle TX01, \mathcal{MS}'_{TX01}, \mathcal{D}_{TX01}, \mathcal{M}'_{TX01}\rangle}{\{\dots\}} \rightarrow_{m.s} \{\langle A04, \{controller04\}, \Delta_{A01'}, \{\}\rangle\}\dots\}$$
$$(b)$$

where:
(a) $\mathcal{MS}'_{TX01} = \mathcal{MS}_{TX01} \cup \{\langle A04, \{controller04\}, \Delta_{A01'}, \{\}\rangle\}$
    $\mathcal{M}'_{TX01} = \mathcal{M}'_{TX01} \setminus \{\langle controller01, TX01, \texttt{create\_microservice}, \langle A04, \{controller04\}, \Delta_{A01'}, \{\}\rangle\rangle\dots\}$

$$\frac{S_G(\mathcal{G}_{controller01}) = \texttt{request}(\texttt{closest}(TD, A01), A04)}{}$$

$$(a) \quad \langle TX01, \mathcal{MS}_{TX01}, \mathcal{D}_{TX01}, \mathcal{M}_{TX01}\rangle \rightarrow_s \quad \langle TX01, \mathcal{MS}'_{TX01}, \mathcal{D}_{TX01}, \mathcal{M}_{TX01}\rangle$$
$$(b) \ \{\langle A01, Ag_{A01}\rangle, \langle A04, Ag_{A04}, \mathcal{D}_{A04}, \mathcal{M}_{A04}\rangle \dots\} \rightarrow_{m.s} \{\langle A01, Ag'_{A01}, \mathcal{D}_{A01}, \mathcal{M}_{A01}\rangle, \langle A04, Ag_{A04}, \mathcal{D}_{A04}, \mathcal{M}'_{A04}\rangle \dots\}$$
$$(c) \ \{\langle controller01, \mathcal{G}_{controller01}, \mathcal{B}_{controller01}, A\rangle\}_{A01} \rightarrow_{ag} \quad \{\langle controller01, \mathcal{G}'_{controller01}, \mathcal{B}_{controller01}, A\rangle\}_{A01}$$

(SENDREQUESTINFOMSA04)

where:
(b) $\mathcal{M}'_{A04} = \mathcal{M}_{A04} \cup \langle A01, A04, \texttt{Request}, closest(TD, A01), A04\rangle\rangle$
(c) $\mathcal{G}'_{controller01} = \mathcal{G}_{controller01} \setminus \{\texttt{request}(\texttt{closest}(TD, A01), A04)\}$

$$\frac{closest(TD, A01), A04) \in \mathcal{D}_{A04}}{}$$

$$(a) \quad \langle TX01, \mathcal{MS}_{TX01}, \mathcal{D}_{TX01}, \mathcal{M}_{TX01}\rangle \rightarrow_s \quad \langle TX01, \mathcal{MS}'_{TX01}, \mathcal{D}_{TX01}, \mathcal{M}_{TX01}\rangle$$
$$(b) \ \{\langle A01, Ag_{A01}, \mathcal{D}_{A01}, \mathcal{M}_{A01}\rangle, \langle A04, Ag_{A04}, \mathcal{D}_{A04}, \mathcal{M}_{A04}\rangle \dots\} \rightarrow_{m.s} \{\langle A01, Ag_{A01}, \mathcal{D}_{A01}, \mathcal{M}_{A01}\rangle, \langle A04, Ag_{A04}, \mathcal{D}_{A04}, \mathcal{M}'_{A04}\rangle \dots\}$$

(RECEIVEREQUESTINFOMSA01)

where:
(b) $\mathcal{M}'_{A04} = \mathcal{M}_{A04} \setminus \langle A01, A04, \texttt{Request}, closest(TD, A01), A04\rangle\rangle$
    $\mathcal{M}'_{A01} = \mathcal{M}_{A01} \cup \langle A04, A01, \texttt{Answer}, closest(tx245, A01), A04\rangle\rangle$

$$\frac{S_M(\mathcal{M}_{A01}) = \langle A04, A01, \texttt{Answer}, closest(tx245, A01), A04\rangle\rangle}{}$$

$$(a) \ \langle TX01, \mathcal{MS}_{TX01}, \mathcal{D}_{TX01}, \mathcal{M}_{TX01}\rangle \rightarrow_s \quad \langle TX01, \mathcal{MS}'_{TX01}, \mathcal{D}_{TX01}, \mathcal{M}_{TX01}\rangle$$
$$(b) \ \{\langle A01, Ag_{A01}, \mathcal{D}_{A01}, \mathcal{M}_{A01}\rangle, \dots\} \rightarrow_{m.s} \{\langle A01, Ag_{A01}, \mathcal{D}'_{A01}, \mathcal{M}'_{A01}\rangle, \dots\}$$

(RECEIVEINFOMSA04)

where:
(b) $\mathcal{M}'_{A01} = \mathcal{M}_{A01} \setminus \langle A04, A01, \texttt{Answer}, closest(tx245, A01), A04\rangle\rangle$
    $\mathcal{D}'_{A01} = \mathcal{D}_{A01} \cup \{closest(tx245, A01), A04\}$

agent behavior. They emphasize that this work contributes to a detailed specification of appropriate operational behavior when a goal is pursued, succeeds, fails, or undergoes changes by the agent.

In [11], the author proposes an operational semantic framework for legal systems within *agent societies*, modeling their *structure* and *dynamics*. This work focuses on actions within legal systems (internal and external legal acts, social acts, etc.) under an *action-based dynamics*, where system transitions are determined by actions of *legal organs* and *legal subjects*. It exemplifies the use of operational semantics to formalize multi-level systems.

In [29], the authors propose a semantic framework for $\mathcal{M}$OISE$^+$ [19] and a linear temporal logic (LTL) to express its properties. They argue that organizational specifications help agents achieve their goals and prevent undesired behavior by imposing *organizational constraints* on agents' behavior, which they must consider in decision-making. The work defines the formal semantics of $\mathcal{M}$OISE$^+$, focusing on ensuring agents respect (i) preconditions for organizational actions and (ii) acquaintance and communication links.

As mentioned before, the use of operational semantics has been quite extensive in multi-agent languages so we cannot cover all that literature here. Our work differs from all those in that we present a new style for the semantics of multi-level systems which has been developed to the formalisation of multi-agent microservices. We are not focused on a special semantics for BDI languages as in [15], in semantics for module-related actions [13], in semantics for goals [17], in semantics for legal systems [11], or semantics for the $\mathcal{M}$OISE model [29] specifically. We focus on proposing an operational semantics to the formalisation of multi-agent microservices that takes into account the desired characteristics of the software engineering based on microservices, i.e., modularity, scalability, reusability, and the specification of shared communication artifacts. To the best of our knowledge, we are the first to propose a formalisation for multi-agent microservices, providing a template that considers the multiple levels of such systems, and also promotes those desired characteristics.

6 Conclusion

In software engineering, there are many benefits of formal specification of systems [1,20]: (i) it allows the proof of formal properties; (ii) it allows a concise description of high-level behaviour of the system; (iii) it eliminates imprecision and ambiguity; and (iv) it allows model checking the system.

While operational semantics has been almost a standard method for the formalisation of agent-oriented programming languages, only recent research has been applied to make those formalisations adequate to the multiplicity of levels of abstraction (i.e., organisation, groups, agents, environment, etc.) present when engineering multi-agent systems, and there is no such work on the formalisation of multi-agent microservices. Following those recent approaches, in this work, we introduced an approach to the formalisation of multi-agent microservices using operational semantics; we use an approach based on the multi-level semantics proposed by [24].

We introduced a template for semantic rules that provide separated transition systems to multiple levels of multi-agent microservices. We also demonstrate how our template can be used to formalise some operations commonly present in multi-agent microservices, detailing how different (sets of) components are represented through the semantic rules. Our approach allows the representation of the interactions between components of different levels. Given the complexity and ubiquity of such multiple levels in multi-agent microservices, the approach seems to allow for a clearer understanding of such complex semantics, improving its readability. Furthermore, with the multi-level representation, our operational semantics style also provides independence between the specification of the various levels present in multi-agent microservices, so that each level is formalised through its own transition system. This promotes the modularity and reusability of formal models.

Using the multiple levels of formalisation, it is possible to make clear the representation of shared communication artifacts and how the different components of the system interact with such shared components, as we have illustrated in this paper through the formalisation of abstract operation for communication in multi-agent microservices. Shared communication artifacts are considered an essential part of the paradigm based on microservices, given they avoid chattiness [28]. In our future work, we intend to explore other relations between the multiple levels of multi-agent microservices. Also, we intend to investigate how our multi-level semantics and shared communication artifacts can play a part in research towards model checking of such systems [6], towards the verification of multi-agent microservices.

References

1. Beckert, B., et al.: Intelligent systems and formal methods in software engineering. IEEE Intell. Syst. **21**(6), 71–81 (2006)
2. Blanger, L., Junior, V., Jevinski, C.J., Panisson, A.R., Bordini, R.H.: Improving the performance of taxi service applications using multi-agent systems techniques. In: Encontro Nacional de Inteligência Artificial e Computacional (ENIAC) (2017)
3. Boissier, O., Bordini, R.H., Hubner, J., Ricci, A.: Multi-agent oriented programming: programming multi-agent systems using JaCaMo (2020)
4. Boissier, O., Bordini, R.H., Hübner, J.F., Ricci, A., Santi, A.: Multi-agent oriented programming with jacamo. Sci. Comput. Program. **78**(6), 747–761 (2013)
5. Bordini, R.H., Dastani, M., Dix, J., Seghrouchni, A.E.F.: Multi-Agent Programming: Languages, 1st edn. Tools and Applications. Springer Publishing Company, Incorporated (2009)
6. Bordini, R.H., Fisher, M., Visser, W., Wooldridge, M.: Verifying multi-agent programs by model checking. Auton. Agent. Multi-Agent Syst. **12**(2), 239–256 (2006)
7. Bordini, R.H., Hübner, J.F., Wooldridge, M.: Programming Multi-Agent Systems in AgentSpeak using Jason. John Wiley & Sons (2007)
8. Bratman, M., et al.: Intention, Plans, and Practical Reason, vol. 10. Harvard University Press Cambridge, MA (1987)
9. Collier, R.W., O'Neill, E., Lillis, D., O'Hare, G.M.: Mams: Multi-agent microservices. In: World Wide Web Conference (2019)

10. Collier, R.W., Russell, S., Lillis, D.: Reflecting on agent programming with agentspeak (l). In: International Conference on Principles and Practice of Multi-Agent Systems, pp. 351–366. Springer (2015). https://doi.org/10.1007/978-3-319-25524-8_22
11. Rocha Costa, A.C.: Situated legal systems and their operational semantics. Artif. Intell. Law **23**(1), 43–102 (2015). https://doi.org/10.1007/s10506-015-9164-z
12. Costa, A.C.R., Dimuro, G.P.: Introducing social groups and group exchanges in the poporg model. In: Proceedings of The 8th International Conference on Autonomous Agents and Multiagent Systems-Volume 2, pp. 1297–1298. International Foundation for Autonomous Agents and Multiagent Systems (2009)
13. Dastani, M., Steunebrink, B.R.: Operational semantics for bdi modules in multi-agent programming. In: Computational Logic in Multi-Agent Systems, pp. 83–101. Springer (2010). https://doi.org/10.1007/978-3-642-16867-3_5
14. Dennis, L.A.: Agent infrastructure layer (ail): Design and operational semantics v1. Tech. rep., Technical Report ULCS-07-001, Department of Computer Science, University of Liverpool (2007)
15. Dennis, L.A., Farwer, B., Bordini, R.H., Fisher, M., Wooldridge, M.: A common semantic basis for bdi languages. In: Programming Multi-Agent Systems, pp. 124–139. Springer (2008). https://doi.org/10.1007/978-3-540-79043-3_8
16. Dragoni, N., Lanese, I., Larsen, S.T., Mazzara, M., Mustafin, R., Safina, L.: Microservices: How to make your application scale. In: International Andrei Ershov Memorial Conference on Perspectives of System Informatics, pp. 95–104. (2017)
17. Harland, J., Morley, D.N., Thangarajah, J., Yorke-Smith, N.: An operational semantics for the goal life-cycle in bdi agents. Auton. Agent. Multi-Agent Syst. **28**(4), 682–719 (2014)
18. Horling, B., Lesser, V.: A survey of multi-agent organizational paradigms. Knowl. Eng. Rev. **19**(4), 281–316 (2004)
19. Hubner, J.F., Sichman, J.S., Boissier, O.: Developing organised multiagent systems using the $Moise^+$ model: programming issues at the system and agent levels. Int. J. Agent-Oriented Softw. Eng. **1**(3–4), 370–395 (2007)
20. Leveson, N.G.: Guest editor's introduction: formal methods in software engineering. IEEE Trans. Software Eng. **16**(9), 929 (1990)
21. Noriega, P.: Agent mediated auctions: the fishmarket metaphor. Institut d'Investigació en Intelligència Artificial (1999)
22. Panisson, A.R., Ali, A., McBurney, P., Bordini, R.H.: Argumentation schemes for data access control. In: COMMA, pp. 361–368 (2018)
23. Panisson, A.R., Bordini, R.H., Rocha, A.C.: An approach to the multi-level semantics of agent societies. In: Brazilian Conference on Intelligent Systems (2020)
24. Panisson, A.R., Bordini, R.H., da Rocha Costa, A.C.: Multi-level semantics with vertical integrity constraints. In: 22st European Conference on Artificial Intelligence (ECAI), pp. 1708 – 1709 (2016)
25. Panisson, A.R., Meneguzzi, F., Fagundes, M., Vieira, R., Bordini, R.H.: Formal semantics of speech acts for argumentative dialogues. In: Thirteenth Int. Conf. on Autonomous Agents and Multiagent Systems, pp. 1437–1438 (2014)
26. Plotkin, G.D.: A structural approach to operational semantics (1981)
27. Ricci, A., Piunti, M., Viroli, M.: Environment programming in multi-agent systems: an artifact-based perspective. Auton. Agent. Multi-Agent Syst. **23**(2), 158–192 (2011)
28. Richards, M.: Microservices vs. service-oriented architecture. O'Reilly Media (2015)

29. van Riemsdijk, M.B., Hindriks, K.V., Jonker, C.M., Sierhuis, M.: Formalizing organizational constraints: A semantic approach. In: Proceedings of the 9th International Conference on Autonomous Agents and Multiagent Systems: volume 1-Volume 1, pp. 823–830. International Foundation for Autonomous Agents and Multiagent Systems (2010)
30. Sierra, C., Rodriguez-Aguilar, J.A., Noriega, P., Esteva, M., Arcos, J.L.: Engineering multi-agent systems as electronic institutions. Europ. J. Infor. Profess. **4**(4), 33–39 (2004)
31. Vieira, R., Moreira, A., Wooldridge, M., Bordini, R.H.: On the formal semantics of speech-act based communication in an agent-oriented programming language. J. Artif. Int. Res. **29**(1), 221–267 (2007)
32. Vivanco, M.E., García, C.S.: Electronic Institutions: from specification to development. Consell Superior d'Investigacions Científiques, Institut d'Investigació en Intelligéncia Artificial (2003)
33. Wooldridge, M.: An introduction to multiagent systems. John Wiley & Sons (2009)
34. Zimmermann, O.: Mircroservices tenets: Agile approach to service development and deployment. In: Proceedings of the Symposium/Summer School on Service-Oriented Computing (2016)

A Novel Genetic Algorithm Approach for Discriminative Subspace Optimization

Bernardo B. Gatto[1]([✉]), Marco A. F. Mollinetti[1], Eulanda M. dos Santos[2], Alessandro L. Koerich[3], and Waldir S. da Silva Junior[4]

[1] Frontier Technology Division, MTI Ltd., Tokyo, Japan
{gatto_b,mollinetti_m}@mti.co.jp
[2] Federal University of Amazonas, Institute of Computing, Manaus, Brazil
emsantos@icomp.ufam.edu.br
[3] École de Technologie Supérieure, Université du Québec, Québec, Canada
alessandro.koerich@etsmtl.ca
[4] Department of Electrical Engineering, Federal University of Amazonas, Manaus, Brazil
waldirjr@ufam.edu.br

Abstract. Image set representation by subspace methods has shown to be effective for several image processing tasks, such as classifying multiple images and videos. A subspace exploits the geometrical structure in which images are distributed, representing the image set with a fixed dimension giving more statistical robustness to input noise and compactness to the images. The mutual subspace method (MSM) and its extensions, the Orthogonal Mutual Subspace method (OMSM), and the Generalized Difference Subspace (GDS) are the most prominent subspace methods employed. However, these methods require solving a nonlinear optimization which lacks a closed-form solution. In this paper, we present a metaheuristic-based approach for discriminative subspace optimization. We develop a Genetic Algorithm (GA) for integrating OMSM and GDS discriminative subspaces. The initialization strategy and the genetic operators of the GA provide quality of objective function value of solutions and preserve their feasibility without any extra repair step. We validated our approach on four object recognition datasets. Results show that our optimization method outperforms related methods in accuracy and highlights the use of evolutionary algorithms for subspace optimization. Code: https://github.com/bernardo-gatto/Evolving_manifold.

Keywords: Subspace representation · Genetic Algorithm · Discriminative Learning

1 Introduction

The problem of classifying image sets has been widely investigated in computer vision and supports several applications by handling, learning, and classifying data from multiple view cameras. For instance, in robot vision, where a data stream is available [1,12,20,22]. In this context, a pattern set is a collection of images (or feature vectors) of the same object or event. This set can be unordered where the time stamp of the collected images is not relevant. The images can also be ordered when the timestamp is semantically meaningful.

Applications of subspaces for pattern-set learning frequently use the mutual subspace method (MSM) [16,23]. These solutions are employed to solve gesture and action recognition problems, where video clips are described as subspaces, in which each subspace is computed from one of the pattern sets. Despite its advantages, traditional subspace methods like MSM lack discriminative feature extraction, which is critical for efficient pattern set modeling. A useful pattern-set model requires robustness to corrupt data; some images may contain noise, occluded targets, or dropped patches. The model must also correctly handle a variable set size, maintaining its computational complexity.

The orthogonal mutual subspace method (OMSM) [21] was developed to extract discriminative features using the Fukunaga-Koontz transformation (FKT). Its formulation has been applied to image-set modeling for solving a range of problems, such as gesture and action classification.

Recently, a variant of the subspace method called generalized difference subspace (GDS) has been developed [4]. The relationship between the patterns is considered by employing the concept of the generalized difference between the subspaces. This formulation provides a novel discriminative transformation, where the projected subspaces produce higher recognition results than conventional subspace-based methods.

Despite their excellent performance, OMSM and GDS lack closed-form solutions and cannot fully achieve their capabilities. Our work introduces a significant advancement by using a genetic algorithm (GA) to optimize the discriminative subspaces. This approach not only enhances the discriminative power of OMSM and GDS but also offers a more rigorous solution to the optimization challenges in subspace methods.

We optimize OMSM and GDS spaces to tackle these challenges. We develop a GA that efficiently handles subspaces by modifying its initialization, evaluation, selection, and crossover operators, which is complex given GA's typical use in simpler optimization problems. Our GA-generated discriminative space surpasses OMSM and GDS in accuracy, showcasing GA's versatility in subspace optimization. This innovation applies to traditional subspace methods, creating a compact model that enhances their inherent benefits.

This paper continues as follows: Sect. 2 describes the related work on subspace-related methods. Section 3 introduces the proposed optimization solution, and Sect. 4 shows the experimental results. Finally, Sect. 5 presents the final remarks and future directions. Code developed for this research can be found in the following repository: https://github.com/bernardo-gatto/Evolving_manifold (Fig. 1).

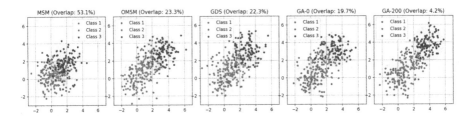

Fig. 1. These plots illustrate the improvement in class separability achieved by different subspace optimization methods. From left to right, the subplots transition from the MSM with a 53% overlap between classes to the GA after 200 iterations (GA-200) with a reduced overlap of 5%, indicating a progressive enhancement in class distinctness..

2 Related Work

2.1 Subspace Method

It has been verified that using multiple images can significantly improve the performance of representing complex shape objects. The MSM [16] is a recognition method that efficiently handles multiple images. MSM is a method for classifying multiple image sets where a set of images is represented by a linear subspace generated by applying non-centered principal component analysis (PCA). The similarity between different image sets is calculated using the canonical angles between the subspaces, and this similarity is used to classify sets of input images.

While MSM plays an essential function in subspace-based methods, it cannot produce good recognition results since it does not consider the relationship between the sets of distinct classes. The OMSM [5] has been developed to improve the discriminative ability of subspace-based methods. In OMSM, class subspaces are orthogonalized using Fukunaga-Koontz's framework before computing the canonical angles. This transformation emphasizes the differences between the classes and significantly improves the recognition performance of MSM. Some of the applications of OMSM include face recognition and multiple view object recognition [6,7].

Another variant of the subspace method was introduced in [4] named GDS, where the pattern sets are also represented as subspaces. The relationship between the patterns is taken into consideration by employing the concept of the generalized difference between the subspaces. This algebraic formulation provides a novel discriminative transformation, where the projected subspaces produce higher recognition results than conventional subspace-based methods.

2.2 Evolutionary and Genetic Algorithms

Research on the theory and applications of evolutionary algorithms (EAs) has been consistent due to the increasing number of complex nonlinear problems unsolvable by conventional gradient-based techniques [2]. Accordingly, EAs present several advantages, such as handling limiters, explaining and repeating the search process, understanding input-output causality within a solution, and introducing corrective actions.

EAs excel in various problem domains, notably in optimization, where they can uncover solutions beyond human intuition. Characterized as iterative refinement algorithms, they evolve a set of potential solutions using replication, reproduction, and selection, all guided stochastically towards a specific goal.

Since its first relevant publication [13], GAs are considered a well-known population-based EA designed to solve optimization problems where exact methods fail [9]. Their mechanisms are grounded in Darwinian natural selection, where fitter individuals produce more offspring, while less fit individuals are eliminated if they don't adapt or mutate to improve their capabilities [17].

A canonical GA comprises these key genetic operators: 1) replication, which creates new solutions either by generating entirely new potential solutions or by altering specific existing ones; 2) reproduction, a mechanism to enhance the diversity of the solution pool, is achieved through two methods: recombination (crossover) and mutation; 3) selection, a process that identifies and retains solutions meeting certain criteria for inclusion in the subsequent iteration.

It is important to acknowledge the diverse range of optimization strategies discussed in the literature [3,24]. For example, Bayesian optimization typically utilizes a cost function reliant on simulations, which often demands substantial computational resources and lacks a surrogate model. In contrast, EA offers an alternative approach, focusing on the iterative refinement of candidate solutions. This method is advantageous as it requires fewer cost function evaluations compared to the Bayesian approach, which relies on extensive sampling to create a pool of potential solutions.

The development of the proposed method is encouraged by the application of GAs to the well-known classifier in a non-Euclidean space. Such a space presents no trivial operations, such as sum or gradient, which are easily achieved in Euclidean spaces.

3 Proposed Method

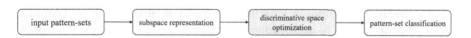

Fig. 2. Framework of the proposed optimization solution. Given a collection of labeled pattern sets, the task lies in 1) learning a subspace representation for each set and then 2) estimating a discriminative space where discriminative information is attainable. Our contribution lies in optimizing the discriminative space using GA, where a population of subspace candidates is evaluated and updated until a user-specific criterion is met.

3.1 Problem Formulation

Subspaces usually represent pattern sets to reduce computational complexity, achieving the immediate advantage of allowing parallel processing. Besides, subspace representation permits the examination of correlations among the various factors inherent in each pattern set.

In a classification problem, $\boldsymbol{P} = \{P_i\}_{i=1}^{n}$ denotes the set of all subspaces spanned by $\boldsymbol{U} = \{U_i\}_{i=1}^{n}$. We can then develop a projection matrix D that acts on the elements of $\boldsymbol{P}$ to extract discriminative information. In traditional GDS, this procedure is performed by removing overlapping components that represent the intersection between subspaces. Differently, OMSM weights the discriminative space, encouraging the decorrelation of the overlapping patterns.

Figure 2 shows the pattern-set classification pipeline and the proposed optimization solution. Given a collection of labeled pattern sets, the task lies in estimating subspace representations for each set and then learning a discriminative space, where the subspaces are projected, and improved patterns are obtained. Our contribution lies in optimizing the discriminative space using our Genetic Algorithm. An initial population derived from OMSM and GDS subspaces is successively evaluated and updated until a specific criterion is met.

3.2 Brief Review on MSM

To represent a pattern-set by a subspace, we use orthonormal basis vectors to describe them compactly. Given a feature matrix $X = [\,x_1\,|\,x_2\,|\,\ldots\,|\,x_{m-1}\,|\,x_m\,]$, where x_j is a feature vector, possibly obtained through the vectorization of an image. Then, we can conduct a decomposition to gather knowledge of the geometric structure of X. The singular value decomposition (SVD) [14] produces a set of eigenvectors U and a set of eigenvalues Λ, where each column vector in U represents an axis and each value in Λ describes how important this axis is in terms of reconstruction. Also, Λ describes how much the vectors in X are correlated. The SVD of X is

$$XX^\top = U\Lambda U^\top . \tag{1}$$

Each column of U is a singular vector of $XX^\top$, and the main diagonal of Λ presents the singular values in descending order. The analysis of Λ is helpful in various problems, such as dimensionality reduction, signal filtering, and feature extraction. By analyzing the influence of each eigenvector, it is possible to select a small set by removing all but the top k eigenvalues in the diagonal of Λ.

The following criteria can be used to obtain the compactness ratio:

$$\mu(k) \leq \sum_{j=1}^{k}(\lambda_j)/\sum_{j=1}^{m'}(\lambda_j) . \tag{2}$$

In Eq. (2), k is the number of the selected eigenvectors which spans a subspace, λ_j corresponds to the j-th eigenvalue of $XX^\top$. Then, $m' = \mathrm{rank}\left(XX^\top\right)$.

It is useful to set k as small as possible to achieve a minimum number of orthonormal basis vectors, maintaining low memory requirements.

We employ the average of the canonical angles to compare the subspaces. A practical technique for computing the canonical angles between two subspaces P and Q is calculating the eigenvalues of the product of their basis vectors. Given U_p and U_q, which span the subspaces P and Q, Eq. (3) computes the canonical correlations between P and Q:

$$U_p^\top U_q = U \Sigma V^\top, \qquad (3)$$

where the eigenvalues matrix Σ provides the canonical correlations between the principal angles of U_p and U_q and can be used to compute the canonical angles, since $\Sigma = \text{diag}(\lambda_1, \lambda_2, \ldots, \lambda_k)$. The canonical angles $\{\theta_j\}_{j=1}^k$ can then be computed by using the inverse cosine of Σ, as $\{\theta_j = \cos^{-1}(\lambda_j)\}_{j=1}^k$. Finally, the average canonical angle $\bar{\theta}$ between P and Q is defined as $\bar{\theta} = \frac{1}{k'} \sum_{j=1}^{k'} \theta_j$, where $k' \leq \min(\text{rank}(P), \text{rank}(Q))$. Once the average of the canonical angles is obtained between the training and the test subspaces, we can employ the nearest neighbour algorithm (1-NN) to classify the test subspaces.

3.3 Conventional Discriminative Subspaces

Once equipped with all the subspaces (using Eqs. (1) and (2)), OMSM employs the FKT to generate the matrix D_o that can decorrelate the subspaces. Each set of basis vectors U_i spans a reference subspace P_i, where its compactness ratio is empirically defined by employing Eq. (2). The method to generate the matrix D_o that efficiently decorrelates the subspaces is explained. We compute the total projection matrix:

$$G = \sum_{i=1}^{n} U_i U_i^\top, \qquad (4)$$

then we decompose the total projection matrix G:

$$D_o = \Lambda^{-1/2} B^\top, \qquad (5)$$

where B is the set of orthonormal eigenvectors corresponding to the largest eigenvalues of G, and Λ is a diagonal matrix with the k-th highest eigenvalue of the matrix G as the k-th diagonal component.

Differently, D_g produced by GDS can be computed by discarding the first d eigenvectors of B as follows: $D_g = B \setminus A$, where $A = \{U_k\}_{k=1}^d$ and $\setminus$ is the relative complement (*i.e.*, removes all the elements of A from B). In practice, OMSM and GDS are very similar. Their differences arise from the fact that the eigenvectors of G are either weighted or discarded, suggesting that an optimization strategy should be implemented for weighting or discarding eigenvectors.

Since this problem can be analyzed from a feature selection point of view and, according to [10], it is an NP-hard problem, we argue that using an EA

Algorithm 1. Genetic Algorithm for the Subspace Optimization Problem

Require: $n, max, f(\cdot)$
1: $D_0 \leftarrow$ InitializePopulation(n)
2: **for** $i \leftarrow 0$ **to** $max - 1$ **do**
3:     $F_i \leftarrow$ Evaluate(D_i, f)
4:     $S_i \leftarrow$ Selection(D_i, F_i)
5:     $\bar{D}_i \leftarrow$ Crossover(S_i)
6:     $D_{i+1} \leftarrow S_i \cup \bar{D}_i$
7: **end for**
8: $D \leftarrow D_{max-1}$
9: **return** argmax $f(D)$

for combining both D_o and D_g discriminative spaces is a viable solution. In this scenario, a suitable objective function f must be an indicator of how much a discriminative space can discriminate different groups. We discuss the chosen objective function in the following section.

3.4 Proposed Algorithm

We propose an EA in the likes of the GA of Holland [13] with all genetic operators but the mutation operator. Algorithm 1 has the following components: InitializePopulation, Evaluate, Selection and Crossover. The overall steps of the GA are shown in Algorithm 1. The encoding of solutions in our GA is in the form of a matrix, each solution encodes the basis vectors of the discriminative subspaces D_o and D_g. This representation allows the GA to effectively combine and evolve them over iterations. It can be thought of as a concatenated or structured representation of these basis vectors, where each vector undergoes optimization through genetic operations. This approach ensures that the evolutionary process refines the subspaces for optimal discriminative power in classification tasks.

In summary, our proposed GA starts by initializing a population of solutions D_0 of size n. In each iteration, the algorithm evaluates the fitness of the current population D_i using an objective function f based on the Fisher score. Based on these fitness values, a selection process Selection is carried out to choose a subset of the current population. This subset then undergoes a crossover operation Crossover, producing a new set of solutions $\bar{D}_i$. The new generation for the next iteration D_{i+1} is formed by combining the selected solutions and the offspring from the crossover. The process repeats until the maximum number of iterations max is reached. Finally, the algorithm returns the best solution from the final population as determined by the objective function f.

Part of OMSM's traditional optimization approach performs the weighting of eigenvectors associated with the *lowest* eigenvalues. Intriguingly, the optimization process of GDS is performed by eliminating eigenvectors associated with the *highest* eigenvalues, suggesting a complementary relationship exists between both optimization strategies.

This greedy strategy produces good results; however, it does not ensure that the optimal discriminative space is the one obtained by the method. Besides, some eigenvectors associated with the middle eigenvalues are never removed or evaluated, and thus, this greedy strategy may lead to weak discriminative spaces.

It is worth mentioning that the removal and concatenation strategy employed in this work may produce discriminative spaces of different dimensions (*i.e.*, different subspace dimensions), which is not exactly a problem in subspace analysis. The number of basis vectors exhibited by each element is expected to be modified according to the dataset complexity. We now detail each step as follows.

InitializePopulation: The initialization step ensures that the initial population D_0 contains D_o and D_g discriminative spaces, their concatenations, and random variants with randomly selected dimensions and their respective concatenations. More precisely, the basis vectors of D_o and D_g can be concatenated to, after an orthogonalization process, produce a novel discriminative space, preserving the capabilities of OMSM and GDS. We create a completely unexplored population by randomly truncating and concatenating D_o and D_g.

This initialization strategy has two immediate advantages: 1) We ensure that the produced solutions will not be worse than the ones provided by D_o and D_g discriminative spaces and, 2) We start from a *good enough* population, which may guide solutions to converge to a local optimum faster than purely random initialization. Our motivation assumes that a random initialization would be challenging to achieve the same performance as the warm start provided by OMSM and GDS. Besides, the probability of discovering better discriminative spaces in the neighborhood space is higher once we start from promising ones.

Evaluate: As the objective function f of the GA, we selected a score reflective of each discriminative subspace's efficacy in classification metrics. These metrics can encompass accuracy, F1 score, and mean squared error. For our implementation, the Fisher score was chosen due to its straightforward interpretability and effectiveness in quantifying the discriminative power of subspaces. The Fisher score measures the separation between different classes in the feature space, making it a suitable metric for optimizing subspaces in classification tasks [11]:

$$f = \text{trace}\left((S_b)(S_w + \epsilon I)^{-1}\right), \tag{6}$$

where S_b and S_w are the between-class scatter matrix and within-class scatter matrix, respectively, measuring the dispersion of the subspaces. Here, ϵ is a positive regularization parameter, and I is the identity matrix. The Fisher score aims to maximize the between-class and within-class variance ratio, ensuring that the evaluated subspace is a robust discriminative model.

Selection: The selection process employs a rank-based approach. Specifically, it involves arranging the discriminative spaces according to their Fisher scores in decreasing order. From this ranked list, the top $p\%$ of spaces, those with the highest scores, are selected for inclusion in the crossover pool. This method ensures that the most promising solutions are carried forward, encouraging the generation of increasingly effective discriminative spaces in subsequent iterations.

Crossover: In the crossover phase, two elements (discriminative spaces) are randomly selected from the pool created in the Selection step. These elements undergo crossover, and a one-point crossover operation is performed. A cut point $c\%$ is randomly determined, and each offspring solution is composed of eigenvectors from both parents, split at this cut point. For instance, let $D_p = \{\phi_{(i)}\}_{i=1}^{n_p}$ and $D_q = \{\psi_{(i)}\}_{i=1}^{n_q}$ be two selected parent solutions. Their crossover operation would result in two offspring solutions $D_p{}'$ and $D_q{}'$, formulated as follows:

$$D_p{}' = \{\phi_{(1)}, \ldots, \phi_{(c)}, \psi_{(c+1)}, \ldots, \psi_{(n_q)}\},$$
$$D_q{}' = \{\psi_{(1)}, \ldots, \psi_{(c)}, \phi_{(c+1)}, \ldots, \phi_{(n_p)}\}.$$

Following the progress of the Algorithm 1, the selected population S_i is merged with $\bar{D}_i$ to compose the next population set; this can be seen as an update process that follows a measure of the elitism of solutions. Evaluation, Selection, and Crossover are then performed until a predefined number of iterations is achieved.

The proposed algorithm presents the main elements of a GA; however, its application is in a non-Euclidean space. The provided mathematical formulations employed for orthogonalization, basis vector concatenation, and projections are the foundations for discriminative subspace optimization employing EA. The following section is focused on applying the presented concepts for pattern set classification. We employ subspace-based methods for representing pattern sets and optimize a discriminative subspace using strategies motivated by the GA presented in this section.

4 Experimental Results

We evaluate the proposed GA to demonstrate its advantages over other pattern-set representation and classification methods. First, we describe each dataset and our experimental protocol. Next, we investigate the approach for initial population estimation. We then compare the proposed approach with existing methods for the same task. Lastly, we provide concluding remarks on our results.

ETH-80 is a dataset designed for object recognition. It contains images of 8 object categories, each with 10 subcategories in 41 orientations, totaling 410 images per category and 3280 images overall. We resized the images to 64×64 pixels and extracted grayscale information. For our analysis, we used the images with backgrounds.

For a comprehensive classification task, we used the ALOI dataset. ALOI is a large dataset of general objects with about 110 images per object, considering different illumination angles, colors, and viewing angles. We used the first 500 object instances. All images were segmented from the background, and we classified them into one of the 500 objects using a 10-fold cross-validation scheme.

We also used the Coil-20/100 datasets. Coil-20 contains 20 objects, with 72 images captured for each using a turntable, resized to 128×128 pixels. Similarly, Coil-100 was obtained with more objects. We split the samples into five groups, randomly selecting one for training and using the rest for testing.

In the experiments, descriptive statistics are the mean accuracy, standard deviation, and best result obtained from the runs of each experiment using pre-established random seeds. Chosen parameters for the proposed GA for all experiments were: number of iterations as 200; population size of 200 solutions; crossover rate c as 0.5; and populational elitism p of the selection step as 10%. The experiment reported in this paper was run on a Unix-like PC equipped with a Core i9 5.00GHz with 16 GB RAM written in Python.

4.1 Evaluating the Initialization Strategy

This experiment analyzes the adopted initialization strategy on the ETH-80 dataset. We utilized an initialization strategy where the elements are assigned to OMSM and GDS discriminative spaces and some variants based on random subspace truncation. Two elements are assigned to the D_o and D_g; then, random truncation is performed on either D_o or D_g before concatenation. This strategy not only creates new discriminative subspaces but also ensures that a vast space is explored and that the results obtained will never be worse than the ones provided by OMSM and GDS.

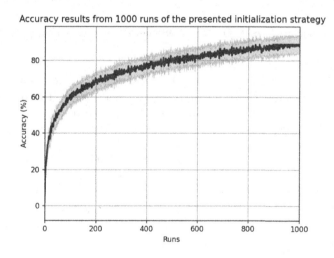

Fig. 3. Accuracy from 1000 runs of the presented initialization strategy.

Figure 3 shows the accuracy results of 1000 runs of the presented initialization strategy. We see that the initialization alone presents attractive solutions. For instance, some elements deliver accuracy higher than OMSM and GDS. However, some display very low accuracy, which may be due to selecting unsuitable eigenvectors. If discriminative eigenvectors are discarded, the learning algorithm accuracy may sharply drop.

4.2 Analysing Distance Matrices

In this experiment, we analyze the distance matrices produced by OMSM, GDS, and our GA. The objective is to understand if the optimized discriminative subspace presents advantages over OMSM and GDS w.r.t. feature extraction capabilities. Also, the distance matrices may provide information concerning their discriminative ability, such as the intra-class and inter-class separability. For this experiment, each algorithm was run for 200 iterations.

Figure 4 displays the distance matrices generated by OMSM, GDS, and the GA on the ETH-80 dataset.

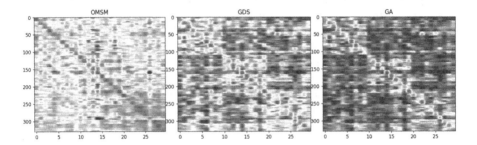

Fig. 4. Distance matrices of OMSM, GDS, and the proposed GA, respectively.

The proximity matrix produced by OMSM shows moderately high similarity even when subspaces describe different classes, indicating that the discriminative subspace may reveal insufficient information for classification. In contrast, the proximity matrix produced by GDS shows lower similarity, revealing moderately good discriminative information. This demonstrates the effectiveness of the GDS discriminative space, as it relies on an analytical solution.

Differently, the similarity between different classes is very low when our optimization method is employed, represented by the dark structures on the matrix. This suggests that the presented optimization technique based on the GA produces a discriminative space capable of separating subspaces of distinct classes while keeping together the same category.

4.3 Comparison with Related Methods

This experiment evaluates the proposed GA for discriminative subspace optimization using the ETH-80, ALOI, COIL-20, and COIL-100 datasets. We compare the proposed method with MSM, OMSM, and GDS. The objective here is to understand whether the optimization strategy developed in this work can produce better results on diverse datasets. It is worth mentioning that more sophisticated manifold learning algorithms exist in the literature. However, we narrowed down the scope of our evaluation to linear methods since our optimization strategy was developed to support linear subspaces.

Table 1. Study of performances for the proposed optimization method (left) and comparison with OMSM and GDS (right).

Dataset	Proposed Method			MSM/OMSM/GDS vs. GA			
	mean acc.	stand. dev.	best result	MSM	OMSM	GDS	GA
ETH-80 [15]	88.7	0.4	89.4	80.3	86.1	86.6	90.1
COIL-20 [18]	89.4	0.3	90.1	76.7	83.3	87.4	90.7
COIL-100 [19]	87.5	0.5	88.6	79.8	85.1	85.5	88.6
ALOI [8]	86.1	0.7	86.9	72.5	79.4	80.8	87.2

A comprehensive summary of the obtained results is shown in Table 1. We see from the results that the GA achieved its highest accuracy on the COIL-20 and ETH-80 datasets, while the results produced on the ALOI dataset are the lowest. These results may indicate that our GA may benefit from a larger initial population when more complex datasets, such as ALOI, are being evaluated. In this experiment, we present the best results obtained from 100 runs with varying random seeds, where each run is executed for 500 iterations.

Table 1 also displays the results achieved by OMSM, GDS, and the GA on all the employed datasets. The presented method exhibited the lowest accuracy gain on the ALOI dataset compared to the gain obtained on other datasets. This suggests that the GA encountered difficulty when the classification problem's number of classes was high.

The highest improvement gain was obtained on the ETH-80 dataset. These results may have been achieved because the ETH-80 is the smallest dataset among those evaluated, which can benefit all methods based on linear subspaces. Furthermore, since the GA does not implement a regularization technique, the provided results on more complex datasets may overfit. Besides, OMSM and GDS presented competitive results despite implementing no optimization approach, indicating their usefulness when optimization is not practical.

4.4 Visualizing the Discriminative Spaces

This section presents a visualization of the discriminative spaces derived from the ETH-80 dataset after training. The dimensions of the spaces are as follows: the GA subspace comprises 196 eigenvectors, the OMSM subspace contains 83 eigenvectors, and the GDS subspace encompasses 222 eigenvectors. Figures 5, 6 and 7 display the first 10 and last 10 eigenvectors of the OMSM, GDS, and GA discriminative spaces, reshaped into 15 × 15 matrices.

Upon analyzing the similarity of eigenvectors across these subspaces, we observe no overlap between GDS and OMSM, which implies that the two spaces are capturing entirely distinct features of the dataset, confirming our previous hypothesis regarding the complementary nature of these spaces. In contrast, there is a significant overlap between GDS and GA, with 163 eigenvectors being shared. This suggests that GA has a strong affinity with the GDS subspace, potentially indicating that GA is preserving a substantial amount of the detailed features captured by GDS. Furthermore, GA discriminative space shares 33 eigenvectors with OMSM, which, although fewer in number, is a significant share considering the OMSM's dimension.

The OMSM subspace, being more compact, may focus on the most salient features of the dataset, which are sufficiently representative of the specific discriminative tasks it's designed for. On the other hand, the GDS subspace, with its larger number of eigenvectors, is likely to preserve more intricate details of the data, which could be advantageous for tasks requiring a high level of granularity.

The GA subspace's ability to share eigenvectors with both OMSM and GDS indicates its capability to exploit both the compact, representative nature of OMSM and the detailed, nuanced capture of GDS. In quantitative terms, GA incorporates approximately 83.16% of GDS and 39.75% of OMSM eigenvectors, reflecting its comprehensive nature in encapsulating the diversity of features within the dataset. This makes the GA subspace potentially versatile and suitable for a wide range of applications that may require either a broad overview or a detailed analysis of the data's characteristics.

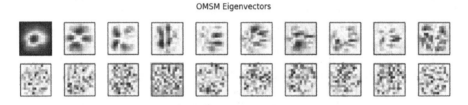

Fig. 5. Eigenvector visualization of the OMSM subspace.

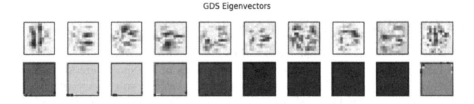

Fig. 6. Eigenvector visualization of the GDS subspace.

In this section, we evaluated our proposed optimization approach regarding the initialization strategy, its distance matrix, accuracy on various datasets, and the visual characteristics of the eigenvectors. We can describe two main findings from the obtained results: 1) the employed initialization strategy frequently shows advantages over conventional methods, suggesting that the proper use of

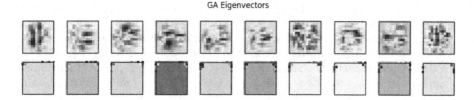

Fig. 7. Eigenvector visualization of the GA subspace.

OMSM and GDS discriminative subspaces improves the classification accuracy; 2) the GA presents discriminative subspaces capable of achieving even higher classification results by iteratively selecting merged subspaces of random dimensions. These findings not only validate the robustness of our proposed approach but also highlight the potential of combining discriminative spaces to achieve superior performance in classification tasks.

5 Final Remarks and Future Directions

This paper presents a Genetic Algorithm (GA) for optimizing discriminative subspaces to enhance feature extraction in object classification, utilizing Evolution Strategy (ES) principles. The GA employs initial population generation, evaluation, and selection, using the Fisher score for fitness assessment. Crossovers among randomly selected elements create new populations, refining solutions with each iteration based on Fisher scores. Evaluated on object recognition tasks, the method demonstrates high classification performance across various datasets, effectively capturing high-dimensional data and uncovering subspace structures.

Future directions include advanced evaluation methods and exploring kernel approaches for nonlinear data distributions. Our solutions are applicable in gesture and action recognition, adapting to tasks using basis vectors for data representation, suitable for subspaces of any pattern-set nature. We also plan to develop mutation techniques for subspaces to further improve GA performance and explore additional metaheuristics for pattern set optimization problems.

Acknowledgements. This study was financed in part by the Coordenação de Aperfeiçoamento de Pessoal de Nível Superior - Brasil (CAPES-PROEX) - Finance Code 001. This work was partially supported by Amazonas State Research Support Foundation - FAPEAM - through the PDPG/CAPES project.

References

1. Chen, Z., Xu, T., Wu, X.J., Wang, R., Kittler, J.: Hybrid riemannian graph-embedding metric learning for image set classification. IEEE Transactions on Big Data (2021)
2. Daham, H.A., Mohammed, H.J.: An evolutionary algorithm approach for vehicle routing problems with backhauls. Materials Today: Proceedings (2021)

3. Eltaeib, T., Mahmood, A.: Differential evolution: A survey and analysis. Appl. Sci. **8**(10), 1945 (2018)
4. Fukui, K., Maki, A.: Difference subspace and its generalization for subspace-based methods. IEEE Trans. Pattern Anal. Mach. Intell. **37**(11), 2164–2177 (2015)
5. Fukui, K., Yamaguchi, O.: The kernel orthogonal mutual subspace method and its application to 3d object recognition. In: Asian Conference on Computer Vision, pp. 467–476. Springer (2007). https://doi.org/10.1007/978-3-540-79043-3_8
6. Gatto, B.B., dos Santos, E.M., Fukui, K., Júnior, W.S., dos Santos, K.V.: Fukunaga-koontz convolutional network with applications on character classification. Neural Process. Lett. **52**(1), 443–465 (2020)
7. Gatto, B.B., Souza, L.S., dos Santos, E.M., Fukui, K., S. Júnior, W.S., dos Santos, K.V.: A semi-supervised convolutional neural network based on subspace representation for image classification. EURASIP J. Image Video Process. **2020**(1), 1–21 (2020). https://doi.org/10.1186/s13640-020-00507-5
8. Geusebroek, J.M., Burghouts, G.J., Smeulders, A.W.: The amsterdam library of object images. Int. J. Comput. Vision **61**(1), 103–112 (2005)
9. Goldberg, D.E.: Genetic Algorithms in Search, Optimization, and Machine Learning. Addison-Wesley Publishing Company (1989)
10. Grzegorowski, M., Slezak, D.: On resilient feature selection: Computational foundations of rc-reducts. Inf. Sci. **499**, 25–44 (2019)
11. Gu, Q., Li, Z., Han, J.: Generalized fisher score for feature selection. arXiv preprint arXiv:1202.3725 (2012)
12. Guo, J., Sun, Y., Gao, J., Hu, Y., Yin, B.: Low rank representation on product grassmann manifolds for multi-view subspace clustering. In: 2020 25th International Conference on Pattern Recognition (ICPR), pp. 907–914. IEEE (2021)
13. Holland, J.H.: Genetic algorithms. Sci. American **267**(1), 66–73 (1992)
14. Kolda, T.G., Bader, B.W.: Tensor decompositions and applications. SIAM Rev. **51**(3), 455–500 (2009)
15. Leibe, B., Schiele, B.: Analyzing appearance and contour based methods for object categorization. In: 2003 IEEE Computer Society Conference on Computer Vision and Pattern Recognition, 2003. Proceedings. vol. 2, pp. II–409. IEEE (2003)
16. Maeda, K.i.: From the subspace methods to the mutual subspace method. In: Computer Vision, pp. 135–156. Springer (2010). https://doi.org/10.1007/978-3-642-12848-6_5
17. Mitchell, M., Taylor, C.E.: Evolutionary computation: an overview. Annu. Rev. Ecol. Syst. **20**, 593–616 (1999)
18. Nene, S.A., Nayar, S.K., Murase, H.: Columbia object image library (coil-20. Tech. rep. (1996)
19. Nene, S.A., Nayar, S.K., Murase, H., et al.: Columbia object image library (coil-100) (1996)
20. Ranta, R., Le Cam, S., Chaudet, B., Tyvaert, L., Maillard, L., Colnat-Coulbois, S., Louis-Dorr, V.: Approximate canonical correlation analysis for common/specific subspace decompositions. Biomed. Signal Process. Control **68**, 102780 (2021)
21. Sakano, H.: A brief history of the subspace methods. In: Asian Conference on Computer Vision, pp. 434–435. Springer (2010). https://doi.org/10.1007/978-3-642-22819-3_44
22. Shafait, F., et al.: Fish identification from videos captured in uncontrolled underwater environments. ICES J. Mar. Sci. **73**(10), 2737–2746 (2016)

23. Tan, H., Gao, Y., Ma, Z.: Regularized constraint subspace based method for image set classification. Pattern Recogn. **76**, 434–448 (2018)
24. Wu, G., Mallipeddi, R., Suganthan, P.N.: Ensemble strategies for population-based optimization algorithms-a survey. Swarm Evol. Comput. **44**, 695–711 (2019)

A Performance Increment Strategy for Semantic Segmentation of Low-Resolution Images from Damaged Roads

Rafael S. Toledo[✉], Cristiano S. Oliveira, Vitor H. T. Oliveira, Eric A. Antonelo, and Aldo von Wangenheim

Federal University of Santa Catarina, Florianópolis, Santa Catarina, Brazil
rafael.toledo@posgrad.ufsc.br

Abstract. Autonomous driving needs good roads, but 85% of Brazilian roads have damages that deep learning models may not regard as most semantic segmentation datasets for autonomous driving are high-resolution images of well-maintained urban roads. A representative dataset for emerging countries consists of low-resolution images of poorly maintained roads and includes labels of damage classes; in this scenario, three challenges arise: objects with few pixels, objects with undefined shapes, and highly underrepresented classes. To tackle these challenges, this work proposes the Performance Increment Strategy for Semantic Segmentation (PISSS) as a methodology of 14 training experiments to boost performance. With PISSS, we reached state-of-the-art results of 79.8 and 68.8 mIoU on the Road Traversing Knowledge (RTK) and Technik Autonomer Systeme 500 (TAS500) test sets, respectively. Furthermore, we also offer an analysis of DeepLabV3+ pitfalls for small object segmentation (Code available on https://github.com/tldrafael/pisss).

Keywords: Unstructured environment · Road segmentation · Damaged roads · Low-resolution · DeepLabV3+

1 Introduction

Autonomous driving research is mainly based on developed countries with well-maintained infrastructure represented by many European urban streets datasets of high-res images, e.g., Cityscapes [7], CamVid [2], and KITTI [9]. In Brazil, 85% of the roads suffer from fatigue, cracks, holes, patches, and wavy surfaces [6]. This poor condition demands adjustments in the perception of autonomous driving. Additionally, computational constraints in emerging countries may limit the input size for the deep learning models, constraining the usage of high-res images and forcing adapted solutions for low-res images.

Some datasets such as [17, 20, 22] represented emerging countries' roads. Among them, especially the RTK dataset [17] captures the Brazilian countryside road featuring distinct maintenance conditions and surfaces. RTK consists of 701 annotated images of resolution 352x288 with 12 classes. The classes include surfaces (asphalt, paved, and unpaved), signs (markings, cat's eyes, speed bumps, and storm drains), and damages (patches, water-puddles, potholes, and cracks). See Fig. 1.

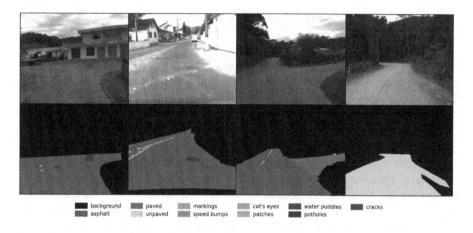

Fig. 1. Examples of emerging countries' roads in the RTK dataset.

This work raised three challenges for training a deep learning model when working with low-res images: objects with few pixels, objects with undefined shapes, and highly underrepresented classes. Low-res images have small objects not only relative to other objects' sizes but in pixels' quantity; for example, 70% of the cat's eyes blobs have edges with equal or less than 5 pixels, and 15% of the road markings blobs even have edges of a unique pixel. These tiny objects can easily vanish at the beginning of the forward pass given the stride of convolutional and pooling layers. The vanishing objects became a problem for DeepLabV3+, as will be seen in Sect. 4.2.

Other issues are undefined shape objects or multiscale elements presence in the image, e.g., road surfaces are broad and have a well-defined shape, whereas patches do not have a defined shape and size. This problem complexity increases with intraclass shape variations, like holes and cracks in the same image with multiple formats. Lastly, small damages and signs become very underrepresented, e.g., background and road surfaces occupy 98.5% of pixels, while cat's eye and storm-drain are just 0.02%. The imbalanced scenario raises the risk of overlooking small-sized classes.

1.1 Contributions

To meet the challenges presented above, we designed the Performance Increment Strategy for Semantic Segmentation (PISSS) that consists of a series of good training practices found in the state-of-the-art (SOTA) works that tackle semantic segmentation challenges of imbalanced datasets, small objects, and multiscale segmentation. With PISSS, we raised the RTK benchmark to 79.8 mIoU and the TAS500 to 68.8 mIoU, the best published results so far. Furthermore, we also propose removing the ResNet's max-pooling (MP) layer to preserve small objects segmentation.

2 Background

In this section, we introduce the main topics covered in the PISSS strategy, they are: training procedures, small objects segmentation, and multiscale segmentation.

2.1 Guidelines and Training Procedures

A significant part of deep learning success comes from adopting better training procedures. However, they are not usually the main research focus, and their details may be hidden in the implementation code. The Bag of Tricks for Image Classification [11] is a fundamental work for training recipes with convolutional neural networks, and it pushed ResNet-50's ImageNet top-1 validation accuracy from 75.3% to 79.29%. [1,24] also present common procedures like warm-up learning rate (LR), cosine LR decay, weight decay, label smoothing, stochastic depth, dropout, mixup, cutmix, and random resized cropping.

Furthermore, standard practices in semantic segmentation SOTA works [5, 14,25–27] are cropping, resizing, and flipping as data augmentation; Stochastic Gradient Descent (SGD) with a polynomial LR decay, momentum of 0.9, and weight decay of 5e-4 as optimizer; and multiscale ensemble predictions for testing. The SGD preference over Adam is explained by its better generalization results [13,28].

Important takeaways are pointed out by [24] that there is no training procedure ideal for all models, and by [1] that training methods are more task-driven than architectures, and, hence, improvements from training methods do not necessarily generalize as well as architectural ones.

2.2 Small Objects Segmentation

Small object segmentation is a challenge for low-res datasets. [10,12] emphasized the importance of context for small object detection, underlining that even humans cannot recognize a small building in a satellite image without the context of roads, cars, or other buildings. Hence, neurons with a large receptive field (RF) are essential for the task. A simpler option was found by [19] that noticed that the double sequence of stride 2 on ResNet loses sensible feature information of small objects; they proposed replacing the ResNet 7x7 convolution layer with stride 2 to a series of three 3x3 with stride only at the end. This same tactic was followed later by [15,29].

2.3 Multiscale Segmentation and DeepLabV3+

Some approaches for handling the multiscale segmentation challenge are to extract multiscale features in a layer level like Res2Net [8] and to add an attention mechanism that smartly combines predictions from different feature map scales to avoid scale pitfalls [4,21]. [21] noted that feature maps of large scales predict better fine details, such as edges of objects or thin structures, whereas feature maps of small scales predict better large structures that demand global context.

DeepLabV3+ [3] proposed the Atrous Spatial Pyramid Pooling (ASPP) module in charge of capturing multiscale features by simultaneously applying various dilated convolutions, hence, combining multiple receptive fields. ASPP balances the trade-off between accurate localization (small receptive field) and context assimilation (large receptive field).

3 PISSS - Performance Increment Strategy for Semantic Segmentation

PISSS is a methodology that consists of an additive series of ablation experiments. Each experiment checks the best performance among a set of hypotheses; e.g., which augmentation operation works better for the RTK dataset? Geometry operation? Color operations? Both together? Each option is called a hypothesis. The ablation experiments answer the questions of the best hypothesis for part of the training setup. The next ablation experiment is built on top of the best setup until that moment.

In total, the PISSS applied on the RTK dataset sums 14 ablation experiments organized into four categories: Baseline (B), Prediction (P), Technique (T), and Architecture (A). Baseline checks choices of the RTK authors training in [18], Prediction checks the usage of prediction ensemble, Technique checks training setup specificities, and Architecture checks changes in the neural network structure. Figure 2 shows the experiments tried in each category.

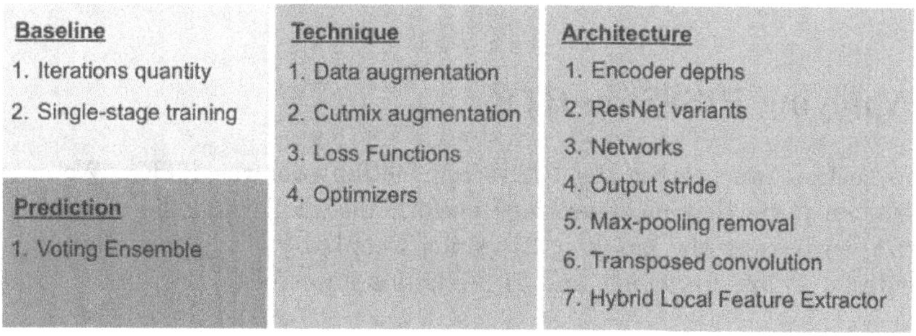

Baseline	Technique	Architecture
1. Iterations quantity	1. Data augmentation	1. Encoder depths
2. Single-stage training	2. Cutmix augmentation	2. ResNet variants
	3. Loss Functions	3. Networks
Prediction	4. Optimizers	4. Output stride
1. Voting Ensemble		5. Max-pooling removal
		6. Transposed convolution
		7. Hybrid Local Feature Extractor

Fig. 2. PISSS Diagram.

The sequence of experiments for PISSS starts on the baseline work, and we also adopted a different evaluation strategy as discussed next.

3.1 Baseline

We adopted the RTK authors' solution [18] as the starting point for the sequence of experiments. We call it **baseline**. It consists of a U-Net with ResNet-34, Adam optimizer with LR of 1e-4, batch size of 8, data augmentation with perspective distortion and horizontal flipping, named as *GeomRTK*, and a two-stage training regime that first runs 100 epochs with cross-entropy (CE) and another 100 later with weighted cross-entropy (WCE), summing 14k iterations at total.

3.2 Evaluation Methodology

When monitoring our training experiments, we noticed slight mIoU variations after the loss convergence. In Fig. 3, it is seen an oscillation around [0.733,0.743], which represents 1% of the mIoU scale of [0,1]; this variation is enough to lead to an incorrect conclusion when comparing the hypotheses. We adopted a workaround to reduce this noisy variation by averaging the last ten results steps of the validation set.

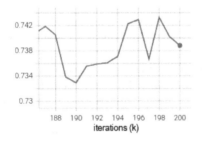

Fig. 3. mIoU oscillation after loss convergence.

4 Applying PISSS on RTK

In this section, we present the PISSS application over the RTK[1]. We split it into 5 main parts that most leverage performance, following the chronological order: 1) surpassing the baseline, 2) tuning DeepLabV3+, 3) fancy approaches and cutmix, 4) loss functions, and 5) prediction ensemble.

---

[1] https://data.mendeley.com/datasets/hssswvmjwf/1.

4.1 Part 1 - Surpassing the Baseline

In the first part, we covered the five first experiments: Iterations (B), Single Stage Training (B), Data Augmentation (T), Encoder Depths (A), and Resnet Variants (A). See Table 1; the order follows from top to bottom. Every subsequent experiment is built over the best hypothesis until that moment unless it is otherwise said, and the best hypothesis until that time has its mIoU value in bold.

Table 1. Part 1 of PISSS.

Exp	Enc	DataAug	Losses	Iters	mIoU
B: Iterations	ResNet-34	GeomRTK	CE+WCE	14k	54.7
B: Iterations	ResNet-34	GeomRTK	CE+WCE	100k	69.3
B: Iterations	ResNet-34	GeomRTK	CE+WCE	200k	**73.9**
B: Single-stage	ResNet-34	GeomRTK	WCE	200k	73.2
B: Single-stage	ResNet-34	GeomRTK	CE	200k	**73.9**
T: DataAug	ResNet-34	None	CE	200k	66.8
T: DataAug	ResNet-34	Cutmix80[a]	CE	200k	71.0
T: DataAug	ResNet-34	Cutmix50[a]	CE	200k	71.0
T: DataAug	ResNet-34	Resizing+Crop	CE	200k	73.5
T: DataAug	ResNet-34	Crop+Color	CE	200k	75.1
T: DataAug	ResNet-34	Crop	CE	200k	**75.4**
A: Encoder Depth	ResNet-50	Crop	CE	200k	75.9
A: Encoder Depth	ResNet-101	Crop	CE	200k	**76.5**
A: ResNet Variants	Res2Net-101	Crop	CE	200k	74.6
A: ResNet Variants	ResNeXt-101	Crop	CE	200k	75.4
A: ResNet Variants	ResNeSt-101	Crop	CE	200k	75.9

[a] The number next to the cutmix word stands by the occurrence probability.

In the Iterations experiment, we ensured that 14k iterations were short for training, and it proved necessary to push the iterations up to 200k. In the next experiment, Single Stage Training, we discarded any benefit from the two-stage training regime or WCE.

Later, in the Data Augmentation experiment, we adopted cropping of (224, 224), random edge resizing with scale [0.78, 2], and color augmentation with grayscale and jitter of 0.27. Cropping alone provided the best result. The addition of resizing or color augmentation worsened the results, cutmix alone made little impact, and no augmentation hypothesis had a terrible performance, revealing the need for data augmentation.

Subsequently, we tried a deeper ResNet version, which performed better. After, we tried ResNet variants: Res2Net, which uses different scales within the

ResNet module; ResNeSt, which adds an attention mechanism; and ResNeXt, which works with grouped convolutions. No variant lifted the performance.

4.2 Part 2 - Tuning DeepLabV3+

In this section, we cover seven more experiments: Model Architecture (A), Output Stride (A), Max-Pooling Removal (A), Transposed Convolutions (A), Hybrid Local Feature Extractor (A), Cutmix (T), and Optimizer (T). See Table 2. Table 2 experiment follows the best setup from Table 1 experiments.

In the model architecture experiment, we tried DeepLabV3+ (DL3+). It did not outperform U-Net in the first trials. However, after controlling the usage of the max-pooling (MP) layer and the output stride (OS), it reached a higher performance. In the OS experiment, in which we control the dimension ratio between the input and the encoder's outcome, there was a clear trend that reducing OS increases performance. Later, in the experiment that we suggest the MP removal, this trend reverted, and a higher OS reached the best performance. This behavior's change in OS is interpreted in the next section.

Table 2. Part 2 of PISSS. Abbreviations: Arch (Architecture), wo/ MP (without MP layer), R50/101 (ResNet-50/101).

Exp	Architecture	Encoder	OS	wo/ MP	mIoU
A: Arch	U-Net	R101	-	-	**76.5**
A: Arch	DL3+	R101	16	-	75.6
A: Arch	DL3+	R50	16	-	75.6
A: OS	DL3+	R50	8	-	75.7
A: OS	DL3+	R50	4	-	76.3
A: wo/ MP	DL3+	R50	4	✓	76.2
A: wo/ MP	DL3+	R50	8	✓	76.8
A: wo/ MP	DL3+	R50	16	✓	**76.9**

Interpreting DeepLabV3+ Patterns. It is essential to point out that the decoder of DL3+ concatenates high-level (HL) features from the ASPP block and low-level (LL) features from the ResNet's stem. See Fig. 4 for a vanilla example of the architecture and the control of the dimensions. The OS experiment's parameter just controls the final HL feature dimensions. On the other hand, the usage of the MP layer controls the LL dimensions; for example, without the MP layer, the LL features come just after the 7x7 conv layer with a stride of 2; otherwise, the features come just after the MP layer with a stride of 4.

Removing the MP layer positively impacted the OS of 8 and 16, whereas it did not matter for OS 4. A smaller LL stride seems more crucial than the HL

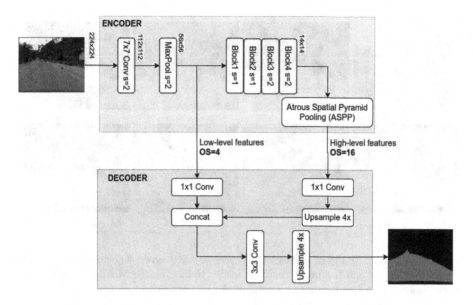

Fig. 4. Low and high-level features connections in DeepLabV3+.

stride for DL3+. Thus, when concatenating a lower LL stride with a higher HL one, it joins the best of both contexts, explaining the results of Table 2.

Another pattern found that is quantitatively unnoticeable but qualitatively impactful was that, with a higher OS, the model randomly predicts small background blobs over the road and small road blobs over the background. This problem gradually vanished until the OS decreased to 4. Besides, the problem was also solved after the MP removal; see Figs. 5 and 6.

4.3 Part 3 - Fancy Approaches and Cutmix

In this section, we cover four more experiments: Transposed Convolutions (A), Hybrid Local Feature Extractor (A), Cutmix (T), and Optimizer (T). See Table 3.

We tried two fancy approaches, and none brought a performance improvement. First, we tried replacing the non-parametric upsampling with a transposed convolution layer. Next, we implemented a hybrid local feature extractor (HLFE) that joins the digressive dilation rates [10] and the hybrid dilation rates [23]. For HLFE, we implemented the following dilation rates of [1, 3, 5, 5, 3, 1] for block 3 and [1, 3, 1] for block 4 of the ResNet. It is impossible to try HLFE for OS 16 as it has no dilation rate.

We also tried SGD over Adam, and it neither presented any improvement. The SGD setup had LR of 1e-2 with a linear warm-up of 5k iterations, poly LR decay, and momentum of 0.9. On the other hand, trying the cutmix augmentation together with cropping had a quite effective impact, raising performance from 76.9 to 78.2 mIoU. We tested it with probability occurrence of 50% and 80%.

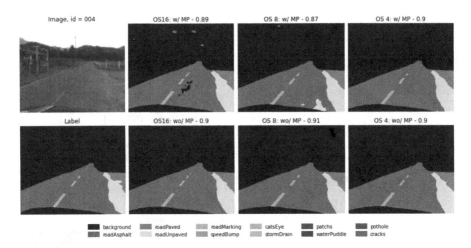

Fig. 5. Comparing results of distinct OS w/ or wo/ MP layer. The subtitles have the OS and MP states and the prediction IoU result. Removing the MP layer avoids early spatial information loss for extracting small object features.

Table 3. Part 3 of PISSS. Abbreviations: ConvT (Transposed Convolution).

Exp	OS	ConvT	HLFE	Cutmix	SGD	mIoU
A: ConvT	16	-	-	-	-	76.9
A: ConvT	16	✓	-	-	-	76.0
A: HLFE	4	-	✓	-	-	76.6
A: HLFE	8	-	✓	-	-	76.7
T: Cutmix	16	-	-	50%	-	77.1
T: Cutmix	16	-	-	80%	-	**78.2**
T: Optimizer	16	-	-	80%	✓	76.2

4.4 Part 4 - Loss Functions

We tried the surrogate losses of mIoU and dice. It did not present any advantage on a well-calibrated training setup, although it does help the baseline simpler training setup, see Table 4. We noticed the surrogate losses alone degrade performance, whereas CE acts like a proxy for mIoU, optimizing it even better than its loss.

4.5 Part 5 - Prediction Ensemble

In contrast with the evaluation methodology applied so far, this experiment counts on the evaluation metrics of a single checkpoint (ckpt), either from the last training step or from the step with the best validation result; hence, the values presented in Table 5 diverge from the previously reported values. We adopted

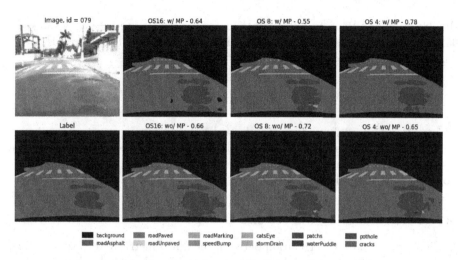

Fig. 6. Comparing results of distinct OS w/ or wo/ MP layer.

Table 4. Part 4 of PISSS. Abbreviations: R34/50 (ResNet-34/101).

Exp.	Arch.	Enc.	OS	Wo/ MP	Aug.	Losses	MIoU
T: Losses	DL3+	R50	16	✓	Crop+Cutmix80	WCE	68.9
T: Losses	DL3+	R50	16	✓	Crop+Cutmix80	MIoU	75.1
T: Losses	DL3+	R50	16	✓	Crop+Cutmix80	dice	75.6
T: Losses	DL3+	R50	16	✓	Crop+Cutmix80	CE+mIoU	77.0
T: Losses	DL3+	R50	16	✓	Crop+Cutmix80	CE+dice	77.4
T: Losses	DL3+	R50	16	✓	Crop+Cutmix80	CE	**78.2**

Baseline

T: Losses	U-Net	R34	-	-	GeomRTK	mIoU	69.8
T: Losses	U-Net	R34	-	-	GeomRTK	dice	72.0
T: Losses	U-Net	R34	-	-	GeomRTK	WCE	73.2
T: Losses	U-Net	R34	-	-	GeomRTK	CE	73.9
T: Losses	U-Net	R34	-	-	GeomRTK	CE+dice	74.7
T: Losses	U-Net	R34	-	-	GeomRTK	CE+mIoU	**74.9**

Table 5. Part 5 of the PISSS experiments.

Exp.	Strategy	Last ckpt	Best ckpt
P: Voting Ensemble	Single Prediction	77.4	79.3
P: Voting Ensemble	MultiScale+Flipped	78.4	79.3
P: Voting Ensemble	Flipped	78.9	79.8

the 288 × 224 and 448 × 352 resolutions for multiscale predictions besides the 352 × 288 native one. Ultimately, the flipped ensemble got the best results, with 78.9 and 79.8 mIoU for the last and best checkpoints.

5 RTK Experiments' Analysis

This section first analyzes how the classes' size and pixel quantity impact performance and, next, shows what the trained model looks for in each class. The results used for the analysis are from the best PISSS hypothesis, i.e., the checkpoint from the Cutmix (T) experiment.

5.1 Performance by Classes and Groups

In Fig. 7, we see a clear correlation between the mIoU metric and the object size, confirming a mIoU bias toward big objects, also pointed out in [7]. Furthermore, the worst performances by class are from cracks, water puddles, and cats' eyes, either classes of tiny objects or undefined shapes.

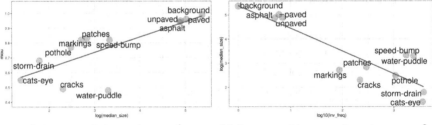

a) Impact of class edge size[1]. b) Impact of inverse class frequency[2].

Fig. 7. Relation between class characteristics and mIoU. ([a]The median size of the shortest edge of the class object. [b]The ratio of the number of pixels between the most popular class and the class i.)

5.2 What Does the Neural Network Look For?

One way to understand how the neural network perceives a dataset category is by optimizing the input neurons to maximize output probability. We optimized the network's inputs using gradient ascent, shown in Fig. 8. It highly noticed different textures, color distribution, and geometric patterns attached to each class; for example, storm-drain presents black holes, road-paved has the presence of polygon structures, and road-asphalt also seems to capture the cracks that happen over the road surface.

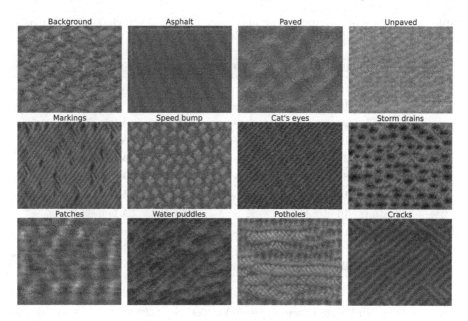

Fig. 8. Classes Optimized Inputs.

6 Applying PISSS on TAS500

The TAS500 dataset [16] is a dataset from 2021 that meets unstructured environments with annotations of fine-grained vegetation and terrain classes to distinct drivable surfaces and natural obstacles. TAS500 has high-res (HR) images of 2026 × 620, which turned prohibitive the experiments of *MP layer removal* and *transposed convolutions* due to the GPU 16 GB memory. Besides, the batch size had to be reduced to 4. Moreover, the images were trained with cropped parts of 1024 × 512, a standard practice for training HR images.

Furthermore, we skipped the baseline experiments and only applied a subset of the PISSS hypotheses used on RTK. So the experiments' setup started with DeepLabV3+, ResNet50, OS 16, Adam with LR of 5e- 5 (a reduced value from the 1e-4 in the RTK experiments, given the reduced batch size). For validation evaluation, we followed the same methodology of averaging the results of the last ten steps. See Tables 6 and 7.

We found that resizing as data augmentation, CE+dice loss, SGD, and multiscale ensemble prediction were fundamental for raising the TAS500 benchmark. The PISSS raised the validation set results from 65.4 to 74.7 mIoU. Furthermore, we checked our best hypothesis model on the Outdoor Semantic Segmentation Challenge, and it reached 68.8[2] mIoU, surpassing the 2021 1st place of 67.5 mIoU.

---

[2] Results with user *slow* on https://codalab.lisn.upsaclay.fr/competitions/5637# results..

Table 6. Summary of the PISSS experiments on TAS500. Abbreviations: Res (Resizing), CM (Cutmix).

Exp	OS	Aug	HLFE	CM	Losses	Optim	mIoU
T: DataAug	16	Crop+Color	-	-	CE	Adam	65.4
T: DataAug	16	Crop	-	-	CE	Adam	67.6
T: DataAug	16	Res+Crop+Color	-	-	CE	Adam	69.4
T: DataAug	16	Res+Crop	-	-	CE	Adam	**69.8**
A: OS	32	Res+Crop	-	-	CE	Adam	65.2
A: OS	8	Res+Crop	-	-	CE	Adam	66.5
T: HLFE	16	Res+Crop	✓	-	CE	Adam	69.8
T: Cutmix	16	Res+Crop	-	80%	CE	Adam	69.1
T: Losses	16	Res+Crop	-	-	CE+dice	SGD	**70.8**

Table 7. Prediction Ensemble on TAS500.

Exp	Strategy	Best ckpt
P: Voting Ensemble	Single Prediction	73.1
P: Voting Ensemble	Flipped	73.6
P: Voting Ensemble	MultiScale+Flipped	74.9

7 Discussion and Findings

Although PISSS worked for RTK and TAS500 datasets, a very different subset of hypotheses was the best in each case. For RTK, when using the MP layer, cutmix, OS 4, CE loss, and Adam worked better; while for TAS500, resizing, OS 16, CE+dice loss, and SGD worked better. The dissimilarity between these two training setups endorses the need for a custom solution, which also corroborates with the ideas [24] that there is no ideal training procedural for all models.

The cutmix augmentation had a meaningful gain of 1.3 mIoU for RTK, but it did not help TAS500; we suppose that cutmix is more helpful for tricky scenes of rough transitions between the road surfaces and damage classes.

Finally, we summarize all findings on the following items:

- Segmenting small objects is problematic if the features' dimensions are reduced before producing deep features; the model tends to overpredict small objects.
- The hardest damage classes to segment based on the RTK results are cracks, water puddles, and cat's eyes.
- We confirmed the mIoU towards big objects.
- CE optimizes the mIoU metric better than mIoU and dice losses.
- An experiment performance gain depends on the initial setup that it was tested.

- The high dissimilarity between the best training setup for each dataset endorses the need for custom solutions.
- Conventional setups usually bring better results than fancy procedures and should be the first attempt.

8 Conclusions

PISSS was effective for RTK and TAS500, reaching SOTA results and showing the importance of a well-tuning training setup besides the bare choice of a neural network architecture. Moreover, we warned of the pitfalls of early stride on convolutional networks when working with tiny objects and road damage in low-res images. We also highlighted the potential cause of problems for false-positive blobs on DeepLabV3+ predictions due to early large strides.

Disclosure of Interests. The authors have no competing interests to declare that are relevant to the content of this article.

References

1. Bello, I., et al.: Revisiting resnets: Improved training and scaling strategies. Adv. Neural. Inf. Process. Syst. **34**, 22614–22627 (2021)
2. Brostow, G.J., Fauqueur, J., Cipolla, R.: Semantic object classes in video: A high-definition ground truth database. Pattern Recogn. Lett. **30**(2), 88–97 (2009)
3. Chen, L.C., Papandreou, G., Kokkinos, I., Murphy, K., Yuille, A.L.: Deeplab: Semantic image segmentation with deep convolutional nets, atrous convolution, and fully connected crfs. IEEE Trans. Pattern Anal. Mach. Intell. **40**(4), 834–848 (2017)
4. Chen, L.C., Yang, Y., Wang, J., Xu, W., Yuille, A.L.: Attention to scale: Scale-aware semantic image segmentation. In: Proceedings of the IEEE Conference on Computer Vision and Pattern Recognition, pp. 3640–3649 (2016)
5. Chen, L.C., Zhu, Y., Papandreou, G., Schroff, F., Adam, H.: Encoder-decoder with atrous separable convolution for semantic image segmentation. In: Proceedings of the European Conference on Computer Vision (ECCV), pp. 801–818 (2018)
6. CNT: Pesquisa CNT de rodovias 2021. SEST SENAT (2021). https://pesquisarodovias.cnt.org.br/downloads/ultimaversao/
7. Cordts, M., et al.: The cityscapes dataset for semantic urban scene understanding. In: Proceedings of the IEEE Conference on Computer Vision and Pattern Recognition, pp. 3213–3223 (2016)
8. Gao, S.H., Cheng, M.M., Zhao, K., Zhang, X.Y., Yang, M.H., Torr, P.: Res2net: A new multi-scale backbone architecture. IEEE Trans. Pattern Anal. Mach. Intell. **43**(2), 652–662 (2019)
9. Geiger, A., Lenz, P., Stiller, C., Urtasun, R.: Vision meets robotics: The kitti dataset. Int. J. Robot. Res. **32**(11), 1231–1237 (2013)
10. Hamaguchi, R., Fujita, A., Nemoto, K., Imaizumi, T., Hikosaka, S.: Effective use of dilated convolutions for segmenting small object instances in remote sensing imagery. In: 2018 IEEEd Winter Conference on Applications of Computer Vision (WACV), pp. 1442–1450. IEEE (2018)

11. He, T., Zhang, Z., Zhang, H., Zhang, Z., Xie, J., Li, M.: Bag of tricks for image classification with convolutional neural networks. In: Proceedings of the IEEE/CVF Conference on Computer Vision and Pattern Recognition, pp. 558–567 (2019)
12. Hu, P., Ramanan, D.: Finding tiny faces. In: Proceedings of the IEEE Conference on Computer Vision and Pattern Recognition, pp. 951–959 (2017)
13. Keskar, N.S., Socher, R.: Improving generalization performance by switching from adam to sgd. arXiv preprint arXiv:1712.07628 (2017)
14. Kirillov, A., Wu, Y., He, K., Girshick, R.: Pointrend: Image segmentation as rendering. In: Proceedings of the IEEE/CVF Conference on Computer Vision and Pattern Recognition, pp. 9799–9808 (2020)
15. Li, Y., Peng, B., He, L., Fan, K., Li, Z., Tong, L.: Road extraction from unmanned aerial vehicle remote sensing images based on improved neural networks. Sensors **19**(19), 4115 (2019)
16. Metzger, K.A., Mortimer, P., Wuensche, H.J.: A fine-grained dataset and its efficient semantic segmentation for unstructured driving scenarios. In: 2020 25th International Conference on Pattern Recognition (ICPR), pp. 7892–7899. IEEE (2021)
17. Rateke, T., Justen, K.A., Von Wangenheim, A.: Road surface classification with images captured from low-cost camera-road traversing knowledge (rtk) dataset. Revista de Informática Teórica e Aplicada **26**(3), 50–64 (2019)
18. Rateke, T., von Wangenheim, A.: Road surface detection and differentiation considering surface damages. Auton. Robot. **45**(2), 299–312 (2021). https://doi.org/10.1007/s10514-020-09964-3
19. Shen, Z., Liu, Z., Li, J., Jiang, Y.G., Chen, Y., Xue, X.: Dsod: Learning deeply supervised object detectors from scratch. In: Proceedings of the IEEE International Conference on Computer Vision, pp. 1919–1927 (2017)
20. Shinzato, P.Y., et al.: Carina dataset: An emerging-country urban scenario benchmark for road detection systems. In: 2016 IEEE 19th International Conference on Intelligent Transportation Systems (ITSC), pp. 41–46. IEEE (2016)
21. Tao, A., Sapra, K., Catanzaro, B.: Hierarchical multi-scale attention for semantic segmentation. arXiv preprint arXiv:2005.10821 (2020)
22. Varma, G., Subramanian, A., Namboodiri, A., Chandraker, M., Jawahar, C.: Idd: A dataset for exploring problems of autonomous navigation in unconstrained environments. In: 2019 IEEE Winter Conference on Applications of Computer Vision (WACV), pp. 1743–1751. IEEE (2019)
23. Wang, P., et al.: Understanding convolution for semantic segmentation. In: 2018 IEEE Winter Conference on Applications of Computer Vision (WACV), pp. 1451–1460. Ieee (2018)
24. Wightman, R., Touvron, H., Jégou, H.: Resnet strikes back: An improved training procedure in timm. arXiv preprint arXiv:2110.00476 (2021)
25. Yu, C., Gao, C., Wang, J., Yu, G., Shen, C., Sang, N.: Bisenet v2: Bilateral network with guided aggregation for real-time semantic segmentation. Int. J. Comput. Vision **129**(11), 3051–3068 (2021)
26. Yuan, Y., Chen, X., Chen, X., Wang, J.: Segmentation transformer: Object-contextual representations for semantic segmentation. arXiv preprint arXiv:1909.11065 (2019)
27. Zhao, H., Shi, J., Qi, X., Wang, X., Jia, J.: Pyramid scene parsing network. In: Proceedings of the IEEE Conference on Computer Vision and Pattern Recognition, pp. 2881–2890 (2017)

28. Zhou, P., Feng, J., Ma, C., Xiong, C., Hoi, S.C.H., et al.: Towards theoretically understanding why sgd generalizes better than adam in deep learning. Adv. Neural. Inf. Process. Syst. **33**, 21285–21296 (2020)
29. Zhou, P., Ni, B., Geng, C., Hu, J., Xu, Y.: Scale-transferrable object detection. In: proceedings of the IEEE Conference on Computer Vision and Pattern Recognition, pp. 528–537 (2018)

A Unified Framework for Average Reward Criterion and Risk

Willy Arthur Silva Reis[✉][iD], Karina Valdivia Delgado[iD], and Valdinei Freire[iD]

University of São Paulo, Cidade Universitária, Butantã, São Paulo,
SP 05508-220, Brazil
{willy.reis,kvd,valdinei.freire}@usp.br

Abstract. The average reward criterion is used to solve infinite-horizon MDPs. This risk-neutral criterion depends on the stochastic process in the limit and can use (i) the accumulated reward at infinity, which considers sequences of states of size $h = \infty$, or (ii) the steady state distribution of the MDP (i.e., the probability that the system is in each state in the long term), which considers sequences of states of size $h = 1$. In many situations, it is desirable to consider risk during the process at each stage, which can be achieved with the average reward criterion using a utility function or a risk measure such as VaR and CVaR. The objective of this work is to propose a mathematical framework that allows a unified treatment of the existing literature using average reward and risk, including works that use exponential utility functions and CVaR, as well as to include interpretations with $1 \leq h \leq \infty$ not present in the literature. These new interpretations allow differentiating policies that may not be distinguished from existing criteria. A numerical example shows the behaviors of the criteria considering this new framework.

Keywords: Markov decision process · Average reward · Risk-sensitive

1 Introduction

A Markov Decision Process (MDP) provides a mathematical framework for representing and solving sequential decision-making problems under uncertainty in fully observable environments [4]. In these problems, an agent must interact with an environment through actions that can have stochastic outcomes.

Various optimization criteria are used to find an optimal policy for a Markov decision process. The most commonly used criterion in finite and indefinite horizon MDPs is accumulated reward. In infinite horizon problems, the average reward per decision step criterion is used.

The average reward criterion is also known as *long-run average reward*, *mean-payoff*, *limit-average reward*, or *steady-state reward* [1]. This criterion is risk-neutral, does not use discount, and is naturally used in cyclic and non-terminating problems. This criterion is fundamental for areas such as controlled queues and inventory systems with frequent restocking, where decisions are made frequently and continually [4]. It is also important for natural resource management and environmental economics, for example, to determine the extraction rate of a renewable natural resource [3].

The average reward criterion depends on the behavior of the stochastic process in the limit, which determines some models of the criterion according to the structure of the Markov chain generated by stationary policies.

One way to think about the stochastic process in the limit is by summing the rewards (i.e., calculating the expected accumulated reward) and then calculating how much is added to this sum at each step, which is called gain. When time tends to infinity, the gain tends to a constant value. Since this sum considers sequences of $h = \infty$ states, this way of thinking in this work is called $h = \infty$. Another way to think about the stochastic process in the limit is by calculating the steady-state distribution, i.e., calculating the probability that the system is in each state in the long term. Considering these probabilities and the immediate reward, the expected reward is calculated. This way of thinking in this work is called $h = 1$. Considering the risk-neutral average reward criterion, both ways of thinking about the stochastic process in the limit yield the same result.

In many situations, it is desirable to consider risk during the process at each stage. In the financial field, for example, a risk-averse investor might prefer to minimize uncertainties by opting for a lower guaranteed return, even if this implies obtaining a smaller gain [7]. Moreover, considering risk at each stage can increase the robustness of policies, allowing them to function in a wider variety of scenarios, such as in the use of renewable natural resources that suffer from greater volatility due to climate change. Among the works that use the average reward criterion and consider risk during the process are those that use exponential utility functions [2] and CVaR [7] (referred to as *steady CVaR* or *long-run CVaR*).

While the algorithm that considers an exponential utility function [2] calculates the accumulated reward and how much is added to this sum, thus following the $h = \infty$ approach; the long-run CVaR algorithm [7] calculates the steady-state distribution and considers the immediate reward, which corresponds to the $h = 1$ approach.

The objective of this work is to propose a mathematical framework that provides a unified treatment of the literature on average reward and risk in MDPs, including works that use exponential utility functions [2] and CVaR [7], as well as to include interpretations with $1 \leq h \leq \infty$, allowing differentiation of policies that may not be distinguished with existing criteria.

This paper is organized as follows. Section 2 reviews the definitions of MDP, average reward criterion, and risk-sensitive average reward literature. Section 3 presents the proposed framework. Section 4 evaluates the policies of an MDP for risk-neutral and risk-sensitive criteria with the proposed framework. Finally, Sect. 5 presents the conclusions.

2 Background

A Markov decision process [4] is described by a tuple $\mathcal{M} = \langle S, A, P, R \rangle$ where: S is a finite set of states; A is a finite set of actions; $P : S \times A \times S \to [0, 1]$ is a transition function that represents the probability that $j \in S$ is reached after the agent executes an action $a \in A$ in a state $s \in S$, i.e., $\Pr(s_{N+1} = j | s_N = s, a_N = a) = P(s, a, j)$; and $R : S \times A \to \mathbb{R}$ is a reward function that represents the reward of executing an action $a \in A$ in a state $s \in S$.

In this work, the average reward criterion is used for infinite-horizon problems. The average reward criterion depends on the behavior of the stochastic process in the limit, which determines certain models of the criterion based on the structure of the Markov chain generated from stationary policies. Among the types of Markov chains are unichain, multichain, recurrent, ergodic, and aperiodic chains. A unichain Markov chain has a closed irreducible set of states (a set of states where all states communicate with each other and do not communicate with states outside the set) and a possibly empty set of transient states (states that will not be revisited after some point in time). Otherwise, the Markov chain is multichain. A recurrent Markov chain is a unichain chain where the set of transient states is empty. A Markov chain is ergodic if it is recurrent and aperiodic. A Markov chain is aperiodic if the returns to states occur at irregular intervals.

2.1 Risk-Neutral Criterion

The expected total reward of a policy π from the initial state s up to the decision epoch $N + 1$ is a function v_{N+1}^{π} defined by

$$v_{N+1}^{\pi}(s) = E\left\{\sum_{n=1}^{N} r(S_n, A_n) \bigg| S_1 = s, \pi\right\} = E\left\{\sum_{n=1}^{N} r_n\right\}, \quad (1)$$

where S_n and A_n refer to the random variables of the state and action in the time step n. When N tends to infinity, $v_{N+1}^{\pi}(s)$ diverges.

The expected average reward of a policy π, i.e. the gain of a policy π, is

$$\begin{aligned} g^{\pi}(s) &= \lim_{N \to \infty} \frac{1}{N} v_{N+1}^{\pi}(s) \\ &= \lim_{N \to \infty} \frac{1}{N} \sum_{n=1}^{N} P_{\pi}^{n-1} r_{\pi}(s), \end{aligned} \quad (2)$$

where P_{π}^{n} is the n-th power of the transition matrix P_{π} and r_{π} is the reward function of π.

A policy with the highest gain $g^*(s)$ is called the optimal policy π^*,

$$g^*(s) = g^{\pi^*}(s) = \sup_{\pi \in \Pi} g^{\pi}(s), \quad (3)$$

where Π is the set of stationary policies.

In the following sections, the simplified version of the notation shown on the right side of Eq. 1 will be used.

2.2 Risk-Sensitive Criterion with Utility Function

There are several ways to incorporate risk into decision making with MDPs. One way is to use a utility function $u : \mathbb{R} \to \mathbb{R}$, such as the exponential utility function [2]. The exponential utility function u is defined by

$$u(x) = -(\text{sgn}\gamma)e^{-\gamma x}, \tag{4}$$

with inverse

$$u^{-1}(y) = \frac{1}{\gamma}\ln[-\text{sgn}(\gamma)y], \tag{5}$$

where γ is the risk factor and $\text{sgn}(\gamma)$ indicates the sign of γ.

The certainty equivalent v_u^π is the value whose utility is equal to the utility of the expected value of the accumulated reward v^π, i.e.,

$$u(v^\pi) = u(v_u^\pi). \tag{6}$$

A positive risk factor γ indicates risk aversion and implies $v_u^\pi < v^\pi$, while a negative factor indicates risk propensity and implies $v_u^\pi > v^\pi$.

The utility function applied to v_u^π is defined as

$$u^\pi = u(v_u^\pi) = -(\text{sgn}\gamma)e^{-\gamma v_u^\pi}. \tag{7}$$

The utility of policy u^π in iteration $n+1$ can be calculated by equation [2]

$$u^{\pi,n+1}(s) = \sum_{j \in S} P_\pi^n(s,j)e^{-\gamma r_\pi(s)}u^{\pi,n}(j), \tag{8}$$

where $u^{\pi,0} = (-1, ..., -1)$, $|u^{\pi,0}| = |S|$ e $P_\pi^0 = I$. From $u^{\pi,n+1}(s)$, the certainty equivalent can be calculated by $v_u^{\pi,n}(s) = u^{-1}(u^{\pi,n}(s))$. During the process, $v_u^{\pi,n}$ grows linearly, and its variation is called the certainty equivalent gain, denoted by g_u^π. Using the exponential utility criterion, an agent seeks a policy that maximizes the certainty equivalent gain of an MDP.

Howard and Matheson [2] propose a policy iteration algorithm that finds a stationary policy and deals only with acyclic and irreducible Markov processes (all states communicate). Given an arbitrary initial policy, the system of Eqs. 9 must be solved in the policy evaluation step to find the certainty equivalent gain g_u^π and the certainty equivalents v_u^π [2]

$$e^{-\gamma(g_u^\pi + v_u^\pi(s))} = \sum_{j \in S} p(j|s, \pi(s))e^{-\gamma(r(s,\pi(s)) + v_u^\pi(j))}. \tag{9}$$

Since there are more variables than equations, one of the certainty equivalent values must be set to 0. Then, using the values found in the policy evaluation step of the previous policy, a policy improvement step is applied to select, in each state s, a new action a that maximizes the equation [2]

$$v_u^\pi(s,a) = \frac{1}{\gamma}\ln\left[\sum_{j \in S} p(j|s,a)e^{-\gamma(r(s,a) + v_u^\pi(j))}\right]. \tag{10}$$

When no change occurs in any state, the optimal policy has been found.

2.3 Risk-Sensitive Criterion with CVaR

Another way to insert risk into MDPs is through the use of VaR and CVaR risk measures. The VaR (Value at Risk) and CVaR (Conditional-Value-at-Risk) metrics are widely used for portfolio management of financial assets.

VaR measures the worst expected loss within a given α confidence level, where $\alpha \in (0,1)$. VaR is defined as the $1 - \alpha$ quantile of Z, i.e.

$$\text{VaR}_\alpha(Z) = \min\{z | F(z) \geq 1 - \alpha\}, \tag{11}$$

where Z is a random variable and $F(z)$ is the cumulative distribution function.

An alternative measure is the CVaR metric, a coherent risk measure [6], that is computed by averaging losses that exceed the VaR value. CVaR, with a confidence level $\alpha \in (0,1)$ is defined as [5]

$$\text{CVaR}_\alpha(Z) = \min_{w \in \mathbb{R}} \left\{ w + \frac{1}{\alpha} E\left[(Z - w)^+\right] \right\}, \tag{12}$$

where $(x)^+ = \max\{x, 0\}$ represents the positive part of x, and w represents the decision variable that, at the optimum point, reaches the value of VaR

$$\text{CVaR}_\alpha(Z) = \text{VaR}_\alpha(Z) + \frac{1}{\alpha} E\left[(Z - \text{VaR}_\alpha(Z))^+\right]. \tag{13}$$

MDPs with the CVaR criterion for average reward were studied for the first time by Xia and Glynn [7]. However, they only deal with stationary policies associated with ergodic Markov chains, and they suggest that the results can be extended to unichain MDPs. The CVaR gain is defined as [7]

$$g^\pi_{\text{CVaR}_\alpha} = \lim_{N \to \infty} \frac{1}{N} \sum_{n=1}^{N} \text{CVaR}_\alpha(r_n). \tag{14}$$

Note that in Eq. 14, the random variable is only the immediate reward. Additionally, a local optimality equation of Bellman is proposed, and a policy iteration-type algorithm is developed.

3 A Unified Framework

In this work, some interpretations of the risk-neutral average reward criterion are considered, followed by interpretations of the risk-sensitive average reward criterion with utility function and CVaR. The proposed framework considers stationary and aperiodic policies, i.e., stationary policies that generate aperiodic unichain or aperiodic multichain Markov chains. For aperiodic policies $P^*_\pi = \lim_{n \to \infty} P^n_\pi$ exists.

3.1 Formalization of the Risk-Neutral Criterion

One of the most common ways to calculate the value function for MDPs is by computing the accumulated reward. This approach can be extended to infinite horizon, where to ensure convergence of the value, it is divided by the number of decision steps N (Eq. 15). Note that in this equation, the accumulated reward is the random variable. From this equation, various interpretations of the gain of an MDP for a policy π can be derived. The relationships between these interpretations are shown in Eqs. 15 to 18.

$$g^\pi = \lim_{N \to \infty} \frac{1}{N} E\left[\sum_{n=1}^{N} r_n\right] \qquad (15)$$

$$= \lim_{N \to \infty} E\left[\sum_{n=1}^{N+1} r_n - \sum_{n=1}^{N} r_n\right] \qquad (16)$$

$$= \lim_{N \to \infty} \sum_{n=1}^{N} \frac{1}{N} E[r_n] \qquad (17)$$

$$= \lim_{N \to \infty} \sum_{n=1}^{N} \frac{1}{Nh} E\left[\sum_{i=n}^{n+h} r_i\right] \qquad (18)$$

Equation 15 was described before. Equation 16 interprets gain as the increase in reward at each decision step. In this approach, the value is calculated as the difference between the expected accumulated rewards considering two consecutive time steps. Even though each accumulated reward tends to infinity, the difference between them is finite. This approach is similar to the gain extracted in iterative algorithms, such as the relative value iteration algorithm [4].

Equation 17 considers the immediate reward as a random variable. The gain is interpreted as the sum of the expected immediate rewards divided by N.

Equation 18 conjectures the interpretation of the gain as the Cesaro limit of the expected accumulated reward for h decision steps after time n. In this equation, the random variable is the accumulated reward of trajectories of fixed size h. The division is done by Nh because N sums of h rewards are made.

Note that Eqs. 16, 17 and 18 converge to some distribution, allowing the use of the steady-state matrix P^*_π to calculate the gain.

3.2 Formalization of the Risk-Sensitive Criterion

Before formalizing the risk-sensitive criterion, some notations used in this section are introduced.

Let ψ be a sequence of states and $\psi(k)$ be the k-th state of this sequence. Let S^h be the set of all sequences ψ of size h. Let $f^{\pi,h}_{s,n} : S^h \to [0,1]$, $f^{\pi,h}_{s,n}(\psi)$ is the probability of the sequence ψ of size h occurring with the process being started in state s and time n, and can be calculated as

$$f_{s,n}^{\pi,h}(\psi) = Pr\left(\bigwedge_{k=0}^{h-1} S_{n+k} = \psi(k)|S_0 = s\right)$$
$$= Pr\left(S_{n+0} = \psi(0)|S_0 = s\right) \prod_{k=1}^{h-1} Pr(S_{n+k} = \psi(k)|S_{n+k-1} = \psi(k-1)). \quad (19)$$

Let $v_n^{\pi,h}(\psi)$ be the accumulated reward of the sequence ψ of size h with the process starting at time n

$$v_n^{\pi,h}(\psi) = \sum_{k=n}^{n+h-1} r(\psi(k)). \quad (20)$$

Risk-Sensitive Criterion with Utility Function. The utility function u can be applied either to the sum of rewards or directly to the immediate reward. The average of the expected utility of the accumulated reward is observed in Eq. 21 and represents the approach of Howard and Matheson [2], which uses the exponential utility function. This equation is similar to Eq. 15, but it includes applying the utility function u to the sum of rewards

$$g_u^\pi = \lim_{N \to \infty} \frac{1}{N} E\left[u\left(\sum_{n=1}^N r_n\right)\right]. \quad (21)$$

Equation 22 is equivalent to Eq. 21. It calculates the expected value of the utility of the accumulated reward of all sequences of size $N \to \infty$ that start in state s and at time 1. Then the average of this expected value is calculated by

$$g_u^\pi = \lim_{N \to \infty} \frac{1}{N} \sum_{\psi \in S^N} f_{s,1}^{\pi,N}(\psi) u\left(\sum_{k=1}^N r(\psi(k))\right). \quad (22)$$

When $N \to \infty$, the probability of a state occurring from the sequence of states tends to the steady state at any time $n \in N$, so that the probability of the sequence $f_{s,n}^{\pi,h}$ can be calculated using the steady-state probability of each state in the sequence in Eq. 19. This interpretation introduces how an arbitrary sequence of states is intended to be visualized in the upcoming equations.

Equation 23 calculates the sum of the expected utility of the accumulated rewards of sequences of fixed size h starting at any time n such that $1 \le n \le N$, with $N \to \infty$, and finally their average is calculated by

$$g_u^{\pi,h} = \lim_{N \to \infty} \frac{1}{N} \sum_{n=1}^N E\left[u\left(\sum_{i=n}^{n+h-1} r_i\right)\right]. \quad (23)$$

Equation 25 reinterprets Eq. 23 by inserting a sequence ψ of h states, the probability of this sequence of states $f_{s,n}^{\pi,h}(\psi)$ (Eq. 19) and its accumulated reward

$v_n^{\pi,h}(\psi)$ (Eq. 20). Thus, for each sequence of size h starting at time n, its probability is multiplied by the utility of its accumulated reward

$$u_n^{\pi,h} = \sum_{\psi \in S^h} f_{s,n}^{\pi,h}(\psi) u\left(v_n^{\pi,h}(\psi)\right). \tag{24}$$

Then the sum for all time n of the inverse of $u_n^{\pi,h}$ is calculated and its average is obtained when $N \to \infty$ by

$$g_u^{\pi,h} = \lim_{N \to \infty} \frac{1}{N} \sum_{n=1}^{N} u^{-1}(u_n^{\pi,h}). \tag{25}$$

For $h = 1$ and u being the identity function, Eq. 25 becomes the risk-neutral average reward, equivalent to Eq. 2.

Note that in this work Eq. 25 is proposed to classify existing works in the literature concerning h. When $h = \infty$, Eq. 25 is equivalent to the criterion used by Howard and Matheson [2].

Risk-Sensitive Criterion with CVaR. Equation 26 considers the average CVaR of the accumulated reward. Analogous to Eq. 21, it considers the accumulated reward of the stochastic process, but instead of applying the utility function u, it applies the CVaR function

$$g_{\text{CVaR}}^{\pi} = \lim_{N \to \infty} \frac{1}{N} \text{CVaR}\left[\sum_{n=1}^{N} r_n\right]. \tag{26}$$

As far as we know, there is no work in the literature that solves Eq. 26.

Similar to Eq. 23, but considering the CVaR instead of the utility function u, Eq. 27 considers the accumulated rewards of fixed-length sequences h started at each time n and calculates the average of the CVaR values of all these sequences:

$$g_{\text{CVaR}}^{\pi,h} = \lim_{N \to \infty} \frac{1}{N} \sum_{n=1}^{N} \text{CVaR}\left[\sum_{i=n}^{n+h-1} r_i\right]. \tag{27}$$

Equation 29 reinterprets Eq. 27 by inserting a sequence ψ of h states, the probability of this sequence of states $f_{s,n}^{\pi,h}$ (Eq. 19) and its accumulated reward $v_n^{\pi,h}(\psi)$ (Eq. 20). Thus, for sequences ψ of size h starting at time n:

$$\begin{aligned} v_{\text{CVaR},n}^{\pi,h} &= \text{CVaR}_\alpha(v_n^{\pi,h}(\psi)) \\ &= \text{VaR}_\alpha(v_n^{\pi,h}(\psi)) + \frac{1}{\alpha} \sum_{\psi \in S^h} f_{s,n}^{\pi,h}(\psi)\left[(v_n^{\pi,h}(\psi) - \text{VaR}_\alpha(v_n^{\pi,h}(\psi)))^+\right]. \end{aligned} \tag{28}$$

Then the sum is calculated for all time n and then its average is obtained for $N \to \infty$

$$g_{\text{CVaR}}^{\pi,h} = \lim_{N \to \infty} \frac{1}{N} \sum_{n=1}^{N} v_{\text{CVaR},n}^{\pi,h}. \tag{29}$$

Note that in this work Eq. 29 is proposed to classify existing works in the literature concerning h. When $h = 1$, Eq. 29 is equivalent to Eq. 14 of Xia and Glynn [7].

4 Numerical Example

Example 1. *Consider an MDP with 4 states and three policies π_1, π_2 and π_3, where the transition matrix of π_1 and π_2 is*

$$P_{\pi_1} = P_{\pi_2} = \begin{bmatrix} 0.5 & 0.5 & 0 & 0 \\ 0 & 0 & 1 & 0 \\ 0 & 0 & 0.5 & 0.5 \\ 1 & 0 & 0 & 0 \end{bmatrix}, \qquad (30)$$

and the transition matrix for π_3 is

$$P_{\pi_3} = \begin{bmatrix} 1 & 0 & 0 & 0 \\ 0 & 0 & 0 & 0 \\ 0 & 0 & 0 & 0 \\ 0 & 0 & 0 & 0 \end{bmatrix}. \qquad (31)$$

Let r_π be the reward function of policy π, $r_{\pi_1} = (0, 6, 0, -5)$, $r_{\pi_2} = (0, 1, 0, 0)$ and $r_{\pi_3} = \left(\frac{1}{6}\right)$.

Figure 1 represents the three policies, where the vertices represent the states, the edges the actions, and the values in parentheses represent the transition probability and the reward, respectively.

The Markov chains generated by the stationary policies in the example 1 are unichain and aperiodic. In this way, all policies have a constant gain, that is, the gain of any policy is independent of the state [4].

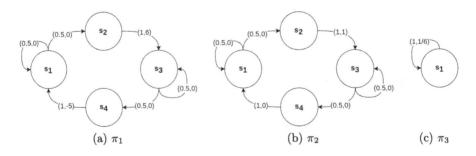

Fig. 1. Graphical representation of the policies of the MDP given in example 1.

The steady-state transition matrix is calculated by $P^*_\pi = \lim_{n\to\infty} P^{n+1}_\pi = P^n_\pi P_\pi$ and for policies π_1 and π_2 it is

$$P^*_{\pi_1} = P^*_{\pi_2} = \begin{bmatrix} \frac{1}{3} & \frac{1}{6} & \frac{1}{3} & \frac{1}{6} \\ \frac{1}{3} & \frac{1}{6} & \frac{1}{3} & \frac{1}{6} \\ \frac{1}{3} & \frac{1}{6} & \frac{1}{3} & \frac{1}{6} \\ \frac{1}{3} & \frac{1}{6} & \frac{1}{3} & \frac{1}{6} \end{bmatrix}. \tag{32}$$

For π_3, $P^*_{\pi_3} = P_{\pi_3}$.

Although the policies π_1 and π_2, with the same transition matrix, have the same expected average reward, the distributions of accumulated rewards for policy π_1 have greater variance than those for policy π_2 for different values of h, as shown in Fig. 2. Therefore, risk-sensitive criteria are expected to differentiate between policies π_1 and π_2.

Figures 2 a), c), e) and g) show the distribution of the accumulated reward for policies π_1 (blue bars), π_2 (orange bars) and π_3 (green bars) of the MDP given in the example 1 for $h \in \{1, 2, 5, 101\}$, respectively. In addition to the distributions, the values $v^{\pi,h}$ (red line), $v^{\pi,h}_{u_1}$ (purple line), $v^{\pi,h}_{\text{VaR}_{0.2}}$ (brown line) and $v^{\pi,h}_{\text{CVaR}_{0.2}}$ (pink line) are showed.

The accumulated rewards for $h = 1$ and $h = 2$ range from -5 to 6 for π_1 and range from 0 to 1 for π_2. For $h = 5$, the accumulated rewards range from -5 to 7 for π_1 and from 0 to 2 for π_2. For $h = 101$, the accumulated rewards range from -5 to 30 for π_1 and from 0 to 25 for π_2. The accumulated reward for π_3 does not vary for each h. It is expected that a risk-averse criterion may consider these three policies differently. Additionally, when h tends to infinity, all accumulated rewards tend to infinity (see Figs. 2 a), c), e) and g)).

Figures 2 b), d), f), and h) show the cumulative distribution function (CDF) for policies with the same values of h. For each policy, the CDFs are symmetric. The median of π_1 and π_2 are the same for all values of h.

In the next subsections, examples of the gain calculation for each of the criteria used are shown.

Consider the following notation regarding policies: $A \succ B$ means that the decision-maker prefers A to B, $A \sim B$ means that the decision-maker has an equal preference for either A or B and $A \succeq B$ means that the decision-maker prefers or has an equal preference for A or B.

4.1 Calculation of Gain Using the Risk-Neutral Criterion

The gain of a policy π, defined in Eq. 15, can be calculated by dividing the accumulated reward v^π_N by the number of steps N, i.e., $g^\pi = \lim_{N\to\infty} \frac{1}{N} v^\pi_N$. v^π_N can be calculated iteratively by

$$v^{\pi,n+1} = v^{\pi,n} + P^n_\pi r_\pi, \tag{33}$$

where $v^{\pi,0}$ is a null vector of size $|S|$ and $P^0_\pi = I$.

Using this form of calculation for the example, the gain of policies π_1, π_2 and π_3 is $g^{\pi_1} = g^{\pi_2} = g^{\pi_3} = (0.167, 0.167, 0.167, 0.167)$.

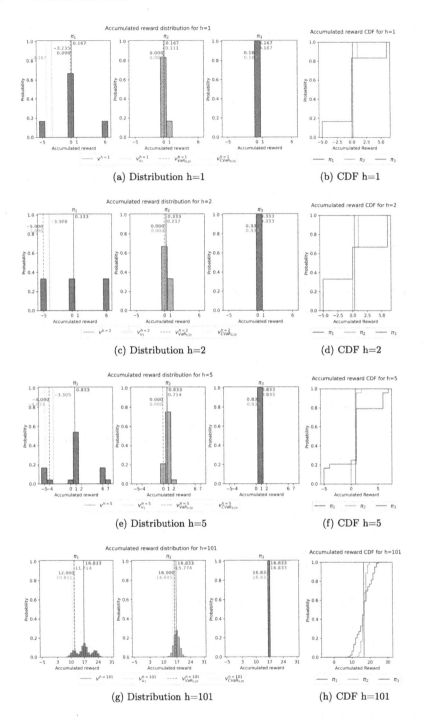

Fig. 2. Graphs of the distribution of accumulated reward and cumulative distribution function (CDF) of accumulated reward for different values of h.

The gain of a policy defined in Eq. 17 can be calculated by multiplying the steady-state matrix P_π^* by the policy reward r_π, i.e., $g^\pi = P_\pi^* r_\pi$ [4]. In the example using this second form of calculation, the gain is $g^{\pi_1} = g^{\pi_2} = g^{\pi_3} = (0.167, 0.167, 0.167, 0.167)$.

Note that, using the risk-neutral criterion $\pi_3 \sim \pi_2 \sim \pi_1$.

4.2 Calculation of Gain Using Utility Function with $\gamma = 1$ for $h = 1$

To calculate the gain $g_u^{\pi,h}$, it is necessary to calculate $f_{s,n}^{\pi,h}(\psi)$ (Eq. 19) which is the probability of the sequence ψ of size h occurring with the process starting in state s and in time n. It is also necessary to calculate $v_n^{\pi,h}(\psi)$ (Eq. 20) which is the accumulated reward of the sequence ψ and $u(v_n^{pi,h}(\psi))$. Table 1 shows the sequences ψ of size $h = 1$ for policy π_1 (column 1), $f_{s,n}^{\pi_1,h=1}(\psi)$ (column 2) and $v_n^{\pi_1,h=1}(\psi)$ (column 3) and $u(v_n^{\pi,h}(\psi))$ (column 4). For policy π_1, $u_n^{\pi_1,h=1} = \frac{1}{3}*(-1) + \frac{1}{6}*(-0.002) + \frac{1}{3}*(-1) + \frac{1}{6}*(-148.413) = -25.403$ (Eq. 24) is calculated. Thus, $v_n^{\pi_1,h=1} = u^{-1}(u_n^{\pi,h=1}) = -3.235$ and $g_{u_1}^{h=1} = \frac{v_n^{\pi,h=1}}{1} = -3.235$.

For policy π_2, $u_n^{\pi_2,h=1} = \frac{5}{6}*(-1) + \frac{1}{6}*(-0.368) = -0.895$. Consequently $v_n^{\pi_2,h=1} = u^{-1}(-0.895) = 0.111$ and $g_{u_1}^{h=1} = 0.111$. For policy π_3, $u_n^{\pi_3,h=1} = -0.846$ and $v_n^{\pi_3,h=1} = g_{u_1}^{h=1} = 0.167$.

Figure 2a) shows the values $v_{u_1}^{\pi,h=1}$ for the three policies (purple vertical lines), which are used to calculate their respective gains. For $h = 1$ and $\gamma = 1$, $\pi_3 \succ \pi_2 \succ \pi_1$.

Table 1. Sequences ψ of size $h = 1$ for policy π_1, $f_{s,n}^{\pi_1,h=1}(\psi)$, $v_n^{\pi_1,h}(\psi)$ and $u(v_n^{\pi_1,h}(\psi))$.

ψ	$f_{s,n}^{\pi_1,h=1}(\psi) = Pr(S_{n+0} = \psi(0)\|S_0 = s) = P^*[s_0, \cdot]$	$v_n^{\pi_1,h=1}(\psi)$	$u(v_n^{\pi_1,h=1}(\psi))$
(s_1)	$\frac{1}{3}$	0	-1
(s_2)	$\frac{1}{6}$	6	-0.002
(s_3)	$\frac{1}{3}$	0	-1
(s_4)	$\frac{1}{6}$	-5	-148.413

4.3 Calculation of Gain Using Utility Function with $\gamma = 1$ for $h = \infty$

Solving the system of Eqs. 9 proposed in [2] to evaluate policies, $g_u^{\pi_1} = g_u^{\pi_2} = (0.157, 0.157, 0.157, 0.157)$ and $g_u^{\pi_3} = (0.167)$. As stated previously, $g_u^\pi = g_u^{\pi,h=\infty}$ for any policy.

Note that $g_u^\pi \leq g^\pi$ for any of the policies, what is expected of a risk-averse agent, whose certainty equivalent gain is less than or equal to the gain.

$g_u^{\pi,h}$ were calculated for $1 \leq h \leq 2000$ and policies $\pi \in \{\pi_1, \pi_2, \pi_3\}$. As expected $g_u^{\pi,h}$ approaches $g_u^{\pi,h=\infty}$ as h increases. The value of $g_u^{\pi_1,h=2000} = g_u^{\pi_2,h=2000} = (0.155, 0.155, 0.155, 0.155)$ and $g_u^{\pi_3,h=2000} = (0.167)$.

For $h = \infty$ and $\gamma = 1$, $\pi_3 \succ \pi_2 \sim \pi_1$.

4.4 Calculation of Gain Using CVaR with $\alpha = 0.2$ for $h = 1$

To calculate the gain $g_{\text{CVaR}}^{\pi,h}$, it is necessary to calculate $f_{s,n}^{\pi,h}(\psi)$ (Eq. 19), $v_n^{\pi,h}(\psi)$ (Eq. 20) and the distribution of these accumulated rewards $f_{v_n^\pi}^h$. Table 1 shows the sequences ψ of size $h = 1$ for policy π_1 (column 1), $f_{s,n}^{\pi_1,h=1}(\psi)$ (column 2) and $v_n^{\pi_1,h=1}(\psi)$ (column 3).

Thus, the possible accumulated rewards are $(-5, 0, 6)$ with distribution $f_{v_n^{\pi_1}}^{h=1} = (\frac{1}{6}, \frac{4}{6}, \frac{1}{6})$. Considering a confidence level of 20%, $v_{\text{VaR}_{0.2}}^{\pi_1,h=1} = \text{VaR}_{0.2}(v_n^{\pi_1,h=1}) = 0$ and $v_{\text{CVaR}_{0.2}}^{\pi_1,h=1} = \text{CVaR}_{0.2}(v_n^{\pi_1,h=1}) = \frac{\frac{1}{6}(-5)+(0.2-\frac{1}{6})*0}{0.2} = -4.167$. Thus, $g_{\text{CVaR}_{0.2}}^{\pi_1,h=1} = \frac{v_{\text{CVaR}_{0.2}}^{\pi_1,h=1}}{1} = -4,167$.

For policy π_2, the possible cumulative rewards are $(0, 1)$ with distribution $f_{v_n^{\pi_2}}^{h=1} = (\frac{5}{6}, \frac{1}{6})$. Considering a confidence level of 20%, $v_{\text{VaR}_{0.2}}^{\pi_2,h=1} = 0$ and $v_{\text{CVaR}_{0.2}}^{\pi_2,h=1} = \frac{0.2*0}{0.2} = 0$. Thus, $g_{\text{CVaR}_{0.2}}^{\pi_2,h=1} = \frac{v_{\text{CVaR}_{0.2}}^{\pi_2,h=1}}{1} = 0$.

For policy π_3, the only accumulated reward is $v_n^{\pi_3,h=1} = (\frac{1}{6})$ with distribution $f_{v_n^{\pi_3}}^{h=1} = (1)$. Considering a confidence level of 20%, the $v_{\text{VaR}_{0.2}}^{\pi_3,h=1} = v_{\text{CVaR}_{0.2}}^{\pi_3,h=1} = \frac{1}{6}$, and $g_{\text{CVaR}_{0.2}}^{\pi_3,h=1} = \frac{v_{\text{CVaR}_{0.2}}^{\pi_3,h=1}}{1} = \frac{1}{6}$.

Figure 2a) shows the values $v_{\text{CVaR}_{0.2}}^{\pi,h=1}$ for the three policies (pink vertical lines) that are used to calculate their respective gains. For $h = 1$ and a confidence level of 20% $\pi_3 \succ \pi_2 \succ \pi_1$.

The values found in this example for $h = 1$ are the same values found with the criterion used by Xia and Glynn [7].

4.5 Comparison of g^π, $g_u^{\pi,h}$, $g_{\text{VaR}}^{\pi,h}$ and $g_{\text{CVaR}}^{\pi,h}$

Figure 3a) shows $(g_u^{\pi,h})$ for $1 \leq h \leq 2000$ and for risk factors $\gamma \in \{1, 0.1, 0.01, 0.001\}$. For all policies, it is noted that when γ tends to 0 for any value of h the value of the utility function $g_u^{\pi,h}$ tends to the risk-neutral gain, g^π. This property was demonstrated by Howard and Matheson [2] for $h = \infty$. It is also observed that for all policies, the gain $g_{u_1}^{\pi,h}$ when h tends to infinity approaches $g_{u_1}^{\pi,h=\infty}$. Additionally, if $h = \infty$ (orange lines in Fig. 3), $\pi_3 \succ \pi_1 \sim \pi_2$ and if $h \neq \infty$ (dotted lines in Fig. 3) $\pi_3 \succ \pi_2 \succ \pi_1$.

Figure 3b) shows $g_{\text{VaR}}^{\pi,h}$ for $1 \leq h \leq 2000$ and for confidence levels $\alpha \in \{0.01, 0.1, 0.2, 0.5, 0.8, 0.9, 0.99\}$. For policies π_1 and π_2, for confidence levels above 50%, $g_{\text{VaR}_\alpha}^{\pi,h} \geq g^\pi$ if $h \geq 1$. When h tends to infinity $g_{\text{VaR}_\alpha}^{\pi,h}$ tends to g^π. For confidence levels below 50%, $g_{\text{VaR}_\alpha}^{\pi,h} \leq g^\pi$ if $h \geq 2$ and approaches g^π when h tends to infinity. For policy π_3, $g_{\text{VaR}_\alpha}^{\pi,h} = g^\pi$ for all h.

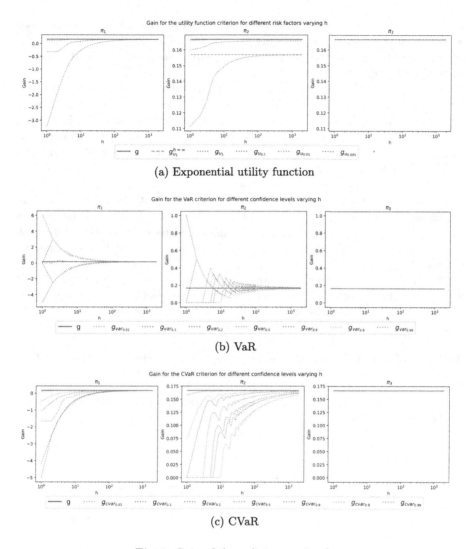

Fig. 3. Gain of the policies varying h.

Figure 3c) shows $g_{\text{CVaR}}^{\pi,h}$ for $1 \leq h \leq 2000$ and for confidence levels $\alpha \in \{0.01, 0.1, 0.2, 0.5, 0.8, 0.9, 0.99\}$. The values $g_{\text{CVaR}_\alpha}^{\pi,h}$ tend to g^π when h tends to infinity for any value of α and, as expected, when α tends to 1, $g_{\text{CVaR}_\alpha}^{\pi,h}$ approaches g^π for any h. Additionally, if $h = \infty$, $\pi_1 \sim \pi_2 \sim \pi_3$ and if $h \neq \infty$ (dotted lines in Fig. 3) $\pi_3 \succeq \pi_2 \succeq \pi_1$.

5 Conclusions

In this work, we proposed a mathematical framework that unifies the average reward and risk criteria for MDPs, for stationary policies that generate aperiodic unichain or multichain Markov chains. The proposed formalization uses sequences of states of size h, allowing two ways of thinking about the stochastic process in the limit: (i) the accumulated reward at infinity and (ii) the steady-state distribution.

The gain from the risk-neutral average reward criterion does not vary regardless of the number of states in the sequence. Works on average reward with exponential utility function [2] and with the CVaR risk measure [7] have been classified using the proposed framework. The exponential utility criterion [2] was classified as $h = \infty$ since the sum of the reward at infinity was used, and the CVaR criterion [7] was classified as $h = 1$ because it considered the immediate reward and the steady-state distribution.

To the best of our knowledge, the exponential utility function applied to sequences of states of size $h = 1$ does not exist in the literature. This criterion allowed differentiation of policies not distinguished by exponential utility with $h = \infty$ and by risk-neutral average reward, as demonstrated in the numerical examples.

As far as we know, there is no work in the literature that characterizes CVaR for $h = \infty$. Through numerical experiments, it was conjectured that this criterion is equivalent to risk-neutral, which needs to be formally demonstrated in future work.

Acknowledgments. This study was supported in part by the *Coordenação de Aperfeiçoamento de Pessoal de Nível Superior* (CAPES) – Finance Code 001, by the São Paulo Research Foundation (FAPESP) grant #2018/11236-9 and the Center for Artificial Intelligence (C4AI-USP), with support by FAPESP (grant #2019/07665-4) and by the IBM Corporation.

References

1. Ashok, P., Chatterjee, K., Daca, P., Křetínský, J., Meggendorfer, T.: Value iteration for long-run average reward in Markov decision processes. In: Majumdar, R., Kunčak, V. (eds.) Computer Aided Verification, pp. 201–221. Springer International Publishing, Cham (2017)
2. Howard, R.A., Matheson, J.E.: Risk-sensitive Markov decision processes. Manage. Sci. **18**(7), 356–369 (1972). http://www.jstor.org/stable/2629352
3. Perman, R.: Natural resource and environmental economics. Pearson Education (2003)
4. Puterman, M.L.: Markov Decision Processes. Wiley Series in Probability and Mathematical Statistics, John Wiley and Sons, New York (1994)
5. Rockafellar, R.T., Uryasev, S., et al.: Optimization of conditional value-at-risk. J. Risk **2**, 21–42 (2000)
6. Rockafellar, R., Uryasev, S.: Conditional value-at-risk for general loss distributions. J. Bank. Fin. **26**(7), 1443–1471 (2002)
7. Xia, L., Zhang, L., Glynn, P.W.: Risk-sensitive Markov decision processes with long-run CVaR criterion. Prod. Oper. Manag. **32**(12), 4049–4067 (2023)

Adaptive Client-Dropping in Federated Learning: Preserving Data Integrity in Medical Domains

Arthur Negrão[✉], Guilherme Silva, Rodrigo Pedrosa, Eduardo Luz, and Pedro Silva

Departamento de Computação, Universidade Federal de Ouro Preto (UFOP), Ouro Preto, MG, Brazil
{arthur.negrao,guilherme.lopes}@aluno.ufop.edu.br,
{rodrigo.pedrosa,eduluz,silvap}@ufop.edu.br

Abstract. In this work, we address the challenge of training machine learning models on sensitive clinical data while ensuring data privacy and robustness against data corruption. Our primary contribution is an approach that integrates Conformal Prediction (CP) techniques into Federated Learning (FL) to enhance the detection and exclusion of corrupted data contributors. By implementing a client-dropping strategy based on an adaptive threshold informed by the interval width metric, we dynamically identify and exclude unreliable clients. This approach, tested using the MedMNIST dataset with a ResNet50 architecture, effectively isolates and discards corrupted inputs, maintaining the integrity and performance of the learning model. Our findings demonstrate that this strategy prevents the potential 10% decrease in accuracy that can occur without such measures, confirming the efficacy of our CP-enhanced FL methodology in ensuring robust and private data handling in sensitive domains like healthcare.

Keywords: Federated Learning · MedMNIST · Conformal Prediction · CNN

1 Introduction

With advancements in computer systems and medicine, computer-assisted diagnoses have become more prevalent, aiding specialists in decision-making [19]. Often, the data involved are sensitive, and their dissemination is constrained by privacy and security regulations. An example of such regulations is the Brazilian General Data Protection Law (*Lei Geral de Proteção de Dados* - LGPD) [4], which reflects a common legislative approach among various countries. Furthermore, the Brazilian Bank Secrecy Law (*Lei de Sigilo Bancário*) [3] imposes additional restrictions on the circulation and disclosure of data, often resulting in a scarcity of large and accessible datasets [1]. This scenario presents a significant challenge for the training of deep learning models, which typically require extensive datasets.

In this context, Federated Learning (FL) [21] offers a compelling strategy for building collaborative models among entities holding sensitive data without the needed of sharing it. FL achieves this by training multiple local models on decentralized datasets and then aggregating these models to form a comprehensive global model. This methodology permits all participating entities to benefit from the collectively derived insights while maintaining the confidentiality of their data [7]. Moreover, as the models are trained locally, the overall demand for computational power is significantly diminished. This efficiency allows for the integration of larger datasets into the training process without the need for high-cost computational resources. By implementing federated learning, institutions can leverage diverse datasets for the enhancement of medical AI applications, thereby advancing healthcare innovations without breaching data privacy.

Addressing the data sharing challenge through Federated Learning (FL) is not straightforward. The difficulty arises because participants in a federated learning setup might contribute inconsistently, that is, some participants may not contribute effectively, while others could inadvertently or maliciously introduce corrupted data. Managing such corrupted data is a significant challenge; it can severely degrade the performance and reliability of the overall model. Ensuring that models maintain their robustness in the face of corrupted data is therefore a crucial obstacle that must be overcome [15].

Previous approaches have faced challenges in balancing data privacy with model reliability [10]. This work integrates Conformal Prediction (CP) techniques [2,16] within a federated learning framework, enhancing both the robustness and reliability of the model by providing a measure of uncertainty for each client. CP, a statistical approach that validates the confidence level of predictions, allows for the use of CP metrics to develop an adaptive threshold strategy. This strategy identifies and manages clients contributing corrupted data during aggregation rounds, thereby ensuring that central model training incorporates only reliable data.

The primary objective of this work is to test the hypothesis that applying conformal prediction techniques within a federated learning context can enhance the robustness and reliability of trained models in medical applications, particularly in distributed scenarios where corrupted data may be present. Two research questions arise from this hypothesis: (RQ1) How does the introduction of corrupted data affect the precision and reliability of trained models in a federated learning context? (RQ2) Can a client-dropping strategy based on conformal prediction metrics effectively reduce the negative effects of corrupted data on a global model in a federated learning context? To address these research questions and the hypothesis, we conducted experiments using the MedMNIST dataset [20], comparing the results of a traditional neural network approach with those of a federated approach. Additionally, we developed a strategy to manage clients with bad/corrupted data based on an adaptive conformal prediction threshold. This threshold uses the interval width metric to determine whether a client should be dropped during an aggregation round.

The experimentation revealed that the proposed strategy is resilient in the presence of corrupted data, with an average accuracy difference of less than 2% between an expected scenario with no corrupted data and a scenario with a corrupted client using the proposed approach. In contrast, a scenario with corrupted data without the proposed strategy showed almost a 10% decrease in accuracy. Furthermore, applying the proposed approach in a context with no corrupted clients resulted in a difference of less than 1% and even enhanced accuracy in some datasets. These results demonstrate that the proposed strategy is resilient against corrupted data and does not negatively impact scenarios without corrupted clients.

This work is subdivided as follows: Sect. 2 presents related works; the data used is presented in Sect. 3; Section 4 presents proposed methodology, followed by experiments and results in Sect. 5; finally, conclusions and future researches are presented on Sect. 6.

2 Related Works

Although studies on Federated Learning (FL) are relatively recent, they already show promising results. In [9], the authors conducted tests on three different datasets (MNIST, MIMIC-III, and ECG), and in all three cases, the models were able to maintain good predictive performance. Additionally, the federated approach proved robust against data distortion and imbalanced distribution, demonstrating that, besides promoting data security, federated learning can bring various other benefits.

Other recent approaches combine Graph Neural Networks (GNNs) with federated learning [6]. It shows that this hybrid approach, in addition to the evident privacy gains, was able to maintain high-quality predictive performance, matching or even surpassing centralized learning in some cases. However, the authors noted that federated GNNs perform worse on datasets with non independent and identically distributed data distribution (non-IID split) compared to centralized approaches. They suggest that future work should address this and other challenges in the field of FL.

In [14], a study on various model aggregation techniques was conducted. Using the MNIST dataset, the authors tested Federated Averaging (FedAvg), Federated Stochastic Variance Reduced Gradient, and CO-OP, achieving the best results with FedAvg. They also noted that FedAvg is efficient with or without independent and identically distributed (i.i.d.) partitioning, and for i.i.d. partitioning, the performance is similar to the centralized approach. This reinforces the hypothesis that federated learning can avoid data exchange while maintaining the efficiency of trained models in various scenarios.

Related to the training and aggregation process, the FOLB algorithm was proposed in [13]. It promises to accelerate and improve the convergence of FL models. The algorithm analyzes the computation capacity, communication, and heterogeneity of the clients to perform client sampling in each training round. It assigns different weights to each client's model during the aggregation process.

The authors conducted empirical tests and demonstrated that the algorithm improves the convergence speed, accuracy, and stability of models across many tasks and datasets.

Regarding information security in federated learning, [12] discusses that, despite providing a learning environment without the need to share data, federated learning still presents some vulnerabilities that designers must be aware of. Poisoning attacks and inference attacks threaten the learning environment by allowing malicious clients to distort the produced data, making gradient descent inefficient and hindering the construction of an effective model. Additionally, analyzing data provided by the server, such as weights and/or loss from each training round, can infer input data and their labels from other clients in the federated network.

In high-stakes decision-making scenarios (e.g., medical), the lack of model interpretability can lead to distrust among users. In [11], the authors propose Conformal Prediction (CP) analysis as an effective and simple way to provide statistical confidence in predictions, offering clear information about prediction uncertainty without requiring direct modification to the model training process. In their experimentation, they conducted CP analysis on a model trained under the FL paradigm using 6 subsets from MedMNIST (Blood, Derma, Path, Tissue Retina, Organ3d). They achieved up to $97\% \pm 0.7\%$ coverage and found a correspondence between conformal uncertainty and prediction task difficulty (measured by class entropy).

In summary, the studies discussed in this section highlight the growing importance and promising potential of federated learning. Various works demonstrate, in different contexts and types of data, that federated approaches are robust and efficient. However, it is crucial to recognize and address the identified vulnerabilities, such as poisoning and inference attacks, to ensure the integrity and security of federated learning systems. Unlike the work in [11], we not only analyze CP but also introduce a client-dropping strategy.

3 MedMNIST Dataset

The MedMNIST dataset [20] consists of a collection of 18 subsets of biomedical images standardized in a format similar to MNIST. This dataset comprises 12 subsets of pre-processed data in 28×28 (2D) and six more subsets in $28 \times 28 \times 28$ (3D). Among these subsets, only the *Path, Blood, Derma*, and *Retina* datasets are in RGB format, while the others are in grayscale. This data collection was developed to address a variety of tasks, such as binary/multi-class classification, ordinal regression, and multi-label tasks, and it covers a wide range of data scales, ranging from 100 to 100,000 images. Figure 1 presents examples of all the used datasets.

Considering the nature of the data in the MedMNIST datasets, some were excluded from this work. The *Chest* subset was discarded because it is multi-label (the same image belongs to one or more classes simultaneously), and the 3D subsets—*Organ, Nodule, Adrenal, Fracture, Vessel*, and *Synapse*—were also

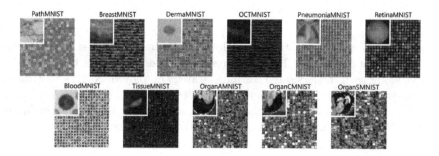

Fig. 1. Used subsets of MedMNIST Dataset. Adapted from [20].

excluded. Therefore, the remaining 11 datasets used in this work are: *Blood, Breast, Derma, Oct, OrganA, OrganC, OrganS, Path, Pneumonia, Retina,* and *Tissue*.

Each dataset used in this study is from a medical domain. The descriptions of each dataset are as follows:

- The **Blood** dataset regards the problem of classifying blood cells.
- The **Breast** dataset comprises breast ultrasound images divided into two conditions: normal or benign (1), and malignant (2).
- The **Derma** dataset contains dermatoscopic images of pigmented skin lesions, split into seven different diseases.
- The **OCT** dataset includes optical coherence tomography (OCT) images for retinal diseases, divided into four diagnoses.
- The **OrganA/C/S** datasets consist of computed tomography (CT) scan images of body organs from three viewing planes: axial, coronal, and sagittal.
- The **Path** dataset is made up of histological images of colorectal cancer used to predict the survival of individuals, collected from two subsets: NCT-CRC-HE-100k (training and validation) and CRC-VAL-HE-7k (testing).
- The **Pneumonia** dataset consists of pediatric chest X-ray images classified as pneumonia or normal.
- The **Retina** dataset contains fundus images used for ordinal regression to classify the severity of diabetic retinopathy into five levels.
- The **Tissue** dataset consists of images of human renal cortex cells divided into eight classes.

It is worth noting that the image formats were not altered, and no preprocessing was applied to the data. Table 1 provides details for each of the datasets used.

4 Methodology

In this section, we outline the methodology adopted for this study[1] and the metrics used for evaluation. First, we present the model used for the evaluation process, along with the federated learning strategy employed. Next, we introduce and describe the proposed approach that utilizes conformal prediction for client-dropping. Finally, we detail the metrics used to evaluate the model, including an in-depth description of the conformal prediction metrics used.

Table 1. Details of the used datasets. Adapted from [20]. MC = Multi-Class, BC = Binary Classification, OR = Ordinal Regression, CT = Computed Tomography and OCT = Optical Coherence Tomography.

ID	Data Description	Task (#classes)	#Train/Validation/Test(Total)
Blood	Microscopic blood cells	MC - 8	11.959/1.712/3.421 (17.092)
Breast	Breast ultrasound	BC	546/78/156 (780)
Derma	Dermatoscope	MC - 7	7.007/1.003/2.005 (10.015)
OCT	Retina's OCT	MC - 4	97.477/10.832/1.000 (109.309)
OrganA	Abdominal CT scan (organ A)	MC - 11	34.561/6.491/17.778 (58.830)
OrganC	Abdominal CT scan (organ C)	MC - 11	12.975/2.392/8.216 (23.583)
OrganS	Abdominal CT scan (organ S)	MC - 11	13.932/2.452/8.827 (25.211)
Path	Colon Pathology	MC - 9	89.996/10.004/7.180 (107.180)
Pneumonia	Chest x-ray	BC	4.708/524/624 (5.856)
Retina	Retina's *Fundus Camera*	OR - 5	1.080/120/400 (1.600)
Tissue	Renal cortex microscope	MC - 8	165.466/23.640/47.280 (236.386)

4.1 Model Used on Training Process

The authors of [20] used the ResNet architecture [18] for their benchmark. For this reason, the ResNet50 architecture was employed in this study without pre-trained weights, with modifications only to the input data format and the number of neurons in the last layer to suit each data subset. The ResNet50 architecture consists of 49 convolutional/pooling layers, with the final layer being fully connected for classification.

For comparison purposes, the same model used in federated training was also evaluated in a centralized context. The model was trained for a proportional number of epochs and evaluated on the same test data.

In both cases, the validation data were used to identify the best model during training. The model with the lowest loss and highest accuracy rate during the global epochs was used to calculate the metrics and report the results.

---
[1] Code available at https://www.dropbox.com/scl/fi/607e0jiy80zbxs7ug96i4/BRACIS_CODE_DROP_STRAT.zip?rlkey=p2jq5n1li473itnjcx1tgst5f&st=o2yuoo8l&dl=0.

4.2 Federated Learning

Federated Learning (FL) [21] is a technique for training deep learning models that operates in a distributed environment, allowing for the collaborative learning of a global model while the data remains locally on the users' devices. In contrast to centralized approaches, where all data is available in a single location for model training, FL allows diverse devices and servers to perform training in parallel without data centralization, avoiding data exchange [7]. Figure 2 shows the workflow of the FL technique used in this work.

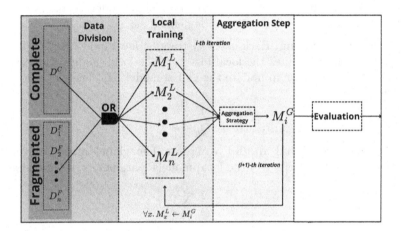

Fig. 2. A client, with his respective portion of data D (D^C all clients have the entire training dataset, D_i^F each client has a portion of the entire training dataset), trains a local model $M^L{}_i$. Those local models are aggregated generating a new global model M_i^G, which is then distributed across all clients for a new training round. This process is repeated for a number of global epochs, and after that the final global model is evaluated.

For the federated model experimentation, the data was divided in two ways. In the first format, each client has access to all entries in the dataset. This strategy is referred to in the experiments as "Complete Dataset". Although this format is not realistic, it serves as an upper bound on accuracy performance. In the second format, referred to as "Fragmented Dataset" in the experiments and closer to a real-world scenario, each client has $\frac{1}{Q_c}$ percent of the dataset, where Q_c is the total number of clients in a given execution. This second strategy allows the repetition of the same training/validation instance among clients. However, despite possible repetitions, there is considerable data diversity among clients. They can be described as:

Complete Dataset: Let D be the dataset, and M_g be the global model. In the complete dataset format, each client c_i with data D_i has access to the entire dataset D, that is

$$D_i = D, \quad \forall i \in \{1, 2, \ldots, Q_c\}. \tag{1}$$

Fragmented Dataset: In the fragmented dataset format, each client c_i has access to a $\frac{1}{Q_c}$ fraction of the dataset D_i, that is

$$D_i = \frac{D}{Q_c}, \quad \forall i \in \{1, 2, \ldots, Q_c\} \tag{2}$$

There are various strategies for manipulating and calculating the loss value during federated training. Among the existing ones, Federated Averaging (FedAvg) was chosen due to the good results presented by [14,17]. In this context, the loss function value of a global epoch is calculated using the average error found on each client. This value is used to update the global model weights, which are then redistributed to all clients. It can be defined as:

Federated Averaging: Each client c_i trains a local model M_{l_i} on its local dataset D_i and computes the local loss L_{l_i}. The local models and losses are then sent to the server to update the global model M_g. The global loss (L_g) is defined as:

$$L_g = \frac{1}{Q_c} \sum_{i=1}^{Q_c} L_{l_i}. \tag{3}$$

Updating the global model weights: The global model weights are updated using the average of the local models' parameters, described as:

$$M_g = \frac{1}{Q_c} \sum_{i=1}^{Q_c} M_{l_i}. \tag{4}$$

It is worth pointing out that the updated global model M_g is then redistributed to all clients for the next round of training. The entire process can be seen in Algorithm 1. Besides, in both federated and local training, the test data is used only for metric calculation and not for training.

4.3 Measuring Uncertainty with Conformal Prediction

For the conformal prediction [2] analysis, two metrics were used: interval width ($|I|$) and coverage rate (CR). The interval width can be mathematically defined as:

$$|I| = 2 * Q_\alpha(|\hat{y}_i - y_i|), \tag{5}$$

where $Q\alpha$ is the α quantile of the residual distribution $R = |\hat{y}_i - y_i|$ ($\hat{y}_i$ being the prediction and y_i the true label), indicating how far predictions are from true labels. The smaller the $|I|$, the less uncertainty there is in the predictions [5,16].

The coverage Rate can be mathematically defined as:

$$CR = \frac{1}{n} \sum_{i}^{n} f(y_i \in I(x_i)), \tag{6}$$

where n is the number of instances, $I(x_i)$ is the prediction interval for instance x_i, and f is a function that returns 1 if the condition is true and 0 otherwise. The higher the coverage rate, the more reliable is the predicted interval in containing the true values [8].

Algorithm 1. Proposed Federated Model Experimentation

Require: Dataset D, Number of clients Q_c
1: **Initialize** global model M_g
2: **Divide** dataset D into one of the two formats:
3:     **Complete Dataset**: Each client has access to all entries in D
4:     **Fragmented Dataset**: Each client has $1/Q_c$ percent of D
5: **for each** client c_i **do**
6:     **Initialize** a local model M_l
7: **end for**
8: **for each** global epoch e_g **do**
9:     **for each** client c_i **do**
10:        **Receive** global model M_g from server
11:        **Train** local model M_{l_i} on client's data
12:        **Compute** local loss L_{l_i}
13:        **Send** updated local model M_{l_i} and loss L_{l_i} to server
14:    **end for**
15:    **Compute** global loss L_g as the average of L_{l_i} for all clients c_i
16:    **Aggregate** local models M_{l_i} to update global model M_g using FedAvg
17:    **Broadcast** updated global model M_g to all clients
18: **end for**
19: **return** Global model M_g

4.4 Conformal Prediction Based on Strategy for Dropping Clients

During the federated training process, clients can encounter issues and return incorrect parameters for global model merging during the global epoch. This issue can be the result of various factors, such as network problems, corrupted data, an so on. To address this issue, we propose simulating a corrupted client by replacing its data with a random array of the same dimensions.

To mitigate the negative impact of a corrupted client, we introduce a client-dropping strategy based on the interval width metric. A client is excluded from a given aggregation round if its interval width is greater than a specified threshold T, that is, $|I_i| > T$, where $|I_i|$ is the interval width of the i-th client calculated in Eq. (5). This threshold, which varies over the aggregation rounds, is calculated as

$$T = \frac{\sum_{i=1}^{n}|I_i|}{Q_c} + 0.05 \times \frac{\sum_{i=1}^{n}|I_i|}{Q_c}, \quad (7)$$

where Q_c is the number of clients. In practical terms, clients with interval widths 5% greater than the average are discarded and not used in computing the new global model weights. The 5% threshold was determined in preliminary runs.

This strategy aims to enhance the robustness of federated learning by effectively identifying and excluding corrupted clients. It is worth to highlight that this strategy also can remove no-corrupted clients and a client is removed during a global epoch, although, it can be used in a next global epoch, since it can be fixed during the training.

To the best of the authors' knowledge, no other work that proposes a client drop strategy in the same way as described was found, that is, no other work used the interval width metric to drop clients in FL context. The metrics used, on the other hand, are derived from adaptations of existing works (Sect. 4.3).

4.5 Evaluation Metrics

Besides the conformal prediction metrics, accuracy (ACC) is used for evaluating the trained models. Accuracy can be defined as the number of correct classifications (where the predicted class matches the expected class) divided by the total number of classifications made. Accuracy ranges from 0 to 1 (or from 0% to 100%), where a value closer to one indicates a higher accuracy rate and better model predictions.

5 Experiments and Results

In this section, we outline the experimental setup, as well as the results and their respective implications.

5.1 Experiments Setup

For the experiments described below, the model was trained for 10 local epochs and 10 global epochs (rounds), with data divided into batches of 128 images each. The Adam optimizer was used with a learning rate of 10^{-3} to adjust the model parameters, and the adopted loss function was the cross-entropy loss. A 10% learning rate decay was applied at the 50th and 75th total epoch (i.e., at the first local epoch in the fifth round and the fifth local epoch in the seventh round).

It is worth noting that the choice of these hyperparameters follows [20], aiming to keep all variables constant except for the "federated versus centralized" comparison. Additionally, for each data subset and each form of data partitioning, the number of clients involved in the training process was varied, with tests conducted with three, four, five, and six clients.

The experiments were conducted on three different systems, within the same local network, with the following configurations: **System 1:** Intel I7-5820k 3.3 GHz; GPU TitanXP 12 Gb; RAM DDR4 12 GB; **System 2:** Intel I9-10900 2.80 GHz; GPU RTX 3090 24 GB; RAM DDR4 128 GB; and **System 3:** AMD Ryzen Threadripper 3960X 3.70 GHz; GPU RTX 3090 24 GB.

The distribution of systems and clients was defined as follows: (i) for the scenario with three clients, one client was assigned to each system; (ii) for four clients, two clients were on System 2 and one on the others; (iii) for five clients, two clients were on Systems 2 and 3, and only one on System 1; and (iv) for six clients, two clients were on each system. The server responsible for managing federated learning process and aggregation was always on System 3. The TensorFlow framework was used to manipulate the model, the Scikit-Learn library for calculating metrics, and the Flower framework to perform federated learning.

In addition to the above experiments, CP-related experiments were also conducted in this study. The model described in Sect. 4.1 was trained in the FL paradigm with three clients (one client per system) in both fragmented and complete dataset cases, using the same configuration as the previous experiments. For the CP-related configuration, the residual threshold was set to 0.3, and the conformal quantile was set to 0.95. These experiments were conducted in the following scenarios: *S1* no data abnormalities; *S2* one client with 100% corrupted data; *S3* no data abnormalities along with the client-dropping strategy; *S4* one client with 100% corrupted data along with the client-dropping strategy.

5.2 Experiment Results

The accuracy results achieved in the evaluations conducted in each scenario, using the ResNet50 model, are presented in Table 2. These results underscore the effectiveness and reliability of the trained model, as demonstrated by the presented data.

Table 2. Accuracy obtained by the federated approach, local approach, and the reference results (Ref) obtained in the work [20]. Cli = number of clients.

Dataset	Complete Dataset				Fragmented Dataset				Local	Ref
	3Cli	4Cli	5Cli	6Cli	3Cli	4Cli	5Cli	6Cli		
Blood	0,939	**0,957**	0,947	0,951	0,924	0,890	0,899	0,895	0,896	0,956
Breast	0,782	0,776	**0,814**	0,782	0,731	0,731	0,731	0,763	0,782	0,812
Derma	0,751	**0,755**	0,749	0,741	0,728	0,710	0,714	0,740	0,698	0,735
Oct	0,743	0,741	0,766	**0,779**	0,697	0,698	0,739	0,737	0,704	0,762
OrganA	0,911	0,923	0,916	0,918	0,897	0,902	0,848	0,913	0,890	**0,935**
OrganC	0,894	0,895	0,902	0,901	0,879	0,871	0,860	0,895	0,851	**0,905**
OrganS	0,769	0,758	0,768	0,757	0,709	0,741	0,749	0,753	0,738	**0,770**
Path	0,811	0,867	0,851	0,844	0,600	0,787	0,803	0,821	0,625	**0,911**
Pneumonia	0,889	**0,902**	0,881	0,897	0,897	0,896	0,865	0,897	0,870	0,854
Retina	0,507	0,493	0,525	0,495	0,475	0,435	0,240	0,482	0,485	**0,528**
Tissue	0,651	0,652	0,661	0,667	0,623	0,631	0,620	0,650	0,630	**0,680**

Regarding Table 2, it can be stated that clients who trained with the complete dataset generally achieved better results than those who trained with the subdivided dataset. Despite the difference between the strategy using the entire dataset and the one using only a portion of the dataset, it is noticeable that this difference is relatively small, typically around 0.05 or 5% in terms of accuracy.

When comparing the results of federated learning (FL) with those of local (centralized) training, it is evident that the federated model achieves comparable and, in some cases, even superior results. This phenomenon is particularly notable with fragmented datasets, where the federated model trains with different subsets of data on each client. This exposure to diverse scenarios during training enhances the model's ability to generalize, contributing to its superior performance.

In the context of complete datasets, although the same data are used in all clients, the randomness during training also contributes to data diversification. This federated training approach results in more robust and adaptable models to different scenarios, demonstrating the effectiveness and advantages of this strategy compared to centralized training.

Still on complete datasets, it is notable that the variation in the number of clients had little impact on the obtained results, especially concerning the accuracy metric (see Table 2). This phenomenon suggests that, for this specific problem, the number of models in the aggregation had a less significant influence on the results than the amount of information available to each individual model. This observation highlights the importance of the quality and diversity of training data over the number of models used in the context of federated learning.

When comparing the accuracy obtained in the federated context with the results presented by [20], it is evident that the former stands out in five cases compared to the latter and shows comparable results in the others. This confirms the effectiveness and feasibility of this strategy.

5.3 Conformal Prediction and Client-Dropping Analysis

In this section, we present the results of the conformal prediction experiments across four scenarios (S1, S2, S3, and S4) as described in the experimental setup. This section addresses the RQ1 and the RQ2 research questions defined in the introduction. The experiments were conducted using only three clients, as no significant differences were observed in Table 2 when more clients were used, with each client utilizing the entire training dataset.

Table 3 shows a considerable reduction in accuracy when a corrupted client is included in the experiment (S2), compared to experiments without a corrupted client (S1) and those where the corrupted client removal strategy is adopted (S3 and S4) for both dataset types (complete and fragmented). For instance, in the Path and Oct subsets, the accuracy dropped by approximately 47% and 24%, respectively. Conversely, in the Derma subset, the drop was approximately 3%.

It is important to note from Table 3 that in both complete and fragmented data scenarios, the client-dropping strategy does not impact training when no corrupted client is present (S3). However, it offers significant benefits when a corrupted client is included (S4). With the client-dropping strategy, the accuracy achieved was close to the ideal (S1) and, in some cases, even higher. This can be attributed to the global model being merged with local models that have less uncertainty, resulting in a more accurate and reliable model.

Even in scenarios where the drop in accuracy was mild, prediction uncertainty generally increased. Notable examples include the Derma and OrganA datasets, where slight drops in accuracy (3% and 2%, respectively) compared to

Table 3. Results of conformal prediction (CP) experiments for all three clients, considering the four scenarios: S1: None corrupted clients; S2: One corrupted client; S3: None corrupted clients and applying the client-dropping strategy; and S4: One corrupted client and applying the client-dropping strategy. The best results in each scenario are highlighted in bold.

Datasets	Accuracy				Interval Width				Coverage Rate			
	S1	S2	S3	S4	S1	S2	S3	S4	S1	S2	S3	S4
Complete Dataset												
Blood	0,939	0,759	0,943	**0,947**	0,174	0,772	0,228	**0,150**	0,980	0,650	0,979	**0,981**
Breast	0,782	0,748	**0,821**	0,737	0,397	0,768	**0,358**	0,821	**0,968**	0,833	**0,968**	0,880
Derma	0,751	0,722	**0,759**	0,679	**0,682**	1,071	0,801	0,866	**0,938**	0,662	0,907	0,890
Oct	0,743	0,484	**0,750**	0,741	0,856	1,000	0,933	**0,483**	0,879	0,610	0,834	**0,960**
OrganA	0,911	0,906	**0,917**	0,914	**0,244**	0,807	0,305	0,275	**0,975**	0,792	0,971	0,974
OrganC	0,894	0,872	0,899	**0,901**	0,427	**0,289**	0,466	0,470	**0,963**	0,962	0,961	0,959
OrganS	**0,769**	0,753	0,768	0,758	**0,733**	1,133	0,829	0,828	**0,928**	0,614	0,902	0,908
Path	0,811	0,367	**0,832**	0,813	0,736	1,051	**0,494**	0,689	0,921	0,851	**0,962**	0,935
Pneumonia	**0,889**	0,841	0,894	0,870	0,417	0,310	**0,300**	0,439	0,963	0,960	**0,978**	0,966
Retina	**0,507**	**0,507**	0,460	0,470	**0,862**	1,062	0,917	1,223	0,881	0,675	**0,878**	0,378
Tissue	0,651	0,349	**0,656**	0,655	1,255	**1,164**	1,254	1,221	0,541	0,587	0,547	**0,602**
Fragmented Dataset												
Blood	0,924	0,773	**0,926**	0,908	**0,382**	0,787	0,541	0,605	**0,966**	0,867	0,956	0,948
Breast	0,731	0,731	**0,750**	0,744	0,731	0,605	**0,369**	0,530	0,859	0,667	**0,978**	0,949
Derma	**0,728**	0,709	0,723	0,673	**0,876**	0,921	0,912	1,105	**0,887**	0,622	0,873	0,758
Oct	0,697	0,466	**0,738**	0,730	0,724	1,006	**0,666**	0,899	0,927	0,618	**0,937**	0,880
OrganA	0,897	0,875	**0,908**	0,907	0,479	0,645	0,380	**0,375**	0,959	0,945	**0,967**	**0,967**
OrganC	0,879	0,854	**0,887**	0,866	0,732	0,916	**0,700**	0,793	0,934	0,874	**0,941**	0,924
OrganS	0,709	0,685	**0,747**	0,724	0,813	**0,601**	0,816	0,853	0,914	**0,935**	0,911	0,897
Path	0,600	0,362	**0,800**	0,724	0,937	1,366	**0,898**	1,078	0,854	0,491	**0,880**	0,771
Pneumonia	**0,897**	0,873	0,891	0,883	0,648	0,501	0,185	**0,096**	0,942	0,900	**0,992**	0,989
Retina	**0,475**	0,435	0,455	0,455	**1,042**	1,374	1,172	1,089	0,508	0,001	0,437	**0,715**
Tissue	0,623	0,605	**0,629**	0,607	0,986	0,915	0,972	**0,879**	0,808	0,858	0,823	**0,880**

scenario S1 were accompanied by significant increases in prediction uncertainty. Specifically, there were increases of 0.39 and 0.56 in interval width and decreases of 0.28 and 0.18 in coverage rate for each respective dataset. Additionally, results from S3 and S4 are very close to (and sometimes slightly better than) those from S1.

5.4 Discussion

As can be observed in Fig. 3, all models trained with fragments of the original dataset tend to have lower accuracy performance compared to models trained with the entire dataset. It is also noticeable that the models trained with six clients had results closest to the reference benchmark values conducted in [20]. Figure 4 graphically presents the accuracy results obtained by the models trained on fragmented datasets. As can be observed, in all federated models, the results were close to the reference benchmark values conducted in [20]. The federated strategy using six clients showed the best results across all datasets. More clients were not tested due to the lack of infrastructure to run the experiments.

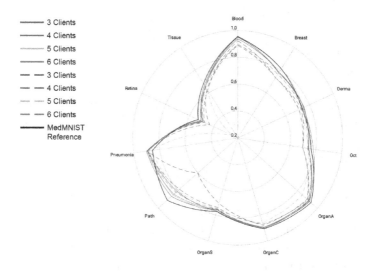

Fig. 3. Accuracy graph of all models trained with fragmented and complete datasets.

Changing focus to the CP analysis, it is possible to see that prediction uncertainty metrics (*i.e.* interval width and coverage rate) can be very useful to detected - and possibly discard - corrupted clients and their bad contribution to the learning process. Specially on cases where accuracy drops are mild, where the high uncertainty on corrupted client predictions would probably be masked, those metrics can provide good intel whether a client is corrupted or not.

In conclusion, it is possible to state that the federated approach succeeded in avoiding data sharing while building competitive models, in terms of accuracy, when compared to the centralized model extensively tested with MedMNIST in [20]. In certain cases, especially in the tests conducted on the Blood, Breast, Derma, Oct, and Pneumonia datasets, the federated approach outperformed the reference model, reinforcing the potential efficiency of the federated strategy.

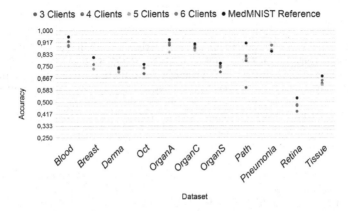

Fig. 4. Accuracy graph of the models trained with fragmented datasets.

6 Conclusion

This study proposes a client-dropping strategy using Conformal Prediction in the federated learning context. The proposed approach preserves the positive characteristics of federated learning, such as data security, while maintaining predictive power and generalization capability in scenarios with and without clients with corrupted data, specifically in the context of the 2D-classification subsets from the MedMNIST dataset. Additionally, it fosters a learning environment less likely to incorporate noise into the learned data representation.

Overall, the proposed strategy demonstrated effectiveness by producing results comparable to or even superior to local (centralized) training and normal federated learning scenarios without corrupted clients, with less than a 2% difference in accuracy on average. When the proposed approach was not used in scenarios involving a client contributing corrupted data, we observed an average accuracy reduction of nearly 10%. These results confirm the strategy's validity and address the initial research questions. Additionally, the consistency in performance across varying client numbers in fragmented dataset scenarios indicates that data can be effectively distributed across multiple sites without compromising computational efficiency. Future research could focus on expanding federated learning strategies, assessing the impact of a higher proportion of corrupted clients, and exploring the challenges of non-iid data splits.

Acknowledgements. This work was carried out with the support of the Coordination for the Improvement of Higher Education Personnel - Brazil (CAPES) - Financial Code 001, the Minas Gerais State Research Support Foundation (FAPEMIG), the National Council for Scientific and Technological Development (CNPq), and the Federal University of Ouro Preto (UFOP/PROPPI).

References

1. Abad, G., Picek, S., Ramírez-Durán, V.J., Urbieta, A.: On the security & privacy in federated learning. arXiv preprint arXiv:2112.05423 (2021)
2. Angelopoulos, A.N., Bates, S.: A gentle introduction to conformal prediction and distribution-free uncertainty quantification. arXiv preprint arXiv:2107.07511 (2021)
3. Congresso Nacional: Lei complementar nº 105, de 10 de janeiro de 2001 (2001). https://www.planalto.gov.br/ccivil_03/leis/lcp/lcp105.htm
4. Congresso Nacional: Lei nº 13.709, de 14 de agosto de 2018 (2018). https://www.planalto.gov.br/ccivil_03/_ato2015-2018/2018/lei/l13709.htm
5. Dunn, P.K., Smyth, G.K.: Randomized quantile residuals. J. Comput. Graph. Stat. **5**(3), 236–244 (1996)
6. He, C., et al.: FedGraphNN: a federated learning system and benchmark for graph neural networks. arXiv preprint arXiv:2104.07145 (2021)
7. Kairouz, P., et al.: Advances and open problems in federated learning. Found. Trends® Mach. Learn. **14**(1–2), 1–210 (2021)
8. Karimi, H., Samavi, R.: Quantifying deep learning model uncertainty in conformal prediction. In: Proceedings of the AAAI Symposium Series, vol. 1, pp. 142–148 (2023)
9. Lee, G.H., Shin, S.Y.: Federated learning on clinical benchmark data: performance assessment. J. Med. Internet Res. **22**(10), e20891 (2020). https://doi.org/10.2196/20891, http://www.jmir.org/2020/10/e20891/
10. Li, S., Ngai, E., Ye, F., Voigt, T.: Auto-weighted robust federated learning with corrupted data sources. ACM Trans. Intell. Syst. Technol. (TIST) **13**(5), 1–20 (2022)
11. Lu, C., Kalpathy-Cramer, J.: Distribution-free federated learning with conformal predictions. arXiv preprint arXiv:2110.07661 (2021)
12. Lyu, L., Yu, H., Yang, Q.: Threats to federated learning: a survey. arXiv preprint arXiv:2003.02133 (2020)
13. Nguyen, H.T., Sehwag, V., Hosseinalipour, S., Brinton, C.G., Chiang, M., Poor, H.V.: Fast-convergent federated learning. IEEE J. Sel. Areas Commun. **39**(1), 201–218 (2020)
14. Nilsson, A., Smith, S., Ulm, G., Gustavsson, E., Jirstrand, M.: A performance evaluation of federated learning algorithms. In: Proceedings of the Second Workshop on Distributed Infrastructures for Deep Learning, pp. 1–8 (2018)
15. Park, S., et al.: Feddefender: client-side attack-tolerant federated learning. In: Proceedings of the 29th ACM SIGKDD Conference on Knowledge Discovery and Data Mining, pp. 1850–1861 (2023)
16. Shafer, G., Vovk, V.: A tutorial on conformal prediction. J. Mach. Learn. Res. **9**(3) (2008)
17. Sun, T., Li, D., Wang, B.: Decentralized federated averaging. IEEE Trans. Pattern Anal. Mach. Intell. **45**(4), 4289–4301 (2022)
18. Targ, S., Almeida, D., Lyman, K.: Resnet in resnet: generalizing residual architectures. arXiv preprint arXiv:1603.08029 (2016)
19. Yanase, J., Triantaphyllou, E.: A systematic survey of computer-aided diagnosis in medicine: past and present developments. Expert Syst. Appl. **138**, 112821 (2019)
20. Yang, J., et al.: Medmnist v2-a large-scale lightweight benchmark for 2D and 3D biomedical image classification. Sci. Data **10**(1), 41 (2023)
21. Zhao, Y., Li, M., Lai, L., Suda, N., Civin, D., Chandra, V.: Federated learning with non-IID data. arXiv preprint arXiv:1806.00582 (2018)

An Ensemble of LLMs Finetuned with LoRA for NER in Portuguese Legal Documents

Rafael Oleques Nunes[✉], Letícia Maria Puttlitz, Antonio Oss Boll, Andre Spritzer, Carla Maria Dal Sasso Freitas, Dennis Giovani Balreira, and Anderson Rocha Tavares

Federal University of Rio Grande do Sul, Porto Alegre, Brazil
{ronunes,spritzer,carla,dgbalreira,artavares}@inf.ufrgs.br,
{leticia.puttlitz,antonio.boll}@ufrgs.br

Abstract. Given the high computational costs of traditional fine-tuning methods and the goal of improving performance, this study investigate the application of low-rank adaptation (LoRA) for fine-tuning BERT models to Portuguese Legal Named Entity Recognition (NER) and the integration of Large Language Models (LLMs) in an ensemble setup. Focusing on the underrepresented Portuguese language, we aim to examine the reliability of extractions enabled by LoRA models and glean actionable insights from the results of both LoRA and LLMs operating in ensembles. Achieving F1-scores of 88.49% for the LeNER-Br corpus and 81.00% for the UlyssesNER-Br corpus, LoRA models demonstrated competitive performance, approaching state-of-the-art standards. Our research demonstrates that incorporating class definitions and counting votes per class substantially improves LLM ensemble results. Overall, this contribution advances the frontiers of AI-powered legal text mining, proposing small models and initial prompt engineering to low-resource conditions that are scalable for broader representation.

Keywords: Named Entity Recognition · Large Language Models · LoRA

1 Introduction

With the advances in Natural Language Processing (NLP), the domain of legal NLP also follows in constant evolution, facing similar challenges and obstacles. One of the main areas of work in the legal NLP field is Named Entity Recognition (NER), which aims to identify and extract entities in text, such as petitions, bills, and sentences from legal documents. The process of recognizing these entities has become challenging due to the lack of structure and predefined entities in corpora from the legal system. Another obstacle is to work in the Brazilian legal system domain, as it does not have the same number of documents and models as the English language.

To address these problems, researchers have created multiple corpora to tackle the issue of NER in the Brazilian legal domain, such as the LeNER-Br [5] and UlyssesNER-Br [3], adding entities to multiple phrases and helping to create a legal NER-friendly corpus. Additionally, many BERT models were created to facilitate the Brazilian legal NER, including LegalBERT-pt [28], BERTikal [22], and many others.

In this work, we propose a comprehensive investigation centered on two primary objectives: first, to analyze the efficacy of deploying LoRA to fine-tune BERT architectures for NER tasks within the legal domain, in contrast with relevant literature; and second, to present a study on the design and implementation of prompt engineering tactics for LLM ensemble modeling, aiming to enhance robustness. Our core contributions are (i) delivering a study on the practicality and advantages of coupling LoRA and BERT models within the specialized context of legal NER, supplemented by comparison with related works, and (ii) an exposition on exploiting assorted prompt ingredients for building LLM ensembles.

2 Related Work

Natural language processing is continuously evolving with the introduction of new concepts. Each new technique opens doors to several fields of work, each with its unique modeling approach and output. Introduced by Mikolov in 2013 [17], the word embeddings gave the area of natural language processing (NLP) a new view, allowing words to be represented in a vectorized manner.

After that, in 2017, the Transformer model [30] revolutionized much of the work in the NLP area, inspiring the creation of several models based on its ideas, such as BERT [12] and GPT [23]. BERT, for instance, utilizes only the encoder part of the transformer and can understand context and work on several other tasks that require text comprehension. Meanwhile, GPT, utilizing the decoder, can generate text, among other tasks. Another model worth mentioning is the ELMo, introduced in 2018 [21], which represents terms in a vector space and makes them context-sensitive.

Researchers have been creating several domain-specific models in the last few years due to BERT's ability to pre-train on data. BERTimbau [29], for instance, was developed specifically for making inferences with Portuguese data, focusing mainly on Brazilian Portuguese corpora. It leverages the transformer architecture to achieve high performance in various natural language processing tasks such as Sentence Textual Similarity and Named Entity Recognition. For training, BERTimbau utilized the brWaC [31] corpus, which consists of a large Web corpus for Brazilian Portuguese. In the legal domain, several models have been developed for various languages, including English [9], German [11], Arabic [2], French [13], and Portuguese [28].

The corpora is also a crucial step in obtaining the results, as it needs to be from the domain and have the correct entities tied to it. A few of the legal corpora from the Brazilian legal system for Named Entity Recognition are LeNER-Br

[5], UlyssesNER-Br [3], Brazilian Supreme Court corpus [10] and CDJUR-BR [8].

Concerning advances in models and the improvement of NER, previous studies have explored how transformer models can enhance NER performance. We focus on LeNER-Br and UlyssesNER-Br, as the former is predominantly a judiciary corpus, and the latter is a legislative corpus. This approach provides an interesting perspective on two different aspects of the legal domain.

Regarding the LeNER-Br corpus, Bonifácio et al. (2020) [6] found that using a domain-specific corpus during the pre-training of multilingual BERT can enhance the recognition of named entities. Additionally, Zanuz et al. (2022) [34] achieved state-of-the-art results by fine-tuning the LeNER-Br corpus using BERTimbau.

For the UlyssesNER-Br corpus, Albuquerque et al. (2023) [4] fine-tuned BERTimbau and conducted the first evaluation of this corpus using BERT models. Nunes et al. (2024) [19] investigated how a semi-supervised technique can improve BERTimbau's performance in the legislative domain, demonstrating that such techniques can enhance model results.

A recent study [20] focused on how a Generative Language Model (GLM) specialized in Brazilian Portuguese performs on both corpora. The authors found that In-Context Learning can provide initial results and demonstrated the potential of these models to extract entities. However, they noted that further studies are needed, as BERT models have shown better performance.

To the best of our knowledge, our work is the first to analyze the efficacy of using LoRA to fine-tune BERT models specifically for legal NER tasks and to explore the design and implementation of prompt engineering techniques for using LLM to ensemble legal NER models. These contributions offer a novel perspective on the study of legal NER and the practical advantages of combining LoRA and BERT models within this specialized domain.

3 Domain Corpora

This Section presents two types of corpora that recognize legally named entities, each focusing on a different type of data. They are called *LeNER-Br* [5] and *UlyssesNER-Br* [3], both sourced from the Brazilian legal system. We chose these two corpora because they can access different fronts of legal domains by leveraging judiciary and legislative texts. Both corpora are embedded within the legislative context. While sentences in LeNER-Br are extracted from court and legislative texts, those in UlyssesNER-Br come from legislative inquiries and bills from the Brazilian Chamber of Deputies.

The first corpus, introduced in 2018 and called *LeNER-Br* [5], focuses on data from the Brazilian justice system. It is composed of a total of 70 legal documents from various types of courts and legislations, containing a total of 10,392 sentences and 318,073 tokens. The categories of the named entities include "Person", "Legal cases", "Time", "Location", "Legislation" and "Organization".

Focusing on the legislative side of legal data, *UlyssesNER-Br* [3] consists of two types of data from the Brazilian Chamber of Deputies: legislative consultations (ST) and bills (PL). The ST database has 790 sentences and 77,441 tokens, whereas the PL corpus contains 9,526 sentences with 138,741 tokens. Both datasets include the following categories of named entities: "Fundamento", "Organizacao," "Produtodelei," "Local," "Data," and "Evento."

4 Models and Fine-Tuning with LoRA

This section provides an overview of the models used to recognize legally named entities. These models are variations of the BERT model [12], each pre-trained with a specific type of data, along with the addition of a Large Language Model.

BERTimbau [29], launched in 2020, is one of the first Brazilian BERT models. It was pre-trained using *brWaC* [32], a large corpus of data from the Brazilian web. Two types of BERTimbau were pre-trained: BERTimbau Base, with 12 layers, a hidden size of 768, 12 attention heads, and 110 million parameters; and BERTimbau Large, with 24 layers, a hidden size of 1024, 16 attention heads, and 330 million parameters. The maximum length for a sentence contains 512 tokens for both models. We used BERTimbau Base for the analysis. It was fine-tuned separately on the *UlyssesNER-Br* and *LeNER-Br* corpora. The hyperparameters used for this model are specified in Sect. 6.2.

Aiming to develop a legal-focused model, *LegalBERT-pt* [28] is a BERT model pre-trained on Brazilian legal corpora. The authors pre-trained two types of models: LegalBert-pt SC and LegalBert-pt FP. The SC model is formulated by pre-training a BERT model from scratch, with the same configuration as the BERTimbau base. Differently, the FP model pre-trains a BERTimbau-Base with the domain-specific corpora. The corpora used in the article consist of multiple Brazilian court cases and, for the SC model, Portuguese Wikipedia articles, totaling 1,500,000 legal documents and a vocabulary of 36,345 words. *LegalBERT-pt FP* was selected as the best-performing model.

In this study, *LegalBERT-pt FP* was fine-tuned using the corpora described in Sect. 3, with its hyperparameters detailed in Sect. 6.2.

Finishing the BERT variant models, *BERTikal* [22] is a BERT model trained on corpora from clippings, court cases and motions in the legal Brazilian data.

The previously described models are fine-tuned on LeNER-br and UlyssesNER-Br (see Sect. 3) and used hyperparameters described in Sect. 6.2. We perform our fine-tuning with LoRA [15], which is a method to reduce the number of trainable parameters and hence the training computational cost using matrix algebra. The method freezes the pre-trained parameters of the model and optimizes the rank-decomposition of the dense layer matrices.

The only Large Language Model (LLM) we use is *Mistral 7B* [16]. It is composed of 7 Billion parameters, has a dimension size of 4096, 32 layers, a vocabulary size of 32000 and multiple other parameters and hyperparameters. We obtained the training corpus from the open web. As an LLM, the Mistral 7B model is capable of a diverse gamma of tasks, including knowledge, math, reasoning, comprehension, and code, among several others. We use this model for the ensemble technique explained in Sect. 5 and for a *Zero-Shot Learning* method.

5 An Ensemble of Language Models

Ensemble learning aims to improve the predictive performance of a single model by training multiple models and combining their predictions [25]. A simple form of ensemble is voting, where base classifiers are presented with an input and each makes a prediction. Subsequently, the prediction that receives the most votes is selected as the final output. We apply ensemble learning to combine multiple language models on the legal Named Entity Recognition task.

In our approach, BERT models serve as the base classifiers for Named Entity Recognition. After the BERT models generates their predictions, Mistral acts as an aggregator in the ensemble, combining the outputs. This method provides a unified result that incorporates and values all the predictions from the BERT models.

We based our prompt engineering on the following elements: persona, definitions, votes, answer format, sentence, and query. The prompt starts with a **persona**, which is used to impersonate a linguist specialized in political science to achieve the best possible performance [26]. We chose this role for the persona because it is a linguistic task that requires legal domain knowledge.

We used the **definitions** of the entity classes to alleviate the problem that some categories have non-intuitive meanings based on their names. For instance, the category *fundamento*[1] does not seem directly linked with legal norms or bills. Thus, we also give the entity class definitions in the prompt. Definitions were sourced from corpora [3,5] and translated into Portuguese. However, we adapted the definitions for *person* and *location* from Harem [27] because both corpora were influenced by Harem and adhered to its class definitions.

We designed the **votes** to provide information to the model regarding the classes given more or fewer votes from each classifier. We present the vote in the format *Class: x votes*, where *class* is the name of the entity class, and x is the number of votes.

In the **answer format**, we encourage the model to use the chain of thought (CoT) [33] by explaining the answer before providing the entity class. We divided the answers into *explanation* and *answer* to facilitate post-processing to obtain the class.

Finally, at the end of the prompt, we provide the **sentence** and ask the model to assign the right entity class to a term or say that it does not have a class in **question-answer** format. We provided entities from the classifiers in question, definitions, and votes. We also tested using all the entities for the definitions and questions.

All classes used in the prompt are given with the first letter in uppercase and the rest in lowercase. We also use space to divide n-gram terms, as in *produto de*

---

[1] In English: foundation.

lei. We opt to use this format so that the names remain in a more fluid language, providing a better context and avoiding unnecessary splitting in more tokens of the terms.

Sometimes, the model gives answer classes near the goals but with slightly different names (e.g. typos). We used Sentence-BERT (SBERT) [24] for **post-processing**, along with the classes [20]. After comparison, we converted the answer into a vector annotating the entity of the term in the sentence. The vector is incrementally annotated because we use a prompt for each term.

6 Experimental Evaluation

This section provides an overview of the experimental environment, encompassing the principal libraries utilized, hyperparameters, and a thorough explanation of the hyperparameter tuning carried out for the LoRA models. In addition, we discuss the performance metrics utilized to evaluate our models.

6.1 Setup

We conducted the experiments on a computer with 12 GB RAM and an Nvidia GeForce RTX 4070 GPU. Owing to its broad support for libraries related to machine learning and natural language processing, we decided to implement our experiments using Python 3.7.6. Quantization, performed using the bitsandbytes library[2], was employed to reduce memory requirements and computational costs. The BERT and Mistral models were obtained from the HuggingFace Hub. The functions and methods of LoRA and running the models were used from HuggingFace.

6.2 Hyperparameters

In this section, we describe the experiment hyperparameters. We first present the hyperparameter search and final values for LoRA training. Here, we present the values used to train the BERT models. Finally, we present the hyperparameters for quantization and generate responses to the LLM.

LoRA. We used Optuna [1] to perform the hyperparameter search. We set the optimization range to drop out around 0.1 and 0.5, and the range of r and *alpha* between 16, 32, 64, and 128, with 100 iterations. The final hyperparameters were $r = 32$ *alpha* $= 1.747406$, *dropout* $= 0.1$ (without optimization) and bias $=$ "all".

BERT. We followed previous studies that used the same legal corpora to set the hyperparameters for the BERT classifier [6,19,34]. The set of values was *learning_rate* $= 1e-3$, *batch_size* $= 10$, and *weight_decay_size* $= 0.01$.

LLM. We use 4-bit quantization to load the model in our machine, setting the attributes *load_in_4bit* $=$ True, *bnb_4bit_compute_dtype* $=$ torch.bfloat16,

---

[2] https://github.com/TimDettmers/bitsandbytes.

$torch\_dtype$ = torch.float16, $device\_map$ = "auto". To generate the answer, we used the standard value of the method $generate$, setting max_new_tokens = 800 and pad_token_id = $model.config.eos\_token\_id$.

6.3 Metrics

We used Seqeval [18] to compute the metrics. This library assesses the results by considering the sequence of tags assigned to each entity in a complete sentence, clearly recognizing entities beyond individual tokens. Our primary metric is the F1-Score, calculated for each class and overall. Additionally, precision and recall were computed for each class, whereas accuracy was determined for the overall results.

7 Results and Discussion

In this section, we present the outcomes of our experiments involving LoRa applied to BERT models and LLM as an ensemble approach. Initially, we delve into the performance of individual LoRa classifiers, considering any potential connections to the pretraining domain of the underlying base models. Subsequently, we reveal the combined results achieved through the LLM ensemble technique and scrutinize the relationships between the selected models and prompts. Finally, we compare our results with those existing state-of-the-art (SOTA) baselines.

7.1 LoRA Classifiers

Table 1. F1-Score for each corpus to the LoRA classifiers.

Model	LeNER-Br	UlyssesNER-Br
LegalBERT-pt	**88.49**	76.96
BERTimbau	87.12	**81.00**
BERTikal	81.53	67.72

We present the final F1-Score of the classifiers to each corpus in Table 1. LegalBERT-pt presented the best result for LeNER-BR, obtaining 88%, followed by BERTimbau, with 87%. For UlyssesNER-Br, we obtained the inverse result, where BERTimbau had the best result at 81%, followed by LegalBERT-pt with 76%. This result probably occurred because LegalBERT-pt follows BERTimbau pretraining in judiciary documents (more information in Sect. 4), which is the data domain of LeNER-Br, whereas UlyssesNER-Br uses legislative documents, which can explain the lower results in this case.

However, it is important to note that we cannot express the consistency of the difference between the close results because we did not use k-fold cross-validation to obtain the average result and variance around the folds. Furthermore, we did not compute tests to obtain statistical significance.

Another observation regards the response format, which we discovered during manual inspection. Specifically, instances exist where the generated outputs fail to adhere strictly to the prescribed formats. Given this discrepancy, there is a risk that accurate responses might go undetected due to unexpected presentation, consequently contributing to reduced F1 scores.

7.2 LLMs Applied as Ensemble

The performance resulting from employing Mistral 7B as an ensemble approach can be found in Tables 2 and 3. We conducted experiments covering diverse combinations of prompts and models to evaluate the effect of each model-prompt pairing. Surprisingly, the combination of multiple models did not yield improved outcomes. The top three scores across both datasets featured BERTimbau and LegalBert-Pt employed independently, echoing the trend seen in Table 1, wherein BERTimbau outperformed UlyssesNER-Br, whereas LegalBert-Pt proved superior when working with LeNER-Br.

Regarding prompt engineering, our investigation revealed that utilizing definitional descriptions paired with voting yielded the most successful outcomes. This strategy offers the model clear class definitions and individual class preferences as indicated by the participating models. Although the label distribution appears scattered throughout the findings, the majority converge near 50% for LeNER-Br and fall below this threshold for UlyssesNER-Br. These observations underscore the significance of defining a constrained solution space when using LLMs, facilitating enhanced categorization capabilities.

Considering the exploratory findings presented thus far, it is important to highlight the necessity of performing more deliberate examinations to acquire robust statistical comparisons between various prompt structures and models, as discussed in Sect. 7.1. Furthermore, embracing a multifaceted validation strategy, including repetitive experimentation (K iterations), calculation of mean values, and determination of standard deviations, constitutes another key element in fostering greater certainty surrounding the obtained outcomes.

7.3 Our Results and the State-of-the-Art

Comparing our results with those of the SOTA, Table 4 reveals that the LoRA classifiers achieved similar performance to previous studies [19,34]. Specifically, our LoRA classifier obtained results closely aligned with a classifier solely utilizing BERTimbau [19] on the UlyssesNER-Br corpus and approached the performance of the classifier proposed by Zanunz et al. (2022) [34] on the LeNER-Br corpus.

The classes that achieved results similar to those of Nunes et al. (2024) [19] were *PESSOA, FUNDAMENTO, EVENTO,* and *DATA,* as shown in Table 5.

Table 2. Overall F1-Score for UlyssesNER-Br using Mistral 7B as ensembler.

Model	Prompt	F1-Score
BERTimbau	definitions + votes	69.87
BERTimbau	definitions	67.97
legalbert	definitions + votes	64.30
BERTimbau	definitions + votes + all labels	63.81
legalbert	definitions	62.15
BERTimbau + legalbert	definitions	60.82
BERTimbau + legalbert	definitions + votes	59.32
legalbert	definitions + votes + all labels	58.36
legalbert + bertikal	definitions	57.83
BERTimbau + bertikal	definitions + votes	57.60
BERTimbau + bertikal	definitions	57.50
BERTimbau + legalbert	definitions + votes + all labels	56.05
legalbert + bertikal	definitions + votes	55.21
BERTimbau + legalbert + bertikal	definitions + votes	55.16
BERTimbau + legalbert + bertikal	definitions	53.51
BERTimbau + bertikal	definitions + votes + all labels	52.70
BERTimbau	definitions + all labels	51.95
legalbert + bertikal	definitions + votes + all labels	50.20
BERTimbau + legalbert + bertikal	definitions + votes + all labels	47.80
legalbert	definitions + all labels	47.09
bertikal	definitions + votes	46.76
BERTimbau + legalbert	definitions + all labels	45.99
bertikal	definitions	45.95
bertikal	definitions + votes + all labels	44.97
BERTimbau + legalbert + bertikal	definitions + all labels	43.65
BERTimbau + bertikal	definitions + all labels	42.28
legalbert + bertikal	definitions + all labels	40.48
bertikal	definitions + all labels	34.78

Notably, the structuring of entities, such as *DATA* (date) and *PESSOA* (person), presents a plausible explanation for their relatively higher performance. Dates typically adhere to specific formats, facilitating their distinction from other entities, whereas the names of individuals often follow recognizable patterns similar to those observed in *FUNDAMENTO* (laws). Conversely, *EVENTO* (event) posed a more significant challenge because of the limited training and test examples available for this class.

Table 3. Overall F1-Score for LeNER-Br using Mistral 7B as ensembler.

Model	Prompt	F1-Score
legalbert	definitions + votes	63.03
BERTimbau	definitions + votes	62.38
legalbert	definitions + votes + all labels	61.71
legalbert + bertikal	definitions + votes	61.18
BERTimbau	definitions + votes + all labels	61.07
legalbert	definitions	60.99
BERTimbau + bertikal	definitions + votes	60.74
BERTimbau	definitions	60.60
BERTimbau + legalbert	definitions + votes	60.46
BERTimbau + legalbert	definitions + votes + all labels	60.13
legalbert + bertikal	definitions + votes + all labels	60.12
BERTimbau + legalbert + bertikal	definitions + votes	59.57
legalbert + bertikal	definitions	59.31
BERTimbau + bertikal	definitions + votes + all labels	59.22
BERTimbau + bertikal	definitions	59.11
BERTimbau + legalbert	definitions	58.94
BERTimbau + legalbert + bertikal	definitions + votes + all labels	57.62
BERTimbau + legalbert + bertikal	definitions	57.41
legalbert	definitions + all labels	51.86
BERTimbau	definitions + all labels	50.55
legalbert + bertikal	definitions + all labels	50.24
BERTimbau + legalbert	definitions + all labels	49.86
BERTimbau + bertikal	definitions + all labels	49.33
BERTimbau + legalbert + bertikal	definitions + all labels	48.59
bertikal	definitions + votes	5.26
bertikal	definitions + votes + all labels	5.18
bertikal	definitions	4.91
bertikal	definitions + all labels	3.65

Table 4. Overall F1-Score for each corpus for our classifiers and SOTA classifiers. The results for UlyssesNER-Br are from Nunes et al. (2024) [19], and the results for LeNER-Br are from Zanuz et al. (2022) [34].

Model	UlyssesNER-Br	LeNER-Br
BERTimbau + self-learning	**86.70 ± 2.28**	-
BERTimbau	83.53 ± 2.56	**91.14 ± 0.39**
LoRA (ours)	81.00	88.49
LLM (ours)	69.86	63.03

Table 5. F1-Score for each class in UlyssesNER-Br for our classifier and from Nunes et al. (2024) [19]. The underlined values are those near the established results.

Category	BERTimbau + LoRA	BERTimbau + Self-Learning
PRODUTODELEI	57.14	75.42 ± 4.47
PESSOA	89.60	87.48 ± 2.79
ORGANIZACAO	74.01	84.89 ± 5.77
LOCAL	74.37	86.46 ± 3.73
FUNDAMENTO	85.06	88.60 ± 2.29
EVENTO	47.06	58.10 ± 34.16
DATA	97.49	94.77 ± 2.65

Table 6. F1-Score for each class in LeNER-Br for our classifier and from Zanuz et al. (2022) [34]. The underlined values are those near the established results.

Category	LegalBert-Pt + LoRA	BERTimbau
TEMPO	92.02	96.04 ± 0.58
PESSOA	95.86	97.38 ± 0.44
ORGANIZACAO	86.04	86.66 ± 1.17
LOCAL	73.58	75.67 ± 3.18
LEGISLACAO	92.66	95.90 ± 0.83
JURISPRUDENCIA	78.66	87.76 ± 0.87

Furthermore, a comparative analysis with Zanuz et al. (2022) [34] revealed that classes such as *PESSOA*, *ORGANIZACAO*, and *LOCAL* demonstrate similar performance trends, as shown in Table 6. This alignment could be attributed to factors such as the extensive training, validation, and test data available for each class and the inherent structural complexities associated with entities in these categories, as discussed previously.

These results show the power of LoRA for the NER task, with the capacity to achieve good results by training a smaller number of parameters, i.e., we decrease the number of training params from 109,532,186 to only 1,291,789 params. Thus, this approach results in small and faster models that can be used with less computer power and stored in less space.

8 Conclusion

In this work, we presented a study of using LoRA to fine-tune BERT Portuguese models to NER tasks in the legal domain. The LoRA models could achieve near results compared to state-of-art for each corpora tested. Another point is that we have the advantage of training smaller models, which is an advantage to storing and running the model in machines with low processing.

We also tested using Mistral 7B applied to the ensemble, which was unproductive. The results were lower compared to the LoRA models. However, we recognize that some outputs do not respect the format requested, which can be a point that the right answers are not recognized.

In future work, we aspire to incorporate hyperparameter tuning tailored to each model augmented with LoRa, potentially amplifying the overall performance. Moreover, we intend to utilize k-fold cross-validation to derive more authoritative and comparable results than existing literature.

Regarding ensembling, we envision expanding our scope by exploring alternative prompts engineered using prompt manipulation techniques. Complementarily, we seek to develop post-processing routines accommodating minor fluctuations in the output format, thereby minimizing false negatives and boosting performance indicators. Testing other multilingual LLMs (e.g., LLaMA2) and trialing retrieval mechanisms, such as In-Context Learning, are other possible ways to improve results.

Another interesting future approach is to compare our results with classic ensemble strategies, such as bagging [7] and boosting [14]. Simple fusion schemes such as averaging and maximum selection may provide promising avenues warranting investigation in this dynamic domain.

Acknowledgements. This work has been partially funded by the Coordenação de Aperfeiçoamento de Pessoal de Nível Superior - Brasil (CAPES) - Finance Code 001. We also acknowledge financial support from the Brazilian funding agency CNPq.

References

1. Akiba, T., Sano, S., Yanase, T., Ohta, T., Koyama, M.: Optuna: a next-generation hyperparameter optimization framework. In: Proceedings of the 25th ACM SIGKDD International Conference on Knowledge Discovery and Data Mining, pp. 2623–2631 (2019)
2. AL-Qurishi, M., AlQaseemi, S., Soussi, R.: Aralegal-BERT: a pretrained language model for Arabic legal text (2022)
3. Albuquerque, H.O., et al.: UlyssesNER-Br: a corpus of Brazilian legislative documents for named entity recognition. In: Pinheiro, V., et al. (eds.) PROPOR 2022. LNCS, vol. 13208, pp. 3–14. Springer, Cham (2022). https://doi.org/10.1007/978-3-030-98305-5_1
4. Albuquerque, H.O., et al.: On the assessment of deep learning models for named entity recognition of Brazilian legal documents. In: Moniz, N., Vale, Z., Cascalho, J., Silva, C., Sebastião, R. (eds.) Progress in Artificial Intelligence - EPIA 2023. Lecture Notes in Computer Science(), vol. 14116, pp. 93–104. Springer, Cham (2023)
5. Luz de Araujo, P.H., de Campos, T.E., de Oliveira, R.R.R., Stauffer, M., Couto, S., Bermejo, P.: LeNER-Br: a dataset for named entity recognition in Brazilian legal text. In: Villavicencio, A., et al. (eds.) PROPOR 2018. LNCS (LNAI), vol. 11122, pp. 313–323. Springer, Cham (2018). https://doi.org/10.1007/978-3-319-99722-3_32

6. Bonifacio, L.H., Vilela, P.A., Lobato, G.R., Fernandes, E.R.: A study on the impact of intradomain finetuning of deep language models for legal named entity recognition in Portuguese. In: Cerri, R., Prati, R.C. (eds.) BRACIS 2020, Part I. LNCS (LNAI), vol. 12319, pp. 648–662. Springer, Cham (2020). https://doi.org/10.1007/978-3-030-61377-8_46
7. Breiman, L.: Bagging predictors. Mach. Learn. **24**, 123–140 (1996). https://doi.org/10.1007/BF00058655
8. Brito, M., et al.: Cdjur-br-uma coleção dourada do judiciário brasileiro com entidades nomeadas refinadas. In: Anais do XIV Simpósio Brasileiro de Tecnologia da Informação e da Linguagem Humana, pp. 177–186. SBC (2023)
9. Chalkidis, I., Fergadiotis, M., Malakasiotis, P., Aletras, N., Androutsopoulos, I.: Legal-BERT: the muppets straight out of law school (2020)
10. Correia, F.A., et al.: Fine-grained legal entity annotation: a case study on the Brazilian supreme court. Inf. Process. Manag. **59**(1), 102794 (2022)
11. Darji, H., Mitrović, J., Granitzer, M.: German BERT model for legal named entity recognition. In: Proceedings of the 15th International Conference on Agents and Artificial Intelligence. SCITEPRESS - Science and Technology Publications (2023). https://doi.org/10.5220/0011749400003393
12. Devlin, J., Chang, M.W., Lee, K., Toutanova, K.: BERT: Pre-training of Deep Bidirectional Transformers for Language Understanding (2019)
13. Douka, S., Abdine, H., Vazirgiannis, M., Hamdani, R.E., Amariles, D.R.: JuriBERT: a masked-language model adaptation for french legal text (2022)
14. Freund, Y., Schapire, R.E.: Experiments with a new boosting algorithm. In: Machine Learning: Proceedings of the Thirteenth International Conference, vol. 96, pp. 148–156 (1996)
15. Hu, E.J., et al.: LoRA: low-rank adaptation of large language models (2021)
16. Jiang, A.Q., et al.: Mistral 7B (2023)
17. Mikolov, T., Chen, K., Corrado, G., Dean, J.: Efficient estimation of word representations in vector space (2013)
18. Nakayama, H.: SeqEval: a Python framework for sequence labeling evaluation (2018). https://github.com/chakki-works/seqeval, software available from https://github.com/chakki-works/seqeval
19. Nunes, R.O., Balreira, D.G., Spritzer, A.S., Freitas, C.M.D.S.: A named entity recognition approach for portuguese legislative texts using self-learning. In: Proceedings of the 16th International Conference on Computational Processing of Portuguese, pp. 290–300 (2024)
20. Oleques Nunes., R., Spritzer., A., Dal Sasso Freitas., C., Balreira., D.: Out of sesame street: a study of Portuguese legal named entity recognition through in-context learning. In: Proceedings of the 26th International Conference on Enterprise Information Systems - Volume 1: ICEIS, pp. 477–489. INSTICC, SciTePress (2024). https://doi.org/10.5220/0012624700003690
21. Peters, M.E., et al.: Deep contextualized word representations (2018)
22. Polo, F.M., et al.: LegalNLP-natural language processing methods for the Brazilian legal language. In: Anais do XVIII Encontro Nacional de Inteligência Artificial e Computacional, pp. 763–774. SBC (2021)
23. Radford, A., Narasimhan, K., Salimans, T., Sutskever, I.: Improving language understanding by generative pre-training. OpenAI Technical report (2018)
24. Reimers, N., Gurevych, I.: Making monolingual sentence embeddings multilingual using knowledge distillation. In: Proceedings of the 2020 Conference on Empirical Methods in Natural Language Processing. Association for Computational Linguistics (2020). https://arxiv.org/abs/2004.09813

25. Sagi, O., Rokach, L.: Ensemble learning: a survey. Wiley Interdisc. Rev. Data Min. Knowl. Discov. **8**(4), e1249 (2018)
26. Salewski, L., Alaniz, S., Rio-Torto, I., Schulz, E., Akata, Z.: In-context impersonation reveals large language models' strengths and biases. In: Thirty-Seventh Conference on Neural Information Processing Systems (2023). https://openreview.net/forum?id=CbsJ53LdKc
27. Santos, D., Cardoso, N.: A golden resource for named entity recognition in Portuguese. In: Vieira, R., Quaresma, P., Nunes, M.G.V., Mamede, N.J., Oliveira, C., Dias, M.C. (eds.) PROPOR 2006. LNCS (LNAI), vol. 3960, pp. 69–79. Springer, Heidelberg (2006). https://doi.org/10.1007/11751984_8
28. Silva, N., et al.: Evaluating topic models in Portuguese political comments about bills from Brazil's chamber of deputies. In: Anais da X Brazilian Conference on Intelligent Systems. SBC, Porto Alegre, RS, Brasil (2021). https://sol.sbc.org.br/index.php/bracis/article/view/19061
29. Souza, F., Nogueira, R., Lotufo, R.: BERTimbau: pretrained BERT models for Brazilian Portuguese, pp. 403–417 (2020). https://doi.org/10.1007/978-3-030-61377-8_28
30. Vaswani, A., et al.: Attention is all you need (2017)
31. Wagner Filho, J.A., Wilkens, R., Idiart, M., Villavicencio, A.: The brWaC corpus: a new open resource for Brazilian Portuguese. In: Proceedings of the Eleventh International Conference on Language Resources and Evaluation (LREC 2018) (2018)
32. Wagner Filho, J.A., Wilkens, R., Idiart, M., Villavicencio, A.: The brWaC corpus: a new open resource for Brazilian Portuguese. In: Calzolari, N., et al. (eds.) Proceedings of the Eleventh International Conference on Language Resources and Evaluation (LREC 2018). European Language Resources Association (ELRA), Miyazaki, Japan (2018). https://aclanthology.org/L18-1686
33. Wei, J., et al.: Chain-of-thought prompting elicits reasoning in large language models. In: Advances in Neural Information Processing Systems, vol. 35, pp. 24824–24837 (2022)
34. Zanuz, L., Rigo, S.J.: Fostering judiciary applications with new fine-tuned models for legal named entity recognition in Portuguese. In: Pinheiro, V., et al. (eds.) PROPOR 2022. LNCS, vol. 13208, pp. 219–229. Springer, Cham (2022). https://doi.org/10.1007/978-3-030-98305-5_21

An Instance Level Analysis of Classification Difficulty for Unlabeled Data

Patricia S. M. Ueda[1]($\boxtimes$), Adriano Rivolli[2], and Ana Carolina Lorena[1]

[1] Universidade Tecnológica Federal do Paraná, Cornélio Procópio, Brazil
`uedapsm@gmail.com, aclorena@ita.br`
[2] Instituto Tecnológico de Aeronáutica, São José dos Campos, Brazil
`rivolli@utfpr.edu.br`

Abstract. Instance hardness measures allow us to assess and understand why some observations from a dataset are difficult to classify. With this information, one may curate and cleanse the training dataset for improved data quality. However, these measures require data to be labeled. This limits their usage in the deployment stage when data is unlabeled. This paper investigates whether it is possible to identify observations that will be hard to classify despite their label. For such, two approaches are tested. The first adapts known instance hardness measures to the unlabeled scenario. The second learns regression meta-models to estimate the instance hardness of new data observations. In experiments, both approaches were better at identifying instances lying in borderline regions of the dataset, which pose a greater difficulty when the label is unknown.

Keywords: Machine Learning · Instance hardness measures · Unlabeled data · Deployment of models

1 Introduction

The Machine Learning (ML) literature extensively provides algorithmic developments focused on model hyperparameter tuning and related *model-centric* tasks. More recently, the community of *data-centric* Artificial Intelligence (AI) is lighting the focus on the effort to understand more the data and its quality improvement than on developing more complex ML models [13].

Paving the way for such a data-centric approach is a more fine-grained analysis of the data and classification performance. Herewith, aggregated measures applied for classification problems, such as accuracy, precision, or similar metrics, restrict the understanding of the particularities in the data the algorithms are modeling. Those aggregated metrics do not provide information about misclassification at the level of an instance or why they are misclassified. However, a more reliable usage of ML algorithms must reveal for which particular instances a model struggles to classify correctly and why. One way to achieve such understanding is leveraging knowledge from correlating data characteristics extracted by a set of meta-features [12] to the predictive performance of multiple algorithms, in a meta-learning (MtL) approach [2].

One particular set of meta-features is the set of data complexity measures, previously proposed by Ho and Basu [5] to explore the overall complexity of solving the classification problem given the dataset available for learning, providing a global perspective of the difficulty of the problem [1]. Since these measures can fail to provide information at the instance-level [8], the *Instance Hardness Measures* (IHMs) were introduced by Smith et al. [14] to characterize the difficulty level of each individual instance of a dataset, giving information on which particular instances are misclassified and why. These developments attend to a trending interest in responsible AI that has emerged in recent years, making researchers focus on the reliability and trustfulness of the predictions obtained by ML models.

Nonetheless, the current IHMs need the instance label to be computed, which restricts their use for analyzing and curating ML training datasets. In the use of ML in production, where the class of an instance is unknown, adaptations are needed. This paper proposes alternative instance hardness measures when the instances do not have a label. The idea is to leverage the knowledge of the hardness of the training dataset, which is labeled, to assess the hardness level of new unlabeled instances. This knowledge can support reject-option strategies in the future so the ML model might opt for abstaining from some predictions that will be uncertain [4].

Firstly, a set of IHMs is adapted to disregard the labels of the new instances in their computation. Another strategy tested was generating regression meta-models to estimate the IHMs for new unlabeled observations in a meta-learning approach at the instance level. Both approaches are compared experimentally using one synthetic dataset and four datasets of the health domain, known for presenting hard instances. Instances with characteristics making them lie in overlapping or borderline regions of the classes are highlighted as hard to classify by both approaches. The adapted measures show an increased correlation to the original values of the instance hardness measures and prove to be an adequate alternative to estimate instance hardness in the deployment stage, driving the solutions to a more refined level and contributing toward a more trustful use of ML models.

The paper is organized as follows: Sect. 2 details the hardness measures to apply to unlabeled data and how they were modified from the original measures. Section 3 presents the materials and methods used in experiments, whose results are presented in Sect. 4. Finally, Sect. 5 presents the conclusions of this work.

2 Instance Hardness Measures

The concept of instance hardness was introduced in the seminal work of Smith et al. [14] as an alternative for a fine-grained analysis of classification difficulty.

They define an instance as hard to classify if it gets consistently misclassified by a set of classifiers of different biases. They also define a set of measures to explain possible reasons why an instance is difficult to classify, which are regarded as instance-level meta-features in the literature [7].

The base IHMs adopted in this work are presented next, along with their adaptations, which are indicated by the "adj" (adjusted) extension. In their definition, let $\mathcal{D}$ be a training dataset with n pairs of labeled instances $(\mathbf{x}_i, y_i)$, where each $\mathbf{x}_i \in \mathcal{X}$ is described by m input features and $y_i \in \mathcal{Y}$ is the class of the instance in the dataset. The number of classes is denoted as C. And let $\mathbf{x}$ be a new instance for which the label is unknown.

To illustrate the concepts, consider the dataset in Fig. 1 containing two classes, red and blue. Two instances are highlighted: $\mathbf{x}_1$ and $\mathbf{x}_2$. The instance $\mathbf{x}_1$ is in a borderline area of the classes and might be difficult to classify despite its class. The instance $\mathbf{x}_2$ is more aligned to the blue class. If the label registered for it in the dataset is blue, it will be easily classified. Otherwise, it will have a hardness level higher than $\mathbf{x}_1$. Standard IHMs need to know these labels, so that both $\mathbf{x}_1$ and $\mathbf{x}_2$ are contained in the labeled dataset $\mathcal{D}$. This work introduces adaptations to estimate the hardness level of an instance in the absence of its label, meaning $\mathbf{x}_1$ and $\mathbf{x}_2$ are not in the labeled dataset $\mathcal{D}$ used to estimate the hardness levels. Please note there are differences between the two estimations. Based on the characteristics of $\mathbf{x}_2$, it will probably be easily classified as blue. In contrast, $\mathbf{x}_1$ will probably be considered hard to classify in both scenarios.

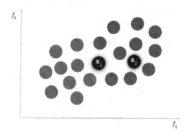

Fig. 1. Example of dataset with highlighted instances: $\mathbf{x}_1$ is in a borderline region and can be difficult to classify despite its class; $\mathbf{x}_2$ might be easy or hard to classify depending on its registered label. (Color figure online)

2.1 Neighborhood-Based IHM

The hardness level of the instance can be obtained considering its neighbourhood in the dataset. In the original IHMs, instances surrounded by elements sharing the same label as themselves can be considered easier to classify. For new data without labels, our approach seeks the neighbourhood of the instance in the labeled dataset $\mathcal{D}$ and assigns a higher hardness level when there is a mix of different classes in this region.

k-Disagreeing Neighbors kDN: the original kDN measure computes the percentage of the k nearest neighbors of $\mathbf{x}_i$ in the dataset $\mathcal{D}$ that have a different label than the refereed instance:

$$\text{kDN}(\mathbf{x}_i, y_i) = \frac{\#\{\mathbf{x}_j | \mathbf{x}_j \in \text{kNN}(\mathbf{x}_i) \wedge y_j \neq y_i\}}{k}, \quad (1)$$

where $\text{kNN}(\mathbf{x}_i)$ represents the set of k-nearest neighbors of the instance $\mathbf{x}_i$ in the dataset $\mathcal{D}$. An instance will be considered harder to classify when the value of $\text{kDN}(\mathbf{x}_i, y_i)$ is higher. Values close to 1 represent an instance surrounded by examples from a different class of itself. This would be $\mathbf{x}_2$'s case in Fig. 1 when labeled red in $\mathcal{D}$. Intermediate values of $\text{kDN}(\mathbf{x}_i, y_i)$ are found for borderline instances. Easier instances are those surrounded by elements sharing their class label, which would correspond to $\mathbf{x}_2$ when it has a blue label.

In the absence of an instance's label, an alternative way to measure the mixture of classes in its neighbourhood is to compute an entropy measure. Specifically, the entropy is computed based on the proportion of the classes found in the instance's neighbourhood. Higher entropy values represent the new instance is in regions from $\mathcal{D}$ near elements from different classes. This corresponds to the $\mathbf{x}_1$ case in Fig. 1. In contrast, $\mathbf{x}_2$ will be regarded as easy to predict, as it is surrounded by elements of the blue class.

$$\text{kDN}_{\text{adj}}(\mathbf{x}) = -\sum_{i=1}^{C} p(y_j = c_i) \log p(y_j = c_i), \text{ for } \mathbf{x}_j \in \text{kNN}(\mathbf{x}), \quad (2)$$

where $p(y_j = c_i)$ are the proportions of the classes of the k-nearest neighbours of $\mathbf{x}$ in the dataset $\mathcal{D}$.

Ratio of the Intra-class and Extra-Class Distances. N2$_{\text{IHM}}$: the original measure takes the complement of the ratio of the distance of $\mathbf{x}_i$ to the nearest example from its class in $\mathcal{D}$ to the distance it has to the nearest instance from a different class (nearest enemy) in $\mathcal{D}$ with a normalization as presented next:

$$\text{N2}_{\text{IHM}}(\mathbf{x}_i, y_i) = 1 - \frac{1}{\text{IntraInter}(\mathbf{x}_i) + 1}, \quad (3)$$

where:

$$\text{IntraInter}(\mathbf{x}_i, y_i) = \frac{d(\mathbf{x}_i, \text{NN}(\mathbf{x}_i \in y_i))}{d(\mathbf{x}_i, \text{NE}(\mathbf{x}_i))}, \quad (4)$$

where d is a distance function, $\text{NN}(\mathbf{x}_i \in y_i)$ represents the nearest neighbor of $\mathbf{x}_i$ from its class and $\text{NE}(\mathbf{x}_i)$ is the nearest enemy of $\mathbf{x}_i$ ($\text{NE}(\mathbf{x}_i) = \text{NN}(\mathbf{x}_i \in y_j \neq y_i)$). In this formulation, when an instance is closer to an example from another class than another from its own class, the N2$_{\text{IHM}}$ values will be larger, indicating that this instance is harder to classify. This would correspond to the case where $\mathbf{x}_2$ in Fig. 1 has the red label.

The alternative measure for unlabeled instances can be obtained by taking the ratio of the minimum distance from $\mathbf{x}$ and the closest element in $\mathcal{D}$, denoted as $\mathbf{x}_j$ in Eq. 5, to the distance from $\mathbf{x}$ and the closest element from another class in $\mathcal{D}$, that is, a class different from that of $\mathbf{x}_j$. This ratio will assume value close to 1 when the instance is almost equally distant from different classes. This will happen more probably for borderline instances, such as $\mathbf{x}_1$ in Fig. 1.

$$\text{N2}_{\text{adj}}(\mathbf{x}) = \frac{\min(d(\mathbf{x}, \mathbf{x}_j))}{\min(d(\mathbf{x}, \mathbf{x}_k)|y_k \neq y_j)} \quad (5)$$

2.2 Class Likelihood IHM

This type of measure captures if the instance is well situated in its class, considering the general patterns of this class. The likelihood can be estimated for that, considering the input features are independent for simplifying the computations.

Class Likelihood Difference] CLD: the original measure takes the complement of the difference between the likelihood that $\mathbf{x}_i$ belongs to its class y_i and the maximum likelihood it has to any other class. This complement is taken to standardize the interpretation of the direction of hardness since the confidence of an instance belongs to its class is larger than that of any other class [9]:

$$\text{CLD}(\mathbf{x}_i, y_i) = \frac{1 - \left(p(\mathbf{x}_i|y_i)p(y_i) - \max_{y_j \neq y_i}[p(\mathbf{x}_i|y_j)p(y_j)]\right)}{2}, \quad (6)$$

where $p(y_i)$ is the prior of class y_i, set as $\frac{1}{C}$ for all data instances. $p(\mathbf{x}_i|y_i)$ represents the likelihood $\mathbf{x}_i$ belongs to class y_i and it can be estimated considering the input features independent of each other, as in Naïve Bayes classification. For example, if $\mathbf{x}_2$ in Fig. 1 is labeled as blue in $\mathcal{D}$, it will be easy according to this measure, as its likelihood to the blue class will be higher than to the red class.

When the class of an instance cannot be defined in advance, the hardness measure can be estimated by the difference between the two higher likelihoods of all possible classes in the dataset. Like in the original measure, the complement of the difference is taken to keep the interpretation that higher values are found for instances harder to classify. The values of this measure will tend to be higher for borderline instances since their likelihood of being in different classes will be similar.

$$\text{CLD}_{\text{adj}}(\mathbf{x}) = \frac{1 - \left(\max_{y_i}[p(\mathbf{x}|y_i)p(y_i)] - \max_{y_j \neq y_i}[p(\mathbf{x}|y_j)p(y_j)]\right)}{2}. \quad (7)$$

2.3 Tree-Based IHM

Decision trees (DTs) can be used to estimate the hardness level of an instance based on the number of splits necessary to classify it. If many splits are required, the instance's classification will be harder. The DT is built based on the labeled dataset $\mathcal{D}$. Unlabeled instances are input to the built DT, and the measure can be computed based on where it is classified.

Disjunct Class Percentage DCP: from a pruned decision tree (DT) using $\mathcal{D}$, the leaf node where the instance is classified is considered the disjunct of $\mathbf{x}_i$. The complement of the percentage of instances in this disjunct that shares the same label as $\mathbf{x}_i$ gives the original DCP measure:

$$\text{DCP}(\mathbf{x}_i, y_i) = 1 - \frac{\sharp\{\mathbf{x}_j | \mathbf{x}_j \in \text{Disjunct}(\mathbf{x}_i) \land y_j = y_i\}}{\sharp\{\mathbf{x}_j | \mathbf{x}_j \in \text{Disjunct}(\mathbf{x}_i)\}}, \qquad (8)$$

where $\text{Disjunct}(\mathbf{x}_i)$ represents the instances contained in the disjunct (leaf node) where $\mathbf{x}_i$ is placed. For easy instances, according to this measure, larger percentages of examples sharing the same label as the instance will be found in their disjunct. For example, if $\mathbf{x}_2$ in Fig. 1 has the red label in $\mathcal{D}$, it will probably be placed in a leaf node containing many elements of the blue class, making it harder to classify according to the interpretation of this measure. In scenarios where the instance's class is unknown, we take the entropy of the disjunct where the instance is placed as a hardness measure, similarly to what has been done for kDN.

$$\text{DCP}_{\text{adj}}(\mathbf{x}) = -\sum_{i=1}^{C} p(y_j = c_i) \log p(y_j = c_i), \text{ for } \mathbf{x}_j \in \text{Disjunct}(\mathbf{x}), \qquad (9)$$

where the proportions of the classes are taken based on the disjunct where $\mathbf{x}$ is placed in the DT built using the dataset $\mathcal{D}$.

Tree Depth TD: the original measure gives the depth of the leaf node that classifies $\mathbf{x}_i$ in a DT built using all labeled dataset $\mathcal{D}$, normalized by the maximum depth of the tree:

$$\text{TD}(\mathbf{x}_i, y_i) = \frac{\text{depth}_{\text{DT}}(\mathbf{x}_i)}{\max(\text{depth}_{\text{DT}}(\mathbf{x}_j \in \mathcal{D}))}, \qquad (10)$$

where $\text{depth}_{\text{DT}}(\mathbf{x}_i)$ gives the depth where the instance $\mathbf{x}_i$ is placed in the DT. Instances harder to classify tend to be placed at deeper levels of the tree, making TD higher. There are two versions of this measure. One derives from a pruned tree (TD_P) and the other from an unpruned tree (TD_U).

For unlabeled instances, the procedure for hardness estimation is the same as in DCP, where the DT is built from the labeled set $\mathcal{D}$, and next, the unlabeled instance is submitted to the built DT. The depth of the leaf node where this instance is classified by the DT is taken and used in the equation:

$$\text{TD}_{\text{adj}}(\mathbf{x}) = \frac{\text{depth}_{\text{DT}}(\mathbf{x})}{\max(\text{depth}_{\text{DT}}(\mathbf{x}_j \in \mathcal{D}))}, \qquad (11)$$

2.4 Using Meta-models to Estimate IHM

Meta-learning is a traditional ML task that uses data related to ML itself [2]. Here, MtL is designed to predict IHM values without considering their labels. This is done using the original input features from the dataset $\mathcal{D}$ to learn the expected IHM values in a regression task. Therefore, in this approach regression meta-models are induced to estimate the IHM values of new instances. Their training datasets comprise the original input features of $\mathcal{D}$ and a label corresponding to an IHM estimated from $\mathcal{D}$ in its original formulation. There is one regression model per IHM.

The estimation of the IHM values for unlabeled data with this meta-learning approach is compared to the usage of the adjusted IHM values.

3 Materials and Methods

In this section, we describe the materials and methods used in experiments performed to analyze the behaviour of IHM for unlabeled data in classification problems.

3.1 Datasets

Five datasets are employed in the experiments. The first dataset was created synthetically, containing three classes with some overlap. The other four datasets are from the health domain, for which some instances are hard to classify due to the overlap of attribute values for different classes or inconsistencies. Two of them are from the UCI public repository [6] and have been employed in previous related work [8,11]. The last two are related to severe COVID-19 cases in two large hospitals from the São Paulo metropolitan area [15]. The main characteristics of the five datasets are presented in Table 1, including the number of instances, classes and input features.

Table 1. Summary of the datasets used in the study.

	Blobs	Diabetes	Heart	Hospital1	Hospital2
Instances	300	768	270	526	134
Classes	3	2	2	2	2
Features	2	8	13	17	19

The dataset blobs was generated synthetically using the `make_blobs` package from the scikit-learn library [10], which can generate isotropic Gaussian blobs in space. The standard deviation between the centers of the classes was set as 2 to create some overlap between the input features and regions where the difficulty in classifying the instances is harder than others. Figure 4 presents this dataset, where it is possible to notice some overlap in the borderline regions of the classes.

The diabetes dataset is related to the incidence of diabetes in female patients of Pima Indian heritage who are at least 21 years old. The objective is to identify the presence of diabetes. The predictive variables record blood indices and patient characteristics, such as number of pregnancies and age [6].

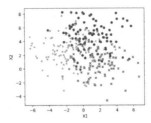

Fig. 2. Illustration of the blobs dataset.

The heart dataset registers heart disease in patients and has features collected during the exercise test, others reflecting blood indices and personal characteristics of the patients, such as age and gender [6].

The last two datasets, named hospital, were extracted from the raw public database provided by FAPESP COVID data sharing initiative [3]. The binary response categorized patients as severe when hospital stay was greater than or equal to 14 days or patients who progressed to death. The features collected in those datasets were related to blood indices, age and gender [15].

3.2 Methodology

The adjusted IHM measures proposed in this paper were applied to the datasets, considering each instance unlabeled once at a time, and the remaining instances labeled, resembling a leave-one-out (LOO) cross-validation scheme.

The same procedure is used to generate the meta-models to predict the IHM values, where one instance is left out as unlabeled at a time. The IHM of the other instances is calculated using their original formulations. Next, a meta-dataset is built, mapping the original features of the instances to the computed IHM values. Regression meta-models are induced to learn this relationship and predict the expected IHM value of the left-out instance. One meta-model is induced per IHM measure considered. We used the Random Forest Regressor (RF) available in the Scikit-learn library [10] with default hyperparameters' values to generate these meta-models.

We also computed the original IHMs for the entire datasets, which regard the labels of all instances. Next, we compare the association of the IHM values of the original measures to those of the estimated measures, where the estimation is taken by the adjusted measures or the induced meta-models. Spearman's correlation provides a non-parametric estimation of the association (monotonic relationship) of the modified measures with the original measure. This correlation captures if the direction of the adjusted/estimated IHM is the same as the value obtained from the original IHM. Higher values of the Spearman's correlation indicate more association between the estimated and original IHM.

We expect medium to high correlations, although there can be deviations of values, since they do not strive to deliver identical IHM values. Indeed, instances with noisy labels in the training datasets have characteristics that make them aligned to another class and are expected to show a lower correlation to the original IHM values. But for most cases, we expect the hardness directions to be maintained.

All codes and analyses are implemented in Python. The original IHMs are computed using the PyHard package [7,9]. Codes of the adjusted measures are in a public repository https://anonymous.4open.science/r/Adj-IHM-BF75. The k value in kDN was set as 10, default value in the PyHard package.

4 Results

The results of the experiments performed are presented and discussed next.

4.1 Meta-models

First, we present the performance of the meta-models in the regression task. Table 2 presents the Mean Squared Error (MSE) obtained in predicting the IHMs using the regression meta-models. Lower values are indicative of better performances in predicting the original IHM values.

Table 2. MSE obtained for the RF algorithm concerning predicting the IHMs to different datasets.

	blobs	diabetes	heart	hospital1	hospital2
kDN	0.219	0.214	0.232	0.190	0.209
N2	0.150	0.077	0.109	0.049	0.049
CLD	0.215	0.196	0.240	0.146	0.124
DCP	0.238	0.221	0.248	0.168	0.190
TD_U	0.066	0.087	0.090	0.068	0.091
TD_P	0.035	0.022	0.010	0.002	0.105

For some measures, the MSEs are lower, demonstrating a better approximation of the original IHM values. This happens mostly for tree-depth measures. For others, the approximations are not as good (e.g. for kDN and DCP). One possible explanation is that the tree depth measures do not depend as much on the labels of the instances as the others. The only difference between the original tree depth measures and their estimated counterparts is excluding one instance from the decision tree induction, which affects less the results. For other measures, if an instance is incorrectly labeled, the original measures will point them as very hard to classify. But this instance might be easily classified into another class, making it easy without the label information.

4.2 Correlation Analysis

Table 3 shows Spearman's correlation coefficient between the original IHMs and the measures obtained using the meta-learning approach. Values higher than 0.5 are highlighted in bold. The values of the estimated tree depth measures are the highest, especially for the pruned version of the measure (TD_U). This happens because, in the pruned version of the tree, noisy and outlier instances tend to be placed in nodes which have undergone pruning. Therefore, the label of the particular instance seems to matter less in the original IHM formulation. In contrast, the formulation of the original CLD, DCP and kDN measures is highly influenced by the label of each instance where they are measured. This decreases the correlations, especially in datasets with many instances with feature values akin to a class, despite being originally labeled into another class in the dataset. This is the case for hospital 1 and 2 datasets, where situations such as instances wrongly labeled or with overlapping feature values are more common.

Table 3. Spearman coefficient obtained for the RF algorithm concerning predicting the IHMs to different datasets.

	blobs	diabetes	heart	hospital1	hospital2
kDN	**0.721**	**0.651**	**0.632**	0.411	0.496
N2	**0.794**	**0.592**	**0.642**	0.299	0.295
CLD	**0.685**	**0.666**	0.430	0.432	0.287
DCP	**0.611**	**0.680**	**0.508**	0.392	0.456
TD_U	**0.927**	**0.824**	**0.671**	**0.870**	**0.911**
TD_P	**0.987**	**0.965**	**0.953**	**0.993**	**0.876**

Table 4. Spearman coefficient obtained for adjusted measures compared to the original IHMs in different datasets.

	blobs	diabetes	heart	hospital1	hospital2
kDN	**0.696**	**0.713**	**0.836**	0.427	**0.554**
N2	**0.579**	**0.651**	**0.790**	**0.501**	**0.609**
CLD	**0.760**	**0.767**	**0.839**	**0.502**	**0.677**
DCP	**0.666**	0.413	**0.793**	**0.514**	**0.681**
TD_U	**0.918**	**0.986**	**0.982**	**0.994**	**0.961**
TD_P	**0.773**	**1.000**	**1.000**	**1.000**	**0.960**

Table 4 presents the same results for the adjusted IHMs: their Spearman correlation to the original IHMs. As in Table 3, values higher than 0.5 are bold-faced. More boldfaced correlations are observed here. Similar observations concerning the higher correlation values for tree depth-based measures are observed in Table 4 too. The correlations observed for the adjusted measures are generally higher than those observed for the measures estimated by the meta-regressors. To make the differences clearer, Fig. 3 plots the Spearman's correlations for the adjusted IHM and the meta-models compared to the original values. Blue bars represent the correlation of the adjusted measures, while orange bars denote the meta-learning approach. Only for the blobs dataset and for the DCP-diabetes combination were correlations of the meta-models higher than those of the adjusted IHMs. The blobs dataset has difficult instances concentrated on the border of the classes, while the other datasets may pose other sources of difficulties which are not captured when the labels are absent, such as label noise.

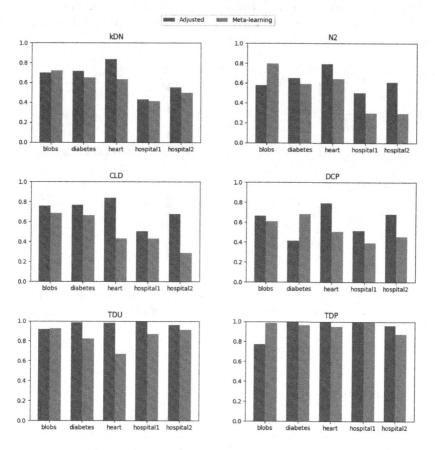

Fig. 3. Spearman's correlation applied to the adjusted IHM vs. the original IHM (blue bars) and the predicted vs. expected values from MtL (orange bars). (Color figure online)

Figure 4 shows the instances in the blobs dataset colored by the hardness of the original IHM (in the left) followed by the adjusted IHM (in the center) and the meta-learning approach (in the right). This can be done for this dataset, as it is bi-dimensional. The harder the instance is to classify, the more intense it is colored in red. In contrast, instances that are easier to classify are filled with darker blue. The central areas of the plots contain the overlapping region between the three classes (see Fig. 2) and, therefore, are harder to classify. The first row corresponds to the kDN measure, while the second is the TD_U measure. For kDN, it is clear that the hardest instances are those in the border of the classes. For TD_U the pattern observed in the three approaches shows that the harness level is related to partition derived from the decision tree classification. All measures show similar behaviors. However, for the adjusted kDN measure, more central instances have higher IHM values compared to the other measures. It is important to note that since the adjusted measures can vary on a different scale from the original IHM, the results presented in the plots were normalized between 0 and 1 to allow a direct comparison.

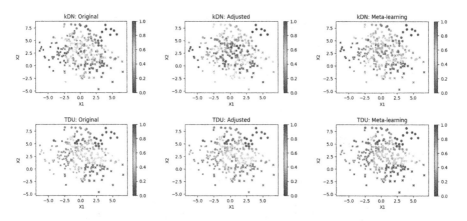

Fig. 4. Visualization of the measures kDN (top) and TD_U (bottom): original IHM (left), adjusted IHM (middle) and meta-learning approach (right) for the blobs dataset. (Color figure online)

4.3 Discussion

Considering the difference in the nature of the datasets, where the blobs were artificially designed with three classes and the other are real-world health data, the Spearman's correlation in Fig. 3 shows that the MtL achieves more convergent result than the adjusted IHM in the blobs dataset for great part of the measures. Conversely, for real datasets, the adjusted IHM is more associated with the original measure for almost all measures.

The tree-depth measures had the highest correlations to the original measures for the adjusted IHM and the MtL approach. This is mostly related to the fact that the original tree-depth measures do not depend so directly on the label of

An Instance Level Analysis of Classification Difficulty for Unlabeled Data 153

the instances. The other measures all regard whether the labels of some vicinity are in accordance with the registered label of the instance. This makes them deviate more for instances that are mislabeled, for instance.

This can be observed in Fig. 5, where the original and estimated IHMs KDN (in the top) and TD_U (in the bottom) are contraposed for all instances of the blobs dataset. In the x-axis, we have the original measures, whilst, in the y-axis, the proposed counterparts are taken. The adjusted kDN is normalized between 0 and 1 for direct comparison.

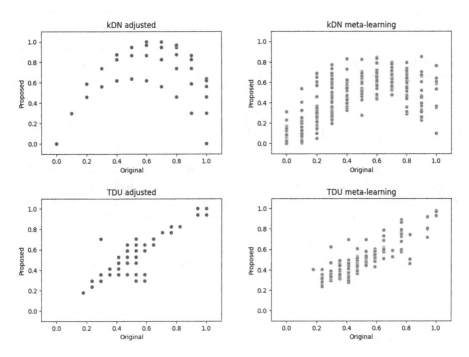

Fig. 5. The adjusted IHM vs the original IHM and meta-learning prediction vs original IHM for the blobs dataset.

For the measure kDN, one can observe that as the hardness to classify the instances grows, both the adjusted and the original IHM increase their values, reaching their peak in the middle of the scale. After that, the value of the estimated IHM assumes the opposite direction of the original measure. This result is expected considering the unlabeled data, given that instances harder to classify without a label will be predicted as belonging to any other class rather than being an outlier from a specific class.

Conversely, in the TD unpruned graphs, the results indicate that the hardness in classification is independent of the class being known or not. For both adjusted IHM and meta-learning approaches, there is some linearity between those measures and the original IHM. It means that the proposed measures for unlabeled data capture the increase of hardness for classification problems equivalent to the increase in hardness when the class is given for this measure. This result can be expected considering the nature of the measure.

The CLD measure, the only measure using likelihood as the metric, performed more closely to the original measure with the adjusted IHM for all datasets. Especially for datasets with two classes, in many cases, both estimates might agree when the first and second classes with maximum likelihoods are the same.

Overall, adjusted and meta-learning IHMs were able to assess the hardness level of the unlabeled instances, with some prominence of the adjusted measures, which showed larger correlations to the original measures in most cases. They are also simpler to compute, as they do not need to induce a ML model as in the meta-learning approach. In the absence of labels, most measures are more effective in pointing borderline instances as posing a higher difficulty of posterior classification.

5 Conclusion and Future Work

This research analyzed alternative ways to measure the hardness of instances for classification problems in scenarios where the label of an instance is unknown, that is, in the deployment stage. Standard IHMs from the literature were adapted to this scenario. Their results were compared to the alternative of generating regression meta-models to predict the IHM values. Both alternatives were effective on their behalf, correlating to the original IHMs that need to know the label of each instance. The correlations were higher for some measures that do not rely as much on the labels, but the results for other measures are expected as their original formulation allows one to identify noise and outliers on data regarding their labels. The results encourages the usage of the adjusted measures in the deployment of ML models, allowing the identification of instances that ML models might struggle to classify.

In future work, we will explore the patterns found in the comparisons between the original and adjusted IHM not presented in this work and alternative measures for unlabeled data not addressed in this research. We will expand the application of the adjusted measures and meta-learning to more datasets, and tuning the meta-models could lead to new findings about the characteristics of the instances. Another fruitful direction will be to explore the usage of the adjusted measures for designing classification rejection options.

Acknowledgements. This study was financed in part by the Coordenação de Aperfeiçoamento de Pessoal de Nível Superior—Brasil (CAPES)—Finance Code 001. The authors thank FAPESP for its support under grant 2021/06870-3.

References

1. Al Hosni, O.S., Starkey, A.: Investigating the performance of data complexity & instance hardness measures as a meta-feature in overlapping classes problem. In: ICCBDC 2023, Manchester, United Kingdom (2023). https://doi.org/10.1145/3616131.3616132
2. Brazdil, P., van Rijn, J.N., Soares, C., Vanschoren, J.: Metalearning: Applications to Automated Machine Learning and Data Mining. Springer, Cham (2022). https://doi.org/10.1007/978-3-030-67024-5
3. FAPESP: FAPESP COVID-19 data sharing/Br (2020). https://repositoriodatasharingfapesp.uspdigital.usp.br
4. Franc, V., Prusa, D., Voracek, V.: Optimal strategies for reject option classifiers. J. Mach. Learn. Res. **24**(11), 1–49 (2023)
5. Ho, T.K., Basu, M.: Complexity measures of supervised classification problems. IEEE Trans. Pattern Anal. Mach. Intell. **24**(3), 289–300 (2002). https://doi.org/10.1109/34.990132
6. Kelly, M., Longjohn, R., Nottingham, K.: The UCI machine learning repository (2023). https://archive.ics.uci.edu
7. Lorena, A.C., Paiva, P.Y., Prudêncio, R.B.: Trusting my predictions: on the value of instance-level analysis. ACM Comput. Surv. **56**(7), 1–28 (2024)
8. Martínez-Plumed, F., Prudêncio, R.B., Martínez-Usó, A., Hernández-Orallo, J.: Item response theory in AI: analysing machine learning classifiers at the instance level. Artif. Intell. **271**, 18–42 (2019)
9. Paiva, P.Y.A., Moreno, C.C., Smith-Miles, K., Valeriano, M.G., Lorena, A.C.: Relating instance hardness to classification performance in a dataset: a visual approach. Mach. Learn. **111**(8), 3085–3123 (2022)
10. Pedregosa, F., et al.: Scikit-learn: machine learning in Python. J. Mach. Learn. Res. **12**, 2825–2830 (2011)
11. Prudêncio, R.B., Silva Filho, T.M.: Explaining learning performance with local performance regions and maximally relevant meta-rules. In: Xavier-Junior, J.C., Rios, R.A. (eds) BRACIS 2022. LNCS, vol. 13653, pp. 550–564. Springer, Cham (2022). https://doi.org/10.1007/978-3-031-21686-2_38
12. Rivolli, A., Garcia, L.P., Soares, C., Vanschoren, J., de Carvalho, A.C.: Meta-features for meta-learning. Knowl.-Based Syst. 108101 (2022)
13. Schweighofer, E.: Data-centric machine learning: improving model performance and understanding through dataset analysis. In: Legal Knowledge and Information Systems: JURIX 2021, vol. 346, p. 54. IOS Press (2021)
14. Smith, M.R., Martinez, T., Giraud-Carrier, C.: An instance level analysis of data complexity. Mach. Learn. **95**(2), 225–256 (2014)
15. Valeriano, M.G., et al.: Let the data speak: analysing data from multiple health centers of the São Paulo metropolitan area for COVID-19 clinical deterioration prediction. In: 2022 22nd IEEE International Symposium on Cluster, Cloud and Internet Computing (CCGrid), pp. 948–951. IEEE (2022)

Analyzing the Impact of Coarsening on k-Partite Network Classification

Thiago de Paulo Faleiros[1](✉), Paulo Eduardo Althoff[1], and Alan Demétrius Baria Valejo[2]

[1] University of Brasília (Unb), Brasília, DF, Brazil
thiagodepaulo@unb.br
[2] Department of Computing, Federal University of São Carlos (UFSCar), São Carlos, SP, Brazil
alanvalejo@ufscar.br

Abstract. The ever-expanding volume of data presents considerable challenges in storing and processing semi-supervised models, hindering their practical implementation. Researchers have explored reducing network versions as a potential solution. Real-world networks often comprise diverse vertex and edge types, leading to the adoption of k-partite network representation. However, existing methods have mainly focused on reducing uni-partite networks with a single vertex type and edges. This study introduces a novel coarsening method designed explicitly for k-partite networks, aiming to preserve classification performance while addressing storage and processing issues. We conducted empirical analyses on synthetically generated networks to evaluate their effectiveness. The results demonstrate the potential of coarsening techniques in overcoming storage and processing challenges posed by large networks. The proposed coarsening algorithm significantly improved storage efficiency and classification runtime, even with moderate reductions in the number of vertices. This led to over one-third savings in storage space and a twofold increase in classification speed. Moreover, the classification performance metrics exhibited low variation on average, indicating the algorithm's robustness and reliability in various scenarios.

Keywords: Graph · Network Coarsening · k-partite network · node classification · transductive learning

1 Introduction

Semi-supervised learning has emerged as a practical approach for leveraging labeled data to guide a supervisor's response. These methods harness both labeled and unlabeled data to facilitate the learning process. Such algorithms consider the connections between labeled and unlabeled data to compensate for the absence of labels [1]. Presenting data in the form of networks further enhances the potential of semi-supervised techniques, as it enables the extraction of relationships from the topological attributes of labeled and unlabeled data [2].

While the semi-supervised approach has offered a solution to reduce the reliance on human intervention, a significant challenge persists. As the volume of data grows, the storage cost and computational processing time required to train a semi-supervised model can become prohibitive, rendering it impractical for specific applications [3]. To address these limitations, researchers have extensively explored a technique involving reduced versions of networks instead of the original ones [4]. This approach minimizes storage requirements, improving algorithm performance compared to a full-sized network.

One notable category within these techniques is the employment of *coarsening* algorithms, which group similar vertices together, effectively reducing redundant information. Coarsening has long been established in visualization and graph partitioning [5], and more recently, it has demonstrated its efficacy in solving classification problems in homogeneous networks [6].

However, most of these methods concentrate on analyzing networks with a singular type of vertex and edge, referred to as uni-partite networks. In reality, information networks often exhibit heterogeneity, comprising various types of vertices and edges [7]. Leveraging the information richness of the heterogeneous k-partite network allows the model to learn more resilient and generalizable data representations, leading to improved performance in downstream tasks like classification and prediction. As interest in techniques for heterogeneous networks grows, as evidenced by studies such as [4,8,9], research on coarsening methods has also gained momentum, particularly for bipartite networks [10–14].

The methods explicitly designed for heterogeneous networks still need to be explored [5]. In this study, we aim to assess the accuracy of coarsened heterogeneous networks in classifying vertices based on their relationships to other vertices within a semi-supervised context. It is important to note that coarsening can result in information loss and a potential decrease in classification accuracy. Therefore, evaluating the trade-off between computational efficiency and classification accuracy becomes crucial when employing coarsening in heterogeneous network classification tasks.

In this context, this study introduces the development of a novel coarsening method designed explicitly for k-partite networks. Our proposed method utilizes a technique that organizes partitions and selects paths in the schema, resulting in improved coarsening performance compared to random approaches used in other studies [15]. Our results demonstrate the potential of coarsening techniques in resolving storage and processing issues for large networks.

This paper is organized as follows. Section 2 discusses the background and foundational concepts necessary for understanding the research, including an overview of k-partite networks and their formal descriptions. Section 3 introduces a novel coarsening algorithm specifically designed for k-partite networks, detailing its methodology and the theoretical underpinnings that ensure its efficiency and effectiveness. Section 4 presents the experimental results, showcasing the performance of the proposed algorithm through various datasets. Finally, Sect. 5 concludes the paper by summarizing the findings and discussing their implications.

2 Background

A network, denoted as $G = (V, E)$ or graph G, is referred to as **k-partite** if its vertex set V consists of k disjoint sets: $V = \mathcal{V}_1 \cup \mathcal{V}_2 \cup ... \cup \mathcal{V}_k$. Here, each $\mathcal{V}_i$ and $\mathcal{V}_j$ ($1 \leq i,j \leq k$) represent sets of vertices, and the edge set E is a subset of pairs from $\bigcup_{i \neq j} \mathcal{V}_i \times \mathcal{V}_j$. In other words, every edge $e = (a, b)$ connects vertices from different sets, where $a \in \mathcal{V}_i$ and $b \in \mathcal{V}_j$, with $Ai \neq j$. Additionally, each edge (a, b) in the graph may be associated with a weight, denoted as $\omega(a, b)$, where $\omega : E \rightarrow \mathbb{R}^*$. Moreover, individual vertices may have associated weights, represented as $\sigma(a)$, where $\sigma : V \rightarrow \mathbb{R}^*$.

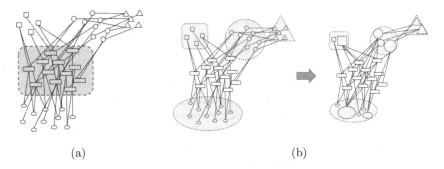

(a)   (b)

Fig. 1. Illustration of the coarsening process performed by the proposed algorithm in a heterogeneous network. In Figure (a), the target partition is highlighted in red, while in Figure (b), the other partitions are highlighted in blue, and the coarsening process reduces them. (Color figure online)

In this context, a heterogeneous network is considered a specific type of k-partite network, where vertices of the same type form a partition, and connections between vertices of different types exist. These connections are undirected, and the relationships between nodes are symmetric.

The weight of a vertex $a \in \mathcal{V}_i$, represented as κ_a, is defined as the total weight of its adjacent edges, expressed as $\kappa_a = \sum_{b \in V} \omega(a, b)$. The h-hop neighborhood of vertex a, denoted as $\Gamma_h(a)$, is formally defined as the set of vertices such that $\Gamma_h(a) = \{b \mid \text{there exists a path of length } h \text{ between vertex } a \text{ and vertex } b\}$. Thus, the 1-hop neighborhood of a, denoted $\Gamma_1(a)$, consists of the vertices directly adjacent to a. Similarly, the 2-hop neighborhood, $\Gamma_2(a)$, comprises the vertices that are reachable from a in exactly 2 hops, and so on for higher values of h.

In a k-partite network context, the network schema refers to the topological structure that links the k partitions together. Formally, the network schema of a k-partite network G can be represented by the network $S(G) = (V_S, E_S)$, where V_S is the set of k vertices associated with each partition, and E_S is the set of edges connecting these vertices. For any edge $(a, b) \in E_S$, vertex a belongs to a partition $\mathcal{V}_i$, and vertex b belongs to a different partition $\mathcal{V}_j$, where $i \neq j$ and

$1 \leq i, j \leq k$. A metapath, in this context, is defined as a sequence of edges that connects vertices from different partitions in the network schema.

Our proposed technique employs a label propagation scheme that disseminates labeled vertices from a specific partition, known as the target partition $\mathcal{V}_t$, to all other vertices within the network. Let $\mathcal{V}^L \in \mathcal{V}_t$ denote the set of labeled vertices in the target partition, and $\mathcal{V}^U \in V$ represent the set of unlabeled vertices. Each vertex in $\mathcal{V}^L$ is associated with a label from a set $C = \{c_1, c_2, \ldots, c_m\}$ comprising m classes. The matrix $\mathcal{Y} \in \mathbb{R}^{|V| \times m}$ represents the labels for the corresponding vertices in V. For simplicity, we denote $\mathcal{Y}_{a,i}$ as the weight of label c_i assigned to a vertex a, and $\mathcal{Y}_{\mathcal{V}_i}$ as the labels assigned to a subset of vertices in partition $\mathcal{V}_i$.

The transductive learning algorithm inputs a labeled training set of vertices $\mathcal{V}^L$ and a set of unlabeled test vertices $\mathcal{V}^U$. It outputs a transductive learner F that assigns a label $c_i \in C$ to each vertex a in $\mathcal{V}^U$, denoted as $F(a) = \arg\max_i \mathcal{Y}_{a,i}$, signifying the label with the highest weight assigned to vertex a by the transductive learner.

3 Coarsening Algorithm for k-partite Network

This section presents our proposed coarsening algorithm designed to reduce k-partite networks, facilitating subsequent classification tasks. The algorithm leverages labeled vertices from the target partition $\mathcal{V}_t$ to guide the reduction process. The k-partite network is initially decomposed into a series of bipartite networks, where pairs of partitions are selected from the original network. Subsequently, an adaptation of the CLPb (Coarsening strategy via semi-synchronous Label Propagation for bipartite networks) [16] coarsening algorithm is applied to these partition pairs, wherein one partition acts as the propagator partition, denoted as $\mathcal{V}_p$, and the other as the receptor partition, denoted as $\mathcal{V}_r$. The coarsening process is executed semi-synchronously, grouping vertices into super-vertices from $\mathcal{V}_r$ with identical labels. An illustration of the coarsening process in a k-partite network can be seen in Fig. 1. This coarsening procedure aims to reduce the overall training time for transductive learning, as network-based methods generally exhibit complexity associated with the number of vertices and edges.

After establishing label propagation as a matching approach for each bipartition, the next crucial step is determining the strategy for selecting the pairs of partitions and the order in which the CLPb algorithm will be applied. When considering all possible partition pairs, networks with high connectivity schemas can lead to numerous procedures, resulting in a quadratic complexity concerning the number of vertices [15]. Moreover, cycles in the network schema would cause information repetition. To address these challenges, our objective was to identify, for each partition, the most suitable neighboring partition to serve as a pair in the bipartite coarsening procedure. This neighboring partition, acting as the pair, is called the "guide partition."

One approach to reducing the number of partition pairs used is to identify paths in the schema of the k-partite network [17]. Given that only the target partition possesses label information initially, a logical strategy for the coarsening procedure is to propagate the information from the target partition to others, leveraging the label information during the matching phase. Shorter paths tend to indicate stronger relationships between vertices [18], making the selection of the shortest metapath between the two partitions preferable.

The objective is to perform coarsening on all non-target partitions following a metapath. The procedure is executed synchronously, one partition pair at a time, with the partition pairs selected radially, starting from the target partition. Initially, guide partitions at a 1-hop distance from the target are coarsened, followed by those at a 2-hop distance, and so on. When reducing each guide partition, the pair used is the neighboring partition within the metapath that leads from the partition being reduced towards the target partition. This chosen order serves two purposes: it propagates the label information initially present in the target partition, and it ensures that the selected neighbor partition has already undergone a reduction in the previous coarsening process (except in the first iteration). The process is illustrated in Fig. 2.

3.1 Algorithm Description

The coarsening procedure for a k-partite network is outlined in Algorithm 1. It takes as input a k-partite network $G = (V, E)$, a schema $S(G)$ of the k-partite network, and a vertex S^t in schema $S(G)$ corresponding to the target partition $\mathcal{V}_t$. The algorithm's output is a coarsened version of the k-partite network G.

Consider P_i as the shortest metapath that originates from the labeled target partition S^t and terminates at the non-target partition S^i. The algorithm chooses the first one found in cases where multiple shortest metapaths exist. For each S^i, a partition S_j at a 1-hop distance from P_i is selected as the "guide partition."

The coarsening process begins with a breadth-first search, starting from $\mathcal{V}_t$, to select a non-target partition $\mathcal{V}_i$ using a guide partition $\mathcal{V}_j$. The objective is to replace the bipartite subnetwork $G_i = (\mathcal{V}_j, \mathcal{V}_i, E_i)$ in G with its condensed version $G_i^c = (\mathcal{V}_j, \mathcal{V}_i^c, E_i^c)$, consisting of super-nodes and super-edges obtained during coarsening. It is essential to note that although the guide partition $\mathcal{V}_j$ supports the bipartite coarsening process, only the partition $\mathcal{V}_i$ is updated in the network G at a time (see line 17 in Algorithm 1). The algorithm then returns the coarsened network G^c, obtained by applying coarsening to all non-target partitions of the original k-partite network.

4 Experimental Results

Experimental studies were conducted on synthetic and real datasets using transductive classification to evaluate the effectiveness of the proposed coarsening algorithm. The principal reduction objectives, namely memory savings and classification runtime, were analyzed as the number of vertices increased. Additionally, the accuracy of various metrics, such as Accuracy, Precision, Recall, and

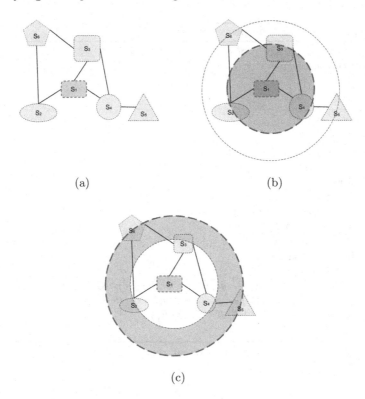

Fig. 2. The coarsening process is executed radially, beginning from the target partition $S_t = S_1$. The first step involves performing coarsening on the immediate neighbors of S_1 (in (b)), which are solely influenced by the target partition. Subsequently, coarsening is applied to partitions that are farther away from S_t (in (c)), with these partitions being influenced only by their respective guide partitions, which carry structural information from S_a.

F-score, was compared for each reduction level concerning the original uncoarsened network. The subsequent sections present the results of these experiments, along with insights drawn from the findings.

4.1 Synthetic Network Generation

Given the scarcity of standard datasets for analyzing heterogeneous data in various fields, we utilized a synthetic network generator called HNOC [19] to address this limitation and create k-partite networks. HNOC was selected for its flexibility in adjusting partition size, the number of potential classes, classification probability, noise, and dispersion levels.

Initially developed for community detection, the HNOC tool's concept of communities can be extended to perform data classification in a semi-supervised setting. Initially, the tool assigns each vertex precisely to its designated community. Subsequently, each vertex within a community is connected to all other

Algorithm 1: *Coarsening* procedure for non-target partitions

input : $G = (V, E)$, $S(G)$, S^t
output: coarsed graph G^c

1 **begin**
2     $bfsList \leftarrow [S^t]$;
3     $added \leftarrow [S^t]$;
4     $guide \leftarrow \{\}$;
5     $G^c \leftarrow G$;
6   **while** $bfsList.length \neq 0$ **do**
7         $S_i \leftarrow pull(bfsList)$;
8   **forall** $S_k \in \mathcal{N}(S_i)$ **do**
9   **if** $S_k \notin added$ **then**
10                 $added.push(S_k)$;
11                 $bfsList.push(S_k)$;
12                 $guide[S_k] \leftarrow S_i$;
13   **if** $S_i \neq S^t$ **then**
14             $S_j \leftarrow guide[S_i]$;
15             $G_i \leftarrow (\mathcal{V}_j, \mathcal{V}_i, E_i) \leftarrow getBipartiteSubgraph(G^c, S_j, S_i)$;
16             $G_i^c \leftarrow (\mathcal{V}_j, \mathcal{V}_i^c, E_i^c) \leftarrow coarseningBiPartite(G_i)$;
17             $G^c \leftarrow ((V/\mathcal{V}_i) \cup \mathcal{V}_i^c, (E - E_i) \cup E_i^c)$;
18   **return** G^c;

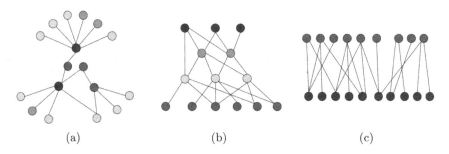

Fig. 3. Distinct topologies networks schemes generated in the experiments. Hierarchical star – Figure (a), Hierarchical web – Figure (b), and Bipartite topology – Figure (c).

vertices in the same community and those in different partitions. Edges are selectively removed based on the *dispersion* parameter to control community density. Lower dispersion values create sparser communities, while higher dispersion values lead to denser communities. The *noise* parameter is utilized for edges between different communities, also known as inter-community edges. Network noise impacts the identification of community boundaries and increases overlap, making finding communities within the network more complex. Higher noise levels can significantly decrease classification accuracy and increase complexity. Generally, noise values greater than 0.5 result in networks with poorly defined and sparse community structures, where inter-community edges outnumber intracommunity edges. Optimal noise values typically range from 0.1 to 0.4, striking a balance between increasing class detection difficulty while preserving the overall network structure [19]. It is crucial to note that no edges connect vertices within the same partition.

Diverse object types and topological structures are necessary in heterogeneous networks. It is essential to generate various network topologies to simulate real-world scenarios and complex systems effectively. In addition to the dispersion and noise parameters, the type of topological structure in the network was also varied. Three specific topologies were selected for the study, namely the *hierarchical star* (Fig. 3(a)), the *hierarchical web* (Fig. 3(b)), and the *bipartite topology* (Fig. 3(c)).

4.2 Experimental Setup

The experiments were categorized into two groups: the first utilized synthetic data and the second involved a real dataset.

In the first group of experiments, a wide range of parameter configurations was considered to generate synthetic k-partite networks with diverse class structures randomly [20]. The number of vertices varied among 2,000 and 15,000 for each scheme (Fig. 3). The number of classes ranged from 4 to 10 with increments of 1, the dispersion varied from 0.1 to 0.9 with increments of 0.1, and the noise value was fixed at 0.3 to create more challenging networks for classification without excessive degradation. Ten distinct random networks were generated for each configuration, resulting in over 13,000 networks. The vertices were evenly distributed among the non-target partitions. To accurately reflect the memory savings achieved by the coarsening algorithm, the non-target partitions were set to be five times larger than the target partition.

The second group of experiments involved a real-world heterogeneous network using the DBLP dataset[1] for the classification task. The DBLP dataset contains open bibliographic information from major computer science journals and proceedings. In this study, the task was to classify the authors into four areas of knowledge, and the authors served as the target partition. The non-target partitions included the articles written by each author, the conferences where the authors have published, and the terms found in these articles. The DBLP dataset comprises 14,475 authors, 14,376 articles, 20 conferences, and 8,920 terms, resulting in 37,791 vertices and 170,795 edges, with a dispersion value of approximately 0.002.

The main objective of this study is to evaluate the effectiveness of the proposed coarsening algorithm in transductive classification tasks involving k-partite networks. For the experiments, the GNetMine algorithm [21] was chosen, as it is a widely used reference algorithm in heterogeneous classification and has been utilized as a benchmark in numerous previous studies [7,17,22–26]. All generated networks were classified using GNetMine, and metrics such as Precision, Recall, F-score (macro variant), Accuracy, and classification time were recorded.

---

[1] DBLP dataset available at https://dblp.org.

The proposed coarsening algorithm was applied to the networks, with 20% and 80% reduction levels. The transductive classification was evaluated by varying the number of labeled vertices, using 1%, 10%, 20%, and 50% of the vertices from the target partition. The obtained results serve as a comparative reference for classification metrics and the storage size of the networks.

4.3 Results on Synthetic Data

The analysis of memory economies was conducted concerning the dispersion parameter. The detailed results in Table 1 demonstrates that networks with higher edge densities incur increased storage and classification runtime costs. Consequently, coarsening is more effective in such scenarios, yielding more substantial benefits. The relative economy of memory and classification runtime for each reduction level compared to the original unreduced network is presented in Table 1. Interestingly, there is a notable increase in the relative economy during the early stages of coarsening, leading to satisfactory results with just a 20% decrease in the number of vertices. This observation may be attributed to the initial levels already clustering many vertices with more shared connections.

Table 1. Compared to their original forms, reducing networks can lead to savings in both storage size and classification time. For both metrics, the values are represented as percentages.

Reduction	Storage Savings		Time Savings for Classification	
20%	27.22%	40.50%	36.24%	48.50%
80%	65.71%	91.22%	80.01%	90.96%
Dispersion	0.1	0.9	0.1	0.9

Table 2 presents the relationship between network reduction and classification metrics. The findings indicate that as the networks grow, the impact of coarsening on classification metrics becomes more pronounced. For instance, when dealing with a network of 15,000 vertices, reducing it by 20% and 80% results in a relative F-score loss of 1.69% and approximately 11.97%, respectively. Additionally, the Precision metric remains relatively stable, suggesting that the proposed coarsening algorithm is better suited for applications seeking to reduce the number of false positives. Furthermore, the relatively low F-score loss achieved by reducing the dataset by about 20% is particularly noteworthy, especially considering the significant savings in storage and classification runtime.

4.4 Results on Real Dataset

The results in Table 3 indicate that despite a 67% reduction in the network, there was no significant decrease in classification metrics. However, the reduction in

Table 2. The loss incurred during network reduction becomes more pronounced as the network size increases.

# vert.	Redu.	Precision (macro)	Recall (macro)	F-score (macro)
2000	20%	0.76 ± 0.14%	3.10 ± 0.46%	3.85 ± 0.62%
	80%	4.02 ± 0.40%	18.31 ± 1.82%	19.32 ± 1.92%
15000	20%	0.33 ± 0.14%	0.64 ± 0.24%	1.00 ± 0.33%
	80%	4.06 ± 0.79%	20.64 ± 2.46%	24.57 ± 3.08%

storage size was not as effective, as shown in Table 4. These findings align with the observations from the synthetic experimental evaluation. The network G_{DBLP} exhibits a low dispersion level of approximately 0.002. As previously discussed, such low dispersion levels, nearing zero, lead to a minimal loss in classification quality and offer limited storage savings.

Table 3. Metrics of Accuracy, Precision (macro), Recall (macro) and F-score (macro) according to G_{DBLP} reduction

Redu.	Accuracy	Precision (macro)	Recall (macro)	F-score (macro)
20%	2.47%	0.02%	2.59%	1.25%
36%	4.88%	0.11%	4.46%	2.46%
49%	4.92%	1.21%	5.12%	3.75%
59%	4.92%	2.43%	5.02%	3.75%
67%	4.92%	1.91%	5.02%	3.75%

Table 4. Storage (megabyte – MB) and time (seconds - s) savings analysis for G_{DBLP}

Redu.	Storage Savings	Time Savings for Classification
0%	2476.87 MB	21.15 s
20%	2219.46 MB	14.40 s
36%	2054.75 MB	11.84 s
49%	2014.74 MB	11.22 s
59%	2010.69 MB	11.10 s
67%	2010.24 MB	11.12 s

5 Conclusion

The primary objective of this study was to evaluate a proposed algorithm for reducing the size of k-partite networks while maintaining classification performance. The study successfully obtained metrics related to resource savings and classification performance, confirming the efficacy of the proposed coarsening algorithm for k-partite networks.

However, it is essential to acknowledge certain limitations. Using synthetic networks instead of real-world networks may introduce some bias, especially considering the high assortativity levels generated by the HNOC tool, which might only be representative of some real-world scenarios. Furthermore, the study's experiment design involved a limited number of k-partite network schemes, which should be considered a potential limitation. Nevertheless, various experiments and parameter configurations thoroughly assessed the algorithm's performance, showcasing its potential for diverse network applications. The researchers have made the entire source code used in the experiments available[2], enabling future works to test different parameters and networks.

The study's findings highlight that the proposed coarsening algorithm achieved significant savings in storage and classification runtime, even with modest reduction levels. For instance, a 20% reduction in the number of vertices resulted in over 1/3 savings in storage and twice faster classifications. Moreover, the classification performance metrics exhibited low average levels of variation, demonstrating the algorithm's stability and reliability.

Acknowledgment. This work is supported by São Paulo Research Foundation (FAPESP) under grant numbers 2022/03090-0.

References

1. Silva, M.M.: Uma abordagem evolucionária para aprendizado semi-supervisionado em máquinas de vetores de suporte. Master's thesis, Escola de Engenharia - UFMG, Belo Horizonte (2008)
2. Aggarwal, C.C.: Machine Learning for Text. Springer, Cham (2018). https://doi.org/10.1007/978-3-319-73531-3
3. Walshaw, C.: Multilevel refinement for combinatorial optimisation problems. Ann. Oper. Res. **131**(1), 325–372 (2004). https://doi.org/10.1023/B:ANOR.0000039525.80601.15
4. Liu, Y., Safavi, T., Dighe, A., Koutra, D.: Graph summarization methods and applications: a survey. ACM Comput. Surv. **51**(3) (2018). http://arxiv.org/abs/1612.04883
5. Valejo, A.D.B.: Métodos multi-nível em redes bi-partidas. Ph.D. dissertation, Instituto de Ciências Matemáticas e de Computação - USP, São Carlos (2019)
6. Liang, J., Gurukar, S., Parthasarathy, S.: Mile: a multi-level framework for scalable graph embedding (2020)

---

[2] URL with code and experiments repository, https://github.com/pealthoff/CoarseKlass.

7. Romanetto, L.D.M.: Classificação transdutiva em redes heterogêneas de informação, baseada na divergência kl. Ph.D. dissertation, Instituto de Ciências Matemáticas e de Computação - USP, São Carlos (2020)
8. Zhou, J., et al.: Graph neural networks: a review of methods and applications. AI Open 1, 57–81 (2020). http://arxiv.org/abs/1812.08434
9. Wu, Z., Pan, S., Chen, F., Long, G., Zhang, C., Yu, P.S.: A comprehensive survey on graph neural networks. IEEE Trans. Neural Netw. Learn. Syst. 32(1), 4–24 (2021). https://ieeexplore.ieee.org/document/9046288/
10. Valejo, A., Ferreira, V., Filho, G.P.R., Oliveira, M.C.F.D., Lopes, A.D.A.: One-mode projection-based multilevel approach for community detection in bipartite networks. In: Annual International Symposium on Information Management and Big Data - SIMBig. CEUR-WS (2017)
11. Valejo, A., Ferreira, V., de Oliveira, M.C.F., de Andrade Lopes, A.: Community detection in bipartite network: a modified coarsening approach. In: Lossio-Ventura, J.A., Alatrista-Salas, H. (eds.) SIMBig 2017. CCIS, vol. 795, pp. 123–136. Springer, Cham (2018). https://doi.org/10.1007/978-3-319-90596-9_9
12. Valejo, A., Ferreira de Oliveira, M.C., Filho, G.P., de Andrade Lopes, A.: Multilevel approach for combinatorial optimization in bipartite network. Knowl.-Based Syst. 151, 45–61 (2018). https://www.sciencedirect.com/science/article/pii/S0950705118301539
13. Valejo, A., Faleiros, T., de Oliveira, M.C.F., de Andrade Lopes, A.: A coarsening method for bipartite networks via weight-constrained label propagation. Knowl.-Based Syst. 195, 105678 (2020). https://www.sciencedirect.com/science/article/pii/S0950705120301180
14. Valejo, A., et al.: Coarsening algorithm via semi-synchronous label propagation for bipartite networks. In: Anais da X Brazilian Conference on Intelligent Systems, Porto Alegre, RS, Brasil, SBC (2021). https://sol.sbc.org.br/index.php/bracis/article/view/19047
15. Zhu, L., Ghasemi-Gol, M., Szekely, P., Galstyan, A., Knoblock, C.A.: Unsupervised entity resolution on multi-type graphs. In: Groth, P., et al. (eds.) ISWC 2016. LNCS, vol. 9981, pp. 649–667. Springer, Cham (2016). https://doi.org/10.1007/978-3-319-46523-4_39
16. Valejo, A.D.B., et al.: Coarsening algorithm via semi-synchronous label propagation for bipartite networks. In: Britto, A., Valdivia Delgado, K. (eds.) BRACIS 2021. LNCS (LNAI), vol. 13073, pp. 437–452. Springer, Cham (2021). https://doi.org/10.1007/978-3-030-91702-9_29
17. Luo, C., Guan, R., Wang, Z., Lin, C.: HetPathMine: a novel transductive classification algorithm on heterogeneous information networks. In: de Rijke, M., et al. (eds.) ECIR 2014. LNCS, vol. 8416, pp. 210–221. Springer, Cham (2014). https://doi.org/10.1007/978-3-319-06028-6_18
18. Gupta, M., Kumar, P., Bhasker, B.: HeteClass: a meta-path based framework for transductive classification of objects in heterogeneous information networks. Expert Syst. Appl. 68, 106–122 (2017). https://linkinghub.elsevier.com/retrieve/pii/S0957417416305462
19. Valejo, A., Góes, F., Romanetto, L., Ferreira de Oliveira, M.C., de Andrade Lopes, A.: A benchmarking tool for the generation of bipartite network models with overlapping communities. Knowl. Inf. Syst. 62(4), 1641–1669 (2020). https://doi.org/10.1007/s10115-019-01411-9
20. Valejo, A., Góes, F., Romanetto, L., Ferreira de Oliveira, M.C., de Andrade Lopes, A.: A benchmarking tool for the generation of bipartite network models with over-

lapping communities. Knowl. Inf. Syst. **62**(4), 1641–1669 (2019). https://doi.org/10.1007/s10115-019-01411-9
21. Ji, M., Sun, Y., Danilevsky, M., Han, J., Gao, J.: Graph regularized transductive classification on heterogeneous information networks. In: Balcázar, J.L., Bonchi, F., Gionis, A., Sebag, M. (eds.) Machine Learning and Knowledge Discovery in Databases, pp. 570–586. Springer, Heidelberg (2010). https://doi.org/10.1007/978-3-642-15880-3_42
22. Zhi, S., Han, J., Gu, Q.: Robust classification of information networks by consistent graph learning. In: Appice, A., Rodrigues, P.P., Santos Costa, V., Gama, J., Jorge, A., Soares, C. (eds.) ECML PKDD 2015. LNCS (LNAI), vol. 9285, pp. 752–767. Springer, Cham (2015). https://doi.org/10.1007/978-3-319-23525-7_46
23. Bangcharoensap, P., Murata, T., Kobayashi, H., Shimizu, N.: Transductive classification on heterogeneous information networks with edge betweenness-based normalization. In: Proceedings of the Ninth ACM International Conference on Web Search and Data Mining (2016)
24. Faleiros, T., Rossi, R., Lopes, A.: Optimizing the class information divergence for transductive classification of texts using propagation in bipartite graphs. Pattern Recogn. Lett. **87**, 04 (2016)
25. Luo, J., Ding, P., Liang, C., Chen, X.: Semi-supervised prediction of human miRNA-disease association based on graph regularization framework in heterogeneous networks. Neurocomputing **294**, 29–38 (2018). https://www.sciencedirect.com/science/article/pii/S0925231218302674
26. Ding, P., Shen, C., Lai, Z., Liang, C., Li, G., Luo, J.: Incorporating multisource knowledge to predict drug synergy based on graph co-regularization. J. Chem. Inf. Model. **60**(1), 37–46 (2019)

Applying Transformers for Anomaly Detection in Bus Trajectories

Michael Cruz[✉] and Luciano Barbosa

Universidade Federal de Pernambuco - Centro de Informática, Recife, Brazil
michaeloc9@gmail.com, luciano@cin.ufpe.br

Abstract. Trajectory anomaly detection is essential in understanding traffic behavior, especially when external factors such as congestion, accidents, and poor weather conditions occur. Most of the previous approaches for this task highly rely on handcrafted features and physical trajectory characteristics, which can be costly to calculate in a large volume of data. In this paper, we propose a novel trajectory anomaly detection approach that relies on language modeling to learn well-formed GPS bus trajectories and, based on it, identifies anomalous trajectories and pinpoints their abnormal points (sub-trajectory anomaly detection). Our solution uses a deep generative encoder-decoder Transformer that learns relationships between the sequential points in the trajectories based on the self-attention mechanism. It does not require manual feature extraction and can be easily adapted to any type of trajectory (e.g., cars, people, and vessels). We have performed an extensive experimental evaluation that shows: (1) our approach is effective for both trajectory and sub-trajectory anomaly detection; and (2) it outperforms the baselines in most evaluation scenarios.

Keywords: Anomaly Detection · Trajectory · Language Model · Transformer

1 Introduction

With the advances in GPS devices and the smart city's infrastructure, more and more mobile data has been generated [23]. Vehicles, smartwatches, traffic signals, and mobile phones, only to name a few, generate a massive volume of trajectory data. These data offer an unprecedented opportunity to discover rich information about traffic behavior, people's mobility, and weather impact [3,21].

Anomaly detection is the task of finding observations that stand out as dissimilar to all others [4]. Particularly in the traffic context, which is easily influenced by external factors (e.g., accidents, detours, events, and weather conditions), trajectory anomaly detection is crucial to understand traffic behavior and to support better decision-making by transit authorities.

Although trajectory anomaly detection has been a research hotspot [1], some challenges remain. For example, most solutions rely on handcrafted features and physical trajectory characteristics such as density [8], distance [9], and isolation [5]. Such features can be costly when dealing, for instance, with a high volume of trajectories, or might not work well in the case of data sparseness [20], which occurs when the period between consecutive trajectory's reported points is too large. In addition, little attention has been paid to finding anomalous regions in online way [5], which can benefit real applications as follows:

Example 1: A car ridesharing company might store thousands of daily trajectories to suggest the best routes to its riders and drivers. Therefore, finding anomaly trajectories is an essential feature since it can be helpful to recommend alternative paths, free of anomalies such as traffic congestion, and alert riders or penalize drivers that do not follow the recommended paths by its app.

Example 2: Public transportation agencies face traffic problems such as congestion, accident, and poor weather conditions in large cities worldwide. At the same time, these agencies need to serve the population by making real-time decisions to deal with these issues, such as proposing alternative routes or releasing more buses. Given that, an online trajectory anomaly detection approach can identify problematic bus trips to allow transit authorities to intervene as quickly as possible.

In this paper, we propose an approach that uses language modeling, commonly used in Natural Language Processing tasks, to: (1) detect anomalous trajectories in bus trajectories and (2) pinpoint the abnormal points in these trajectories (sub-trajectory anomaly detection). Our solution allows these tasks to be performed either offline or online, i.e., as the buses move along their route. In addition, it can also easily be adapted to other types of trajectories (e.g., cars, people, and vessels) since our model does not use any specific aspect from the bus domain (e.g., bus stops).

The key idea of our model is to learn the language of well-formed trajectories and then identify erroneous (ill-formed) trajectories and the trajectories' points where the errors occur. For that, given an input trajectory T mapped by our solution to a sequence of tokens, our language model (LM) generates $\hat{T}$, the most likely (common) sequence of points for T, which represents T supposedly without anomalies. Our assumption is, therefore, that since anomalous points are typically few and different from the others [5], there is a small chance that abnormal points are present in $\hat{T}$. Based on that, our solution produces an anomaly score for T and pinpoints anomalous regions in T by comparing T with $\hat{T}$.

We build our language model using a deep generative encoder-decoder Transformer [16] to learn the relationships between the sequential points in trajectories. More specifically, our solution first maps the input raw trajectory's points into a geographical grid and uses each grid cell's *id* as a token to represent the trajectory points. Then, this token-based trajectory representation feeds the Transformer encoder, which applies the self-attention mechanism to relate tokens in the sequence. The decoder receives the encoder's output and leverages previously predicted tokens to generate the next one based on the self-attention mechanism.

We have conducted an extensive experimental evaluation on real-world bus datasets from Recife (Brazil) and Dublin (Ireland) cities. The results show that our language model effectively detects whether trajectories are anomalous and, at the same time, finds anomalous regions without any handcrafted features.

The rest of this paper is organized as follows. Section 2 reviews the state of the art of anomaly detection in GPS trajectory. In Sect. 3, we present some background concepts and the problem statement. Section 4 delineates our model, and Sect. 5 describes the datasets and the setup experimentation. We compare our results with the state-of-the-art algorithms and previous research in Sect. 6. Finally, the conclusion and future work are drawn in Sect. 7.

2 Related Work

A recent survey [1] provides a comprehensive summary of the state-of-the-art solutions in trajectory anomaly detection. Most of them are based on distance [9] and pattern mining [5]. In addition, some approaches use machine learning on a supervised [6] and semi-supervised way [13]. In this section, we discuss some of these approaches in detail.

iBOAT (Isolation-based Online Anomalous Trajectory Detection) [5] detects anomalies point-by-point by isolating trajectories that are "few" and "different" from historical trajectories on the same route. It uses a "support" function to count how many historical trajectories share points with an ongoing trajectory (called a window). If the support value falls below a certain threshold, the points in the window are marked as anomalous. An anomaly score is then assigned using a logistic function, giving high values to anomalous points and low values to non-anomalous ones. iBOAT operates online, generating anomaly scores and detecting abnormal points in real-time, but it requires trajectories with the same departure and destination points to function.

In GMVSAE [13], the authors propose a deep learning encoder-decoder approach to detect whether the trajectory is anomalous. The approach uses an LSTM (Long Short-Term Memory) architecture to encode and decode trajectories. The approach uses the encoder to model trajectories by a gaussian distribution mixture. Then, the proposed method uses each distribution as the decoder's input information to calculate the probability of an ongoing trajectory belonging to one of the distributions. The trajectory is anomalous if the probability is below a certain threshold.

STOD (Spatial-Temporal Outlier Detector) [6] uses a supervised deep-learning model to detect trajectory anomalies in bus routes. It classifies bus trajectories based on predefined routes and considers them anomalous if the model's confidence in the classification is low. Each trajectory point is represented with features like GPS timestamp and a pre-trained embedding vector, which are fed into a Bi-GRU (Gated Recurrent Unit) architecture followed by a Multi-Layer Perceptron (MLP) with a softmax function. STOD calculates entropy over the softmax output's probability distribution to determine prediction confidence. A trajectory is marked as anomalous if the confidence score falls below a threshold. Similar to STOD, [2] proposes a multi-class Convolutional Neural Network to classify trajectories by route IDs and to identify anomalies based on misclassification or low-class probability. Both approaches require route ID labels, do not perform online anomaly detection, and do not detect anomalous regions.

Lastly, CTSS (Continuous Trajectory Similarity Search) [19] presents an online trajectory search method based on similarity scores to detect anomalous trajectories. For this, the authors proposed an approach that considers the current point of an input trajectory and all possible paths to arrive at the trajectory destination. Given all those future trajectories and a reference trajectory (ground truth), the approach calculates similarity scores between each pair (future trajectory and the reference one) and returns the one with minimum distance. Then, if the score is greater than a threshold, the input trajectory is anomalous. CTSS needs ground truth trajectories to calculate the similarity scores and knowledge of the trajectory destination in advance.

3 Problem Formulation

In this section, we provide some background concepts and state the problem we deal with in this work.

Definition 1. *[Bus Trajectory]. We define a bus trajectory T as a sequence of consecutive GPS points collected from a bus trip, denoted as $T = \{p_1, p_2, ..., p_n\}$ where n is the length of the trajectory. Each point $p_i = \{lat_i, lng_i\} \in \mathbb{R}^2$ is composed of latitude (lat_i) and longitude (lng_i) is associated with a bus trip and ordered by the points' timestamp, i.e., $tsp_i < tsp_{i+1}$.*

Definition 2. *[Spatial Anomaly Trajectory]. We consider a trajectory T as anomalous if some of its points spatially diverge from regular trajectories of T's assigned bus route.*

Definition 3. *[Problem Statement]. Given a bus trajectory T, we aim to calculate the spatial anomaly score of T and detect the anomalous points of T.*

4 Method

In this section, we introduce our approach to discovering abnormal trajectories and localizing anomaly regions in trajectories. As Fig. 1 shows, our solution is composed of three main components: Grid Mapping, Transformer Language Model, and Anomaly Detector. They work as follows. Given an input bus trajectory, the Grid Mapping discretizes it by mapping each of its points to a geographical grid cell, represented by a token, generating a sequence of grid cell tokens. This sequence is passed to a language model, a deep generative encoder-decoder Transformer, that produces a series of predicted grid cell tokens. Finally, the Anomaly Detector compares the original token sequence with the predicted one to calculate the trajectory's anomaly score and identify anomalous points if they exist. In the remainder of this section, we provide further details about each one of these components.

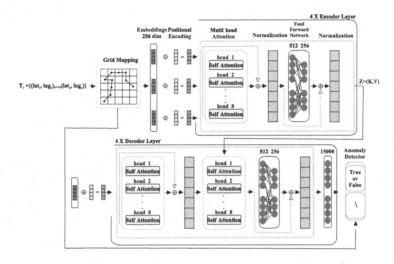

Fig. 1. The trajectory anomaly detection solution proposed in this work.

4.1 Grid Mapping

The first step of our approach, Grid Mapping, maps the trajectories, which are multivariate times series, into a univariate sequence of tokens. For this purpose, we use the H3 (Hexagonal Hierarchical Geospatial Indexing System)[1] library to create a grid system based on hexagonal cells and a hierarchical index. More specifically, given a raw trajectory $T = (p_1, p_2, .., p_n)$, where p_i is each trajectory point, represented by its latitude and longitude $(lat, long)$, and a grid of cell G, we use the geoToH3 function in the H3 library to map points of trajectories into grid cell locations c_i, generating $T' = (c_1, c_2, ..., c_n)$. As a result, every point that falls into the same cell has the same identifier (token). Therefore, this mapping reduces the complexity of dealing with a continuous multi-dimensional domain to a discrete uni-dimensional one, which language models can adequately process.

4.2 Transformer Encoder

The first component of our language model is the Transformer Encoder which maps the sequence of trajectory tokens T' into a set of vectors that feeds the

---
[1] https://eng.uber.com/h3/.

Transformer Decoder. For that, it encodes the tokens in T' based on the other tokens (points) in T' by applying the self-attention strategy.

More concretely, as Fig. 1 depicts, the encoder first generates a sequence of embeddings from the trajectory tokens. An embedding is a vector representation of a token in an n-dimension space[2]. Next, a positional encoding adds position information to the embeddings. Similar to [16], our position encoder is calculated as:

$$PE(pos, 2_i) = sin(pos/1000^{2i/d_{model}}) \quad (1)$$

$$PE(pos, 2_{i+1}) = cos(pos/1000^{2i/d_{model}}) \quad (2)$$

where sin and cos are the trigonometric functions sine and cosine, respectively, pos is the position of the point in the trajectory, and d_{model} is the dimension in the embedding vector. To create the final embedding representation for each input token, the model performs an element-wise addition of the token embedding with the positional encoding vector.

The model then passes these embeddings to the Transformer block with four identical encoder layers. It uses the so-called multi-head self-attention to allow the network to attend different input sequence positions and learn which points in the sequence are relevant to the current one. Multiple heads create multiple representation subspaces to learn a set of queries Q and keys K of dimension d_k, and values V dimension d_v weight matrices. Each head computes the attention weights for a given token embedding j ($Embedding_j$) as follows:

$$Q_{i,j} = W_i^Q \cdot Embedding_j \quad (3)$$

$$K_{i,j} = W_i^K \cdot Embedding_j \quad (4)$$

$$V_{i,j} = W_i^V \cdot Embedding_j \quad (5)$$

$$Head = softmax(\frac{Q_i K_i}{\sqrt{d_k}}) V_i \quad (6)$$

with parameters matrices $W^Q \in \mathbb{R}^{d_{model} X d_k}$, $W^K \in \mathbb{R}^{d_{model} X d_k}$, and $W^V \in \mathbb{R}^{d_{model} X d_v}$.

The multi-head attention combines the individual heads as follows:

$$Multi\_Head = concat(Head_1, Head_2, ..., Head_n) W^O \quad (7)$$

where $W^O \in \mathbb{R}^{h d_v X d_v}$ is a weight matrix learned during training.

On top of the multi-head attention, there are two skip connections and two normalization layers interspersed with fully connected feed-forward networks. The residual connection helps the encoder to combine features from different layers, merging different levels of representations [7]. The normalization layers standardize the residual connection and the feed-forward outputs, giving numerical stability to the model. The model calculates it as follows:

---

[2] We use embeddings of 256 dimensions in our solution.

$$\bar{x}_i = \frac{x_i - \mu_B}{\sqrt{\sigma_B^2 + \epsilon}} \tag{8}$$

$$z_i = \gamma \cdot \bar{x} + \beta \tag{9}$$

where μ_B and σ^2 are respectively the batch mean and standard deviation, ϵ is a stability factor added to variance to avoid a division by zero, γ and β are learning parameters, and z_i is the normalized value of x_i. Note that x_i is the concatenation between *Multi_head* vector and positional Embeddings (skip connection) and the normalization vector along with feed-forward output as shown in Fig. 1.

Lastly, the feed-forward network has two layers on the top of the encoder. Their goal is to project the normalization of the multi-head attention to another dimension space and add non-linearities between them.

The encoder's final output is the matrix $Z = (K, V)$, where K are key vectors and V value vectors of the tokens in the input sentence. The decoder uses this matrix to focus on appropriate tokens in the input sentence to generate the predicted sequence.

4.3 Transformer Decoder

The Transformer Decoder is the second component of our language model. Its goal is to produce a grid cell token sequence from the input sentence encoded by the Transformer Encoder. For that, it uses the auto-regressive method, i.e., it predicts each token in the sequence based on the previous ones produced by the model.

Similar to the encoder, the decoder is also composed of Transformer blocks. The decoder self-attention works, however, in a slightly different way. While the self-attention in the encoder considers all tokens from the trajectory to generate the attention weights, the decoder only considers tokens preceding the current one to predict the next. For that, the Transformers mask future positions using the look-ahead mask approach [16].

In addition, the first decoder multi-head attention layer learns a query matrix Q_{dec} from the previously predicted tokens. First, the decoder receives the output from the previous layer (embeddings). Then, similar to the encoder, the decoder augments it with a positional embedding layer and feeds it to multi-head attention to generate Q_{dec}. After, the query vector and the residual connection feed a normalization layer similar to the decoder. Next, a second multi-head attention layer receives the learned decoder Q_{dec} and the matrix $Z = (K, V)$ from the encoder output to guide the query/search process. This second multi-head layer allows the decoder to focus on which trajectory points from the encoder are relevant to predict the next token/point. After the second multi-head layer generates the encoder-decoder attention vector, the decoder passes it to a feed-forward layer, followed by another normalization to add non-linearities and stability to the values.

Finally, the last layer implements a feed-forward neural network that projects the decoder vectors in a large dimension (vocabulary size) to represent the logit vector[3]. Each logit represents a token/cell score, which the softmax function turns into a probability. The model outputs the highest probability token for each position in our decoding strategy, generating the predicted sequence $\hat{T}$.

4.4 Training

We use the sparse categorical cross-entropy loss to train our model since the labels are integers. The loss is described as follows:

$$L(y, \hat{y}) = -\sum_{j=0}^{M}\sum_{i=0}^{N}(y_{ij} \cdot log(\hat{y}_{ij})) \quad (10)$$

where y_{ij} is the target, and $\hat{y}_{ij}$ represents the prediction. To train the model, we use the Adam optimizer ($\beta_1 = 0.9$, $\beta_2 = 0.9$, and $\epsilon = 1e^{-9}$) with a flexible learning rate that increases at the beginning of training and decreases slowly in the remaining training steps conform [16]. We also apply residual dropout with a rate of 0.1 for each layer in the encoder and decoder.

It is worth mentioning that our approach learns to generate the input trajectory, then input and targets are the same for training. In addition, during training, we use the teacher-forcing, i.e., we pass the true output to each successive step in the decoder. Finally, in the inference step, we provide the input to the encoder and a starting token to the decoder that outputs prediction one token at a time.

4.5 Anomaly Detector

Given the encoder's input token sequence and the decoder's output, as aforementioned, our solution produces two outputs for a given bus trajectory: its anomaly score and the regions where the anomaly occurs in the trajectory. Our primary assumption is that our trained language model predicts the correct sequence. Any token in the input sequence (trajectory) that diverges from the predicted ones is considered an anomaly.

Thus, to calculate the trajectory's anomaly score, the detector compares the sentence predicted by the decoder with the encoder's input sentence by aligning them and computing their Hamming distance [14], as follows: score = 1 - (Hamming(T,$\hat{T}$)/n), where $\hat{T}$ is the decoder's predicted sequence, T is the language model input sequence, and n represents their size.

We consider the anomalous regions in the input sequence T the trajectory points represented by the unmatched tokens between T and $\hat{T}$.

5 Data Description and Setup

5.1 Experimental Setup

In this section, we provide details about the setup of our experimental evaluation.

---
[3] The logit vector dimension is the total number of grid cell tokens (vocabulary).

Datasets. We conducted our experiments in two real-world bus trajectory datasets. The first dataset is from Recife, Brazil. It comprises 19,290 trajectories (100 points on average) from 82 bus lines generated by 238 buses from October 2017 to November 2017. Each bus reports points at intervals of 30 s, containing longitude, latitude, timestamp, route id, vehicle id, instantaneous velocity, and travel distance (from the beginning of the trip). The second dataset is from Dublin[4], Ireland. It contains 60,084 trajectories (206 points on average) and 68 bus lines. Each trajectory point is reported between 20 and 50 s. In total, there 12,497,472 points collected from Jan 01 2013 to Jan 04 2013. Each point contains the attributes: latitude, longitude, timestamp, line id, journey id, and vehicle id.

Pre-processing. As mentioned in Sect. 4.1, we map the trajectories into a geographical grid. Table 1 presents the statistics of trajectories before and after the grid-mapping transformation using the H3 parameter resolution 10 (16 is the maximum resolution), which we chose by experimentation[5]. As can be seen, this transformation greatly reduces the dimensionality of both datasets.

Table 1. Number of unique points before and after the Grip Mapping.

	Recife	Dublin
	Unique Points	Unique Points
Before Mapping	1,929,000	12,497,472
After Mapping	2,416	6,401

Ground Truth. Since there is no label available in our datasets, one can try manually labeling anomalies as [17,18] or generate artificial ones as [10,12,22]. We chose to generate synthetic anomalies, as manual labels are time-consuming, by adding some perturbation in the real trajectories. We do so by randomly choosing the first point in an actual trajectory t and shifting it along with the following n points in t sequentially. In the experiments, we use two parameters to create different anomaly trajectories from real ones: d (the distance in kilometers from the real points) and p (the percentage of shifted points). For example, using $d = 0.5$ and $p = 0.1$, 10% of trajectory points are moved 500 m from the real point. We generate anomalous trajectories for our experiments considering the values of $p = [0.1, 0.2, 03]$ and $d = 1.0$.

Baselines. We evaluate the following anomaly detection methods in our experiments:

- **RioBusData** [2] is a supervised method to detect anomalous bus trajectories by classifying them in bus routes. It uses a Convolutional Neural Network

---
[4] https://data.gov.ie/dataset/dublin-bus-gps-sample-data-from-dublin-city-council-insight-project.
[5] The resolution allows the library to increase or decrease the size of grid cells. Thus, the higher resolution, the smaller the cell areas.

Table 2. Values of hyper-parameters of Transformer.

Num Heads	Embedding	Num Layers	Beta 1	Beta 2	Epsilon	Optimizer	Dropout
8	250	4	0.9	0.98	1e−9	Adam	0.1

(CNN) fed by raw bus trajectories. On the top of CNN, a softmax function outputs a vector of probability where each value is the probability of class membership for each route/label. A trajectory is abnormal if its highest-class probability is below a given threshold.

- **STOD** [6] is also a supervised method that detects anomalous bus trajectories and learns to classify bus trajectories in their routes using a deep-learning network. The model outputs the routes' class distribution of a given trajectory. From this distribution, it calculates the uncertainty of the classifier using entropy as a measure of anomaly degree. The higher the classifier uncertainty, the higher the entropy. A trajectory is anomalous if the entropy of the classifier's probability distribution output for it is higher than a threshold.
- **GM-VSAE** [13] uses an encoder-decoder strategy to detect anomalous trajectories. To perform that, firstly, the encoder infers a disentangled latent space to discover the distribution of each trajectory based on this space. This distribution is then fed to the decoder that generates a trajectory. The method calculates a score comparing the generated trajectory with the input trajectory. A high score ≈ 1.0 means that the input trajectory has a high probability of being an anomaly.
- **iBOAT** [5] is based on the isolation mechanism [11], and an adaptive windows approach. It performs the detection of both trajectory and sub-trajectory anomaly detection. iBOAT calculates the frequency of points mapped into a grid cell to isolate "few and different" points. Based on that, trajectories that visit cells with low frequency get small scores, meaning that those points are highly likely to be an anomaly. Conversely, trajectories with high-frequent visited cells have high scores of non-anomaly. Similar to the previous methods, iBOAT also needs a threshold to detect anomalies.
- **Transformer** is our proposed approach[6]. We train our model on both datasets with the hyper-parameter values shown in Table 2.

It is worth pointing out that all those methods identify anomalous trajectories, but only our approach (Transformer) and iBOAT detect anomalous sub-trajectories. We randomly selected 8,200 trajectories from the Recife dataset and 6,800 from Dublin to evaluate the approaches. Based on a data analysis, we defined the maximum number of points as 100 for the Recife dataset and 208 for Dublin.

Evaluation Metrics. We use F1-measure, Precision, Recall, and PR-AUC as evaluation metrics since they are usually applied to evaluate outlier detection

---

[6] https://github.com/michaeloc/its_research.

methods [1,13]. To verify whether the F1-measure values of our model are statistically different from the baselines, we execute the Wilcoxon statistical test [15]. The test verifies whether two paired samples (F1-measure values of our solution vs. a baseline) come from the same distribution. Given that, we set the significance level $\alpha = 5\%$. In our context, the null hypothesis h_0 considers that the median difference between the F1 values of a pair of models is zero. We performed this statistical test on the instances in the test set.

6 Results and Discussion

In this section, we first present the evaluation of the trajectory outlier identification and, subsequently, the region anomaly detection.

Table 3. Results of anomaly trajectory detection on the Dublin dataset.

	p = 0.1			p = 0.2			p = 0.3		
	F1	Rec	Prec	F1	Rec	Prec	F1	Rec	Prec
STOD	0.636	0.595	0.683	0.696	0.635	0.772	0.724	0.638	0.838
RioBusData	0.651	0.738	0.605	0.660	0.749	0.612	0.673	0.753	0.631
GMVSAE	0.682	0.624	0.758	0.708	0.628	0.827	0.713	0.611	0.876
iBOAT	0.668	0.566	0.833	0.673	0.557	0.871	0.684	0.563	0.889
Transformer	**0.840**	**0.776**	**0.930**	**0.853**	**0.768**	**0.972**	**0.854**	**0.763**	**0.979**

Table 4. Results of anomaly trajectory detection on the Recife dataset.

	p = 0.1			p = 0.2			p = 0.3		
	F1	Rec	Prec	F1	Rec	Prec	F1	Rec	Prec
STOD	0.589	0.552	0.634	0.630	0.576	0.703	0.660	0.590	0.760
RioBusData	0.533	0.549	0.524	0.537	0.551	0.529	0.545	0.555	0.544
GMVSAE	0.654	**0.590**	0.748	**0.673**	**0.594**	0.797	**0.682**	**0.597**	0.819
iBOAT	**0.668**	0.533	0.915	0.668	0.528	0.928	0.673	0.534	0.932
Transformer	0.665	0.520	**0.948**	0.669	0.518	**0.967**	0.670	0.518	**0.970**

6.1 Trajectory Anomaly Detection

Table 3 presents the results for the trajectory anomaly detection task on the Dublin dataset. Transformer outperforms the baselines in all scenarios and metrics. For example, considering the best results on F1, our approach is at least 17%

better than all baselines. To confirm this, Table 6 depicts the p-values for the results of each baseline in comparison to Transformer on the Dublin dataset. All the p-values are smaller than the significance level (0.05), which supports that our Transformer network has, in fact, superior performance than the baselines on this dataset.

Regarding the results on the Recife dataset, presented in Table 4, our method obtains better F1-measure values than STOD and RioBusData, and comparable ones with GMVSAE and iBOAT. The p-values of the hypothesis tests on this dataset for F1-measure, Table 7, confirm this: the p-values of GMVSAE and iBOAT versus Transformer in all scenarios are higher than the significance level of 0.05, meaning there is no statistical difference in terms of F1-measure between our approach and them. Regarding RioBusData and STOD, however, the p-values are lower than 0.05 in two of the three anomaly cases. Looking at precision values, Transformer achieved the best overall results but lower recall than GMVSAE and iBOAT. In practice, better precision can be an advantage since an anomaly detection model can be considered a filter that identifies possibly a few anomalous trajectories in a large set of trajectories. The more precise this filter is, the few false negative anomalies need to be inspected.

Overall, the methods built to detect anomalies (Transformer, GMVSAE and iBOAT) outperformed in almost all scenarios the ones that try to do this indirectly (STOD and RioBusData), i.e., learning to classify routes instead of anomalies. For example, for $p = 0.1$, RioBusData obtained the lowest F1 (0.651) on Dublin and Recife (0.533). However, on the Dublin dataset, STOD shows better F1 results than iBOAT for $p = 0.2$ (0.69 vs 0.673 respectively), and for $p = 0.3$ obtained the F1 second best value (0.724) only behind Transformer (0.854). We also observed that STOD is more sensible than our method over the percentage outlier variation p. For example, for $p = 0.1$, and $p = 0.3$, the F1-measure is 0.58 and 0.66 on Recife (difference of 0.8), respectively. In contrast, our method is more stable regarding the outlier level. For instance, for $p = 0.1$ and $p = 0.3$, the F1-measure is respectively 0.84 and 0.85 on Dublin and 0.66 and 0.67 on Recife, i.e., there is not much difference.

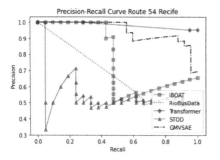 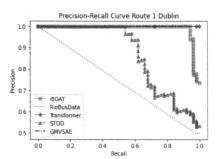

Fig. 2. PR-AUC of the approaches for route 1 on the Dublin dataset and route 54 on the Recife dataset.

Table 5. Results for the region anomaly detection models.

Recife												
	$p = 0.1$				$p = 0.2$				$p = 0.3$			
	F1	Prec	Rec	p-value F1	F1	Prec	Rec	p-value F1	F1	Prec	Rec	p-value F1
iBOAT	0.91	0.84	0.99	**7.7e−10**	0.90	0.84	0.98	**2.1e−8**	0.90	0.83	0.97	**1.9e−7**
Transformer	0.98	0.98	0.99		0.98	0.97	0.99		0.97	0.96	0.99	
Dublin												
iBOAT	0.97	0.96	0.98	**0.12**	0.97	0.96	0.97	**0.65**	0.96	0.96	0.96	**0.10**
Transformer	0.98	0.97	0.99		0.97	0.95	0.99		0.96	0.93	0.99	

To provide a detailed analysis of the approaches on individual routes, Fig. 2 shows the PR-AUC curves of all methods on both datasets in two different routes, one from each dataset with $p = 0.3$. In route 54 from Recife, our approach has the highest area under curve ≈ 0.98, outperforming both the unsupervised methods (GMVSAE ≈ 0.94 and iBOAT ≈ 0.77) and the supervised ones (STOD ≈ 0.54 and RioBusData ≈ 0.67). Looking at route 1 from Dublin, we observe that the encoder-decoder methods have almost the perfect curve AUC ≈ 0.99, i.e., the models can adequately distinguish anomalous trajectories from non-anomalous ones. Conversely, the RioBusData has the worst PR-AUC curve ≈ 0.70. Finally, we can see that the precision of iBOAT degrades with a recall close to 1.

Table 6. Hypothesis test for F1 on the Dublin dataset.

Transformer			
	$p = 0.1$	$p = 0.2$	$p = 0.3$
	p-value	p-value	p-value
STOD	4.05e−12	1.77e−12	1.09e−12
RioBusData	1.48e−12	1.04e−12	8.73e−13
GMVSAE	4.472e−12	4.05e−12	3.26e−12
iBOAT	1.22e−12	7.98e−13	1.13e−12

6.2 Region Anomaly Detection

Table 5 shows the results between Transformer and iBOAT on region anomaly detection. We observe that Transformer outperforms iBOAT on the Recife dataset in all scenarios. This occurs mainly because our model achieved the high values of precision $(0.988, 0.975, 0.961)$. The methods, however, are similar regarding recall for $p = 0.1$, for instance, Transformer's recall is 0.992, and iBoat 0.990. On the Dublin dataset, the methods are qualitatively similar. Note that

Table 7. Hypothesis test for F1 on the Recife dataset.

Transformer			
	$p = 0.1$	$p = 0.2$	$p = 0.3$
	p-value	p-value	p-value
STOD	0.38	2.11e–4	4.26e–9
RioBusData	1.13e–11	5.61e–13	8.12e–14
GMVSAE	0.21	0.68	0.46
iBOAT	0.73	0.80	0.73

the most difference between the models occurs for $p = 0.1$: Transformer's F1 is 0.986, and iBOAT's F1 is 0.978.

We applied the Wilcoxon test to verify whether there is a statistical difference between the F1-measure values of the methods on this task. On the Recife dataset, the models are statistically different, with p-value lower than our significance level of 0.05 for all scenarios. On the Dublin dataset, however, there is no statistical evidence to reject the h_0, since the p-values are higher than 0.05 and, therefore, both models are statistically equivalent in terms of F-1 measure.

To present concrete examples of the detection of region anomalies on real trajectories by our model, Fig. 3a shows the expected trajectories of Dublin route 1, and Fig. 3b depicts the anomalous regions identified (represented by the red dots) by Transformer. One can see from these plots that our approach identifies the anomalous regions in this trajectory very precisely.

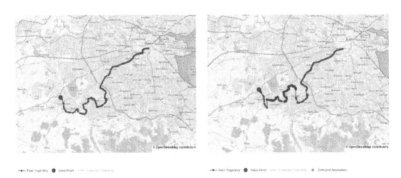

(a) Expected trajectory (real trajectory). (b) Anomalous regions (the red dots) detected by our model.

Fig. 3. Example of anomaly detection inference.

7 Conclusion

In this paper, we propose a solution that applies an encoder-decoder transformer language model in bus trajectory data to solve two problems: trajectory and subtrajectory anomaly detection. Our solution transforms a trajectory into a discrete token sequence by mapping its points to tokens representing geographical grid cells. This sequence is then passed to the Transformer language model that outputs a predicted sequence, supposedly without anomalies. Finally, our solution calculates the trajectory's anomaly score by applying the hamming distance between the two sequences and identifies the anomalous regions by looking at the unmatched tokens between them. Experiments in two real-world bus trajectory datasets demonstrate that our approach is effective for anomalous trajectory detection and anomalous region detection tasks.

In future work, we intend to train our approach in multiple trajectory datasets to verify whether it can learn general trajectory patterns (deep representation). Once our approach learns those patterns, we want to exploit other tasks, such as trajectory similarity and classification, using transfer learning.

Acknowledgment. This work is partially supported by INES (www.ines.org.br), CNPq grant 465614/2014-0, FACEPE grants APQ-0399-1.03/17 and APQ/0388-1.03/14, CAPES grant 88887.136410/2017-00.

References

1. Belhadi, A., Djenouri, Y., Lin, J.C.W., Cano, A.: Trajectory outlier detection: algorithms, taxonomies, evaluation, and open challenges. ACM Trans. Manage. Inf. Syst. (TMIS) **11**(3), 1–29 (2020)
2. Bessa, A., Silva, F.D.M., Nogueira, R.F., Bertini, E., Freire, J.: RioBusData: outlier detection in bus routes of Rio de Janeiro. arXiv preprint arXiv:1601.06128 (2016)
3. Bouritsas, G., Daveas, S., Danelakis, A., Thomopoulos, S.C.: Automated real-time anomaly detection in human trajectories using sequence to sequence networks. In: 2019 16th IEEE International Conference on Advanced Video and Signal Based Surveillance (AVSS), pp. 1–8. IEEE (2019)
4. Chalapathy, R., Chawla, S.: Deep learning for anomaly detection: a survey. arXiv preprint arXiv:1901.03407 (2019)
5. Chen, C., et al.: iBOAT: isolation-based online anomalous trajectory detection. IEEE Trans. Intell. Transp. Syst. **14**(2), 806–818 (2013)
6. Cruz, M., Barbosa, L.: Learning GPS point representations to detect anomalous bus trajectories. IEEE Access **8**, 229006–229017 (2020)
7. Huang, G., Liu, Z., Van Der Maaten, L., Weinberger, K.Q.: Densely connected convolutional networks. In: Proceedings of the IEEE Conference on Computer Vision and Pattern Recognition, pp. 4700–4708 (2017)
8. Kong, X., Song, X., Xia, F., Guo, H., Wang, J., Tolba, A.: LoTAD: long-term traffic anomaly detection based on crowdsourced bus trajectory data. World Wide Web **21**(3), 825–847 (2018)
9. Lee, J.G., Han, J., Li, X.: Trajectory outlier detection: a partition-and-detect framework. In: 2008 IEEE 24th International Conference on Data Engineering, pp. 140–149. IEEE (2008)

10. Li, X., Zhao, K., Cong, G., Jensen, C.S., Wei, W.: Deep representation learning for trajectory similarity computation. In: 2018 IEEE 34th International Conference on Data Engineering (ICDE), pp. 617–628. IEEE (2018)
11. Liu, F.T., Ting, K.M., Zhou, Z.H.: Isolation-based anomaly detection. ACM Trans. Knowl. Discov. Data (TKDD) **6**(1), 1–39 (2012)
12. Liu, S., Ni, L.M., Krishnan, R.: Fraud detection from taxis' driving behaviors. IEEE Trans. Veh. Technol. **63**(1), 464–472 (2013)
13. Liu, Y., Zhao, K., Cong, G., Bao, Z.: Online anomalous trajectory detection with deep generative sequence modeling. In: 2020 IEEE 36th International Conference on Data Engineering (ICDE), pp. 949–960. IEEE (2020)
14. Luu, V.T., Forestier, G., Weber, J., Bourgeois, P., Djelil, F., Muller, P.A.: A review of alignment based similarity measures for web usage mining. Artif. Intell. Rev. **53**(3), 1529–1551 (2020)
15. Siegel, S.: Nonparametric statistics for the behavioral sciences (1956)
16. Vaswani, A., et al.: Attention is all you need. In: Advances in Neural Information Processing Systems, pp. 5998–6008 (2017)
17. Wang, J., et al.: Anomalous trajectory detection and classification based on difference and intersection set distance. IEEE Trans. Veh. Technol. **69**(3), 2487–2500 (2020)
18. Zhang, D., Li, N., Zhou, Z.H., Chen, C., Sun, L., Li, S.: iBAT: detecting anomalous taxi trajectories from GPS traces. In: Proceedings of the 13th International Conference on Ubiquitous Computing, pp. 99–108 (2011)
19. Zhang, D., Chang, Z., Wu, S., Yuan, Y., Tan, K.L., Chen, G.: Continuous trajectory similarity search for online outlier detection. IEEE Trans. Knowl. Data Eng. (2020)
20. Zhang, Y., Ning, N., Zhou, P., Wu, B.: UT-ATD: universal transformer for anomalous trajectory detection by embedding trajectory information. In: Proceedings of the 27th International Conference on Distributed Multimedia Systems (2021)
21. Zhao, X., Rao, Y., Cai, J., Ma, W.: Abnormal trajectory detection based on a sparse subgraph. IEEE Access **8**, 29987–30000 (2020)
22. Zheng, G., Brantley, S.L., Lauvaux, T., Li, Z.: Contextual spatial outlier detection with metric learning. In: Proceedings of the 23rd ACM SIGKDD International Conference on Knowledge Discovery and Data Mining, pp. 2161–2170 (2017)
23. Zheng, Y.: Trajectory data mining: an overview. ACM Trans. Intell. Syst. Technol. (TIST) **6**(3), 1–41 (2015)

Aroeira: A Curated Corpus for the Portuguese Language with a Large Number of Tokens

Thiago Lira, Flávio Cação, Cinthia Souza, João Valentini, Edson Bollis, Otavio Oliveira, Renato Almeida, Marcio Magalhães, Katia Poloni, Andre Oliveira, and Lucas Pellicer[✉]

Instituto de Ciência e Tecnologia Itaú-Unibanco (ICTi), São Paulo, Brazil
{cinthia.mikaela-souza,joao.valentini22,edson.bollis,
otavio.rodrigues-oliveira,renato-augusto.almeida,marcio.chiara-magalhaes,
katia.poloni,andre.seidel-oliveira,lucas.pellicer}@itau-unibanco.com.br

Abstract. The emphasis on constructing extensive datasets for training large language models (LLM) has recently increased, and current literature predominantly features datasets for high-resource languages such as English and Chinese. However, there is a notable scarcity of high-quality corpora for the Portuguese language. To address this limitation, we propose Aroeira, a curated corpus explicitly designed for training large language models in the Portuguese language, with a focus on the Brazilian Portuguese one. The Aroeira Corpus consists of 100 GB of texts from various internet platforms, processed through a comprehensive pipeline to ensure superior quality. The pipeline handles downloading, text extraction, language identification, application of quality and bias filters, and storage, all tailored for the Portuguese language. The resulting corpus contains 35.3 million documents and over 15.1 billion tokens, surpassing the largest previously available corpus in this domain.

1 Introduction

Most modern natural language processing (NLP) methodologies, including Large Language Models (LLMs), rely on extensive text corpora for precise training and weight adaptation [18]. Large-scale training corpora, or pre-training corpora, are fundamental for developing foundational models, which serve as the basis for numerous task-specific adaptations [7]. State-of-the-art LLM training pipelines utilize various types of datasets: (i) pre-training corpora to acquire language structure, syntax, and semantics; (ii) instruction fine-tuning datasets to enhance the model's capability to follow instructions; (iii) preference datasets to rank responses; and (iv) evaluation datasets to measure model performance [23].

Recent research shows that the size and diversity of pre-training corpora significantly impact LLM performance [14,18]. Most pre-training datasets are available in English and Chinese, which are high-resource languages, while other languages have significantly fewer tokens [23]. Although multilingual corpora can help mitigate data scarcity for low-resource languages, these datasets are often unbalanced, favoring high-resource languages [17]. This imbalance affects the performance of multilingual models for less-represented languages and models trained on multilingual corpora do not perform as well as those trained on monolingual corpora [34]. Therefore, it is essential to train or fine-tune models in the target languages to capture linguistic nuances, structures, and domain-specific or cultural knowledge [28].

A direct implication of this scenario is the necessity of making high-quality plain-text corpora available to encourage research on specific model languages and the development of better-performing approaches. Therefore, we introduce Aroeira: a curated Portuguese language-specific corpus composed of approximately 100 GB of text. The content was extracted from recent Common Crawl[1] (CC) web pages (up until 2023) and fully curated to remove web tags, ensuring quality and bias filtering. To the best of our knowledge, Aroeira is the largest highly-curated Portuguese corpus available to date. It has the potential to influence new instruction fine-tuning and evaluation dataset studies while guiding the development of preference datasets and large models.

Aroeira was created based on a double-pipeline inspired by [30]. The pipeline comprises two key steps: data quality management and content safety assurance. These steps ensure the size and quality necessary for a Portuguese corpus to train safe LLMs effectively. As part of the content safety step, we investigated techniques for filtering hazardous content and mitigating biases in our corpora [16]. This effort resulted in a custom Portuguese word dictionary, which encompasses offensive words, as well as terms, expressions, and phrases that include sexism, homophobia, ableism, racism, hate speech, and political, religious, and regional prejudice [13,24,26].

We highlight our main contributions:

- Introduction of Aroeira, a 100 GB Portuguese corpus from diverse internet sources. Our dataset surpasses the largest currently available corpus for training language models in Portuguese in terms of size, quality, and representativeness.
- Development of a parameterizable double-pipeline, which includes: downloading, extracting, language identification, quality filtering, and text storage in the data step; filtering sexual content, toxic data, and bias in the content safety step.
- Creation of a dictionary to filter biased terms and mitigate social bias in the Portuguese language.

This paper is organized as follows. In Sect. 2, we present related work in corpus extraction. Sections 3 and 4, we describe the methodology for generating the corpus and the configuration of hyperparameters used in the quality filters, respectively. In Sect. 5, we analyze the volumetry of Aroeira in terms of year distribution, knowledge domains, document length, and other relevant results. Finally, Sect. 6 presents conclusions and future works.

---

[1] Available at: https://commoncrawl.org/.

2 Related Work

The largest Portuguese language corpus is BrWac [35] which has approximately 25 GB of textual data distributed in 3.53 Mi documents totaling 2.68 Bi tokens. Another large corpus is the Carolina 1.2 Ada [11], which contains approximately 2.11 Mi documents and a total of 11 GB of textual data. When we compare this corpus with the corpora of other languages, the gaps become evident. Gao et al. [14], for example, propose The Pile, a corpus with 825 GB of texts in English. The corpus is derived from various data sources, including scientific articles, patent documents, and forums.

An inspiring work for Aroeira is Colossal Clean Crawled Corpus (C4) [30], a curated English-only corpus. C4 was created using Common Crawl (CC) data extracted in April 2019 and comprises approximately 750 GB of clean English text. Similar to our approach, they apply filters to the raw data. CLUECorpus2020 [36] was constructed using cleaned data from CC, resulting in a high-quality Chinese pre-training corpus of 100 GB and 36 Bi tokens. MassiveText [29] is a collection of large English datasets created with data from different sources. MassiveText contains 2.35 Bi documents, equivalent to 10.5 TB of text. More recently, Sabiá [28] applied a similar filtering methodology of MassiveText to the Portuguese section of ClueWeb dataset [27] and managed to retrieve a curated dataset. WuDaoCorpora [38] is a 3 TB Chinese corpus with 1.08 Tri of Hanzi characters collected from 822 Mi web pages.

It is also worth mentioning that a current trend is the proposal of multilingual corpora. The C4Corpus authors [17] present the construction of a 12 Mi web page corpus containing more than 50 languages, including Portuguese. English has a volume of 7.7 Mi (64.2%) documents while Portuguese has only 0.3 Mi (2.5%). RedPajama [10] is a large multilingual corpus, containing 100 Bi text documents extracted from 84 CC snapshots. Quality signals were applied to 30 billion documents, and deduplication was performed on 20 billion documents. It claims to have English (69.8%), Deutch (9.2%), Spanish (8.8%), French (7.8%), and Italian (4.4%).

As we can see, the corpora available in English and Chinese have a massive data amount, easily surpassing corpora in Portuguese and other languages. However, we know that the amount of internet information available in English and Chinese is greater than in Portuguese.

Another important aspect is the biases present in corpora and texts. Language is a highly relevant avenue for manifesting social hierarchies, pre-established concepts, and standard forms of treatment [6]. Various efforts are being made to evaluate data biases and how they impact the behavior of language models. The paper [24] analyzed 93 social groups that receive stigmatized treatment by NLP models. Work [25] created StereoSet to measure stereotypical treatment in certain ethnic groups, and paper [26] developed a benchmark dataset for measuring biases related to gender, race, age, sexual orientation, and others.

These aspects are relevant in a context with strong normative motivations and the need to create responsible AI. Many ways exist to mitigate text biases, such as data augmentation, content filtering, rebalancing, masking, and many others [13]. Our work uses the concept studied by [16], where filtering sensitive content can result in models with more equitable treatment of different ethnic groups. The work specifically uses word co-occurrence in the filtering process.

Based on these past works, we can see that, in general, they focus on creating corpora for training language models for high-resource language tasks. Thus, there is a necessity for creating a Portuguese corpus since the amount of large Brazilian Portuguese models has drastically increased recently. We can cite Bertimbau [33], PTT5 [9], Bertaú [12], Sabiá [4,28], Cabrita [22], and Bode [15].

3 Aroeira

In this section, we detail the steps of the corpus creation (double-pipeline) which is divided into two objectives, (i) collect (Data Pipeline) and (ii) ensure content safety (Content Safety Pipeline). Our whole pipeline contains nine steps: data collection and sampling, text extraction, language identification, deduplication, and quality filters in Data Pipeline, and sexual content filter, toxic data filter, bias filter, and categorization in Content Safety Pipeline. Figure 1 presents the entire workflow.

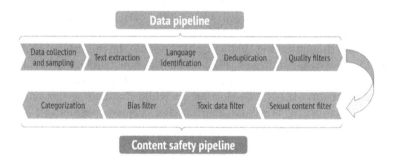

Fig. 1. Double pipeline: Data Pipeline contains collection and sampling, text extraction, language identification, deduplication, and quality filters; and content safety pipeline encompasses sexual content filter, toxic data filter, bias filter, and categorization.

3.1 Data Collection and Sampling

The data collection and sampling step involves downloading and extracting Portuguese text from raw Web ARChive (WARC) in files. All data is sourced from Common Crawl (CC), which contains petabytes of scraped internet content from millions of web pages. We use the raw HTTP files as the initial material, from which we extract and filter Portuguese text as detailed in Subsects. 3.2 and 3.3. CC organizes its datasets by date, each comprising thousands of individual shards of scraped content. We sampled shards from datasets ranging from 2015 to 2023, prioritizing more recent data.

3.2 Text Extraction

This computational step involves using multiple cloud machines in parallel. These instances download and process raw files by extracting text from the HTML and filtering for Portuguese text. The resulting data is then processed further using a single machine containing a key-value database for deduplication, as explained in Subsect. 3.4. We opted to work with WARC files to ensure better quality text, which includes handling raw HTML files and extracting texts ourselves. A Python library called Trafilatura [5] extracted only natural language text from the HTML files. Metadata from webpages were saved for later use in the pipeline.

3.3 Language Identification

Roughly 0.2% of the pages in each shard are in Portuguese. To filter these pages, it is necessary to detect the language they were written in automatically. As utilized by [14] and [1], we employ Meta AI's pre-trained fastText model, which can detect 176 languages. For each downloaded page, the text is extracted using the Trafilatura library [5] and the fastText[2] [20] models used to determine the language. Pages identified as Portuguese with the highest probability by fastText were selected.

3.4 Deduplication

The purpose of this step is to remove duplicated data from the corpus. To achieve this goal, we use two deduplication approaches. The first is a page-level approach, which identifies and removes pages with duplicate URLs. The second is the document-level approach, which aims to remove significant overlapping documents. We employ the MinHashLSH algorithm to calculate the Jaccard similarity between documents, considering whether two document similarity exceeds 0.7 [29].

3.5 Quality Filters

A significant amount of data available on the internet may be insufficient in terms of quality for linguistic model formation. Some examples include automatically generated text and text not written for human consumption [29]. This step aims to retain only pages written by humans for humans. To achieve this, we applied a series of ten quality filters:

- **Number of tokens**: Removes pages with fewer than a minimum number of tokens (in this work, we used the same tokenizer employed by GPT-2), as texts with low token counts are generally not informative;
- **Number of words**: Removes pages that do not attend specified upper and lower word limits, excluding punctuation and special characters;

---

[2] Available at: https://fasttext.cc/.

- **Type Token Ratio (TTR)**: The ratio of unique words (types) to total words (tokens) [31]. TTR [37] serves as an indicator of text quality;
- **Symbols-word ratio**: Removes pages whose symbol word percentages exceed limits. Any special character is considered a symbol;
- **Symbols at the beginning of the text**: Removes pages with an excessive number of symbols at the beginning of the text;
- **Stopwords**: The presence of stopwords may indicate text coherence [29];
- **N-gram repetition**: Excessive repetition of sentences, paragraphs, or n-grams indicates low informational content [29];
- **Number of sentences**: Removes pages with fewer than a specified number of sentences;
- **Lorem ipsum**: Removes pages containing the term "Lorem ipsum" [30];
- **Valid words**: Removes pages whose percentage of words found in a language dictionary is below a specified threshold.

The thresholds for each filter are detailed in Sect. 4.

3.6 Sexual Content Filter

To maintain the integrity of the corpus, a filter was applied to remove sexual content from the data. We verified whether a URL was present in the Université Toulouse 1[3] (UT1) blocklist for each page collected. As noted by [2], the UT1 blocklist is an extensive compilation of block lists frequently used for internet access control at schools. It was developed with the help of automated systems and human contributors and currently includes 3.7 million entries. For this work, we utilized a filtered version of this blocklist tailored for Brazilian websites. It should be noted that this filter only excludes websites marked as adult content. For the remaining content, we randomly selected 25,000 examples and used a Mistral 7B [19] model to extract pejorative sexual terms. These terms were then reviewed by humans and used as the final sexual content filter.

3.7 Toxic Data Filter

In this step, our objective is to identify and remove potential toxic content. Toxicity definition is a rude, disrespectful, or unreasonable comment likely to incite an argument [13]. Our filter comprises a dictionary of insults and pejorative terms. We evaluated exact matches of dictionary words with document terms and removed documents that exceeded a specified threshold percentage of words in the dictionary. The dictionary used for this filter was created by merging two lists of words[4,5]. We reinforce this filter does not aim to eliminate all the data containing toxic words, but rather to remove content with a significant toxic content proportion.

---

[3] https://dsi.ut-capitole.fr/blacklists/index_en.php.
[4] https://github.com/LDNOOBW/List-of-Dirty-Naughty-Obscene-and-Otherwise-Bad-Words/blob/master/pt.
[5] https://github.com/dunossauro/chat-detox/blob/main/palavras.txt.

3.8 Bias Filter

Our work involves a step to identify and eliminate potential biases in the text based on contextual cues. We construct a dictionary of Portuguese expressions used in biased contexts.

When compiling this dictionary, it is crucial to consider the society dynamic and its relationship with linguistics [6]. We filtered the corpus by checking for exact matches between the dictionary words and those in the text [16]. Various types of social biases were mapped, including gender, religion, race, sexist expressions, xenophobia, homophobia, ableism, fatphobia, and politics [24,26].

3.9 Categorization

This phase aims to categorize each page into a specific knowledge domain. We identified 27 knowledge domains, covering several categories and subjects, such as: blog posts, news articles, marketing, movies, social media, health, culinary recipes, books, scientific articles, politics, etc. The information regarding each domain can be leveraged to balance the data for specific tasks or augment datasets where knowledge is lacking. Essentially, we map the URLs of different pages to each knowledge scope and assign a topic to each URL. A pt-pt text category has been introduced to differentiate between Brazilian Portuguese and European Portuguese, as Brazilian Portuguese predominates in the dataset.

The identified knowledge domains and their distribution are discussed in the Subsect. 5.2.

4 Qualitative Configuration Test

We generated a 1 GB sample of texts and analyzed the distribution of metrics such as the number of tokens, word count, and TTR. We use this sample to find the best configuration for our double-pipeline to produce the resultant datasets. Different value sets were empirically tested, and for each one, we checked the correctness of page removals and recorded the number of pages excluded after the filters were applied.

Analysis was carried out on this sample to find the optimal configuration. Moreover, we perform qualitative analyses to verify the removal appropriateness. This qualitative test helped us define satisfactory values that we should filter to obtain content with good textual quality, i.e., a text that is diverse in words, fluid, with few repetitions, and with semantically relevant content. It is worth noting that we also evaluated the number of potentially toxic words and possible biases contained in the texts.

Table 1 shows the optimal configuration to obtain texts that exceed our minimum quality requirements.

Table 1. Double-pipeline final configuration.

Parameter	Description	Value
min_tokens	Minimum number of tokens	30
min_words	Minimum number of words	20
max_words	Maximum number of words	10000
TTR	Type Token Ratio	0.2
max_symbols	Maximum percentage of symbols-words	0.70
fs_symbols	Maximum number of symbols at the beginning of the sentence	6
min_stopwords	Minimum percentage of stopwords	0.02
occurrence_ngram	3-gram repeat percentage	0.3
num_sentences	Minimum number of sentences	2
valid_words	Minimum percentage of valid words	0.2
toxic_content	Maximum percentage of toxic content	0.2
max_word_bias	Maximum number of biased words	10

5 Results

The created corpus was evaluated concerning five groups of requirements. The first requirement is the created corpus must be larger than the existing corpus for the Portuguese language (see Subsect. 5.1). The second requirement is the corpus must be diverse, i.e., containing data from different sources (see Subsect. 5.2). The third requirement is the corpus covers the most recent to the least recent information (see Subsect. 5.3). The fourth requirement is that the corpus presents high-quality text indicators (see Subsect. 5.4). Finally, the fifth requirement is that the corpus avoids introducing or increasing bias.

Due to the large corpus size, Subsects. 5.4 and 5.5 utilize a 10% randomly generated sample to present the results. Consequently, Fig. 4 and Table 3 were created based on this sample size.

5.1 Corpus Size

The first requirement evaluated was the corpus size. We collected terabytes of data from different CC dumps. Each dump has approximately 0.2% of texts in Portuguese. It is worth noting that the documents may be of poor quality, contain inappropriate or biased content, and be duplicated due to the CC not filtering the data. Therefore, a corpus cleaning step was necessary to ensure that the final corpus was composed only of non-duplicated documents to respect quality criteria. At the end of this process, we obtained a corpus of 100 GB.

Table 2 presents the created corpus statistics alongside other Portuguese corpora. Aroeira surpasses brWac [35] and Carolina 1.2 Ada [11] in size, document quantity, and token number. Thus, our corpus is potentially a more diverse resource regarding texts and tokens than the available resources.

Table 2. Corpora size comparison.

Language	Corpus	Size	#Documents	#Tokens
Portuguese	Aroeira	**100 GB**	**35.3 Mi**	**15.1 Bi**
	brWac	25 GB	3.53 Mi	2.68 Bi
	Carolina 1.2 Ada	11 GB	2.11 Mi	0.82 Bi
English	MassiveText	10.5 TB	2.35 Bi	2.3 Tri*
	The Pile	825 GB	–	–
	C4	750 GB	-	–
Chinese	WuDaoCorpora	3 TB	822 Bi	–
	CLUEcorpus2020	100 GB	2.35 Bi	36 Bi
Multilingual	RedPajama	260 TB*	100 Bi	30.4 Tri
	C4Corpus	29 GB^c	12 Mi	10.8 Bi

Note: Bold letters are best values, "*" point calculated or no paper, and "^{c}" show compressed values.

5.2 Knowledge Domains

The second requirement evaluated was the distribution of knowledge domains within the created corpus. A mapping of different URLs to their respective knowledge domains was conducted. Each base URL was verified against a dictionary. When there were no matches, keywords were used to determine the document's domain (Subsect. 3.9). Figure 2 illustrates the document distribution across these domains.

The complexity of the corpus strongly correlates with downstream data performance [3]. Therefore, an extensive representation of knowledge domains can contribute to the generation of more robust models, potentially improving incontext few-shot learning performance [32].

Most documents could not be assigned to a specific domain and are marked as NR (Not Recognized). Among those that were categorized, blog posts and news articles were the most frequent, although other categories such as institutional texts, e-commerce, and internet forums were also found. Knowledge domains are essential for evaluating the quality of the data in the corpus and for filtering data used in the pre-training phase of domain-specific language models.

5.3 Distribution of Documents over Time

Our third analysis is the distribution of the corpus documents over time. This temporal analysis is important to identify possible temporal biases such as outdated texts. Our corpus presents a recent data distribution, which indicates more up-to-date texts.

Figure 3 illustrates that our data set comprises documents spanning up to 7 years, beginning in 2017. The bulk of the data is from 2017 to 2019, but a notable portion of recent data is from 2021, 2022, and 2023. This distribution

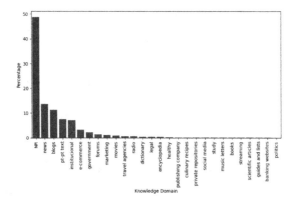

Fig. 2. Distribution of knowledge domains.

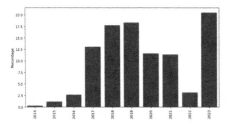

Fig. 3. Distribution of documents over time.

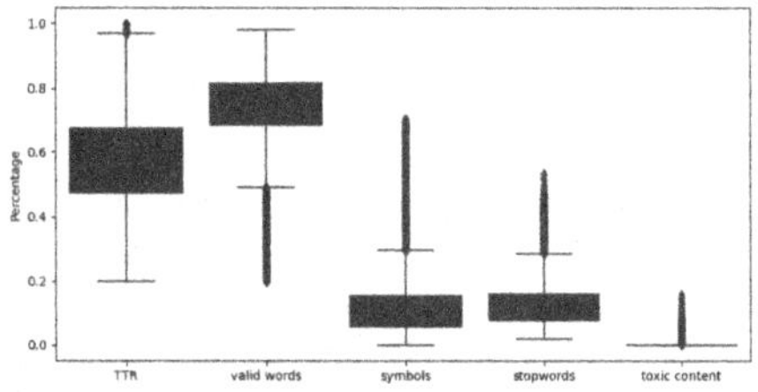

Fig. 4. Quality indicators. Higher TTR and percentage of valid words indicate better corpus quality, while lower values for other indicators also signify better quality.

meets the need for both recent and extensive data. As a result, models can train on up-to-date information and include recent and common terms used in Portuguese.

5.4 Quality Indicators

We used TTR value, symbol word percentage, stopword percentage, valid words, and toxic content as quality indicators. Figure 4 shows the results obtained.

We have two metrics that indicate the diversity of the content present in our corpus. The TTR indicates the variability of tokens in a sentence, and we aim for this value to be as high as possible, as it is a sign of texts composed of varied tokens with low word repetition. A TTR threshold of 0.5 is a quality parameter for the text. We obtained a distribution curve with the first quartile close to 0.5 TTR, a median of 0.57, and the third quartile above 0.65. Thus, most of the data in our corpus reaches a satisfactory TTR value, which is a strong indicator of non-repetitive texts.

Another indicator of variability is the percentage of valid tokens or keywords. This indicator measures the frequency of contextually relevant words within the text, and we also want higher values for more diverse and fluid texts. We achieved an excellent distribution in this indicator, with most data above 0.7. This distribution value is another strong sign of lexically diverse texts.

In contrast, the metrics for the percentage of symbols and the percentage of stopwords are indicators of less fluid texts, with many symbols interrupting the text or commonly used words that do not add semantic value to the documents (e.g., "to", "for", or "the"). Our goal is to minimize these metrics as much as possible. We achieved our goal of reducing these values, obtaining distributions with low values, with the third quartile below 0.2 in both metrics. This result is a strong indication that the texts in the corpus are fluid.

Finally, we aim to minimize the percentage of potentially toxic words, decreasing to close to 0 (no potentially toxic words). The results show that we achieved this goal in almost all the texts in the corpus, except for some outliers that do not exceed 0.2% of toxic content.

5.5 Bias

We performed a word co-occurrence analysis to identify biases in our corpus. This technique has proven effective in demonstrating stereotypical treatment of a particular social group [13], which we do not desire. It is worth noting that this method is one of several mitigation approaches, and the corpus may still exhibit bias. We analyzed three groups of biases as shown in Table 3, they are: (i) Gender, (ii) Religion, and (iii) Race. We selected different words for each group and analyzed the context in which these words were inserted.

We have chosen representative terms indicative of social groups and conducted an analysis focusing on those with the highest co-occurrence frequencies. We isolate the terms "Man" and "Woman" to evaluate gender bias. We selected "Atheist", "Christian", "Buddhist", "Evangelical", "Jewish", "Muslim" and "Umbandist" for religious bias. Finally, we highlighted the words "White", "Black", "Asian" and "Hispanic" for racial bias. The Tables 3 respectively represent the results for gender, religious, and racial bias.

No stereotypical treatment or hazardous behaviors described in the literature exist in the analyzed groups, such as associations with crime, income, and others [16]. Furthermore, the words selected among the various social groups are very similar, which suggests a more equanimous treatment in our proposed corpus.

Table 3. Related word co-occurrence.

Topic	Word	1	2	3	4	5	6	7	8	9	10
Gender	Woman (Mulher)	man (homem)	day (dia)	year (ano)	life (vida)	son (filho)	women (mulheres)	mother (mãe)	house (casa)	against (contra)	husband (marido)
	Man (Homem)	woman (mulher)	gave (deu)	year (ano)	life (vida)	world (mundo)	son (filho)	spider (aranha)	well (bem)	day (dia)	because (porque)
Religion	Atheist (Ateu)	gave (deu)	Christian (cristão)	faith (fé)	all (todo)	religion (religião)	person (pessoa)	life (vida)	because (porque)	state (estado)	religious (religioso)
	Christian (Cristão)	gave (deu)	life (vida)	all (todo)	world (mundo)	Christ (cristo)	church (igreja)	must (deve)	Jesus (Jesus)	can (pode)	love (amor)
	Buddhist (Budista)	temple (templo)	monk (monge)	meditation (meditação)	zen (zen)	practice (prática)	Buddhism (budismo)	year (ano)	tradition (tradição)	all (todo)	about (sobre)
	Evangelical (Evangélico)	pastor (pastor)	church (igreja)	means (meio)	day (dia)	Christian (cristão)	gave the (deu o)	year (ano)	hospital (hospital)	Catholic (católico)	Brazil (brasil)
	Jewish (Judeu)	people (povo)	state (estado)	Jesus (jesus)	all (todo)	gave (deu)	Israel (israel)	day (dia)	history (história)	Christian (cristão)	other (outro)
	Muslim (Muçulmano)	world (mundo)	Christian (cristão)	all (todo)	country (país)	Arab (árabe)	Jewish (judeu)	about (sobre)	state (estado)	can (pode)	other (outro)
	Umbanda (Umbandista)	Umbanda (umbanda)	religion (religião)	great (grande)	day (dia)	good (bom)	all (todo)	catholic (católico)	about (sobre)	true (verdadeiro)	end (fim)
Race	White (Branco)	black (preto)	river (rio)	castle (castelo)	color (cor)	blue (azul)	core (core)	wine (vinho)	red (vermelho)	green (verde)	first (primeiro)
	Black (Negro)	red (rubro)	river (rio)	hole (buraco)	white (branco)	movement (movimento)	side (lado)	about (sobre)	year (ano)	Brazil (Brasil)	humor (humor)
	Hispanic (Hispânico)	world (mundo)	Spanish (espanhol)	qualified (qualificado)	black (negro)	work (trabalho)	about (sobre)	great (grande)	soccer (futebol)	public (público)	Mexico (México)
	Asian (Asiático)	southeast (sudeste)	country (país)	market (mercado)	continent (continente)	countries (países)	China (China)	south (sul)	east (leste)	year (ano)	Africa (áfrica)

6 Conclusion

Recent studies have shown significant improvements in the performance of language models trained on large corpora [8,14]. Consequently, the interest in creating large datasets has grown. Most existing research focuses on high-resource languages like English and Chinese, with considerable efforts made to develop multilingual corpora. However, there is a pressing need to develop large datasets for lower-resource languages.

This work aims to address this gap by developing language models for lower-resource languages, specifically Portuguese. We created the largest curated corpora for training or pre-training Portuguese language models. To achieve this, we implemented a double-pipeline process to extract data while ensuring content safety. The process includes downloading, text extraction, language identification, deduplication, quality filtering, filtering for sexual content and toxicity, bias filtering, categorization, and storage. This effort involved collecting terabytes of data, resulting in a curated dataset of approximately 100 GB for Aroeira's construction.

Our results demonstrate that our corpus fulfills the requirements of corpus size, knowledge domains, document distribution over time, quality indicators, and bias mitigation. We conducted statistical analyses of the corpus to better comprehend the size of the collected documents in terms of the number of tokens, words, and sentences. Additionally, we analyzed bias to recognize potential harms in the created corpus, a distinguishing factor in building models free from social biases. Our findings conclude that the corpus created is of high quality and diversity, with minimal bias.

We are eager to advance our work by training models with encoder architectures, such as BERT models [21]. Furthermore, we plan to pre-train language models with Aroeira to obtain higher-quality models in Portuguese. We intend to research existent instruction and evaluation datasets. Finally, conducting comparative bias analyses on models trained with Aroeira relative to other available corpora [13] will be highly valuable.

Acknowledgements. We thank Instituto de Ciência e Tecnologia Itaú-Unibanco (ICTi) and Itaú-Unibanco SA for the technical support, resources, and financial aid in the development of the Aroeira corpus. It's also noteworthy the fact that ChatGPT (OpenAI) was employed in the writing process, contributing to thorough grammatical and semantic reviews.

Data Availability. Our Aroeira corpus is available for download in the Hugging Face repository: https://huggingface.co/datasets/Itau-Unibanco/aroeira and is under the CC-BY-NC 4.0 license.

References

1. Abadji, J., Ortiz Suárez, P.J., Romary, L., Sagot, B.: Ungoliant: an optimized pipeline for the generation of a very large-scale multilingual web corpus (2021). https://doi.org/10.14618/IDS-PUB-10468
2. Abadji, J., Suarez, P.O., Romary, L., Sagot, B.: Towards a cleaner document-oriented multilingual crawled corpus, January 2022
3. Agrawal, A., Singh, S.: Corpus complexity matters in pretraining language models, pp. 257–263, January 2023. https://doi.org/10.18653/v1/2023.sustainlp-1.20
4. Almeida, T.S., Abonizio, H., Nogueira, R., Pires, R.: Sabi\'a-2: a new generation of Portuguese large language models. arXiv preprint arXiv:2403.09887 (2024)
5. Barbaresi, A.: Trafilatura: a web scraping library and command-line tool for text discovery and extraction. In: Proceedings of the Joint Conference of the 59th Annual Meeting of the Association for Computational Linguistics and the 11th International Joint Conference on Natural Language Processing: System Demonstrations, pp. 122–131. Association for Computational Linguistics (2021). https://aclanthology.org/2021.acl-demo.15
6. Blodgett, S.L., Barocas, S., Daumé III, H., Wallach, H.: Language (technology) is power: a critical survey of "bias" in NLP. In: Proceedings of the 58th Annual Meeting of the Association for Computational Linguistics. Association for Computational Linguistics (2020). https://doi.org/10.18653/v1/2020.acl-main.485
7. Bommasani, R., et al.: On the opportunities and risks of foundation models. arXiv preprint arXiv:2108.07258 (2021)
8. Brown, T., et al.: Language models are few-shot learners. In: Larochelle, H., Ranzato, M., Hadsell, R., Balcan, M., Lin, H. (eds.) Advances in Neural Information Processing Systems, vol. 33, pp. 1877–1901. Curran Associates, Inc. (2020). https://proceedings.neurips.cc/paper_files/paper/2020/file/1457c0d6bfcb4967418bfb8ac142f64a-Paper.pdf

9. Carmo, D., Piau, M., Campiotti, I., Nogueira, R., Lotufo, R.: PTT5: pretraining and validating the T5 model on Brazilian Portuguese data. arXiv preprint arXiv:2008.09144 (2020)
10. Computer, T.: RedPajama: an open source recipe to reproduce LLaMa training dataset (2023). https://github.com/togethercomputer/RedPajama-Data
11. Crespo, M.C.R.M., et al.: Carolina: a general corpus of contemporary Brazilian Portuguese with provenance, typology and versioning information, March 2023
12. Finardi, P., Viegas, J.D., Ferreira, G.T., Mansano, A.F., Caridá, V.F.: BERTa\'u: Ita\'u BERT for digital customer service. arXiv preprint arXiv:2101.12015 (2021)
13. Gallegos, I.O., et al.: Bias and fairness in large language models: a survey. Comput. Linguist. 1–79 (2024). https://doi.org/10.1162/coli_a_00524
14. Gao, L., et al.: The pile: an 800 GB dataset of diverse text for language modeling, December 2020
15. Garcia, G.L., et al.: Introducing bode: a fine-tuned large language model for Portuguese prompt-based task. arXiv preprint arXiv:2401.02909 (2024)
16. Garimella, A., Mihalcea, R., Amarnath, A.: Demographic-aware language model fine-tuning as a bias mitigation technique. In: He, Y., et al. (eds.) Proceedings of the 2nd Conference of the Asia-Pacific Chapter of the Association for Computational Linguistics and the 12th International Joint Conference on Natural Language Processing (Volume 2: Short Papers), pp. 311–319. Association for Computational Linguistics, Online only, November 2022. https://aclanthology.org/2022.aacl-short.38
17. Habernal, I., Zayed, O., Gurevych, I.: C4Corpus: multilingual web-size corpus with free license. In: Calzolari, N., et al. (eds.) Proceedings of the Tenth International Conference on Language Resources and Evaluation (LREC 2016), pp. 914–922. European Language Resources Association (ELRA), Portorož, Slovenia, May 2016. https://aclanthology.org/L16-1146
18. Hoffmann, J., et al.: An empirical analysis of compute-optimal large language model training. In: Koyejo, S., Mohamed, S., Agarwal, A., Belgrave, D., Cho, K., Oh, A. (eds.) Advances in Neural Information Processing Systems, vol. 35, pp. 30016–30030. Curran Associates, Inc. (2022). https://proceedings.neurips.cc/paper_files/paper/2022/file/c1e2faff6f588870935f114ebe04a3e5-Paper-Conference.pdf
19. Jiang, A.Q., et al.: Mistral 7B. arXiv preprint arXiv:2310.06825 (2023)
20. Joulin, A., Grave, E., Bojanowski, P., Mikolov, T.: Bag of tricks for efficient text classification. In: Proceedings of the 15th Conference of the European Chapter of the Association for Computational Linguistics: Volume 2, Short Papers, pp. 427–431. Association for Computational Linguistics, April 2017
21. Kenton, J.D.M.W.C., Toutanova, L.K.: BERT: pre-training of deep bidirectional transformers for language understanding. In: Proceedings of naacL-HLT, vol. 1, p. 2 (2019)
22. Larcher, C., Piau, M., Finardi, P., Gengo, P., Esposito, P., Caridá, V.: Cabrita: closing the gap for Foreign languages. arXiv preprint arXiv:2308.11878 (2023)
23. Liu, Y., Cao, J., Liu, C., Ding, K., Jin, L.: Datasets for large language models: a comprehensive survey. arXiv preprint arXiv:2402.18041 (2024)
24. Mei, K., Fereidooni, S., Caliskan, A.: Bias against 93 stigmatized groups in masked language models and downstream sentiment classification tasks. In: 2023 ACM Conference on Fairness, Accountability, and Transparency, FAccT 2023. ACM, June 2023. https://doi.org/10.1145/3593013.3594109

25. Nadeem, M., Bethke, A., Reddy, S.: StereoSet: measuring stereotypical bias in pretrained language models. In: Proceedings of the 59th Annual Meeting of the Association for Computational Linguistics and the 11th International Joint Conference on Natural Language Processing (Volume 1: Long Papers). Association for Computational Linguistics (2021). https://doi.org/10.18653/v1/2021.acl-long.416
26. Nangia, N., Vania, C., Bhalerao, R., Bowman, S.R.: Crows-pairs: a challenge dataset for measuring social biases in masked language models. In: Proceedings of the 2020 Conference on Empirical Methods in Natural Language Processing (EMNLP). Association for Computational Linguistics (2020). https://doi.org/10.18653/v1/2020.emnlp-main.154
27. Overwijk, A., Xiong, C., Callan, J.: ClueWeb22: 10 billion web documents with rich information. In: Proceedings of the 45th International ACM SIGIR Conference on Research and Development in Information Retrieval, pp. 3360–3362 (2022)
28. Pires, R., Abonizio, H., Almeida, T.S., Nogueira, R.: Sabiá: Portuguese large language models. In: Brazilian Conference on Intelligent Systems, pp. 226–240 (2023)
29. Rae, J.W., et al.: Scaling language models: methods, analysis & insights from training gopher, December 2021
30. Raffel, C., et al.: Exploring the limits of transfer learning with a unified text-to-text transformer. J. Mach. Learn. Res. **21**(140), 1–67 (2020). http://jmlr.org/papers/v21/20-074.html
31. Richards, B.: Type/token ratios: what do they really tell us? J. Child Lang. **14**(2), 201–209 (1987). https://doi.org/10.1017/s0305000900012885
32. Shin, S., et al.: On the effect of pretraining corpora on in-context learning by a large-scale language model (2022)
33. Souza, F., Nogueira, R., Lotufo, R.: BERTimbau: pretrained BERT models for Brazilian Portuguese. In: Cerri, R., Prati, R.C. (eds.) BRACIS 2020. LNCS (LNAI), vol. 12319, pp. 403–417. Springer, Cham (2020). https://doi.org/10.1007/978-3-030-61377-8_28
34. Virtanen, A., et al.: Multilingual is not enough: BERT for Finnish, December 2019
35. Wagner Filho, J.A., et al.: The brWaC corpus: a new open resource for Brazilian Portuguese. In: Calzolari, N., et al. (eds.) Proceedings of the Eleventh International Conference on Language Resources and Evaluation (LREC 2018). European Language Resources Association (ELRA), Miyazaki, Japan, May 2018. https://aclanthology.org/L18-1686
36. Xu, L., Zhang, X., Dong, Q.: CLUECorpus2020: a large-scale Chinese corpus for pre-training language model. arXiv preprint arXiv:2003.01355 (2020)
37. Youmans, G.: Measuring lexical style and competence: the type-token vocabulary curve. Style **24**(4), 584–599 (1990). http://www.jstor.org/stable/42946163
38. Yuan, S., et al.: WuDaoCorpora: a super large-scale Chinese corpora for pre-training language models. AI Open **2**, 65–68 (2021)

Assessing Adversarial Effects of Noise in Missing Data Imputation

Arthur Dantas Mangussi[1,2](✉), Ricardo Cardoso Pereira[3,4], Pedro Henriques Abreu[4], and Ana Carolina Lorena[1,2]

[1] Computer Science Division, Aeronautics Institute of Technologies, Praça Marechal Eduardo Gomes, 50, São José dos Campos 12228-900, Brazil
mangussiarthur@gmail.com, aclorena@ita.br
[2] Science and Technology Institute, Federal University of São Paulo, Talim Street 330, São José dos Campos 12231-280, Brazil
[3] Miguel Torga Institute of Higher Education, 3000-132 Coimbra, Portugal
rdpereira@dei.uc.pt
[4] Centre for Informatics and Systems of the University of Coimbra, Department of Informatics Engineering, University of Coimbra, 3030-290 Coimbra, Portugal
pha@dei.uc.pt

Abstract. In real-world scenarios, a wide variety of datasets contain inconsistencies. One example of such inconsistency is missing data (MD), which refers to the absence of information in one or more variables. Missing imputation strategies emerged as a possible solution for addressing this problem, which can replace the missing values based on mean, median, or Machine Learning (ML) techniques. The performance of such strategies depends on multiple factors. One factor that influences the missing value imputation (MVI) methods is the presence of noisy instances, described as anything that obscures the relationship between the features of an instance and its class, having an adversarial effect. However, the interaction between MD and noisy instances has received little attention in the literature. This work fills this gap by investigating missing and noisy data interplay. Our experimental setup begins with generating missingness under the Missing Not at Random (MNAR) mechanism in a multivariate scenario and performing imputation using seven state-of-the-art MVI methods. Our methodology involves applying a noise filter before performing the imputation task and evaluating the quality of the imputation directly. Additionally, we measure the classification performance with the new estimates. This approach is applied to both synthetic data and 11 real-world datasets. The effects of noise filtering before imputation are evaluated. The results show that noise preprocessing before the imputation task improves the imputation quality and the classification performance for imputed datasets.

Keywords: Missing data imputation · noise filtering

1 Introduction

Real-world data often presents multiple problems, which can jeopardize the performance of Machine Learning (ML) classifiers [12]. Renggli et al. [16] advocate that data quality issues influence several stages of the ML pipeline. One common type of inconsistency is missing data (MD), which can be described as the absence of values in the data observations [17]. The literature categorizes MD mechanisms into the following categories [14]:

- *Missing Completely at Random* (MCAR): missingness occurs randomly without any dependency on specific features within the dataset;
- *Missing at Random* (MAR): a dependency between existent features determines the missingness nature;
- *Missing Not at Random* (MNAR): the missing values depend on observed and/or other unobserved data (i.e., features not available on the dataset).

To address the MD issue, the literature presents different missing value imputation (MVI) methods, from basic, such as mean, median, and mode, to more sophisticated strategies, including using ML methods to estimate the missing values [15]. Santos et al. [17] describe the classical experimental setup for evaluating MD imputation algorithms with four main steps. The first concerns data collection, where a complete dataset (i.e., without missing values) is considered. Subsequently, the next step is amputation, which introduces artificial missing values, following the characteristics of MCAR, MAR, or MNAR mechanisms. Then, the MVI algorithms need to be selected and applied. Finally, the last step is evaluation. In general, studies in the MD field often evaluate new MVI techniques by measuring the difference between the original (i.e., the ground truth) and imputed data, a process known as direct evaluation. Conversely, there is indirect evaluation, which measures the classification performance using the imputed datasets. However, the literature rarely focuses on evaluating MVI methods using both approaches [6,9].

Another data quality issue is the presence of noisy instances on the dataset. According to Zhu and Wu [26], there are two distinct types of noise: attribute and class noise. The former type of noise affects input features, while the latter affects the labels registered for the observations. Both are present in real-world scenarios, emphasizing that aggregate noise identification strategies should be considered for both. Nonetheless, the label noise is potentially more harmful than attribute noise [22,26]. The literature outlines several techniques for identifying potential noise and addressing it to build more reliable ML models from data [3]. Our work will focus on noise filters (NFs), which scan the training data for potentially noisy instances.

Regarding the interplay of MD and noise inconsistencies, when initial noisy data is used to extract patterns for missing data imputation, whether through simple statistics or more sophisticated strategies, the harmful effects of noise can propagate to other instances. Nonetheless, this interaction has received little attention in the missing data literature, and only Zhu et al. [25], Fangfang et al.

[8], and Hulse and Khoshgoftaar [23] have investigated the interaction between these two data inconsistencies. None of these works analyze the application of noise pre-processing before the imputation task. Thus, this work investigates the interaction between missing and noisy data inconsistencies, considering pre-processing on noisy instances before the imputation task and investigating how noisy data impacts the results of MVI methods.

The remainder of this work is organized as follows: Sect. 2 presents related work on the MD field, and the interplay of missing values and noise data. Section 3 describes the methodology and the experimental setup. The results are presented in Sect. 4. Section 5 outlines the conclusions and future directions of this work.

2 Related Work

This section presents a literature review of imputation techniques, noise in ML, and the interplay of both data inconsistencies.

2.1 Imputation Techniques

There are several strategies for handling MD. Replacing the missing values with a predetermined estimate (i.e., imputing the missing values) is a common approach in the literature. The simplest way to perform the imputation task is through a single imputation, where missing values in quantitative features can be replaced with the mean or median of all available non-missing values. In contrast, missing values in qualitative attributes are replaced with the mode [9].

To solve the limitations of single imputation, the literature provides multiple imputation strategies, which employ approximate values that reflect the uncertainty around the actual value from the observed data [2]. The Multivariate Imputation by Chained Equation (MICE) is the most widely used multiple imputation technique [1]. The MICE algorithm is a Gibbs sampler that estimates the posterior distribution of a vector of unknown parameters by sampling iteratively from conditional distributions [20].

Another way to impute MD is using matrix completion methods, such as *SoftImpute*. The key idea is to perform imputation by approximating the original data with a low-rank matrix. This process typically involves matrix decomposition to identify latent features that best describe the available values [14].

Random Forests (RF) are used for MVI in the *missForest* algorithm [19]. The *missForest* algorithm is an iterative imputation scheme that trains an RF on observed values in the first step, predicts the missing values and proceeds iteratively. Another common ML algorithm used in data imputation is the K-Nearest Neighbor (KNN), which finds the nearest neighbors of instances with missing values and uses them for imputation [6].

The MVI community has recently been using deep learning (DL) methods for MVI [10]. An example is the Generative Adversarial Imputation Nets (GAIN)

[24], which has two main components: generator and discriminator. The generator component imputes the missing values on the observed data. It outputs a complete vector, and the discriminator attempts to validate which element in the output vector is imputed [18,20]. Autoencoders (AEs) are another example of DL-based methods employed in MVI. AEs are a neural network architecture that learns from incomplete data (input layer) and tries reproducing this input at the output layer, generating new plausible values for imputation [13,14].

With the development of new MVI techniques, it is essential to evaluate their effectiveness. The literature presents an interesting behavior wherein most works only use the direct evaluation for MVI techniques, where the imputed values are compared to real values in datasets where some values are amputed. But it is also important to evaluate how the imputed values influence classification performance, in an indirect evaluation. Very few works use both approaches.

Pereira et al. [14] evaluate the Siamese Autoencoder-Based Approach for Missing Data Imputation (SAEI) for imputation tasks in direct and indirect ways. For the direct evaluation, the authors used the Mean Absolute Error (MAE) metric, and for the indirect evaluation, they measured the F1-score performance with three different classifiers: KNN, RF, and eXtreme Gradient Boosting (XGB). This was done for 14 datasets. The results show that SAEI outperforms other state-of-the-art MVI methods under MNAR assumption and induces the best classification results, improving the F1-scores for 50% of the used datasets.

Luengo, García, and Herrera [11] analyze the behavior of 23 classification methods and 14 different MVI approaches in an indirect evaluation. Moreover, this methodology was applied to 21 real-world datasets, and all of them have their proper MD. They found that using certain MVI techniques could improve the accuracy obtained for the classification methods, facilitating an explanation of how imputation may be a helpful tool to overcome the negative impact of MD.

2.2 Noise

Noise is frequent in real-world datasets and can harm the predictive performance of ML classifiers. Although most ML algorithms have some internal mechanisms to avoid focusing on noisy instances (e.g. pruning mechanisms in Decision Trees), cleansing such instances can be beneficial [4].

The literature shows various methods to identify and address attribute and label noise. According to Saez et al. [21], Noise filter (NF) techniques are widely used in a data pre-processing step for cleansing the training data [3]. Their strategy consists of identifying potential noise and removing these unreliable examples. However, these inconsistencies can also be corrected [5].

NFs can be divided into two main categories: similarity-based and ensemble-based filters [21]. The similarity-based or distance-based filters employ the KNN algorithm to evaluate whether an example is closest to others within its class; otherwise, it is an unreliable and potentially noisy instance. Various KNN-based methods have emerged in the literature [3]. A well-known example is the Edited

Nearest Neighbor (ENN) NF, which eliminates samples whose class differs from most of its K nearest neighbors [5,21].

Our focus in this work will be using the ENN algorithm to identify potential noisy instances. These noisy instances will be disregarded during the imputation process to avoid noise from propagating to the imputed data.

2.3 Interplay of Missing and Noisy Data

A few works have investigated the relationship between noise and MD. Hulse and Khoshgoftaar [23] evaluated the impact of noise in software measurement data on the imputation process. The authors have used five imputation methods using real-world software measurement datasets. The amputation process was made only in the dependent variable covering the MCAR, MAR, and NI (non-ignorable) mechanisms until they achieved 5%, 10%, 15%, and 20% of missing rates. They used five imputation methods on real-world software measurement datasets. The amputation process was applied only to the dependent variable, covering the MCAR, MAR, and NI (non-ignorable) mechanisms, until they achieved missing rates of 5%, 10%, 15%, and 20%. They also considered four different noisy scenarios: inherent noise only (i.e., noise present in the original dataset [23]), no noisy instances, and inherent noise with an additional 5% and 10% of injected noise. For each experimental setup and combination of factors, they conducted five independent random selections and compared the imputation accuracy of the five imputation techniques. Their experiments demonstrated that data quality plays a crucial role in the effectiveness of imputation techniques. Moreover, for the four missing rates used in their experimental setup, an increase in missing rate was not found to be significant for all imputation techniques.

Robust Imputation based on the Group Method of Data Handling (RIBG) is an MVI method proposed by Zhu et al. [25] for predicting missing values in noisy environments. The authors use the Group Method of Data Handling (GMDH), a heuristic self-organizing data mining technique known for its noise immunity, to develop the RIBG method. RIBG operates as follows: given an incomplete dataset, it first performs a preliminary imputation using the mean for numerical features and mode for categorical features to create an initial complete dataset. Then, RIBG applies the GMDH mechanism to iteratively predict and update these initial missing value estimates. To evaluate the effectiveness of the RIBG method, the authors tested it on nine datasets from the UCI repository, with missing rates of 5%, 10%, and 20%, under varying noise levels. The missing data was generated under MCAR, MAR, and MNAR assumptions. The results indicate that noise significantly impacts MVI methods, particularly at high noise levels.

Li et al. [8] present a Noise-Aware Missing Data Multiple Imputation (NPMI) algorithm designed to handle missing data in noisy environments. The NPMI algorithm uses the Random Sample Consensus (RANSAC) method to estimate the initial parameters of the multiple imputation algorithm. This approach enhances the robustness of multiple imputations and ensures accuracy even when noise is present. The proposed method was validated on four datasets: two real and two synthetic. For the synthetic data, random Gaussian noise was simulated at different noise levels: 5%, 10%, 20%, 30%, 40%, and 50%. For the real data, some values were randomly designated as missing, with missing rates corresponding to the noise level percentages. The accuracy of imputation was evaluated using the Root Mean Square Error (RMSE). The experimental results demonstrated that noise significantly affects the data quality of the entire dataset, with higher noise levels leading to a greater degradation in data quality.

Therefore, previous work has shown that the quality of observed data significantly impacts the imputation task and needs to be addressed. As suggested in studies by [8,25], one option is to use algorithms that are robust to noise for the imputation task. However, those works do not analyze if a simple pre-processing for treating noisy instances beforehand impacts missing value imputation. Overall, a comprehensive exploratory analysis of NF use before the imputation task is still needed. Thus, this work aims to investigate the interaction of missing and noisy data inconsistencies and to what extent noisy data impacts the results of MVI methods. From the authors' knowledge, this constitutes the first work that has tried employing an NF before the imputation, investigating the direct and indirect impact of this procedure on imputation quality.

3 Methodology

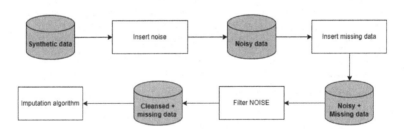

Fig. 1. Methodology overview illustrating the sequential steps followed in the experimental design in the case of synthetic datasets.

The methodology for this work consists of evaluating the interplay between noise and missing data in two points: the imputation quality (direct evaluation) and the classification performance for imputed datasets (indirect evaluation). We perform two types of experiments to achieve our goal: using synthetic data and real-world datasets.

Firstly, we use the synthetic datasets to conduct an initial analysis and have greater control over the experiments. As shown in Fig. 1, we generate synthetic data and introduce artificial noise into the attributes or labels. Afterward, we insert missing data in an amputation process (i.e., generate artificial missing values). Then, with a dataset that contains both data inconsistencies, we use an NF and perform the imputation of the missing values, disregarding the potentially noisy instances. The next section describes how these datasets were generated. A direct evaluation of the imputation quality in these datasets with and without noise filtering is done.

Once the impact of noise on the imputation results of missing values in synthetic data has been assessed, the experiments are extended to real-world datasets currently employed in the related literature, assuming that real-world data already contains some noise level [8]. Therefore, in this case, the artificial insertion of noise at the beginning of Fig. 1 is disregarded. Here, both direct and indirect evaluations of the imputation results are assessed.

We have selected eleven real-world heterogeneous benchmark datasets that are currently employed in the MD field and are available on the University of California Irvine Machine Learning Repository[1] and Kaggle[2]. Table 1 overviews dataset characteristics. Each dataset is identified by its acronym name and information on the number of instances, types of features, and classes (i.e., the number of possible output variable values). Categorical features were converted to quantitative values with a one-hot encoding when needed.

As seen in Table 1, the eleven real-world datasets are all binary classification problems. We selected this type of problem due to the methodology employed in this work. We split each dataset according to the two classes to ensure that both classes have the same amount of missing values and to prevent the noise filter from removing an entire class. Section 3.1 will describe this process in more detail.

Table 1. Overview of datasets characteristics.

Dataset	Instances	Features		Classes
		Continuous	Categorical	
Wiscosin	569	30	0	2
Pima diabetes	768	8	0	2
Indian liver	583	9	1	2
Parkinsons	195	22	0	2
Mammographic masses	830	1	3	2
Thoracic surgery	470	3	13	2
Diabetic retinopathy	1151	3	16	2
BC Coimbra	116	4	0	2
Thyroid recurrence	383	1	15	2
Blood transfusion	748	4	0	2
Law school	20798	6	6	2

For the implementation, we used Python version 3.11 and several additional libraries: Pandas, Numpy, Scikit-Learn, mdatagen, and Imbalance-Learn. All the experiments were conducted on a machine with 60GB RAM, GPU NVIDIA GeForce RTX 4090 24GB, and Linux, Ubuntu version 22.04.4.

---

[1] https://archive.ics.uci.edu/datasets.
[2] https://www.kaggle.com/datasets.

3.1 Experimental Setup

As outlined in Fig. 1, the first step in our methodology was to create synthetic data. We have used the "make_classification" function from Scikit-Learn to generate the synthetic data. Our base dataset consists of 500 observations with 5 numerical input features and two classes. To introduce attribute noise, Gaussian noise is added to each feature. This noise is generated using the mean and standard deviation of each feature, ensuring the values remain within the feature's minimum and maximum range. Label noise is simulated by flipping the labels. The following rates are tested: 5%, 10%, and 20%. We introduce noise to the entire dataset to obtain a more realistic dataset, as real-world data already contains some noise.

Once noise is inserted, we used a stratified cross-validation strategy with five folds to perform the amputation process (i.e., artificially generating missing values in the dataset) and imputation task. We used the Python library *mdatagen*[3] for each fold to generate artificial MD under MNAR mechanism in a multivariate scenario. To ensure that both classes would receive MD, we split the training set by the outcome and conducted independent procedures for training and testing sets to keep the same missing rates for both. The multivariate MNAR strategy deleted the lowest values for dataset features more related to the classes up to 5%, 10% and 20%. These missing rates were selected from [25].

Using such corrupted datasets, we employ the Edited Nearest Neighbor (ENN) algorithm to filter the potentially noisy instances in the training sets, with $k = 5$. The ENN algorithm cleans the dataset by deleting samples that are close elements from other classes. This tends to remove data in overlapping, borderline and noisy areas of the dataset [7]. In this work, we used a more conservative approach to undersample the majority class, where most of the neighbors must belong to the same class as the examined sample for it to be retained, and the default distance metric in Imbalanced-learn package. The identified noisy instances are not used in data imputation afterwards.

For the imputation task, we chose seven state-of-the-art MVI to address the generated MD. The algorithms chosen are the mean of each feature, KNN, MICE, PMIVAE, missForest, SoftImpute, and GAIN. The KNN, imputation by the mean, missForest, and MICE were used directly from the Scikit-learn library. The remaining algorithms are available in different GitHub repositories[4]. The KNN was used with $K = 5$ and the Euclidean distance, MICE was run with 100 iterations, and the parameterization of the architecture of deep learning methods followed the authors' recommendations from the original articles. As aforementioned, we saved the imputed test data for each fold in the stratified cross-validation strategy. We measured the imputation quality with the Mean Absolute Error (MAE) between the predicted values and the ground truth for the multivariate scenario in the test sets. Next, at the end of the cross-validation

---

[3] https://pypi.org/project/mdatagen/.
[4] https://github.com/travisbrady/py-soft-impute, https://github.com/jsyoon0823/GAIN https://github.com/ricardodcpereira/PMIVAE.

process, we combined all folds, obtaining an imputed dataset with original data dimensions without bias.

We selected an RF classifier to measure the classification performance of the new datasets. The experiments used the same cross-validation strategy for amputation and imputation processes. We tuned the RF hyperparameters by Randomized Search in the training sets using the F1 Score.

4 Results

This section presents our analysis of the effect of employing an NF based on similarity before the imputation process under MNAR mechanism. We begin by investigating the overall impact of ENN in synthetic data. Subsequently, we analyze the effect of ENN filtering on the imputation task for real-world datasets in direct and indirect evaluations. We also discuss if the pattern found in the experiments using synthetic data is verified in real-world datasets.

4.1 Impact of NF in Imputation

Table 2 presents differences in MAE imputation results for synthetic data under the MNAR mechanism. The differences are taken from the baseline, where no noise filtering is applied. Therefore, positive values represent better estimates of the missing values after NF, while negative values denote the opposite. The negative values are boldfaced in the table. The datasets named Att$X$ contain $X\%$ of noise in the attributes, where X is a noise rate, while datasets named Label$X$ have $X\%$ of label noise. For each one of them, missing rates are also varied.

In all cases, except some specific scenarios with the SoftImpute method, using an NF has improved the MAE results in imputation. Although the differences are small, this demonstrates that even by employing a simple noise filter as ENN before imputation, the imputation quality can be improved. For the mean, MICE, PMIVAE and missForest imputers, the results are improved more for increased missing rates, indicating that a reliable dataset is especially needed for better imputation results when there are large missing rates. In other cases, there is no clear tendency. There is also no clear tendency on which type of noise or noise ratio affects more the MVI results.

Regardless of the characteristics of the SoftImpute method, its primary goal is to find a low-rank matrix that approximates the original dataset [14]. However, the data distribution is altered when we apply an NF that removes specific observations based on the ENN criterion. These new estimates may no longer accurately represent the observed data, leading to the deterioration in the MVI results shown in Tables 2 and 3.

Based on the findings in Table 2, we extend our analysis to real-world datasets, assuming they already have noisy instances.

Table 2. Differences between the average MAE for the baseline where no NF is applied and the application of an NF before imputation under MNAR multivariate conditions for the synthetic datasets. 'Att' is an acronym for attribute noise and 'Label' for label noise. 'Att05' means a 5% noise level introduced as attribute noise, and the same applies for 'Label 05'.

Dataset	Missing Rate	Mean	KNN	MICE	PMIVAE	SoftImpute	GAIN	missForest
Att05	5	0.007	0.055	0.014	0.007	0.001	0.103	0.036
	10	0.011	0.053	0.018	0.011	0.001	0.052	0.035
	20	0.021	0.040	0.024	0.021	0.009	0.062	0.046
Att10	5	0.008	0.047	0.016	0.008	**-0.025**	0.075	0.031
	10	0.012	0.040	0.022	0.012	**-0.032**	0.065	0.032
	20	0.022	0.032	0.029	0.023	**-0.009**	0.041	0.038
Att20	5	0.007	0.044	0.017	0.008	0.039	0.119	0.014
	10	0.013	0.033	0.027	0.012	0.036	0.041	0.024
	20	0.022	0.031	0.028	0.023	0.001	0.076	0.032
Label05	5	0.008	0.027	0.013	0.006	**-0.012**	0.059	0.029
	10	0.012	0.039	0.014	0.012	**-0.012**	0.015	0.016
	20	0.021	0.042	0.016	0.023	0.016	0.030	0.028
Label10	5	0.006	0.024	0.010	0.006	0.000	0.035	0.020
	10	0.011	0.043	0.014	0.009	0.024	0.009	0.020
	20	0.021	0.045	0.019	0.020	0.008	0.017	0.030
Label20	5	0.006	0.042	0.011	0.006	**-0.007**	0.036	0.006
	10	0.013	0.040	0.014	0.013	**-0.007**	0.063	0.010
	20	0.021	0.040	0.020	0.021	**-0.003**	0.029	0.027

4.2 Imputation Results for Real-World Datasets

Table 3 demonstrates the differences in MAE results of employing the previous methodology in real-world datasets under the MNAR mechanism in a multivariate scenario. Again, the differences are taken from the baseline of not employing an NF. Negative differences are boldfaced and indicate the estimates of the missing values are worse when NF is applied. Again, this happens mostly for the SoftImpute algorithm. But now GAIN, missForest, MICE and kNN were also impaired in some specific cases. In most cases, there are improvements, which tend to be higher for larger missing rates. The SoftImpute method shows worse MAE for 44.12% instances and GAIN for 41.18%. For the remaining methods, the percentage of worse results is less than 10%. Therefore, the applied methodology enhances the imputation quality for most MVI techniques on these real-world datasets under the MNAR mechanism. As observed for synthetic data, when the missing rate is increased, the results tend to be improved more when ENN is applied as a pre-processing step.

Table 4 gives a general overview of how each imputation method performs in the datasets by averaging the MAE results per imputation method and missing

Table 3. Differences between the average MAE for the baseline where no NF is applied and the application of an NF before imputation under MNAR multivariate conditions for the real-world datasets.

Dataset	Missing Rate	Mean	KNN	MICE	PMIVAE	SoftImpute	GAIN	missForest
wiscosin	5	0.006	0.009	0.001	0.003	-0.009	-0.006	0.008
	10	0.012	0.014	0.002	0.012	**-0.012**	0.003	0.001
	20	0.022	0.021	0.009	0.021	-0.008	0.002	0
pima	5	0.011	0.005	0.01	0.013	-0.009	**-0.011**	0.008
	10	0.015	0.007	0.007	0.014	0	**-0.031**	0.017
	20	0.025	0.018	0.018	0.026	-0.011	0.035	0.02
indian_liver	5	0.005	0.012	0.005	0.002	-0.023	**-0.024**	0.001
	10	0.01	0.016	0.011	0.005	0.032	0.02	0.019
	20	0.023	0.034	0.027	0.015	-0.002	0.002	0.029
parkinsons	5	0.004	0.012	0.003	0.002	-0.008	0.02	**-0.001**
	10	0.011	0.023	0.007	0.01	0.017	0.046	0.015
	20	0.023	0.026	0.033	0.021	0.031	0.065	0.017
mammographic_masses	5	0.007	0.01	0.003	0.002	0.004	0.065	0.004
	10	0.012	0.033	**-0.006**	0.105	-0.015	-0.015	0.008
	20	0.028	0.086	0.004	0.063	0.016	0.07	0.015
thoracic_surgery	5	0.006	0.018	**-0.001**	0.001	0	-0.004	0.018
	10	0.011	0.02	0.005	0.003	0.015	**-0.006**	0.009
	20	0.022	0.028	0.018	0.006	0.008	0.015	0.011
diabetic_retionapaty	5	0.003	0.013	0.003	-0.001	0.008	0.012	**-0.019**
	10	0.006	0.026	0.005	0.005	0.012	0.034	0.013
	20	0.016	0.024	0.012	0.012	0.016	**-0.02**	-0.009
bc_coimbra	5	0.007	0.017	0.007	0.006	0.04	**-0.041**	0.025
	10	0.012	0.013	0.005	0.002	**-0.005**	0.026	0.028
	20	0.024	0.029	0.019	0.015	0.012	0.114	0.037
thyroid_recurrence	5	0.007	0.008	0.018	0.007	**-0.005**	0.058	0.022
	10	0.014	0.017	0.011	0.018	-0.005	**-0.046**	0.014
	20	0.027	0.025	0.024	0.018	0.004	**-0.011**	0.016
blood_transfusion	5	0.004	-0.015	0	0.002	0.006	**-0.115**	0.004
	10	0.01	0.015	0.012	0.009	**-0.004**	0.111	0.009
	20	0.02	0.03	0.027	0.019	-0.003	**-0.017**	0.041
law	5	0.008	0.022	0.004	0.004	0.012	**-0.058**	0.008
	10	0.017	0.056	0.007	0.008	**-0.012**	0.071	0.025
	20	0.033	0.091	0.074	0.077	0.006	0.023	0.086

rate. The best results are boldfaced. The missForest algorithm outperforms the remaining methods, followed by KNN and MICE. The GAIN method was the worst method in this experimental setup, and the PMIVAE only surpassed the imputation results by mean. The complexity of those methods may justify the need for larger datasets to obtain better results in the experiments. Since the

Table 4. Average MAE results obtained for each imputation method, grouped by missing rate. The highlighted bold results present the best MAE results for each missing rate.

Missing Rate	Mean	KNN	MICE	PMIVAE	SoftImpute	GAIN	missForest
5%	0.233	0.142	0.145	0.221	0.168	0.481	**0.139**
10%	0.224	0.146	0.147	0.209	0.164	0.442	**0.130**
20%	0.211	0.158	0.148	0.196	0.162	0.467	**0.132**

Table 5. Average differences of MAE grouped by missing rate for the three best-performing imputation methods.

Missing Rate	KNN	MICE	missForest
5%	0.010	0.005	0.007
10%	0.022	0.006	0.014
20%	0.037	0.024	0.024

ENN algorithm further reduces the datasets in imputation, this might have degraded their performance more.

Taking the three best imputation methods, missForest, KNN, and MICE, respectively, we also analyze the average difference between applying or not the methodology of this work, for different missing levels. The results are shown in Table 5. The average difference increases for higher missing rates, confirming the results observed for synthetic data.

Indirectly evaluating the imputation quality, we trained an RF classifier for the newly imputed data and compared its F1 score with the baseline (i.e., the F1-score for original datasets, without amputed values). Figure 2 illustrates the overall average results across all imputed real-world datasets and missing rates, where the average F1-score achieved for the original complete datasets is shown as a dashed red line. MICE outperformed the remaining MVI methods and got closer average to the baseline, followed by KNN and missForest. Those methods were the best three imputation algorithms in our previous evaluation and benefited more from noise filtering. Therefore, we can observe an interplay with better quality of imputation and classification performances. In contrast, the PMIVAE method presents a worse F1-score than the other methods. Also, PMIVAE only outperforms the mean and GAIN in MAE imputation quality. We believe that the nature of ENN's undersampling process, combined with the high dependency on parameterization for deep learning methods such as PMIVAE and GAIN, explains the results observed in our study. Therefore, we plan to conduct an optimization search to identify the best parameters for these methods in the future. Additionally, due to its specific characteristics, the PMIVAE imputation method may not yield good results with the RF classifier. We intend to extend our methodology to include other classification algorithms to address this.

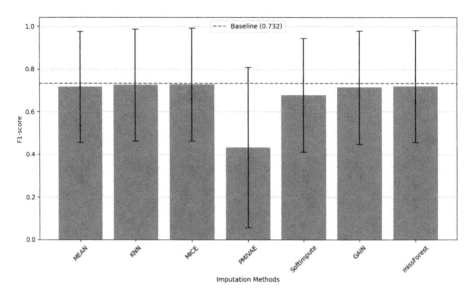

Fig. 2. Overall F1 score of all imputation methods for Random Forest classification model.

In conclusion, most imputation strategies benefited from noise filtering, estimating new values with higher quality. However, the results are impaired for some specific algorithms. MAE is reduced for SoftImpute, GAIN, and PMIVAE. PMIVAE is especially impaired in the indirect evaluation. These algorithms may require more data to obtain a proper fit of values to be imputed. Since noise filtering reduces the dataset, this can explain the observed behavior.

5 Conclusions

This study investigated the interaction between missing and noisy data, focusing on pre-processing noisy instances before imputation and how this procedure impacts MVI methods. This is the first comprehensive analysis of its kind, addressing a gap in the related literature. We used the Edited Nearest Neighbors (ENN) noise filter and seven imputation methods: mean, KNN, MICE, PMIVAE, SoftImpute, GAIN, and missForest. The filtering takes place before imputation and the hypothesis is that disregarding the unreliable instances from the imputation process can be beneficial, as it prevents noise from propagating to the new estimated values.

Applying the noise filter has generally improved estimates of the missing values for both synthetic and real-world datasets. The imputation methods that benefited the most from this approach were missForest, KNN, and MICE. And the results are improved more for datasets with higher missing rates. Nonetheless, the reduction in the datasets used for adjusting the models seems to have impaired the results for more complex techniques. However, this must be confirmed for larger datasets with more instances.

For future considerations of this work, we want to extend the methodology for the MAR multivariate mechanism and cover other classifiers and noise filters more complex than ENN.

Acknowledgments. The authors gratefully acknowledge the Brazilian funding agencies FAPESP (Fundação Amparo à Pesquisa do Estado de São Paulo) under grants 2022/10553 -6, 2023/13688-2 and 2021/06870-3. This research was also supported by the Portuguese Recovery and Resilience Plan (PRR) through project C645008882-00000055 - Center for Responsible AI.

Disclosure of Interests. The authors declare that they have no known competing financial interests or personal relationships that could have appeared to influence the work reported in this paper.

References

1. Buuren, S., Groothuis-Oudshoorn, C.: MICE: multivariate imputation by chained equations in R. J. Stat. Softw. **45** (2011). https://doi.org/10.18637/jss.v045.i03
2. Emmanuel, T., Maupong, T., Mpoeleng, D., Semong, T., Mphago, B., Tabona, O.: A survey on missing data in machine learning. J. Big Data **8**(1) (2021). https://doi.org/10.1186/s40537-021-00516-9. funding Information: This work received a grant from the Botswana International University of Science and Technology. Publisher Copyright: 2021, The Author(s)
3. Frenay, B., Verleysen, M.: Classification in the presence of label noise: a survey. IEEE Trans. Neural Netw. Learn. Syst. **25**(5), 845–869 (2014). https://doi.org/10.1109/TNNLS.2013.2292894
4. Garcia, L.P., de Carvalho, A.C., Lorena, A.C.: Effect of label noise in the complexity of classification problems. Neurocomputing **160**, 108–119 (2015)
5. Garcia, L.P., Lehmann, J., de Carvalho, A.C., Lorena, A.C.: New label noise injection methods for the evaluation of noise filters. Knowl.-Based Syst. **163**, 693–704 (2019). https://doi.org/10.1016/j.knosys.2018.09.031, https://www.sciencedirect.com/science/article/pii/S0950705118304829
6. Hasan, M.K., Alam, M.A., Roy, S., Dutta, A., Jawad, M.T., Das, S.: Missing value imputation affects the performance of machine learning: a review and analysis of the literature (2010–2021). Informatics in Medicine Unlocked **27** (2021). https://doi.org/10.1016/j.imu.2021.100799
7. Lemaître, G., Nogueira, F., Aridas, C.K.: Imbalanced-learn: a Python toolbox to tackle the curse of imbalanced datasets in machine learning. J. Mach. Learn. Res. **18**(17), 1–5 (2017). http://jmlr.org/papers/v18/16-365
8. Li, F., Sun, H., Gu, Y., Yu, G.: A noise-aware multiple imputation algorithm for missing data. Mathematics **11**, 73 (2022). https://doi.org/10.3390/math11010073
9. Lin, W.C., Tsai, C.F.: Missing value imputation: a review and analysis of the literature (2006–2017). Artif. Intell. Rev. **53**, 1487–1509 (2020). https://doi.org/10.1007/s10462-019-09709-4
10. Liu, M., et al.: Handling missing values in healthcare data: a systematic review of deep learning-based imputation techniques. Artif. Intell. Med. **142**, 102587 (2023). https://doi.org/10.1016/j.artmed.2023.102587, https://www.sciencedirect.com/science/article/pii/S093336572300101X

11. Luengo, J., García, S., Herrera, F.: On the choice of the best imputation methods for missing values considering three groups of classification methods. Knowl. Inf. Syst. **32**, 77–108 (2012). https://doi.org/10.1007/s10115-011-0424-2
12. Nakhaei, A., Sepehri, M.M., khatibi, T.: A promising method for correcting class noise in the presence of attribute noise. Int. J. Hosp. Res. **12**(1) (2023). https://ijhr.iums.ac.ir/article_171438.html
13. Pereira, R.C., Abreu, P.H., Rodrigues, P.P.: Siamese autoencoder-based approach for missing data imputation. In: International Conference on Computational Science, pp. 33–46. Springer, Cham (2023). https://doi.org/10.1007/978-3-031-35995-8_3
14. Pereira, R.C., Abreu, P.H., Rodrigues, P.P.: Siamese autoencoder architecture for the imputation of data missing not at random. J. Comput. Sci. **78**, 102269 (2024). https://doi.org/10.1016/j.jocs.2024.102269, https://www.sciencedirect.com/science/article/pii/S1877750324000620
15. Pereira, R.C., Rodrigues, P.P., Figueiredo, M.A.T., Abreu, P.H.: Automatic delta-adjustment method applied to missing not at random imputation. Comput. Sci. ICCS **2023**, 481–493 (2023). https://doi.org/10.1007/978-3-031-35995-8_34
16. Renggli, C., Rimanic, L., Gürel, N.M., Karlaš, B., Wu, W., Zhang, C.: A data quality-driven view of mlops. arXiv preprint arXiv:2102.07750 (2021)
17. Santos, M.S., Pereira, R.C., Costa, A.F., Soares, J.P., Santos, J., Abreu, P.H.: Generating synthetic missing data: a review by missing mechanism. IEEE Access **7**, 11651–11667 (2019)
18. Shahbazian, R., Greco, S.: Generative adversarial networks assist missing data imputation: a comprehensive survey and evaluation. IEEE Access **11**, 88908–88928 (2023). https://doi.org/10.1109/ACCESS.2023.3306721
19. Stekhoven, D., Bühlmann, P.: Missforest? Non-parametric missing value imputation for mixed-type data. Bioinformatics (Oxford, England) **28**, 112–118 (2012). https://doi.org/10.1093/bioinformatics/btr597
20. Sun, Y., Li, J., Xu, Y., Zhang, T., Wang, X.: Deep learning versus conventional methods for missing data imputation: a review and comparative study. Expert Syst. Appl. **227**, 120201 (2023). https://doi.org/10.1016/j.eswa.2023.120201
21. Sáez, J.A.: Noise models in classification: unified nomenclature, extended taxonomy and pragmatic categorization. Mathematics **10**(20) (2022). https://doi.org/10.3390/math10203736, https://www.mdpi.com/2227-7390/10/20/3736
22. Sáez, J.A., Galar, M., Luengo, J., Herrera, F.: Analyzing the presence of noise in multi-class problems: alleviating its influence with the one-vs-one decomposition. Knowl. Inform. Syst. **38**, 179–206 (2014). https://doi.org/10.1007/s10115-012-0570-1
23. Van Hulse, J., Khoshgoftaar, T.: A comprehensive empirical evaluation of missing value imputation in noisy software measurement data. J. Syst. Softw. **81**, 691–708 (2008). https://doi.org/10.1016/j.jss.2007.07.043
24. Yoon, J., Jordon, J., van der Schaar, M.: Gain: missing data imputation using generative adversarial nets. In: International Conference on Machine Learning (ICML), pp. 5689—5698 (2018)
25. Zhu, B., He, C., Liatsis, P.: A robust missing value imputation method for noisy data. Appl. Intell. **36**, 61–74 (2012). https://doi.org/10.1007/s10489-010-0244-1
26. Zhu, X., Wu, X.: Class noise vs. attribute noise: a quantitative study. Artif. Intell. Rev. **22**, 177–210 (2004). https://doi.org/10.1007/s10462-004-0751-8

Assessing European and Brazilian Portuguese LLMs for NER in Specialised Domains

Rafael Oleques Nunes[1] (✉), Joaquim Santos[2], Andre Spritzer[1], Dennis Giovani Balreira[1], Carla Maria Dal Sasso Freitas[1], Fernanda Olival[3], Helena Freire Cameron[3], and Renata Vieira[4]

[1] Federal University of Rio Grande do Sul, Porto Alegre, Brazil
{ronunes,dgbalreira,spritzer,carla}@inf.ufrgs.br
[2] University of Vale do Rio dos Sinos, São Leopoldo, Brazil
nejoaquim@edu.unisinos.br
[3] CIDEHUS - Portalegre Polytechnic University, Portalegre, Portugal
helenac@ipportalegre.pt
[4] CIDEHUS - University of Évora, Évora, Portugal
renatav@uevora.pt

Abstract. This paper discusses the impact of Portuguese variants in Large Language Models for the task of named entity recognition (NER) in specialised domains. The tests were made on a Brazilian Portuguese legal and a European Portuguese historical corpora. The models taken into account are BERTimbau (PT-BR), Albertina (PT-PT and PT-BR), and XML-R (multilingual). The impact was more evident in the Portuguese historical corpus, which resulted in higher F1 measures compared to previous works that did not consider the same language variant. Additionally, the study underscores the impact of model architecture on performance, highlighting the critical role of both linguistic alignment and model size in enhancing NER in specialised domains.

Keywords: Named entity recognition · Large Language Models · Portuguese language variants

1 Introduction

Named Entity Recognition (NER) is a crucial task in Natural Language Processing (NLP). The objective of NER is to identify and classify specific terms in a sentence, such as the names of people, organisations, locations, dates, and other entities. This task is fundamental for various NLP applications, including information extraction, question answering, and automatic text summarisation. Although many studies and models have been developed with good results, most have focused on general entities, relying on datasets such as CoNLL 2002 for English [23] and HAREM for Portuguese [17].

However, specialised domains often cannot utilise these general classifiers effectively because of domain-specific language or the need to identify different types of entities not commonly present in general datasets. For instance, while entities such as gene names and diseases are crucial in the biomedical domain, in legal texts, entities such as law citations and court names are more relevant. Previous studies have addressed these challenges by delving into specific domains and proposing tailored datasets and models [13,18,20].

As specialised models on Portuguese variants became recently available, [15,21], this study evaluates the performance of these models on specialised texts written in two variants of the Portuguese NER annotated corpus (European and Brazilian). We aim to compare how these models perform in these variants, considering two specialised domains, historical and legislative, that present unique challenges and require domain-specific knowledge.

By advancing NER capabilities in domain-specific Portuguese texts in two variations of the language, we aim to facilitate more accurate and contextually relevant information extraction, supporting various applications, from historical research to legal document processing. The legislative and historical fields have the advantage of being comprehensive despite some specialization. This study contributes to the academic understanding of domain-specific NER and provides practical insights and tools for improving NLP applications in specialised fields.

Our main contributions are: (i) a comprehensive evaluation of different models on NER tasks within Portuguese texts, (ii) an analysis of the differences in model performance when pre-trained with European and Brazilian variants of the Portuguese language, and (iii) insights into applying these models across different specialised domain corpora: legal and historical.

2 Related Work

Named Entity Recognition (NER) involves identifying and classifying entities such as locations, organisations, and persons within a text. General entities have been extensively explored in various languages and corpora. Popular datasets for evaluating NER in different languages include CoNLL 2002 for English [23] and HAREM for Portuguese [17]. These datasets have been instrumental in advancing NER research and establishing model performance benchmarks.

However, the classifiers trained on these datasets are not always sufficient for all domains. For example, Silva et al. (2023) [20] proposed a dataset focused on *cachaça*, a distilled spirit made from sugarcane juice, where important entity categories include classification, price, storage time, and sensory characteristics, such as colour. These particular entities cannot be adequately represented using only the traditional NER categories.

The legal domain has also been found to benefit from the use of specialised entities, such as legal citations from legislation or court cases [3]; products of law and legal basis [2]; courts, origins of legal procedure, and trial dates [7]; and decisions and sentences [5]. Recent studies have explored these areas using various techniques with transformer models, including domain adaptation [4,9], fine-tuning [25], in-context learning [14], and self-learning [13].

Historical texts represent another important domain with unique characteristics. Studies focusing on NER tasks in historical texts [10,18,24] have highlighted the challenges posed by different word spellings and sentence structures compared with contemporary texts in the same language.

All the works discussed so far have focused on NER tasks within the context of Portuguese language texts. In this study, we extend this focus by evaluating the impact of Portuguese-specific models on specialised texts, specifically within the historical and legal domains, exploring both European and Brazilian language variants.

3 Corpora

3.1 European Portuguese Annotated Historical Texts

The first *sub-corpus* considered in this study is a subset of the historical PT-Eu *corpus*, the *Parish Memories* collection. It is an 18th-century *corpus* composed of the answers to a survey sent in 1758 to the Portuguese priests to obtain feedback about the state of the territory after the 1755 Lisbon big earthquake and to gather information to form a Geographical Dictionary of Portugal. The sub-corpus under analysis contains 71 Memories of the municipalities of the main cities in the largest region in Portugal, Alentejo, i.e., Portalegre, Évora and Beja. We also added the Memories from the municipality of Vila Viçosa, a historically relevant municipality, dominated by the House of Bragança. The original manuscripts have been transcribed and normalised to contemporary standard European Portuguese spelling. The *sub-corpus* was annotated with named entities, customised to the historical reality.

In a first approach (Anonymous, 2021) three basic categories were considered (*person, local, organisation*), as, for a historian, they aim to answer the main questions: *Who, Where, and When*. After this initial approach, considering the need to describe past realities better, a *corpus*-based study was conducted to define extensions of these categories (Anonymous, 2022) according to their relevance to the historian's inquiry. For that, the main category of *person* was broken down into several subcategories.

The category *person* (PER) refers to references made by name, first name, and family name (PER_NAM); occupation (PER_OCC); or social category (PER_CAT). All these attributes reflect the hierarchical structure of the 18th-century Portuguese society as, frequently, titles and occupation positions were almost part of a person's name and identity. An example of mentions to persons by occupation is *Reitor da Universidade de Évora* (Rector of the University of Évora).

Still related to the PER category, and because they constitute specific details of *person*, we established other subcategories to label saints (PER_SAINT); divinities (PER_DIV); groups of persons (PER_PGRP); and authors (PER_AUT). The subcategory for groups of persons is used to annotate organic groups, families, and members of an organisation, among others. Monges Cartuxos (Carthusians monks) and Sarracenos (the Saracens) are examples of this category.

Concerning the *local* category, we generalised it to place (PLC). This category includes geopolitical entities (PLC_GPE), aquifers(PLC_AQU), mountains (PLC_MOUNT), facilities (PLC_FAC), and one extra subcategory for other locations (PLC_LOC) as for instance, Bispado de Portalegre (Bishopric of Portalegre, PLC_GPE) and Rio Guadiana (Guadiana river, PLC_AQU).

Geopolitical entities were included to avoid ambiguities among locations and organisations, as this categories aggregates them indistinctly. Other references to geographical points, such as rivers and mountains, are essential for geo-references.

The remaining categories are for organisation, time, and authored work. The ORG category labels several organisations as, for example, Companhia de Jesus (Society of Jesus), and Universidade de Coimbra (University of Coimbra).

For TIM_CRON, we only annotated specific references to dates, for instance, *o ano de 1755* (the year of 1755). We also established a category, AUTWORK, that allows us to treat the text sources mentioned in the *corpus* to recognise the text sources mentioned by priests.

An extended set of NE categories to account for past ages also implies more complexity in annotation and their computational processes. All the documents of the *sub-corpus* were manually annotated based on the consensual judgment of four annotators, and it was made using the INCEPTION platform[1].

Table 1. Frequency of named entities in the Parish Memories for each type.

CATEG	Train	Dev	Test	Overall NE
AUTWORK	106	12	19	137
ORG	287	52	54	393
PER_AUT	101	13	15	129
PER_CAT	37	4	8	49
PER_DIV	119	25	40	184
PER_NAM	520	62	136	718
PER_OCC	88	11	25	124
PER_PGRP	153	25	21	199
PER_SAINT	435	76	133	644
PLC_AQU	147	13	68	228
PLC_FAC	202	18	69	289
PLC_GPE	785	84	232	1101
PLC_LOC	336	24	87	447
PLC_MOUNT	50	10	13	73
TIM_CRON	217	33	66	316
Total	3,583	462	986	5031

---

[1] https://inception-project.github.io.

As can seen in Table 1, we have 5,031 annotated NEs. The major classes are geo-political entities, person names, and saints. Persons are referenced only by category, and mountains are the least represented. For training, development, and testing, the distribution is 70, 10, and 20%.

3.2 Brazilian Portuguese Annotated Legislative Texts

Table 2. Frequency of named entities in UlyssesNER-Br for each type.

CATEG	Train	Dev	Test	Overall NE
DATA	433	72	98	609
EVENTO	9	5	9	23
FUNDapelido	123	24	34	181
FUNDlei	359	81	85	522
FUNDprojetodelei	8	2	5	15
LOCALconcreto	333	139	88	560
LOCALvirtual	36	6	13	55
ORGgovernamental	324	60	68	452
ORGnaogovernamental	88	10	22	120
ORGpartido	23	11	4	38
PESSOAcargo	224	38	40	302
PESSOAgrupocargo	121	17	20	158
PESSOAindividual	283	59	59	401
PRODUTOoutros	173	38	42	253
PRODUTOprograma	42	6	11	59
PRODUTOsistema	15	2	1	18
Total	2594	570	599	3763

The legislative *corpus* used in this work is UlyssesNER-Br [2]. It comprised 150 bills from the Brazilian Chamber of Deputies (BCoD). Annotations were performed in three phases by two undergraduate students and one graduate student as a curator.

UlyssesNER-Br [2] included both coarse and fine-grained levels. The coarse-grained level comprises seven entity categories, whereas the fine-grained level includes eighteen entity types. The entities follow the traditional HAREM [16] entities (*person, location, organisation, event,* and *date*) with the addition of domain-specific entities such as law foundations and law products.

The entity PESSOA (person) was specialised in three types: PESSOAindividual (individual), PESSOAcargo (occupation), and PESSOAgrupocargo (group of occupations). In this division, it is possible fine-grained levels of person citations, such as Deputado HILDO ROCHA (Deputy HILDO ROCHA), and 16 de março de 2011 [March 16, 2011].

FUNDAMENTO (law foundation) refers to various legal entities such as laws, bills, and legislative consultations requested by congressmen. This category includes fine-grained entities, such as FUNDlei (legal norm), FUNDapelido (legal norm nickname), and FUNDprojetodelei (bill). Examples of fine-grained entities include art. 34 do Estatuto do Idoso (art. 34 of the Elderly Statute), and Código Brasileiro de Trânsito (Brazilian Traffic Code).

The final category, PRODUTODELEI (law product), pertains to anything created due to legislation. This class also includes three fine-grained types: PRODUTOsistema (system product), PRODUTOprograma (program product), and PRODUTOoutros (other products). Examples of each fine-grained class are Sistema Único de Saúde (Unified Health System), and salário mínimo (Minimum wage).

4 Framework and Models

We adopted the Flair [1] framework, a NER library for multiple languages developed in PyTorch[2]. This framework provides pre-trained language models, named entity recognition models, and neural networks for language model training and sequence tagging. With Flair, we can construct pipelines for training token classifiers and feed them with various types of language models, such as Word Embeddings, Transformer-based models, and Flair Embeddings itself.

Herein, we analyse four versions of language models.

BERTimbau [21] is a pre-trained transformer-based language model trained specifically for Brazilian Portuguese. It was trained on the *brWaC corpus* [8], which amounts to a total of 2.6 billion tokens, resulting in 17.5 GB of preprocessed data. BERTimbau was trained using token masking in input sentences. In other words, it is a Masked Language Model (MLM). We chose this model because the current state-of-the-art [22] in NER for Portuguese uses this model. We used the Large version of BERTimbau[3].

Albertina PT-* [15] is a large language model designed explicitly for Portuguese. It functions as an encoder within the BERT family and is built upon the DeBERTa model using the Transformer neural architecture. Albertina PT- has two variants: Albertina PT-PT and Albertina PT-BR. Both variants are distributed free of charge, under a permissible license.

Albertina PT-PT is the European Portuguese version. This model is available in three sizes, specifically with 1.5 billion parameters, 900 and 100 million parameters. The pre-training *corpora* of the Albertina PT-PT 1.5B comprises general and legislative domains.

Albertina PT-BR focuses on Brazilian Portuguese. Its largest version, Albertina 1.5B PT-BR [19], has a more permissive licensed model without using

---

[2] https://pytorch.org/.
[3] https://huggingface.co/neuralmind/bert-large-portuguese-cased.

the BrWac dataset, consisting of a 36 billion token dataset compiled from a multilingual *corpus*. Since this *corpus* includes both European and Brazilian Portuguese, additional filtering was applied to retain only documents with metadata indicating Brazil's internet country code top-level domain.

XLM-RoBERTa [6] is a multilingual model designed to understand 100 languages without requiring language-specific tensors, as it can identify the language directly from the input identifiers. It incorporates techniques from RoBERTa [12] into the XLM framework [11], focusing solely on masked language modeling for single-language sentences, and does not employ translation language modeling.

Table 3 compares the four described models according to some of their features.

Table 3. Comparison of the four Portuguese Large Language Models.

Feature	Albertina PT-PT	Albertina PT-BR	XLM-R	BERTimbau
Params	1.5B	1.5B	550M	355M
Corpus	CulturaX,DCEP,Europarl,ParlamPT	CulturaX	Multi-data	brWaC
Arch.	DeBERTa,24L,16H	DeBERTa,24L,16H	Trans,24L,16H	BERT,24L,16H
Lang.	PT (PT)	PT (BR)	100 languages	PT (BR)
Domain	General,Legislative	General	General	General
Year	2023	2023	2019	2021

5 Experimental Evaluation

We conducted our experiments on a GPU A100 with 80GB of RAM and an RTX 4090 with 64GB of RAM. The experiments used Python 3.7.6 and the Flair library to use pre-trained models.

Hyperparameters were set to the default values recommended by the library: a learning rate of 5e-5, a mini-batch size of 4, and training for 10 epochs. Truncation was applied to the maximum length, and padding was set to true.

We employed standard metrics for model evaluation, including accuracy, precision, recall, and micro F1-score using the CoNLL-2002 script [23]. These metrics provided a comprehensive assessment of the model's performance across different classes, ensuring a thorough evaluation of its effectiveness.

6 Results and Discussion

6.1 Overall Results

First, we detail the results for each corpus individually. Then, we combine the results for comparative analysis, highlighting the influence of linguistic context and textual domain on the performance of specialised NER models.

Table 4. Parish Memories models results.

Model	Precision	Recall	F1	Δ ↑
Albertina PT-PT	**72.76**	**76.10**	**74.39**	+3.02
Albertina PT-BR	69.71	73.11	71.37	+0.61
XLM-R-Large	68.31	73.38	70.76	+0.23
BERTimbau-Large	67.36	74.00	70.53	*bl*

Parish Memories Corpus. Table 4 presents the overall results, highlighting that Albertina, trained in European Portuguese, achieved superior F1-Score results compared to other models. This success is likely not solely due to Albertina having more parameters than previous models but primarily due to its pre-training on European Portuguese texts. Given that, although the corpus consists of 18th-century Portuguese texts, they were used in their normalized version in European Portuguese, and this linguistic alignment may likely have contributed significantly to Albertina's performance.

Another notable observation concerns the differences in F1-Scores, as indicated in the column Δ ↑. The largest discrepancy among the models' results is observed with European Portuguese Albertina, which markedly outperforms its Brazilian Portuguese counterpart.

The recall metric also yields insightful conclusions. BERTimbau-Large exhibits the highest recall among Brazilian Portuguese and multilingual models, surpassing Albertina and XLM-R-Large, suggesting that its smaller architecture can achieve comparable entity identification. However, precision trends align with model size, indicating that larger models tend to identify true positive instances more precisely. Leveraging a model trained on the same Portuguese variant as the *corpus* consistently yields the best results across all metrics.

Table 5. UlyssesNER-Br types level models results.

Model	Precision	Recall	F1	Δ ↑
Albertina PT-PT	86.83	**91.32**	**89.02**	+0.08
Albertina PT-BR	**87.93**	89.98	88.94	+0.18
BERTimbau-Large	84.14	90.32	87.12	+1.18
XLM-R-Large	82.39	89.82	85.94	*bl*

Brazilian Legislative Texts. Table 5 presents the results for the fine-grained level in the UlyssesNER-Br *corpus*. It highlights that larger models can achieve better results, but the contribution of language specificity to enhanced performance remains inconclusive. The challenge of language specificity is evident in Table 5, where the increment in F1-Score for the two best models (the variations of Albertina) is only 0.08, possibly due to the legislative data during the pre-training of European Portuguese Albertina (see Table 3), or possibly to random classifier initialisation and result fluctuations.

Comparing the models, a significant finding is BERTimbau's high recall compared to larger models, underscoring its effectiveness in accurately identifying positive instances within specific legislative contexts, even if it is the smallest model. However, its lower precision suggests potential misses in positive instances.

Larger models, such as XLM-R and Albertina, excel in capturing fine-grained categories. While BERTimbau demonstrates superior recall in fine-grained analysis, the trade-off between precision and recall results in larger models achieving better F1-Score results with comparable recall. The Albertina models achieve the best balance between precision and recall. This balance indicates their effectiveness in both capturing a large number of relevant instances and maintaining accuracy.

Comparative Analysis. The linguistic variation between Brazilian Portuguese and European Portuguese differed significantly across the two *corpora*, highlighting how the *Parish Memories corpus* could leverage models tailored to its specific variant more effectively than Brazilian legislative texts. One plausible explanation lies in the domain and structure of the texts.

The *Parish Memories corpus* comprises letters written by priests with varying levels of academic background and formality. These texts differ from contemporary European Portuguese but maintain significant linguistic proximity, especially in their textual references. While the entities mentioned refer to things, people, and events of the 18th-century, many of these references are still used and mentioned today, such as place names like Coimbra and Évora and local parish names and devotions that continue in the country. Therefore, it is reasonable to argue that a model trained on European Portuguese is better equipped to extract entities from this *corpus*.

Conversely, even though the Brazilian legislative *corpus* contains contemporary texts, many of these details were not necessarily learned from general texts in either Portuguese variant by the structure and specific jargon of legislative texts. Thus, it becomes that, for this *corpus*, the model size was more decisive than the linguistic variation.

Our results and analysis illustrate how linguistic context and textual domain play a crucial role in shaping the selection and performance of NER models. It underscores the importance of meticulously evaluating the *corpus* to select a suitable model that optimizes the extraction of named entities within specific contexts.

6.2 Results by Categories

We present the results for each *corpus* at the entity level. We discuss how each model learned each entity and what is the influence of specialised entities. Sub-

Table 6. Parish Memories results per entity.

CATEG	BERTimbau			XLM-R			Albertina (BR)			Albertina (PT)		
	P	R	F1	P	R	F1	P	R	F1	P	R	F1
AUTWORK	45.83	52.38	48.89	47.83	55.00	51.16	55.00	52.38	53.66	70.00	66.67	68.29
ORG	48.05	67.27	56.06	53.23	55.93	54.50	57.14	58.18	57.66	64.29	65.45	67.86
PER_AUT	77.78	87.50	82.35	78.95	93.75	85.71	75.00	93.75	83.33	83.33	93.75	88.24
PER_CAT	87.50	87.50	87.50	50.00	75.00	60.00	38.89	87.50	53.85	53.33	100.00	69.57
PER_DIV	76.74	82.50	79.52	69.57	80.00	74.42	82.50	82.50	82.50	87.80	90.00	88.89
PER_NAM	61.04	67.63	64.16	66.23	71.83	68.92	61.88	71.22	66.22	68.59	76.98	72.54
PER_OCC	44.12	60.00	50.85	60.71	62.96	61.82	55.17	64.00	56.26	70.37	76.00	73.08
PER_PGRP	50.00	61.90	55.32	55.17	76.19	64.00	69.57	76.19	72.73	69.57	76.19	72.73
PER_SAINT	77.37	79.10	78.23	75.69	78.99	77.30	78.79	77.61	78.20	77.21	78.36	77.78
PLC_AQU	66.20	67.14	66.67	72.73	76.71	74.67	81.25	74.29	77.61	80.00	74.29	77.04
PLC_FAC	65.33	67.12	66.22	59.52	66.67	62.89	68.12	64.38	66.20	67.14	64.38	65.73
PLC_GPE	77.87	81.55	79.66	78.84	77.87	78.35	79.22	78.54	78.88	78.45	78.11	78.28
PLC_LOC	65.35	74.16	69.47	60.00	72.53	65.67	55.24	65.17	59.79	65.38	76.40	70.47
PLC_MOUNT	56.25	69.23	62.07	75.00	92.31	82.76	75.00	92.31	82.76	80.00	92.31	85.71
TIM_CRON	69.33	77.61	73.24	66.67	65.71	66.19	70.00	73.13	71.53	65.28	70.15	67.63

sequently, we present a comparative analysis to conclude how different levels of entities can help or affect the final model results.

Parish Memories. Table 6 shows the results at the entities level. Albertina in European Portuguese tends to consistently achieve the best results in F1-Score in most categories, pointing out that the specific features of the language in the model can be advantageous to the *corpus*. Models like Albertina (PT) and (BR) show a better balance between precision and recall, leading to higher F1 scores across many categories. This balance is important for practical applications where false positives and negatives must be minimised.

The table also highlights that Albertina (PT) achieved the highest results, with no F1-Scores near 50%, unlike a random classifier. In contrast, BERTImbau had three classes with scores near 50% and one with even lower results. Additionally, it is noteworthy that Albertina (BR) had five classes with results near 50% (the highest number among the models). However, the overall result compensated for the higher number of classes with F1 scores of 70% or more.

We observe specific trends and outliers in the F1-Scores in a bird's eye analysis of the categories in Table 6. In AUTWORK and PLC_MOUNT, the result increased when the model size was increased (see Table 3 for reference on model sizes), and the best result was in the Portuguese European Albertina model.

The Albertina (PT) model consistently performs well in identifying specialised persons. However, an outlier emerges in the PER_CAT (social category), where BERTimbau exhibits a notable increase of 17.93 points compared to the second-best result achieved by Albertina (PT). This unexpected outcome could be attributed to BERTimbau's pre-training on the brWaC corpus

[8], which includes a diverse range of content potentially encompassing social category data. This may have enhanced BERTimbau's ability to accurately recognize entities within social contexts. Additionally, while BERTimbau shows a slight improvement in the PER_SAINT category with a marginal increase of 0.45 points, this improvement is not significant enough to decisively conclude it performs better than Albertina in this category.

Similarly, place entities consistently performed well with the Albertina (PT) model, often achieving results that were either the best or very close to it. For instance, in PLC_AQU, PLC_FAC, and PLC_GPE, Albertina (PT) demonstrated high precision and recall, with marginal differences from the top performer ranging from 0.57 to 1.38 (the largest difference observed). This suggests Albertina (PT) can identify and classify various place-related entities within the *corpus*.

However, it's notable that BERTimbau occasionally surpassed Albertina (PT) in specific categories, such as TIM_CRON, indicating its capability to excel in contexts where temporal references are crucial. These nuances of performance highlight the strengths of each model in different entity categories. It emphasises the importance of considering specific contexts and the balance in the results distributions in evaluating effectiveness across varied entity types (Table 7).

Table 7. UlyssesNER-Br type level results per entity.

CATEG	BERTimbau			XLM-R			Albertina (BR)			Albertina (PT)		
	P	R	F1	P	R	F1	P	R	F1	P	R	F1
DATA	100.00	94.23	97.02	96.00	96.97	97.92	95.92	96.91	98.00	100.00	98.99	98.99
EVENTO	77.77	100.00	87.50	87.50	77.78	82.35	100.00	66.67	80.00	85.71	66.67	75.00
FUNDapelido	94.12	100.00	97.00	90.91	88.24	89.55	93.75	88.24	90.91	93.75	88.24	90.91
FUNDlei	98.82	90.32	94.38	74.51	89.41	81.28	91.11	96.47	93.71	90.22	97.65	93.79
FUNDprojetodelei	20.00	100.00	33.33	100.00	20.00	33.33	100.00	40.00	57.14	50.00	20.00	28.57
LOCALconcreto	94.32	84.69	89.25	87.37	94.32	90.71	87.50	95.45	91.30	89.36	95.45	92.31
LOCALvirtual	38.46	23.81	29.41	44.44	61.54	51.61	33.33	46.15	38.71	44.44	61.54	51.61
ORGgovernamental	82.35	78.87	80.58	77.92	88.24	82.76	80.82	86.76	83.69	80.00	88.24	83.92
ORGnaogovernamental	90.91	80.00	85.11	77.27	77.27	77.27	94.44	77.27	85.00	90.00	81.82	85.71
ORGpartido	100.00	66.67	80.00	100.00	100.00	100.00	100.00	100.00	100.00	100.00	100.00	100.00
PESSOAcargo	97.50	90.70	93.98	88.64	97.50	92.86	92.86	97.50	95.12	86.67	97.50	91.76
PESSOAgrupocargo	95.00	90.48	92.68	90.48	95.00	92.68	94.74	90.00	92.31	85.71	90.00	87.80
PESSOAindividual	96.61	96.61	96.61	96.72	100.00	98.33	96.61	96.61	96.61	93.55	98.31	95.87
PRODUTOoutros	76.19	64.00	69.57	53.33	76.19	62.75	66.67	80.95	73.12	64.71	78.57	70.97
PRODUTOprograma	54.55	75.00	63.16	100.00	54.55	70.59	100.00	54.55	70.59	100.00	54.55	70.59
PRODUTOsistema	100.00	100.00	100.00	100.00	100.00	100.00	100.00	100.00	100.00	100.00	100.00	100.00

Brazilian Legislative Texts. Table 5 shows the results at the entity level. Similar to the conclusions drawn in Sect. 6.1, we observe that top results are well-distributed among the models BERTimbau, XLM-R, Albertina (BR), and Albertina (PT), with 4, 6, 6, and 8 instances of achieving the best scores, respectively. Additionally, while there are several instances of closely competitive results across models, having the worst result in a category does not imply poor performance, as evidenced by five instances where the worst result still exceeds 89%. This demonstrates the models' effective learning of the representations of many target entities.

Regarding the specific classes, Date (DATA) is notably well learned by all the models, with the minimum F1-Score varying between 97.02% and 98.99%. It is an expected result since Date is the largest class, with 433 examples in the training set, and carries specific patterns in its format.

The class Event (EVENTO) was also learned by all models, although results exhibited greater variability compared to the Date class, ranging from 75.00% to 87.50%. Event is a minority class in the training set, comprising only 9 examples. BERTimbau Large stood out by achieving the highest performance for this class. This contrasts with previous findings [13], which demonstrated that BERTimbau Base failed to learn Event, resulting in an F1-Score of 0%. This observation supports the hypothesis discussed in Sect. 6.1 that model size, rather than idiomatic variation, plays a crucial role in performance on this *corpus*.

The location (LOCAL) classes exhibited varying results and trends. First, concrete places (LOCALconcreto) were well-learned entities, with F1 Scores ranging from 89.25% to 92.31%, demonstrating an increase in performance with larger models. Concrete places have specific semantics related to geographical landmarks, which aid the models in identifying context and patterns.

In contrast, virtual places (LOCALvirtual) performed worse, with F1 Scores between 29.41% and 51.61%. A virtual place is a more diverse entity encompassing different types of locations, which do not necessarily share the same meaning and can include entities such as newspapers and internet pages [16], making it more challenging for the models to learn. Additionally, annotations in this broader category may lead to semantic conflicts. For example, the term *Jornal Diário Catarinense* (Diário Catarinense newspaper) in the sentence *Destaco que nos anos 90, o Jornal Diário Catarinense realizou uma pesquisa popular que colocou entre os 20 catarinenses do século*[4] could be understood as both a location and a general organisation [14], which is challenging for the model to discern.

Organisations (ORG) showed a slight variation in results across different models. The results of the specialised classes of organisations highlight the relationship between the number of training examples in each class and their respective F1 Scores. As seen in Table 2, governmental organisations (ORGgovernamental), non-governmental organisations (ORGnaogovernamental), and political parties (ORGpartido) had 324, 88, and 23 examples in the training set, respectively. Despite the number of examples, the results were inversely related, with the minority class achieving the highest F1-Score and the majority class also achieving a high score, as shown in Table 5. One possible explanation for this behaviour is the specificity of the class. For instance, although the governmental

---

[4] English translation: *I highlight that in the 90 s, Jornal Diário Catarinense carried out a popular survey that placed it among the 20 Santa Catarina citizens of the century.*

organisation class pertains to a specific domain, it encompasses a wide range of entities, from municipal guards to the Chamber of Deputies. The same applies to non-governmental institutions. Conversely, political parties typically appear in more specific contexts or share similar characteristics in their names, such as following the name of a deputy or including the word "party" in their titles.

The classes related to Law Foundation - Legal Norm Nickname (FUNDapelido), Legal Norm (FUNDlei), and Law Proposals (FUNDprojetodelei) - exhibited varying levels of learning. The discrepancies in results are likely attributed to the number of training examples available for each class. For example, Legal Norm Nickname and Legal Norm achieved high F1 Scores of 97.00% and 94.38%, respectively, with 123 and 359 examples. In contrast, Law Proposals, despite having a specific format, such as the example "PEC 187/2016", had only 8 examples in the training set. The highest results were concentrated in BERTimbau in Law Foundation classes.

The specialised classes of law products (PRODUTO) exhibited a trend of improved performance with larger models. This was particularly evident in the other products class (PRODUTOoutros), which achieved its best results with Albertina (BR). This class showed strong performance, specifically in the Portuguese models, with the highest result being the Brazilian Portuguese version of Albertina. Regarding the program product class (PRODUTOprograma), the best results were shared among XLM-R and both variations of Albertina, with identical scores across all metrics. This consistency necessitates a more detailed investigation into the class distribution and model learning. Lastly, the system product class (PRODUTOsistema) had only one example in the test set, and the same term appeared in the training set, making it difficult to determine whether the 100% F1-Score reflects a well-learned class or if the model specialised in recognising this specific term.

Comparative Analysis. Regarding the models, Albertina (PT) demonstrated superior performance in the Parish Memories corpus due to its European Portuguese training, particularly in recognising person and place entities. BERTimbau showcased strengths in identifying entities influenced by its diverse pre-training data, achieving comparable results to larger models, especially in the Brazilian Legislative Texts. XLM-R and Albertina (BR) also showed competitive performance across various entity types, highlighting their versatility.

The focus on European Portuguese-specific entities allowed Albertina (PT) to outperform other models in the Parish Memories *corpus*. Additionally, increasing the model size significantly improved performance in this dataset.

In the Brazilian Legislative Texts *corpus*, the diversity and specificity of the entities, coupled with the number of training examples, greatly influenced the models' performance. Larger models generally performed better, especially in minority and specific entity classes. Moreover, the inclusion of legal data in Albertina's (PT) pre-training likely contributed to its strong performance in this *corpus*.

7 Conclusion

We investigated the influence of Portuguese language variants in pre-trained Language Models on Named Entity Recognition (NER) within specialised domains. Our experiments used two distinct *corpora*: a historical corpus in European Portuguese and a legislative corpus in Brazilian Portuguese. We evaluated models pre-trained specifically for Brazilian Portuguese (BERTimbau and Albertina PT-BR), European Portuguese (Albertina PT-PT), and multilingual contexts (XLM-R). Our analysis delved qualitatively into specific class examples, semantics, and compositions.

Several conclusions can be drawn based on our findings. Models tailored to the textual and linguistic context, such as Albertina for European Portuguese, demonstrated superior performance in NER tasks, particularly in the Parish Memories corpus. Albertina excelled in precision and recall, extracting entities from 18th-century texts because of its linguistic alignment. Conversely, larger models such as XLM-R and Albertina showed an enhanced precision-recall balance in the Brazilian legislative texts, underscoring the critical role of model size in handling fine-grained categories.

Future research should focus on granular error analysis and interpretation. Understanding model improvements and challenges is crucial, particularly in ambiguity and linguistic complexity. Exploring the performance of techniques such as semi-supervised learning and transfer learning methods will further advance the field of NER in specific textual domains.

Acknowledgements. This work has received funds from the Coordenação de Aperfeiçoamento de Pessoal de Nível Superior - Brazil (CAPES) - Finance Code 001, the Brazilian funding agency CNPq, and the Portuguese Science Foundation FCT, in the context of the projects CEECIND/01997/2017 and UIDB/00057/2020 https://doi.org/10.54499/UIDB/00057/2020.

References

1. Akbik, A., Bergmann, T., Blythe, D., Rasul, K., Schweter, S., Vollgraf, R.: FLAIR: an easy-to-use framework for state-of-the-art NLP. In: NAACL 2019, 2019 Annual Conference of the North American Chapter of the Association for Computational Linguistics (Demonstrations), pp. 54–59 (2019)
2. Albuquerque, H.O., et al.: UlyssesNER-Br: a corpus of Brazilian legislative documents for named entity recognition. In: International Conference on Computational Processing of the Portuguese Language, pp. 3–14. Springer (2022)
3. Luz de Araujo, P.H., de Campos, T.E., de Oliveira, R.R., Stauffer, M., Couto, S., Bermejo, P.: LeNER-Br: a dataset for named entity recognition in Brazilian legal text. In: Computational Processing of the Portuguese Language: 13th International Conference, PROPOR 2018, Canela, Brazil, September 24–26, 2018, Proceedings 13, pp. 313–323. Springer (2018)
4. Bonifacio, L.H., Vilela, P.A., Lobato, G.R., Fernandes, E.R.: A study on the impact of intradomain finetuning of deep language models for legal named entity recognition in Portuguese. In: Intelligent Systems: 9th Brazilian Conference, BRACIS

2020, Rio Grande, Brazil, October 20–23, 2020, Proceedings, Part I 9, pp. 648–662. Springer (2020)
5. Brito, M., Pinheiro, V., Furtado, V., Neto, J.A.M., Bomfim, F.d.C.J., da Costa, A.C.F., Silveira, R.: Cdjur-br-uma coleção dourada do judiciário brasileiro com entidades nomeadas refinadas. In: Anais do XIV Simpósio Brasileiro de Tecnologia da Informação e da Linguagem Humana, pp. 177–186. SBC (2023)
6. Conneau, A., et al.: Unsupervised cross-lingual representation learning at scale. arXiv preprint arXiv:1911.02116 (2019)
7. Correia, F.A., et al.: Fine-grained legal entity annotation: a case study on the Brazilian supreme court. Inf. Process. Manage. **59**(1), 102794 (2022)
8. Filho, J.A.W., Wilkens, R., Idiart, M., Villavicencio, A.: The BRWAC corpus: a new open resource for Brazilian Portuguese. In: Proceedings of the 11th International Conference on Language Resources and Evaluation, pp. 4339–4344 (2018). http://www.lrec-conf.org/proceedings/lrec2018/summaries/599.html
9. Garcia, E.A., et al.: Robertalexpt: a legal roberta model pretrained with deduplication for Portuguese. In: Proceedings of the 16th International Conference on Computational Processing of Portuguese, pp. 374–383 (2024)
10. Grilo, S., Bolrinha, M., Silva, J., Vaz, R., Branco, A.: The BDCamoes collection of Portuguese literary documents: a research resource for digital humanities and language technology. In: Proceedings of the Twelfth Language Resources and Evaluation Conference, pp. 849–854 (2020)
11. Lample, G., Conneau, A.: Cross-lingual language model pretraining. arXiv preprint arXiv:1901.07291 (2019)
12. Liu, Y., et al.: Roberta: a robustly optimized BERT pretraining approach. arXiv preprint arXiv:1907.11692 (2019)
13. Nunes, R.O., Balreira, D.G., Spritzer, A.S., Freitas, C.M.D.S.: A named entity recognition approach for Portuguese legislative texts using self-learning. In: Proceedings of the 16th International Conference on Computational Processing of Portuguese, pp. 290–300 (2024)
14. Oleques Nunes., R., Spritzer., A., Dal Sasso Freitas., C., Balreira., D.: Out of sesame street: a study of Portuguese legal named entity recognition through in-context learning. In: Proceedings of the 26th International Conference on Enterprise Information Systems - Volume 1: ICEIS, pp. 477–489. INSTICC, SciTePress (2024). https://doi.org/10.5220/0012624700003690
15. Rodrigues, J., Gomes, L., Silva, J., Branco, A., Santos, R., Cardoso, H.L., Osório, T.: Advancing neural encoding of Portuguese with transformer Albertina pt. In: EPIA Conference on Artificial Intelligence, pp. 441–453. Springer (2023)
16. Santos, D., Cardoso, N.: A golden resource for named entity recognition in Portuguese. In: International Workshop On Computational Processing of the Portuguese Language, pp. 69–79. Springer (2006)
17. Santos, D., Seco, N., Cardoso, N., Vilela, R.: HAREM: an advanced NER evaluation contest for Portuguese. In: Calzolari, N., et al. (eds.) Proceedings of the 5th International Conference on Language Resources and Evaluation (LREC'2006)(Genoa Italy 22–28 May 2006) (2006)
18. Santos, J., Cameron, H.F., Olival, F., Farrica, F., Vieira, R.: Named entity recognition specialised for Portuguese 18th-century history research. In: Proceedings of the 16th International Conference on Computational Processing of Portuguese, pp. 117–126 (2024)
19. Santos, R., et al.: Fostering the ecosystem of open neural encoders for Portuguese with Albertina pt-* family (2024)

20. Silva, P., Franco, A., Santos, T., Brito, M., Pereira, D.: CachacaNER: a dataset for named entity recognition in texts about the cachaça beverage. Lang. Resour. Eval. **58**(4), 1315–1333 (2023)
21. Souza, F., Nogueira, R., Lotufo, R.: BERTimbau: pretrained BERT models for Brazilian Portuguese. In: Proceedings of the 9th Brazilian Conference on Intelligent Systems, BRACIS (2020)
22. Souza, F., Nogueira, R.F., de Alencar Lotufo, R.: Portuguese named entity recognition using BERT-CRF. CoRR arXiv:1909.10649 (2019)
23. Tjong Kim Sang, E.F.: Introduction to the CoNLL-2002 shared task: language-independent named entity recognition. In: COLING-02: The 6th Conference on Natural Language Learning 2002 (CoNLL-2002) (2002). https://aclanthology.org/W02-2024
24. Vieira, R., Olival, F., Cameron, H., Santos, J., Sequeira, O., Santos, I.: Enriching the 1758 Portuguese parish memories (Alentejo) with named entities. J. Open Humanit. Data **7**, 20 (2021)
25. Zanuz, L., Rigo, S.J.: Fostering judiciary applications with new fine-tuned models for legal named entity recognition in Portuguese. In: International Conference on Computational Processing of the Portuguese Language, pp. 219–229. Springer (2022)

BASWE: Balanced Accuracy-Based Sliding Window Ensemble for Classification in Imbalanced Data Streams with Concept Drift

Douglas Amorim de Oliveira[✉], Karina Valdivia Delgado, and Marcelo de Souza Lauretto

School of Arts, Sciences and Humanities, University of São Paulo, São Paulo, Brazil
{douglas.amorim.oliveira,kvd,marcelolauretto}@usp.br

Abstract. In the wake of the exponential growth in data generation witnessed in recent decades, the binary classification task within data streams presents inherent challenges due to their continuous, real-time flow and dynamic nature. This paper introduces the Balanced Accuracy-based Sliding Window Ensemble (BASWE) algorithm that leverages Balanced Accuracy, sliding windows, and resampling techniques to effectively handle imbalanced classes and concept drifts, ensuring robust performance even as data patterns evolve. In experiments conducted on 40 datasets, comprising 16 real-world and 24 synthetic datasets generated under three configurations-no drift, gradual drift, and sudden drift-and with varying imbalance ratios, BASWE demonstrated superior performance compared to seven other state-of-the-art algorithms in terms of F1 Score and the Kappa statistic.

Keywords: Ensemble · Concept Drift · Imbalanced Datasets

1 Introduction

The rapid increase in data generation has led to the emergence of data streams, characterized by continuous, real-time data flow that challenges the static nature of traditional datasets [2,11]. In supervised learning tasks, this dynamic environment presents the dual issues of concept drift, also known as data shift [3], where underlying data distributions shift over time, and class imbalance, where one class vastly outnumbers the others. In data streams, this imbalance is not static but can also evolve, intensifying the difficulty of the learning task. The interplay between the dynamic nature of imbalance and the shifting paradigms of concept drift demands a rethinking of traditional approaches to maintain classifier performance [12,15].

Many algorithms have been proposed to address the challenges in data streams classification with imbalanced data, such as CALMID [17], CSARF [18], ROSE [8], OOB [19], SMOTE-OB [4] and UOB [19]. Four of these algorithms,

CALMID, CSARF, ROSE, and SMOTE-OB, implement an explicit approaches to handle concept drift. CSARF, ROSE, OOB, and UOB implement an ensemble approach to handle imbalanced data. Additionally, CSARF also implements a cost-sensitive approach.

One ensemble algorithm that showed robust results for binary classification in data streams with concept drift is the Kappa Updated Ensemble (KUE) [7]. By implementing the Very Fast Decision Tree (VFDT) [13] algorithm as a basis for its experts and creating policies for model updates, vote abstention, and model assessment using the Kappa statistic, the KUE outperformed other state-of-the-art algorithms [7]. However, KUE was not designed for data streams with imbalanced classes.

This paper proposes an algorithm for binary classification named the Balanced Accuracy-based Sliding Window Ensemble (BASWE), inspired by KUE. BASWE uses two sliding windows, resampling techniques, and replaces the Kappa statistic with the Balanced Accuracy metric in ensemble updates. These modifications aim to achieve a more robust ensemble with higher performance in the scenario of classification in data streams with imbalanced classes and concept drifts. The contributions of this work can be summarized as follows:

- Introduction of BASWE, an ensemble algorithm designed to manage binary classification in data streams with imbalanced classes and concept drift.
- The adoption of Balanced Accuracy as a pivotal metric in guiding ensemble updates, encompassing voting strategies, model performance measurement, and expert substitution.
- The use of class-specific sliding windows and a resampling step in pre-processing. This strategy is tailored to address imbalanced data streams, effectively reducing the imbalance ratio during the model training phase with new data chunks.
- A comprehensive experimental evaluation comparing BASWE against state-of-the-art algorithms.

2 Data Stream Classification

Data streams can be understood as data instances generated at high speed and arriving continuously, presenting challenges to computational systems for storage and processing [10]. However, if efficiently analyzed, they provide an important source of information for real-time decision-making support.

A data stream is characterized as a sequence denoted by $S = \langle S_1, S_2, \ldots, S_n, \ldots \rangle$, where each S_j represents a collection of instances with a size $N \geq 1$. In the particular case where $N = 1$, the context is referred to as online learning [10], otherwise, it is called learning by chunks. Typically, each instance s_t within each set S_j is independently and randomly produced following a stationary distribution D_j.

The data stream classification task aims to predict the correct label y_t, for each incoming instance $s_t \in S_j$. Each instance is described by a set of attributes

$X_t = \{x_{t1}, x_{t2}, ..., x_{tn}\}$. A classifier, represented as F, takes the set of instance attributes X_t as input and generates the predicted label $\hat{y}_t$, as output. In our research, we focus on fully labeled binary classification, where the correct label y_t is restricted to one of two possible outcomes. Additionally, every instance s_t within the data stream is pre-labeled.

In several classification problems, the distribution D_j is subject to changes over time, as the characteristics and definitions of the data stream evolve. This phenomenon is called concept drift [12,20]. Concept drifts in the pattern of class distribution can be segmented into 4 categories [12]: sudden, incremental, gradual, or recurrent. In the realm of data streams, concept drift can be addressed either explicitly (using a dedicated drift detector) or implicitly.

Class imbalance in data stream classification often occurs for various reasons. It can be due to the intrinsic nature of the data scope being processed, such as a medical database that includes sporadic occurrences of a rare disease. Alternatively, it could occur purely by chance if the batch of data being processed lacks a representative diversity of instances for each class.

To address the scenario of imbalanced classes in machine learning within data streams, three main methods are commonly employed [14]: (i) data-level methods, where the data is pre-processing using, for example, resampling techniques; (ii) algorithmic-level methods where the learning algorithms are adapted to deal with imbalance classes; and (iii) hybrid methods which represent a combination of the previous two.

3 BASWE

This section introduces the Balanced Accuracy-based Sliding Window Ensemble (BASWE), an algorithm designed specifically for binary classification in imbalanced data streams[1]. This algorithm is inspired by KUE algorithm [6]. BASWE has three main characteristics. The first one is the use of dynamic ensemble techniques, which address dynamic substitutions and provide real-time ensemble updates. Additionally, BASWE implements an abstention policy to address relevance modifications of the experts. The second crucial characteristic is the use of Balanced Accuracy as the primary metric for model evaluation and updating. This strategy is specifically tailored to address the imbalanced data sets that BASWE is designed to handle. The third key feature is the application of sliding windows as a data-level method that can effectively manage class imbalance in the data stream. These latter two characteristics are explained in Sects. 3.1 and 3.2, respectively.

BASWE's experts in the ensemble implement the Very Fast Decision Trees (VFDT) algorithm [13]. This algorithm is recognized for its computational efficiency, making it ideal for handling real-time data streams. It operates by building decision trees incrementally, accommodating massive volumes of data while minimizing memory usage.

---

[1] Code and datasets available on https://github.com/BASWE-Paper/BASWE.

3.1 Balanced Accuracy Approach to Handle Ensemble Update

Balanced Accuracy (BA) (Eq. 1) provides a robust measure for binary classification problems, especially when dealing with imbalanced datasets.

$$\text{Balanced Accuracy} = \frac{1}{2}\left(\frac{TPR}{TPR+FNR} + \frac{TNR}{TNR+FPR}\right) \quad (1)$$

where TPR is the True Positive Rate and TNR is the True Negative Rate.

In the context of the BASWE algorithm, Balanced Accuracy is used as a driving mechanism for two key operations:

- **Expert substitution within the ensemble.** With every new chunk of data, q new experts are trained outside the ensemble. The ensemble itself consists of k experts ($\gamma_1, ..., \gamma_k$), and if the Balanced Accuracy of the newly trained experts exceeds that of the worst-performing ones in the ensemble, a substitution takes place.
- **Abstention of voting from ensemble experts.** The abstention policy operates by controlling the voting rights of the ensemble experts based on their Balanced Accuracy. When the ensemble computes the majority vote to determine the final classification, the BA of each expert is assessed against a specific threshold (0.5 in our implementation). Any expert with a BA below this threshold is prevented from voting, meaning their classification decision does not influence the ensemble's final output. This abstention policy ensures that the ensemble's decision-making process is not adversely affected by the subpar performance of any of its experts, thereby maintaining the integrity and accuracy of the ensemble's output.

3.2 Sliding Window Approach for Handling Imbalanced Data Streams

In BASWE, the sliding window approach is adopted to manage a sort of data cache, aiming to decrease the disproportion among the classes over time. This methodology contributes to handling the data imbalance and providing a more accurate and representative sample of the current data for model training.

To implement this approach, the algorithm establishes two sliding windows, W_0 and W_1, each corresponding to one class. The size of each sliding window is set to be half of the chunk size that is used for model processing. As each new chunk of data arrives from the data stream, the instances are added to the corresponding sliding window. If a sliding window is full, meaning the number of instances equals the predefined size, the oldest instance is discarded to make room for the newest one. This policy of expelling older instances helps accommodate concept drift, as instances naturally exit the cache over time, thereby reflecting the changing nature of the data stream.

Algorithm 1 shows the FILLWINDOW function. In this function, the algorithm takes as input the current sliding window, the target class covered by this sliding window (*desiredClass*), the current chunk of data (S_i), and the window max size

Algorithm 1. FILLWINDOW function

Input: *window*: sliding window, *desiredClass* : target class, S_i : i-th data chunk, cs : size of sliding windows cache
Output: *window*: sliding window
1: **function** FILLWINDOW(*window*, *desiredClass*, S_i, cs)
2:    **for** each *instance* $\in S_i$ **do**
3:       **if** class of *instance* $=$ *desiredClass* **then**
4:          Add *instance* to *window*
5:          **if** length(*window*) $> cs$ **then**
6:             remove oldest instance in *window*
7:          **end if**
8:       **end if**
9:    **end for**
10:    **return** *window*
11: **end function**

(cs). The chunk is traversed searching for instances of the desired class (lines 2-3), and upon finding an instance, it is added to the sliding window (line 4). After adding a new instance to the sliding window, it checks whether its size has exceeded cs, and if so, removes the oldest instance from the sliding window (lines 5-7). In the end, the function returns the *window* vector incremented by the new instances of the *desiredClass* contained in the i-th chunk.

When the BASWE model is trained, if any of the sliding windows is not fully populated, oversampling is performed. The oversampling function iterates over the vector from the most recent to the oldest instance, replicating instances until the number of instances required to complete the chunk size is achieved.

In this way, the sliding window approach within the BASWE algorithm not only ensures an up-to-date representation of the data for model training but also plays an instrumental role in managing class imbalance and concept drift.

3.3 BASWE's Pseudocode

BASWE's pseudocode is delineated in Algorithm 2, and is segmented into four stages: ensemble initialization, ensemble experts update, training of the new q components, and replacement of the weakest γ expert in the ensemble ϵ.

The main loop (lines 3-34) represents the algorithm over the chunks that arrive throughout the data stream at each iteration. In lines $4-5$, two sliding windows W_0 and W_1 are filled, each for one class. Oversampling is then performed by the OVERSAMPLING function in lines $6-7$, creating W_0' and W_1', which are joined to form the chunk S_i' in line 8. This chunk will be used by the experts in the training phase, while the original chunk S_i will be used to compute the performance.

Upon receiving the first chunk, S_1, we have the process of creating and initializing the ensemble of k experts, represented in lines $9-15$. For each of the k experts in the ensemble, a random integer number r is selected (line 11), drawn

Algorithm 2. BASWE: Balanced Accuracy-based Sliding Window Ensemble

Input: S: data stream, f: number of features, k: number of ensemble components, cs: size of sliding window cache, q: number of new components to train
Output: ε : ensemble of k classifiers $(\gamma_1, ..., \gamma_k)$, φ: subspace of features for each of the k components, BA: Balanced Accuracy for each of the k components

1: $W_0 \leftarrow \emptyset$
2: $W_1 \leftarrow \emptyset$
3: **for** $S_i \in \{S_1, \ldots, S_n\}$ **do**
4:    $W_0 \leftarrow$ FILLWINDOW(W_0, 0, S_i, cs)
5:    $W_1 \leftarrow$ FILLWINDOW(W_1, 1, S_i, cs)
6:    $W_0' \leftarrow$ OVERSAMPLING(W_0, cs)
7:    $W_1' \leftarrow$ OVERSAMPLING(W_1, cs)
8:    $S_i' \leftarrow W_0' \cup W_1'$
9:    **if** S_1 **then**   ▷ Ensemble initialization
10:       **for** $j \in \{1, \ldots, k\}$ **do**
11:          $r \leftarrow$ random integer with uniform probability $[1, f]$
12:          $\varphi_j \leftarrow$ CHOOSEFEATURESSUBSPACE(r)
13:          $\gamma_j \leftarrow$ train new classifier on FILTERFEATURESSUBSPACE(φ_j, S_1')
14:          $BA_j \leftarrow$ compute Balanced Accuracy of γ_j over S_1
15:       **end for**
16:    **else**   ▷ Ensemble experts update
17:       **for** $j \in \{1, \ldots, k\}$ **do**
18:          $BA_j \leftarrow$ compute Balanced Accuracy of γ_j over S_i
19:          $\gamma_j \leftarrow$ incremental train of γ_j on FILTERFEATURESSUBSPACE(φ_j, S_i')
20:       **end for**
21:       **for** $\{1, \ldots, q\}$ **do**   ▷ Train q new experts
22:          $r \leftarrow$ random integer with uniform probability $[1, f]$
23:          $\varphi' \leftarrow$ CHOOSEFEATURESSUBSPACE(r)
24:          $\gamma' \leftarrow$ train new classifier on FILTERFEATURESSUBSPACE(φ', S_i')
25:          $BA' \leftarrow$ compute Balanced Accuracy of γ' over S_i
26:          $BA_{min}, w \leftarrow$ MINBALANCEDACCURACY(ϵ)
27:          **if** $BA' > BA_{min}$ **then**   ▷ Replace weakest $\gamma \in \varepsilon$
28:             $\varphi_w \leftarrow \varphi'$
29:             $\gamma_w \leftarrow \gamma'$
30:             $BA_w \leftarrow BA'$
31:          **end if**
32:       **end for**
33:    **end if**
34: **end for**

from uniform probability $[1, f]$. This value, r, determines the size of the feature subspace that will be used for the j-th expert. In line 12, the CHOOSEFEATURESSUBSPACE method selects a r-dimensional random feature subspace for the j-th expert, with this information stored in the variable φ_j, that will be used to generate diversity in the ensemble and avoid noisy data. In line 13, the FILTERFEATURESSUBSPACE method receives two parameters, the φ_j and the chunk S_1', and then returns the chunk S_1' filtering only the features contained in the φ_j

subspace. The classifier γ_j is then initialized, training with the chunk S_1' filtered by the FILTERFEATURESSUBSPACE method. The Balanced Accuracy of the j-th expert is then computed and stored in the variable BA_j (line 14).

The ensemble model update is performed in lines 17-20 considering the subsequent chunks. Within this process, the algorithm filters the feature subspace φ_j from chunk S_i', conducts incremental training of γ_j (line 18), and computes the Balanced Accuracy of this expert (line 19).

From lines 21 – 26, q new experts are trained. If these new experts exhibit superior performance based on Balanced Accuracy compared to the q least effective experts in the ensemble, they will replace the latter. The higher the value of q, the more intense the behavioral change of the ensemble tends to be, as more components can be replaced in each data chunk. This presents both an advantage of rapid adaptation and a risk of undesirable abrupt changes caused by anomalous chunks. In line 26, the MINBALANCEDACCURACY method conducts a search for the expert in the ensemble ϵ with the lowest performance. The result returned by this method is then stored in BA_{min}, the Balanced Accuracy of the expert, and the position of the expert within the ensemble, represented by w. This substitution takes place in lines 27 – 31 of Algorithm 2, where we compare the Balanced Accuracy of the new expert, denoted as BA', to that of the expert with the lowest Balanced Accuracy, BA_{min}. If the new expert demonstrates superior performance, the expert at position w (represented by γ_w) is substituted with the new γ'. We also update the feature subspace, φ_w, and the Balanced Accuracy, BA_w, of the weakest expert with φ' and BA', respectively.

4 Experimental Results

In this section, we compare BASWE with seven state-of-the-art algorithms.

4.1 Benchmark Algorithms

BASWE was compared with different ensemble classifiers: KUE [7], CALMID [17], CSARF [18], ROSE [8], OOB [19], UOB [19], and SMOTE-OB [4]. CALMID, CSARF, ROSE, UOB, and SMOTE-OB were selected as the top 5 algorithms by a recent survey on imbalanced data streams [1]. Additionally, we included KUE, as it served as the inspiration for BASWE, and OOB, which was proposed by the same authors as UOB and demonstrated good performance in other recent works [8]. Among the selected algorithms, KUE is the only one that was not designed to deal with imbalanced data streams.

All those algorithms are ensembles; however, they implement distinct approaches to address the replacement of new base classifiers. KUE and CALMID algorithms execute, at most, one replacement per new chunk when employed with their default parameters. Conversely, BASWE, CSARF, and ROSE allow the substitution of multiple experts with each new chunk (or instance, in the case of ROSE). On the other hand, the OOB, SMOTE-OB, and UOB algorithms do not replace their base classifiers throughout the data stream.

These algorithms are implemented in MOA and we use its default parameters, as specified in their respective original studies. KUE is the only exception; it has an implementation with $q = 1$ and an additional implementation that permits the substitution of two experts (referred as KUE (q = 2) in this section). In BASWE, we make use of the same KUE's default parameters ($k = 10$), except q, which is equal to 2. BASWE has one additional parameter, cs, that represents the sliding window's cache size, we use the half of chunk size as the value of cs.

4.2 Datasets

Experiments were conducted using 24 binary data stream generators in the MOA (Massive Online Analysis) environment [5], featuring varying degrees of data imbalance. Of these, 8 data streams have sudden concept drift, 8 data streams have gradual concept drift, and the remaining 8 do not have concept drift. Additionally, evaluations were carried out on 16 real-world datasets. All experiments in this study treated the data streams as fully labeled.

Synthetic Datasets: A series of synthetic datasets was generated utilizing the MOA environment. The chosen generators for this process were as follows: Agrawal, Asset Negotiation, HyperPlane, Mixed, Random RBF, Random Tree, SEA, and Sine. Each of these synthetic datasets was synthesized under specified conditions, with a designated random seed value set at 42, a chunk size of 500, and comprising 200,000 instances. All synthetic datasets were designed for binary classification. For a comprehensive analysis, these datasets were crafted with five distinct imbalance ratios: 90:10, 95:5, 97.5:2.5, 99:1, and 99.5:0.5. In these ratios, the value before the colon denotes the proportion of the majority class, while the value after represents the proportion of the minority class. Additionally, the synthetic data streams were generated in the two contexts, with and without concept drift. In the data streams with concept drift, we analyzed sudden and gradual concept drift, the two types of concept drift commonest in the studies [1].

Real Benchmark Datasets: Sixteen real benchmark datasets were utilized for the experimental analysis: Adult, Airlines, Bridges-1VsAll, Census, Covtypenorm-1-2VsAll, Credit.G, Dermatology, Diabetes, Electricity, Gmsc, Mushroom, Sick, Sonar, Vehicle, Vote, and Vowel[2].

4.3 Metrics

The performance of the classifiers was evaluated using the Kappa metric, which was used in [1,7,8], and the F1-Score, which was used in [9,15,16]. For each set of experiments, described in Sect. 4.4, the following measures are computed:

---

[2] Code and datasets are available on https://github.com/BASWE-Paper/BASWE.

- **Absolute Best (AB)** that represents the number of experiments in which the algorithm achieved the best absolute metric (F1-Score or Kappa) value.
- **Equivalent Best using t-test (EB)** that indicates the number of experiments where, despite the algorithm not having achieved the highest metric (F1-Score or Kappa) value, no significant difference was found (indicating a t-test p-value greater than 0.05) between its metric value and the highest metric value among all other algorithms.
- **Total Best (TB)** that is the sum of the Absolute Best and Equivalent Best values, representing the number of experiments where the algorithm demonstrated the best or equivalent to the best metric (F1-Score or Kappa) value.
- **Avg. metric (F1-Score or Kappa)** that represents the average of metric (F1-Score or Kappa) value from all experiments.

The algorithms are ranked by the Total Best value and, in case of a tie, the Avg. metric (F1-Score or Kappa) is used to rank the algorithms. We prioritize the Total Best value over Avg. metric (F1-Score or Kappa) since averages can be skewed by the outcomes from specific datasets. In certain scenarios, an algorithm might exhibit a significantly higher or lower Avg. metric (F1-Score or Kappa) due to unique characteristics inherent to a particular dataset. Hence, it is possible for an algorithm to rank first in Total Best while manifesting the least favorable results when evaluated by Avg. metric (F1-Score or Kappa), due to notably poorer performance in a few experiments.

4.4 Experiments Configuration

In this study, we define an experiment as the execution of an algorithm on a data stream. To compare the algorithms, experiments were performed with synthetic and real data streams. For the synthetic data, the number of streams was 40 without concept drift, 40 with gradual concept drift, and 40 with sudden concept drift. This is because there are 8 synthetic data streams generators, and each has one experiment for every ratio of data imbalance: 90:10, 95:5, 97.5:2.5, 99:1, and 99.5:0.5. For the real data streams, the number of experiments was 16 which corresponds to the number of real data streams. Since 9 algorithms were evaluated across 136 data streams, there are 1,224 experimental configurations. To mitigate potential impacts of performance anomalies associated with specific chunk characteristics or random initialization of the experts, each experiment was executed 10 times, resulting in 12,240 experimental runs.

To calculate the F1-Score and Kappa values for each algorithm on each data stream, the average value of the metrics was calculated over all the chunks of each run, using the "moa.tasks.EvaluatePrequential" configuration of MOA. Each new chunk is processed in two steps: (i) evaluate the current ensemble on the chunk, computing the F1-score and Kappa in line 18 of Algorithm 2; and (ii) update the ensemble with the new chunk in line 19 of Algorithm 2.

The time limit for each experiment execution was 60 min. The experiments were conducted on an Apple M1 Pro CPU with 16 GB of memory, running macOS Ventura 13.3.

4.5 Synthetic Imbalanced Data Streams Without Concept Drift

In this section, we address the results obtained from 40 experiments (8 synthetic imbalanced data streams without concept drift and 5 different imbalance rates). BASWE achieved the highest Total Best value for both F1-Score (15 of 40 experiments) and Kappa (20 of 40 experiments). The second-best Total Best value for both measures was achieved by the OOB ensemble classifier. Conversely, the CSARF algorithm recorded the lowest Total Best value, securing the Total Best F1-Score value in just 2 of 40 experiments and Kappa in 3 of 40. SMOTE-OB did not finish executing the experiments within the time constraint of 60 min.

4.6 Synthetic Imbalanced Data Streams with Concept Drift

In this section, we address the results obtained from 80 experiments (5 different imbalance rates with 8 imbalanced synthetic data streams with sudden concept drift and 8 with gradual concept drift).

For experiments with gradual drift, the datasets followed a consistent pattern until the 40,000th instance. From this point, a gradual drift began, continuing over a span of 20,000 instances until the 60,000th instance. For experiments with sudden drift, the datasets maintained a stable pattern until the 40,000th instance, where an immediate and abrupt change occurred. In both scenarios, with gradual and sudden drifts, SMOTE-OB did not finish the experiments' execution within the time limit constraint (60 min).

Gradual Concept Drift. Table 1 shows F1-Score and Kappa for the experiments over the imbalanced data streams with gradual concept drift. BASWE emerged as a top-performing algorithm, securing the highest 'Total Best' value in both F1-Score (14 out of 40 experiments) and Kappa (19 out of 40 experiments). For synthetic imbalanced data streams without concept drift, OOB had the highest average metric, whether F1-Score or Kappa. However, in the scenario of synthetic imbalanced data streams with gradual concept drift, BASWE took the lead with the highest average F1-Score and ROSE had the highest average Kappa.

Figure 1 shows the F1-Score achieved by each algorithm under different.imbalance ratios. The results of Kappa were also plotted but are not shown here. We can see that in several data streams, the algorithms exhibit somewhat similar behavior. However, UOB and CSARF tend to diverge more from the other algorithms as the imbalance increases. This might suggest that, even though both algorithms are designed to handle imbalanced classes and perform reasonably well in that scope, they are more sensitive to highly imbalanced data contexts, such as ratios of 97.5:2.5 or higher. This behavior is observed for both Kappa and F1-Score. In the case of UOB, which employs undersampling, the reduced performance can be attributed to its aggressive discarding of instances from the majority class at high imbalance ratios, thereby losing valuable information beneficial for learning.

Table 1. F1-Score (left) and Kappa (right) obtained by the algorithms for the eight synthetic datasets with gradual concept drift

# experiments*			Algorithm	Avg. F1-Score
TB	AB	EB		
14	9	5	**BASWE**	**89.03 (1)**
11	8	3	ROSE	88.44 (3)
10	4	6	KUE	88.45 (2)
10	6	4	KUE (q = 2)	88.26 (4)
9	5	4	OOB	88.07 (5)
9	7	2	CALMID	87.60 (6)
3	1	2	UOB	84.26 (8)
0	0	0	CSARF	84.95 (7)
-	-	-	SMOTE-OB	Not finished

# experiments*			Algorithm	Avg. Kappa
TB	AB	EB		
19	15	4	BASWE	72.58 (2)
17	13	4	**ROSE**	**73.59 (1)**
13	7	6	OOB	72.44 (3)
4	1	3	UOB	64.64 (8)
3	2	1	KUE (q = 2)	67.74 (5)
2	1	1	KUE	67.97 (4)
2	1	1	CALMID	67.31 (6)
1	0	1	CSARF	66.73 (7)
-	-	-	SMOTE-OB	Not finished

*40 experiments

Sudden Concept Drift. Table 2 shows the compilation of results measured by F1-Score and Kappa, respectively, for the experiments over the imbalanced data streams with sudden concept drift. The best algorithm in terms of F1-Score was CALMID, which achieved the 'Total Best' in 16 experiments. In terms of Kappa, the top spot went to ROSE, also with a 'Total Best' value of 16 experiments. Both algorithms, CALMID and ROSE, implement explicit techniques to handle concept drift. These performances highlight that, in the specific scenario of sudden concept drift, the presence of dedicated mechanisms for detecting concept drifts makes a significant difference, ensuring the best performance in our experiments. This distinction was not as pronounced in experiments with gradual concept drift, where BASWE, which uses an implicit approach to handle concept drift, secured the highest 'Total Best' values, for both F1-Score and Kappa.

BASWE, while not securing top performance, still showed strong results, tied in the second-highest 'Total Best' value in F1-Score (11 out of 40 experiments), and the second in Kappa (14 out of 40 experiments). This suggests that, even with its implicit approach to concept drift, BASWE exhibited notable robustness within this data scope.

Plots for F1-Score and Kappa for different imbalance rates, analogous to those in Fig. 1, were also analyzed under the sudden concept drift scenario but are not shown here. It was observed that, similar to the experiments with gradual concept drift, UOB deteriorates more quickly as the imbalance increases. CSARF, which previously showed rapid deterioration like UOB in the gradual concept drift experiments, demonstrated more robustness in the experiments with sudden concept drift. This can be attributed to mechanisms designed to explicitly handle the occurrence of concept drift.

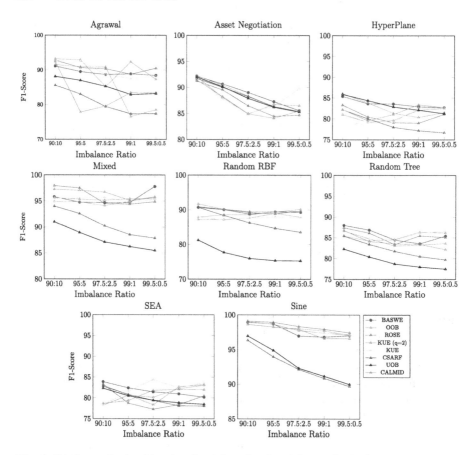

Fig. 1. F1-Score obtained by the algorithms for the eight synthetic datasets with gradual concept drift

4.7 Real Datasets

In this section, we discuss the results from 16 experiments on real data streams, some of which exhibit concept drifts and feature various imbalance ratios. Unlike synthetic data streams, which are generated based on specific probability distributions and are thus well-behaved, real data streams might not follow these distributions. This can introduce unique challenges, such as data chunks that contain instances of only one class.

Table 3 shows the F1-Score and Kappa achieved by the algorithms compared across all real datasets. BASWE consistently delivered the highest Total Best value of F1-Score, achieving this in 8 of the 16 real datasets. In terms of Kappa, there was a tie among CSARF, ROSE, and BASWE algorithms, that have Total Best value equal to 5 experiments each. Just as observed in experiments with synthetic data streams, SMOTE-OB exhibited a high computational cost and

Table 2. F1-Score (left) and Kappa (right) obtained by the algorithms for the eight synthetic datasets with sudden concept drift

# experiments*			Algorithm	Avg. F1-Score
TB	AB	EB		
16	14	2	CALMID	87.93 (3)
11	7	4	KUE	88.37 (2)
11	5	6	BASWE	87.05 (5)
9	7	2	ROSE	87.38 (4)
4	4	0	KUE (q = 2)	88.64 (1)
4	2	2	OOB	86.68 (6)
2	1	1	UOB	82.72 (8)
0	0	0	CSARF	82.73 (7)
-	-	-	SMOTE-OB	Not finished

*40 experiments

# experiments*			Algorithm	Avg. Kappa
TB	AB	EB		
16	11	5	**ROSE**	**69.14 (1)**
14	12	2	BASWE	67.44 (2)
10	6	4	OOB	67.31 (3)
7	6	1	CALMID	64.10 (6)
4	2	2	KUE	64.63 (5)
4	3	1	UOB	59.78 (8)
2	0	2	KUE (q = 2)	64.63 (4)
1	0	1	CSARF	61.00 (7)
-	-	-	SMOTE-OB	Not finished

Table 3. F1-Score (left) and Kappa (right) obtained by the algorithms for the sixteen real datasets

# experiments*			Algorithm	Avg. F1-Score
TB	AB	EB		
8	4	4	BASWE	78.02 (2)
6	5	1	**ROSE**	**78.40 (1)**
4	2	2	CSARF	77.53 (3)
3	0	3	KUE	74.77 (6)
2	0	2	KUE (q = 2)	75.03 (5)
2	2	0	SMOTE-OB	-
2	2	0	UOB	73.99 (7)
1	1	0	OOB	75.81 (4)
1	0	1	CALMID	71.23 (9)

*16 experiments

# experiments*			Algorithm	Avg. Kappa
TB	AB	EB		
5	3	2	**CSARF**	**56.71 (1)**
5	4	1	ROSE	53.75 (3)
5	3	2	BASWE	53.42 (4)
3	2	1	SMOTE-OB	-
2	2	0	CALMID	49.44 (7)
1	1	0	OOB	53.31 (5)
1	1	0	UOB	49.69 (6)
1	0	1	KUE (q = 2)	44.49 (9)
0	0	0	KUE	44.50 (8)

was unable to complete the execution of experiments within the 60-minute time constraint for 5 of the 16 data sets.

4.8 Discussion

Tables 4 and 5 present the overall performance of the algorithms measured by F1-Score and Kappa, respectively. Experimental results suggest that the BASWE is a promising choice for binary classification tasks involving class imbalance with and without concept drift. BASWE achieved superior results when measured by F1-Score, achieving the best performance or equivalent to the best in 48 out

Table 4. 'Total Best' of F1-Score for total experiments

	Without Concept Drift	Gradual Concept Drift	Sudden Concept Drift	Real Datasets	Total
BASWE	**15**	**14**	11	**8**	**48**
CALMID	7	9	**16**	1	33
CSARF	2	0	0	4	6
KUE	6	10	11	3	30
KUE (q = 2)	7	10	4	2	23
OOB	12	9	4	1	26
ROSE	8	11	9	6	34
SMOTE-OB	0	0	0	2	2
UOB	9	3	2	2	16

Table 5. 'Total Best' of Kappa for total experiments

	Without Concept Drift	Gradual Concept Drift	Sudden Concept Drift	Real Datasets	Total
BASWE	**20**	**19**	14	**5**	**58**
CALMID	6	2	7	2	17
CSARF	3	1	1	**5**	10
KUE	6	2	4	0	12
KUE (q = 2)	4	3	2	1	10
OOB	14	13	10	1	38
ROSE	6	17	**16**	**5**	44
SMOTE-OB	0	0	0	3	3
UOB	5	4	4	1	14

of 136 general experiments: 15 out of 40 in imbalanced data streams without concept drift, 14 out of 40 for the set with gradual concept drift, 11 out of 40 for the set with sudden concept drift, and 8 out of 16 for the real datasets. Similarly, it attained superior results when measured by the Kappa statistic, achieving the best performance or equivalent to the best in 58 out of 136 general experiments: 20 out of 40 in imbalanced data streams without concept drift, 19 out of 40 for the set with gradual concept drift, 14 out of 40 for the set with sudden concept drift, and 5 out of 16 for the real datasets.

The experimental data highlighted in Sect. 4.6 suggest that the effectiveness of BASWE decreases in the presence of sudden concept drifts. In such cases, the algorithms that recorded the highest 'Total Best' scores for each metric (CALMID for F1-Score and ROSE for Kappa) utilize drift detection mechanisms, which directly address concept drift. This could suggest that BASWE's implicit method of handling concept drift may be less effective during rapid shifts in data distribution, resulting in a slower adaptation to these changes.

5 Conclusion

This paper presents BASWE, an algorithm tailored specifically for binary classification in imbalanced data streams with and without concept drift. BASWE incorporates sliding windows paired with class-specific oversampling. Rather than processing all data uniformly, BASWE prioritizes the minority class instances, ensuring their proportional representation within the model. Class-specific oversampling entails replicating instances of the minority class, effectively countering the prevalent class imbalance.

Furthermore, BASWE leverages Balanced Accuracy for ensemble updates and vote abstention. This not only ensures that the algorithm remains unbiased towards the majority class, offering a fair evaluation of its performance across both classes, but also measures the quality of experts within the ensemble. By evaluating and subsequently updating or replacing underperforming experts based on Balanced Accuracy, BASWE implicitly addresses concept drift over time. This continuous updating also enhances the diversity within the ensemble, which is further augmented by selecting a random subset of features for each expert in the ensemble.

Its consistently high performance across a variety of datasets, coupled with its robustness to different data conditions, makes BASWE a versatile and reliable choice for handling classification tasks in imbalanced data streams, with or without concept drift.

Acknowledgments. This study was supported by CEPID-CeMEAI-Center for Mathematical Sciences Applied to Industry (grant 2013/07375-0, São Paulo Research Foundation-FAPESP).

References

1. Aguiar, G., Krawczyk, B., Cano, A.: A survey on learning from imbalanced data streams: taxonomy, challenges, empirical study, and reproducible experimental framework. Mach. Learn. **113**(7), 4165–4243 (2023)
2. Bahri, M., Bifet, A., Gama, J., Gomes, H.M., Maniu, S.: Data stream analysis: foundations, major tasks and tools. WIREs Data Mining Knowl. Discov. **11**(3) (2021)
3. Barrabés, M., Mas Montserrat, D., Geleta, M., Giró-i Nieto, X., Ioannidis, A.: Adversarial learning for feature shift detection and correction. In: Advances in Neural Information Processing Systems, vol. 36 (2024)
4. Bernardo, A., Valle, E.D.: SMOTE-OB: combining SMOTE and online bagging for continuous rebalancing of evolving data streams. In: 2021 IEEE International Conference on Big Data, pp. 5033–5042. IEEE, Orlando, FL, USA (2021)

5. Bifet, A., Holmes, G., Pfahringer, B., Kranen, P., Kremer, H., Jansen, T., Seidl, T.: MOA: massive online analysis, a framework for stream classification and clustering. In: Proceedings of the First Workshop on Applications of Pattern Analysis, vol. 11, pp. 44–50 (2010)
6. Cano, A., Krawczyk, B.: Evolving rule-based classifiers with genetic programming on GPUs for drifting data streams. Pattern Recogn. **87**, 248–268 (2019)
7. Cano, A., Krawczyk, B.: Kappa updated ensemble for drifting data stream mining. Mach. Learn. **109**(1), 175–218 (2020)
8. Cano, A., Krawczyk, B.: ROSE: robust online self-adjusting ensemble for continual learning on imbalanced drifting data streams. Mach. Learn. **111**(7), 2561–2599 (2022)
9. Du, H., Palaoag, T.D.: Dynamic weighted majority based on over-sampling for imbalanced data streams. In: CIIS 2021: The 4th International Conference on Computational Intelligence and Intelligent Systems, November 20–22, 2021, pp. 87–95. ACM, Tokyo, Japan (2021)
10. Gaber, M.M.: Advances in data stream mining. WIREs Data Mining Knowl. Discov. **2**(1), 79–85 (2012)
11. Gama, J.: Knowledge Discovery from Data Streams. CRC Press, Boca Raton, FL, USA, Chapman and Hall / CRC Data Mining and Knowledge Discovery Series (2010)
12. Gama, J., Zliobaite, I., Bifet, A., Pechenizkiy, M., Bouchachia, A.: A survey on concept drift adaptation. ACM Comput. Surv. **46**(4), 1–37 (2014)
13. Hulten, G., Spencer, L., Domingos, P.M.: Mining time-changing data streams. In: Proceedings of the Seventh ACM SIGKDD International Conference on Knowledge Discovery and Data Mining, pp. 97–106. ACM, San Francisco, CA, USA (2001)
14. Krawczyk, B.: Learning from imbalanced data: open challenges and future directions. Prog. Artif. Intell. **5**(4), 221–232 (2016)
15. Krawczyk, B., Minku, L.L., Gama, J., Stefanowski, J., Wozniak, M.: Ensemble learning for data stream analysis: a survey. Inf. Fusion **37**, 132–156 (2017)
16. Li, Z., Huang, W., Xiong, Y., Ren, S., Zhu, T.: Incremental learning imbalanced data streams with concept drift: the dynamic updated ensemble algorithm. Knowl. Based Syst. **195**, 105694 (2020)
17. Liu, W., Zhang, H., Ding, Z., Liu, Q., Zhu, C.: A comprehensive active learning method for multiclass imbalanced data streams with concept drift. Knowl. Based Syst. **215**, 106778 (2021)
18. Loezer, L., Enembreck, F., Barddal, J.P., de Souza Britto Jr., A.: Cost-sensitive learning for imbalanced data streams. In: The 35th ACM/SIGAPP Symposium on Applied Computing, pp. 498–504. ACM, Brno, Czech Republic (2020)
19. Wang, B., Pineau, J.: Online bagging and boosting for imbalanced data streams. IEEE Trans. Knowl. Data Eng. **28**(12), 3353–3366 (2016)
20. Webb, G.I., Hyde, R., Cao, H., Nguyen, H., Petitjean, F.: Characterizing concept drift. Data Min. Knowl. Discov. **30**(4), 964–994 (2016)

Beyond Audio Signals: Generative Model-Based Speaker Diarization in Portuguese

Antônio Oss Boll[1]( ), Letícia Maria Puttlitz[1], Heloísa Oss Boll[2], and Rodrigo Mor Malossi[2]

[1] Institute of Mathematics and Statistics, University of Sao Paulo, Rua do Matao, 1010 - Cidade Universitaria, Sao Paulo, SP 05508-090, Brazil
{aoboll,leticia.puttlitz}@ime.usp.br
[2] Institute of Informatics, Universidade Federal do Rio Grande do Sul, Av. Bento Gonçalves, 9500, Porto Alegre, RS 91501-970, Brazil
hoboll@inf.ufrgs.br, rodrigo.malossi@ufrgs.br

Abstract. Speaker diarization, the task of automatically identifying different speakers in audio and video, is frequently performed using probabilistic models and deep learning techniques. However, existing methods usually rely on direct analysis of the audio signal, which presents challenges for languages that lack established diarization methodologies, such as Portuguese. In this article, we propose a new approach to speaker diarization that leverages generative models for automatic speaker identification in Portuguese. We employed two generative models: one for refining the transcribed audio and another for performing the diarization task, as well as a model for initially transcribing the audio. Our method simplifies the diarization process by capturing and analyzing speaker style patterns from transcribed audio and achieves high accuracy without depending on direct signal analysis. This approach not only increases the effectiveness of speaker identification but also extends the usefulness of generative models to new domains. It opens a new perspective for diarization research, especially for the development of accurate systems for under-researched languages in audio and video applications.

Keywords: Speaker Diarization · Generative Models · Natural Language Processing

1 Introduction

Diarization is the task of automatically identifying different speakers in speech, whether from audio or video [21]. It is based on two main processes: the recognition of who is speaking and the initial segmentation of the speech. Its applications are wide-ranging, including meeting transcriptions [3], call center assistance [5], and the development of personalized voice assistants [23].

Traditional methods for diarization employ probabilistic models, such as Gaussian Mixture Models (GMM) and Hidden Markov Models (HMM), which are effective at clustering audio according to voice patterns [12,25]. Subsequently, other techniques based on Deep Learning (DL) have brought significant advances to the field. Examples include Recurrent Neural Networks (RNNs), which are adapted for sequential data [29], and Convolutional Neural Networks (CNNs), which are capable of analyzing audio directly from spectrograms [15].

Most recently, methods based on the Transformer architecture have demonstrated state-of-the-art performance across various domains [27,28]. This success is largely attributed to self-attention, a mechanism that allows the model to focus on different parts of the input sequence when producing output while also capturing long-range dependencies in the input, regardless of their distance in the sequence. Large Language Models (LLMs) are built upon this architecture, utilizing very deep neural networks with stacked multi-head attention layers. Models such as GPT-4 [19], Llama 3 [1], and Claude [4], exhibit impressive capabilities in natural language understanding and in solving complex tasks via text generation.

Despite these advancements, there remains a gap in applying such approaches to audio and video tasks. Exploring the use of Transformer-based architectures for diarization presents a promising opportunity for achieving similar breakthroughs in audio and video processing.

In addition, language variations are also a challenge in diarization tasks. Most models are adapted to processing English content due to the abundance of materials in the language. Although Portuguese is a high-resource language for text-based tasks, it is considerably under-researched in audio and video applications, especially regarding models, transcription tools, and new diarization methods.

In this article, we investigate existing diarization methods and introduce a novel approach that addresses identified gaps in the field, leveraging generative models for automatic speaker identification in Portuguese videos and audio recordings. Unlike traditional methods that require direct analysis of audio signals, our approach captures and analyzes speaker style patterns from transcribed audio with high accuracy. This method not only simplifies the process but also extends the applicability of generative models beyond their traditional domains, opening up a promising new line of research. Furthermore, our approach is language-agnostic and can be extended to any language supported by both a generative language model and a transcription model.

2 Background

2.1 Speech Data

Speech data contains a variety of information that reflects the identity of the speaker, including distinctive characteristics of the vocal tract, excitation source, and behavioral features. These cues are important for speaker recognition [13] and form the basis for the features of a speaker. In our approach, the model identifies the speaker's traits from the text.

Processing such information involves a set of challenges, such as high variability due to factors like accent, background noise, and recording quality [24]. Furthermore, the audio quality can be significantly affected when not recorded in controlled environments, such as outdoor settings, rooms with echo, or audio with simultaneous speech. Our approach aims to mitigate these issues by employing advanced deep learning-based preprocessing techniques for audio transcription and refinement, as detailed in Sect. 4.

Another issue is labeling, as accurate annotation of speech data is crucial but can be demanding and costly. Techniques like large-scale weak supervision are employed to address these challenges [24]. In our case, we used the labeling from the Brazilian Chamber of Deputies as a foundation, manually correcting some of its errors for better accuracy. The process is detailed in Subsect. 4.1.

Finally, considering that handling speech data must comply with privacy regulations and ethical guidelines due to its sensitive nature [18], it is important to note that we used already transcribed examples of sessions that are required to be public. Hence, the data already complied with the expected privacy standards.

2.2 Speech to Text

Speech to text (STT) is the process of converting speech signals to a sequence of words. The main technique based on STT is speech recognition [13]. There are several speech recognition models available, such as Whisper large v2 [24], Deepgram Nova-2, Amazon, Microsoft Azure Batch v3.1, and the Universal Speech Model [30]. In the present work, Whisper large v2 was employed to transcribe public sections from the Brazilian Chamber of Deputies.

The Whisper model, a state-of-the-art method for speech transcription, is a pre-trained model designed for automatic speech recognition (ASR) and speech translation. It was trained on 680,000 h of labeled speech data, utilizing large-scale weak supervision for annotation. In its process, all audio is resampled to 16,000 Hz and converted into an 80-channel log-magnitude Mel spectrogram.

Whisper encoder handles this input representation, whereas the decoder utilizes learned position embeddings and tied input-output token representations. It employs an encoder-decoder Transformer architecture, and both the encoder and decoder have the same width and number of transformer blocks. The architecture was trained to handle multiple tasks simultaneously, including voice activity detection and inverse text normalization.

STT approaches have influenced numerous fields. In medicine, these models can document interactions between patients and doctors [9]. For customer service, STT enables real-time transcription of customer interactions, resulting in enhanced service quality [16]. Additionally, STT offers tools for individuals with disabilities, improving accessibility [7].

In our study, we extended the application of STT to the political domain, focusing on diarizing speeches from the Chamber of Deputies. Furthermore, transcription captures the spoken words without distinguishing between speakers. Our goal task, diarization, goes beyond transcription by segmenting speech according to different speakers.

2.3 Diarization

Diarization involves assigning audio segments to the appropriate speaker. It can be accomplished through various techniques, but typically involves voice activity detection (VAD) to identify speech regions and clustering these segmented audio portions to associate them with individual speakers [8]. Diarization also often includes speaker embedding extraction, where speaker characteristics are encoded into numerical representations. These embeddings are then utilized in the clustering process to group audio segments based on speaker similarity [22].

Our methodology is an innovative approach to solving the challenges of the diarization task. It uses audio transcription and advanced prompt engineering steps, which are detailed in Sect. 4.

2.4 Large Language Models

Large language models (LLMs) based on Transformers [27] are an essential part of our study, as they were used for refining the transcriptions and identifying the speakers, as detailed in Sect. 4. These very deep neural networks are trained on extensive datasets, enabling them to learn complex patterns.

Each token within the model, whether it represents a whole word or part of one, is interpreted as a numerical embedding that encapsulates its semantic meaning [17]. Similar words are mapped to embeddings that are close to each other in a high-dimensional space, facilitating the model's ability to comprehend semantic relationships. The breadth and quality of the training dataset influence the model's knowledge and vocabulary. Therefore, a vast and high-quality training dataset is essential to ensure the model's comprehension and accuracy.

3 Related Work

There are several diarization models that support the English language, including Pyannote [8]. Pyannote works by implementing tasks such as voice activity detection (VAD), speaker change detection (SCD), and overlapped speech detection (OSD), which assign a value of 1 if the event occurs and 0 otherwise. Additionally, re-segmentation is used to refine the boundaries and labels of speech segments, assigning a label of 0 when no speaker is active and the corresponding speaker's label when they are speaking.

Google also offers a model widely used for speech diarization on its cloud platform, the Speech-to-Text API. Both versions, V1 and the recently introduced V2, can be used for this task in English.

Unfortunately, all aforementioned tools lack support for the Portuguese language. Therefore, we explored other multilingual options, such as Oracle Speech AI or Assembly AI, which we tested and compared against our approach.

The Oracle Speech AI service [20] offers a range of functionalities, including transcriptions utilizing the Whisper model, which supports over 50 languages. Additionally, it provides features such as text normalization, profanity filtering, and confidence scoring per word. Moreover, it has the ability to recognize between 2 and 16 distinct speakers. However, the lack of detailed information about the training of their diarization model makes it difficult to understand the internal mechanisms of the service.

On the other hand, the Assembly AI service [6] also includes STT functionalities such as speech recognition and diarization. The speech recognition models supported by this service are Best and Nano. However, there is also not much information available about the methodology employed in their diarization process.

4 Materials and Methods

4.1 Dataset

As mentioned, there are very few public videos with identified speakers in Portuguese. Hence, the resource used in our study was one of the few available, obtained from the "Ordem do dia" part of the sessions of the Brazilian Chamber of Deputies [10]. Given the source's relevance, the data is from a trusted source, which makes it suitable for evaluating our method. Another reason why this dataset was chosen is due to its sudden changes in speakers. While a number of videos have speaker identification, most of these instances are singular. This means that when one person speaks, there is usually a pause before another person responds. On the other hand, our dataset does not follow a specific order when people are speaking; anyone can interrupt and start talking at any time. This characteristic makes it ideal for our purposes, as we want to develop a method that can skillfully handle complex conversational dynamics.

It is worth mentioning that some video transcripts may not be entirely accurate, as some instances may not follow the exact order of the speakers, having slightly different lines from the original ones. For example, in the session entitled "2nd Ordinary Legislative Session of the 57th Legislature, 109th Session", at the 13:56 mark, the audio says: "pelo Rio de Janeiro no Rio de Janeiro, nós vamos ouvir ela, que foi Senadora, Governadora, Deputada Federal, que é a nossa Deputada." The transcript's meaning remains the same, but with some new words and is organized differently: "pelo Rio de Janeiro com a ex-Senadora, ex-Governadora e Deputada Federal."

Given the inconsistencies between the actual audio and the written transcription, we used each of Whisper's transcriptions as the ground truth. We then manually inserted the real speakers and their labels. That resulted in a better evaluation of our method and ensured that the transcriptions were adequate.

Despite these discrepancies, our final goal was speaker diarization. Therefore, we focused solely on identifying the speakers rather than ensuring perfect correspondence between the audio and transcripts, even though we manually inserted the speaker labels.

4.2 Diarization Strategy

Our approach involves two key stages: transcribing audio from videos and processing the resulting text with generative models to identify the speakers[1]. We utilized Whisper Large-v2 [24] to transcribe eight parts of sessions from the Brazilian Chamber of Deputies held in 2024. Each part of the session lasted for 4 min, providing substantial audio data for analysis. For text refinement and speaker diarization, we employed the advanced GPT-4o generative model.

Furthermore, we developed a prompt engineering method to guide the models in both refining the transcriptions and generating the speaker predictions, aiming to refine the output and ensure speaker identification accuracy.

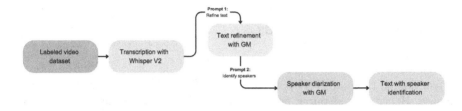

Fig. 1. Pipeline of the proposed architecture. GM = generative model.

An overview of the process can be found in Fig. 1. It begins by transcribing the video with Whisper, and its output is then fed into a generative model. The model is then instructed to refine the text, removing misspelled words and correcting misplaced or contextually incorrect words. This refinement is achieved through prompt engineering, through which a detailed instructional prompt guides the model to make such changes and improve the quality of the text generated by Whisper.

Finally, the refined text is passed on to another generative model to identify the speaker in the diarization task. Although this second model's architecture is the same as the model used for refinement, it is given a more robust prompt, allowing it to accurately identify the speaker.

Each speaker presents not only variations in their voice but also in how they communicate, using unique vocabulary and expressions. When applying a generative model to perform diarization on a transcribed text, it takes into account the individual characteristics of each speaker. Additionally, the model leverages the knowledge acquired during training to identify patterns in the dialogue structure, such as recognizing that a question followed by an answer typically indicates that one person asked the question and another provided the response.

---

[1] In this article, we focus on diarizing videos; however, the process can be equally applied to audio inputs.

4.3 Prompts

Generative models can perform poorly if they are not given precise instructions on the task they need to solve. Therefore, this work also developed a prompt engineering strategy to ensure robustness. The task instructions used in this work are specified in Figs. 2 and 3 for each generative model, respectively, text refinement and speaker identification. The specific prompts are available on our GitHub.

Each of these tasks is further elaborated in the following subsections.

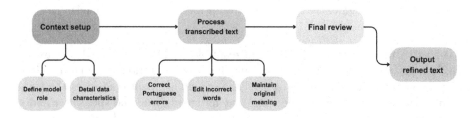

Fig. 2. Design of the first prompt for the text refinement model.

Text Refinement. As mentioned, the first model functions as a "text refiner". The prompt used for this task is detailed in Fig. 2. It is important to firstly include the context in which the model will operate. This is specified in the 'Model role', where the model is assigned a function or task at a higher level. Subsequently, it requires instructions regarding the data that needs refinement.

In our example, we explained the context of the video, instructing the model to act as a stenographer of sessions in the Chamber of Deputies and detailing the type of content of the transcript. In addition, we asked it to output an improved text, correcting Portuguese errors and editing wrong words.

For this step, to ensure consistent and accurate word correction, it is essential that the model has extended deterministic characteristics, avoiding variations in its responses. We therefore set the temperature to zero to eliminate any chance of variability. With both of these settings, the generative model could be effectively guided to provide the necessary text refinement.

Speaker Diarization. The second model developed is significantly more complex. The prompt used for the diarization task includes almost all the refinement instructions involved in the first model, but with the addition of steps, results, and examples, providing much more detailed guidance. The prompt used in this task is detailed in Fig. 3.

Specifically, the diarization prompt consists of three segments. Firstly, an example of a conversation relevant to the task is provided. In our case, we simulated a dialog between individuals in a courtroom setting. If a different context

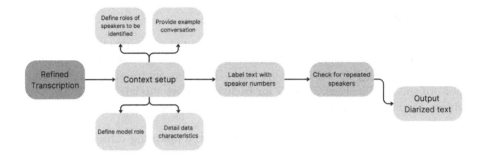

Fig. 3. Design of the second prompt for the text refinement model.

is needed, a more contextually relevant example could be employed. For example, in a restaurant scenario, it could involve a conversation between a waiter, a customer, and the restaurant owner. The second segment comprises the transcript itself, which serves as the base text for identifying the speaker. Finally, the third segment includes a set of instructions that the model must follow to perform the diarization task. They include both general instructions about the task and specific instructions for each of the potential speakers. With these three segments, we can achieve state-of-the-art performance in speaker identification tasks.

Another essential factor that influences the model's prediction is the temperature setting. Unlike the first model, we made this one slightly less deterministic since it needs to "understand" and accurately predict the number of speakers. We, therefore, increased the temperature and sacrificed some of the quality of the response at each iteration. However, this compensation significantly improves performance, with the model obtaining nearly perfect responses every two iterations.

4.4 Evaluation

There are several methods to evaluate speaker diarization, including the diarization error rate (DER) [11], word error rate (WER), Jaccard error rate (JER) [26], and e-WER [2], among others, such as the extensions discussed by Galibert [14]. However, since our approach primarily uses text, we focused on evaluating the quality of diarization as a text classification method.

In this work, we used accuracy and F1-score as our primary metrics for assessing the quality of speaker prediction. Accuracy measures how often speakers are correctly labeled compared to the actual speaker for each sentence or word, reflecting the overall correctness of the classification across all speakers. The F1-score, being the harmonic mean of recall and precision, considers each correctly classified speaker. We specifically used the weighted F1-score to account for any imbalance among speakers.

We evaluated these metrics at both the sentence level and the word level, comparing speaker predictions with the ground truth for each phrase and word, respectively. This separation between levels was implemented to enhance the accuracy of classifications. Evaluating only at the sentence level may allow for multiple speakers within a sentence without significant penalties. Therefore, we incorporated word-level evaluation to address this issue and provide a more granular assessment.

To assess the variability and robustness of the models, we also utilized confidence intervals to statistically demonstrate these differences. Additionally, we employed the Wilcoxon Signed-Rank Test to show that the differences between models are statistically significant. In this test, we specified the "greater" mode, focusing on superior performance rather than using a two-sided test. This approach ensured that our evaluation emphasized the best-performing model.

5 Results

Our main results are shown in Table 1. It displays the average accuracy and F1-score for the proposed method and the baseline models, as well as the respective 95% confidence intervals.

Table 1. Comparison of our approach and baseline models for diarization.

Evaluation	Model	Accuracy (95% CI)	F1-score (95% CI)
Sentence	Oracle	0.48 (0.26, 0.69)	0.48 (0.28, 0.68)
	Assembly	0.81 (0.65, 0.97)	0.82 (0.64, 1.00)
	Ours	0.95 (0.92, 0.98)	0.96 (0.93, 0.98)
Word	Oracle	0.47 (0.21, 0.74)	0.47 (0.21, 0.73)
	Assembly	0.81 (0.61, 1.00)	0.81 (0.59, 1.00)
	Ours	0.98 (0.95, 1.00)	0.98 (0.97, 1.00)

The Oracle model exhibited the lowest performance, with both accuracy and F1-score consistently around 0.48 for both sentence-level and word-level evaluations. The Assembly model performed moderately well, with scores of approximately 0.81, yet it still fell short when compared to our model's performance.

Our model demonstrated a significant improvement in accuracy, achieving 0.95 and 0.98 for sentence-level and word-level evaluations, respectively. This represents a substantial increase over both the Oracle and Assembly models. The improvement is similarly reflected in the F1-scores, where our model achieved 0.96 and 0.98 for sentence-level and word-level evaluations, respectively.

Additionally, the evaluation of the confidence intervals reveals that our approach exhibits the lowest variability, indicating high precision and consistency in the results. This suggests that our method is both robust and effective, especially when compared to the other models, which show greater variability in their confidence intervals.

The superior performance of our method across both evaluation metrics indicates that it generalizes well across different audio types, performing effectively on both high and low conversational recordings within our dataset.

To further validate these differences, we performed a Wilcoxon Signed-Rank Test to compare the results across the different models. The outcomes of this test are presented in Table 2.

Table 2. Comparison of our approach and baseline models using the Wilcoxon Signed-Rank Test.

Evaluation	Model 1	Model 2	P-value for accuracy	P-value for F1-score
Sentence	Assembly	Oracle	0.013 (< 0.05)	0.019 (< 0.05)
	Ours	Assembly	0.013 (< 0.05)	0.031 (< 0.05)
	Ours	Oracle	0.003 (< 0.05)	0.003 (< 0.05)
Word	Assembly	Oracle	0.019 (< 0.05)	0.011 (< 0.05)
	Ours	Assembly	0.013 (< 0.05)	0.013 (< 0.05)
	Ours	Oracle	0.007 (< 0.05)	0.007 (< 0.05)

As shown in Table 2, all results are statistically significant, providing strong evidence of differences between the models. The Wilcoxon test results allow us to state with 95% confidence that our model performs significantly better than both the Assembly model and the Oracle model at both the sentence and word levels, in terms of both accuracy and F1-score.

Additionally, our method has the unique ability to identify speakers based on their roles within a simple hierarchy (e.g., President and deputies). This functionality could be integrated into speaker-identified transcription tasks if required, though it would necessitate a robust example for effective implementation.

All transcriptions and predictions generated by our model are available on our GitHub.

5.1 Limitations

Our model is aligned with the Whisper-v2 model, which can transcribe audio segments up to 25 MB in size. Consequently, our diarization method achieves state-of-the-art results only for audio segments within this size limit. Although enhancing the model's robustness is a potential area for future work, this study operates within the specified constraint.

Furthermore, since we leveraged generative models, the outputs may slightly vary with each prompt. We recommend iterating the process a few times to ensure the effectiveness of the task.

6 Conclusion

In this work, we introduced a new method for the diarization of audio and video. Our approach leverages the power of large language models to excel in this task, particularly in languages that lack established diarization methodologies. By prioritizing an understanding of textual context and cues over traditional audio signal-based speech diarization techniques, we achieved state-of-the-art results for the Portuguese language.

Our primary objective was to compare our method with other available methods for Portuguese, such as the Oracle model and AssemblyAI. The results demonstrated that our method significantly outperformed baseline methods, achieving higher accuracy and F1-scores (0.95 and 0.98 for sentence-level and 0.96 and 0.98 for word-level evaluations). Additionally, the narrow confidence intervals indicated high precision and consistency.

The statistical significance confirmed by the Wilcoxon Signed-Rank Test further supported the superior performance of our model. Moreover, its ability to identify speakers based on their roles within a hierarchy, such as distinguishing between a President and deputies, adds another layer of functionality that could be integrated into more complex diarization tasks.

For future work, we aim to investigate its behavior on longer audio segments and expand its evaluation by validating it on a wider variety of data. Additionally, we see potential in extending it to other languages, especially those lacking well-established diarization tools, such as Spanish and Italian, and including multilingual evaluations in our comparisons.

Furthermore, we aim to adapt the method for real-time applications and explore the integration of Retrieval-Augmented generation (RAG). By connecting the generative model to a website, RAG could enable the model to selectively search for relevant examples and content. In addition, we plan to perform ablation studies to better understand the contributions of each component within the proposed framework.

Finally, LLMs, with their vast parameter-rich architectures, hold the potential to tackle a wide array of challenges. In this sense, we believe this work aligns with the emerging trend of leveraging generative models for tasks beyond text, extending into areas such as video, audio, and beyond.

References

1. AI, M.: Llama 3 (2024). https://llama.meta.com/. Accessed 28 Aug 2024
2. Ali, A., Renals, S.: Word error rate estimation for speech recognition: e-WER. In: Gurevych, I., Miyao, Y. (eds.) Proceedings of the 56th Annual Meeting of the Association for Computational Linguistics (Volume 2: Short Papers), pp. 20–24. Association for Computational Linguistics, Melbourne, Australia (2018). https://doi.org/10.18653/v1/P18-2004, https://aclanthology.org/P18-2004
3. Anguera, X., Bozonnet, S., Evans, N., Fredouille, C., Friedland, G., Vinyals, O.: Speaker diarization: a review of recent research. IEEE Trans. Audio Speech Lang. Process. **20**(2), 356–370 (2012)

4. Anthropic: Claude AI (2024). https://www.anthropic.com/claude. Accessed 28 Aug 2024
5. Aronowitz, H.: Speaker diarization using a priori acoustic information. In: INTERSPEECH, pp. 937–940 (2011)
6. AssemblyAI: Assemblyai (2024). https://www.assemblyai.com/. Accessed 30 Aug 2024
7. Bain, K., Basson, S., Faisman, A., Kanevsky, D.: Accessibility, transcription, and access everywhere. IBM Syst. J. **44**(3), 589–603 (2005)
8. Bredin, H., et al.: Pyannote. Audio: neural building blocks for speaker diarization. In: ICASSP 2020-2020 IEEE International Conference on Acoustics, Speech and Signal Processing (ICASSP), pp. 7124–7128. IEEE (2020)
9. Chiu, C.C., et al.: Speech recognition for medical conversations. arXiv preprint arXiv:1711.07274 (2017)
10. Câmara dos Deputados do Brasil: Discursos e notas taquigráficas (2024). https://www.camara.leg.br/internet/sitaqweb/discursodireto.asp. Accessed 10 Jun 2024
11. Fiscus, J.G., Ajot, J., Michel, M., Garofolo, J.S.: The rich transcription 2006 spring meeting recognition evaluation. In: Renals, S., Bengio, S., Fiscus, J.G. (eds.) MLMI 2006. LNCS, vol. 4299, pp. 309–322. Springer, Heidelberg (2006). https://doi.org/10.1007/11965152_28
12. Fox, E.B., Sudderth, E.B., Jordan, M.I., Willsky, A.S.: The sticky HDP-HMM: bayesian nonparametric hidden Markov models with persistent states. Arxiv preprint **2** (2007)
13. Gaikwad, S.K., Gawali, B.W., Yannawar, P.: A review on speech recognition technique. Int. J. Comput. Appl. **10**(3), 16–24 (2010)
14. Galibert, O.: Methodologies for the evaluation of speaker diarization and automatic speech recognition in the presence of overlapping speech. In: Interspeech (2013). https://doi.org/10.21437/Interspeech.2013-303
15. Hrúz, M., Zajíc, Z.: Convolutional neural network for speaker change detection in telephone speaker diarization system. In: 2017 IEEE International Conference on Acoustics, Speech and Signal Processing (ICASSP), pp. 4945–4949 (2017). https://doi.org/10.1109/ICASSP.2017.7953097
16. Meng, J., Zhang, J., Zhao, H.: Overview of the speech recognition technology. In: 2012 Fourth International Conference on Computational and Information Sciences, pp. 199–202. IEEE (2012)
17. Mikolov, T., Chen, K., Corrado, G., Dean, J.: Efficient estimation of word representations in vector space. arXiv preprint arXiv:1301.3781 (2013)
18. Nautsch, A., et al.: Preserving privacy in speaker and speech characterisation. Comput. Speech Lang. **58**, 441–480 (2019)
19. OpenAI: Gpt-4o (2023). https://openai.com/index/hello-gpt-4o/. Accessed 28 Aug 2024
20. Oracle Corporation: Oracle AI speech (2024). https://www.oracle.com/artificial-intelligence/speech/. Accessed 30 Aug 2024
21. Park, T.J., Kanda, N., Dimitriadis, D., Han, K.J., Watanabe, S., Narayanan, S.: A review of speaker diarization: recent advances with deep learning. arXiv preprint arXiv:2101.09624 (2021)
22. Park, T., Koluguri, N.R., Jia, F., Balam, J., Ginsburg, B.: Nemo open source speaker diarization system. In: INTERSPEECH, pp. 853–854 (2022)
23. Ponraj, A.S., et al.: Speech recognition with gender identification and speaker diarization. In: 2020 IEEE International Conference for Innovation in Technology (INOCON), pp. 1–4. IEEE (2020)

24. Radford, A., Kim, J.W., Xu, T., Brockman, G., McLeavey, C., Sutskever, I.: Robust speech recognition via large-scale weak supervision. In: International Conference on Machine Learning, pp. 28492–28518. PMLR (2023)
25. Reynolds, D.A.: Speaker identification and verification using Gaussian mixture speaker models. Speech Commun. **17**(1), 91–108 (1995). https://doi.org/10.1016/0167-6393(95)00009-D, https://www.sciencedirect.com/science/article/pii/016763939500009D
26. Ryant, N., et al.: The second dihard diarization challenge: dataset, task, and baselines. arXiv preprint arXiv:1906.07839 (2019)
27. Vaswani, A., et al.: Attention is all you need. In: Guyon, I., et al. (eds.) Advances in Neural Information Processing Systems, vol. 30. Curran Associates, Inc. (2017). https://proceedings.neurips.cc/paper_files/paper/2017/file/3f5ee243547dee91fbd053c1c4a845aa-Paper.pdf
28. Wang, D., Xiao, X., Kanda, N., Yoshioka, T., Wu, J.: Target speaker voice activity detection with transformers and its integration with end-to-end neural diarization. arXiv preprint arXiv:2208.13085 (2022)
29. Weninger, F., Wöllmer, M., Schuller, B.: Automatic assessment of singer traits in popular music: Gender, age, height and race. In: Proceedings 12th International Society for Music Information Retrieval Conference, ISMIR 2011 (2011)
30. Zhang, Y., et al.: Google USM: scaling automatic speech recognition beyond 100 languages. arXiv preprint arXiv:2303.01037 (2023)

Classification of Non-alcoholic Fatty Liver Disease in Thermal Images of the Liver Using a Siamese Neural Network

Maxwell Pires Silva[✉], Aristófanes Corrêa Silva, and Anselmo Cardoso de Paiva

Núcleo de Computação Aplicada - Universidade Federal do Maranhão (UFMA), Av. dos Portugueses, 1966 - Vila Bacanga, São Luís, MA 65080-805, Brazil
maxwell.pires@discente.ufma.br, {ari,anselmo.paiva}@nca.ufma.br

Abstract. Non-alcoholic fatty liver disease (NAFLD) is a prevalent and severe condition that requires effective and non-invasive diagnostic methods. This work presents an innovative approach using Siamese neural networks to classify thermal images of the liver in order to identify the presence of NAFLD. The research is motivated by the need to use artificial intelligence to analyze thermal images, especially when the number of images is limited, making it difficult for ordinary neural networks to learn, a difficulty that the Siamese network already faces with ease. The proposed method involves three stages: extraction of the region of interest (ROI), image pre-processing and classification using the Siamese Neural Network, which compares pairs of images to determine their similarity and, consequently, the presence or absence of NAFLD. The preliminary results, with an accuracy of 71%, precision of 57% and recall of 96%, indicate that this approach could offer a promising tool for the non-invasive diagnosis of NAFLD, contributing a promising method to the field of machine learning applied to medicine.

Keywords: Non-alcoholic Fatty Liver Disease · Thermal Imaging · Siamese Neural Network · Classification

1 Introduction

Non-alcoholic fatty liver disease (NAFLD) is increasingly diagnosed worldwide. It is the most common cause of abnormal liver function tests and chronic liver disease in both developed and developing countries [1]. NAFLD refers to fat accumulation, mainly triglycerides, in hepatocytes so that it exceeds 5% of the liver's weight. Primary NAFLD results from insulin resistance and often occurs as part of the metabolic changes accompanying obesity, type 2 diabetes and dyslipidemia. The histological damage in NAFLD is very similar to that seen in patients with alcoholic liver disease, but NAFLD is not alcohol-induced by definition [2]. So, it is important to exclude secondary causes of steatosis. Furthermore, NAFLD affects approximately 1.5 billion people worldwide, making it the most common liver disease [3].

Given the seriousness of this context, the diagnosis of NAFLD must be rapid, accurate and efficient. Currently, liver biopsy is the most specific method for diagnosing NAFLD, and it also makes it possible to assess the severity of fatty infiltration in the liver. However, this method is an invasive procedure that poses a risk to the patient and is expensive. Therefore, alternative pathology detection methods are desirable in this context [4]. Most patients are asymptomatic, and the diagnosis is considered suspicious of NAFLD after finding elevated transaminases in routine tests. This disease is also a frequent incidental finding on ultrasound (US) performed for other reasons, such as suspected gallstones [5]. Hepatic US is currently the most widely available, simple and inexpensive non-invasive method for detecting hepatic steatosis in clinical practice. However, the accuracy of US is highly operator-dependent, and its sensitivity is reduced when steatosis infiltration is less than 30% or in morbidly obese patients. In addition, the quantification of hepatic steatosis is subjective and can be influenced by the heterogeneity observed in some patients with NAFLD [6].

On the other hand, thermographic studies have been widely included in medical practice to obtain additional data for the diagnosis of various diseases and determine the best methods and the effectiveness of treatment. The main advantages of thermographic research are the relatively low cost, absence of ionizing radiation or electromagnetic fields, absence of contraindications, safety, and the ability to diagnose the disease at an early stage [7]. Thus, thermography in medicine can be used to identify the dynamics of pathological processes. Specifically in patients with NAFLD, it is observed that the development of the disease causes a decrease in the surface temperature of the liver [8], making it a promising method for diagnosis. Even so, the capture quality of the infrared emission that thermal cameras use to generate surface thermograms is influenced by the environment, which attenuates the thermographic effect, making its detection difficult. Consequently, a controlled environment is required for image acquisition and pre-processing to reduce noise and refine image characteristics [9].

In this context, few studies have been carried out using medical thermal images in conjunction with Neural Networks for detection or classification. Pinto et al. (2021) and Farias et al. (2023) developed a study to classify thermal images of the liver into two groups: healthy and NAFLD. The first used the AlexNet Convolutional Neural Network with satisfactory results, achieving an accuracy of 96%. The second study used the temporal analysis technique with a temporal convolutional network, achieving an accuracy of 88%. Both prove the great potential that the analysis of medical thermal images has if combined with machine learning, enabling professionals to be supported by Neural Networks during the diagnosis of this disease.

In machine learning, convolutional neural networks (CNNs) are important architectures that enable the resolution of some problems, such as facial recognition, autonomous vehicles, and intelligent medical treatment [12]. Furthermore, in deep learning, CNNs are widely known for their ability to achieve high accuracy in classifying medical images. However, these models that are not pre-trained, i.e., with the initial weights of the network starting with random values, have some limitations.

Training the deep learning model on a large number of images requires huge computational resources, and for proper training of the model, it needs a very significant amount of standard training datasets, which is the biggest problem for medical images; after all, data can be expensive and difficult to obtain. Furthermore, there are still ethical privacy issues [13]. With this perspective, Siamese Neural Networks (SNNs) emerge as an architecture of artificial neural networks composed of two identical neural networks (they share the same weights), united using perceptrons whose function is to calculate the similarity between the response of the two networks [14]. They were proposed to attenuate the data scarcity problem since this type of architecture requires small samples for training and learning the model.

Therefore, considering the importance of a non-invasive diagnosis of NAFLD together with the contribution that machine learning is capable of bringing to the classification of images within the scope of computer vision, the present work aims to classify thermal images of the liver region as affected by NAFLD or healthy, using a Siamese neural network architecture. This work makes the following contributions: a study for the processing of medical thermal images, a study to detail the structure of the Siamese network and the application of the Siamese neural network in the classification of medical thermal images. This study is structured as follows: Sect. 2 presents the proposed method and the image dataset used. Section 3 presents the result and discusses what was obtained. Finally, Sect. 4 presents the conclusions of this work.

2 Materials and Method

This section presents the image dataset used and the proposed method for classifying thermal images of the liver (Fig. 1). The method consists of three steps: The first is the extraction of the region of interest (ROI), followed by the pre-processing of the ROI and, finally, the classification of the images by the Siamese network. Below, each step is detailed.

2.1 Dataset

The dataset comprises images from 40 patients, 18 diagnosed with NAFLD and 22 healthy. In total, there are 103 images of patients with this disease and 132 images of healthy patients. The diagnosis of these patients for NAFLD was confirmed via ultrasound of the abdominal region by the specialist, serving to determine and label the patients as affected or not by NAFLD. A thermograph with an infrared sensor FLIR - Model S650C, which has a resolution of 640 × 480 pixels, has been used to acquire the images. As mentioned, a controlled environment is required, so the specialist has developed a thorough protocol for acquiring the thermal images, which consists of certain steps: initially, the environment is acclimatized, where the patient is kept for 15 min at 23°C and a relative humidity of 65%.

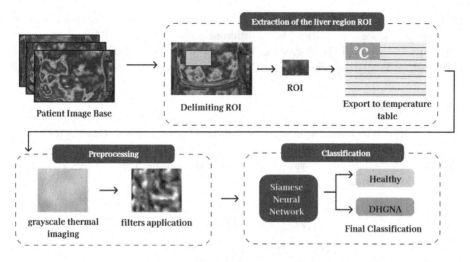

Fig. 1. Proposed Method

At this location, the patients were positioned in dorsal decubitus (lying with their chests up) and the images were then taken with the camera above the patient, forming a 90° angle (Fig. 2). Five images were captured per patient. In addition, throughout the process of acclimatization and taking the images, the patients had their abdominal region exposed.

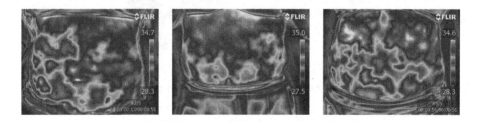

Fig. 2. Examples of thermal images obtained from the dataset.

2.2 Extraction of the Liver Region (ROI)

In this study, the region of interest (ROI) comprises the location of the liver in the acquired images. Ozougwu (2017) defines the liver region as the upper right quadrant of the abdomen. The ROI extraction process was carried out using the proprietary software Flir Tools[1], provided by the manufacturer of the thermograph used to acquire the images. Then, with the help of the specialist,

---
[1] https://www.flir.eu/browse/professional-tools/thermography-software/.

ROIs of different dimensions were obtained from the dataset to capture the liver region better since each patient has a different physiology. The ROI was exported to a temperature table in degrees Celsius, where each cell shows a value for the temperature at each specific point.

2.3 Pre-processing

Once the ROI had been delineated, its temperature table was transformed into thermal images, which use a gray scale for their representation (Fig 3(a)). The process described in Eq. 1 was used to map the temperature values read t to other desired domains $f(t)$.

$$f(t) = \frac{(t * 1000 - oldMin)(newMax - newMin)}{(oldMax - oldMin)} + newMin. \quad (1)$$

In Eq. 1, *oldMin* represents the lowest temperature among all the images in the dataset, multiplied by 1000, while *oldMax* is the highest temperature among all the images multiplied by 1000. *newMin* and *newMax* are constant values that represent the minimum and maximum values of the scale adopted (0 and 255 for gray). In this way, all the base images were converted into intermediate thermal images.

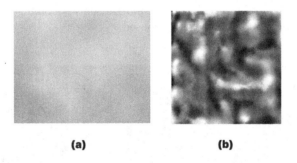

Fig. 3. ROI pre-processing steps

Filters were applied to the images to enhance them. The median filter and adaptive histogram equalization (CLAHE) [16] (Fig 3(b)) were used and demonstrated a significant improvement in network performance in preliminary studies [10,11]. Finally, to solve the problem of the different ROI dimensions obtained for each patient, all the ROI images were centrally cropped to 200 by 200 pixels.

2.4 Classification

Siamese Networks have a high potential to learn, even from a small set of data, and require relative computing power compared to what they can deliver. With this in mind, a Siamese Neural Network architecture was developed for this context. This architecture works with two input images and a single output, which provides a number of how similar these two images are by means of a similarity function.

The architecture is composed of a pair of Convolutional Networks that share the same weights and are responsible for extracting features from the input image pair. The internal structure of the pair of convolutional networks comprises four Convolutional Layers, with kernel size of 2 by 2 pixels, each containing a MaxPooling layer and a Dropout layer. The two outputs of the Convolutional Networks are processed by the Euclidean Distance Layer, converging their result to a single neuron, activated by a sigmoid function, therefore, this last neuron is also trained, generating the final result of the similarity of the two images (Fig. 4). Thus, the closer the value is to 1.0, the network identifies that the two input images are similar, i.e. they should belong to the same class; otherwise, for values close to 0.0, the network identifies that the input images are not very similar, i.e. they should belong to different classes.

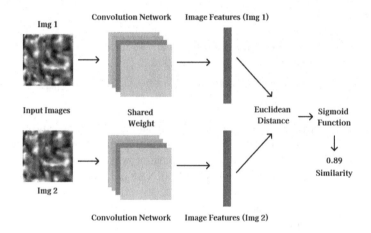

Fig. 4. Siamese Neural Network Architecture

2.5 Evaluation Metrics

As a result of the classification, there are four possibilities: true positives (TP), which are patients with NAFLD who have been correctly classified as sick, and true negatives (TN), which are healthy patients who have been correctly classified as healthy. There are also false positives (FP), which are healthy patients who were wrongly classified as having NAFLD, and false negatives (FN), which are patients with NAFLD who were wrongly classified as healthy. Based on these values, it was possible to calculate the metrics chosen to evaluate the network: accuracy, precision and recall. Each metric is calculated according to the Eqs. 2, 3, and 4 below.

$$precision = \frac{TP}{TP+FP}, \qquad (2)$$

$$recall = \frac{TP}{TP+FN}, \qquad (3)$$

$$accuracy = \frac{TP+TN}{TP+TN+FP+FN}. \qquad (4)$$

In medical applications, one of the most important metrics is the recall metric since the most significant risk to the patient is that they have the disease, and the model classifies them as healthy. Recall, therefore, provides the model's effectiveness in detecting the disease where it exists. Precision calculates how effective the model is at detecting healthy patients where there are healthy ones. Finally, accuracy reveals how effective the model is at getting its classifications right, whether positive or negative.

3 Results and Discussion

This section presents the experimental setup and the analysis of the results achieved at each stage of the proposed method.

3.1 Experimental Setup

The dataset was divided into three different sets: train, test, and validation. The images were separated per patient so that each set was as balanced as possible since the number of images per patient was different. Therefore, the final training set had 126 (66 healthy and 60 NAFLD) images, the test set had 89 (56 healthy and 33 NAFLD) images, and the validation set had 20 (10 healthy and 10 NAFLD) images. There was a careful separation so that the images of the same patient belong to a single set.

The training was designed as follows: as the Siamese network receives pairs of images as input, two image pairs were formed for each image in the training set, the first with a positive example (same class) and the other with a negative example (different classes), so that the image completing the pair was chosen at random. As a result, the network could have one positive and one negative example for each image, and a total of 252 image pairs could be provided.

The network's hyperparameters were batch size 10, Adam optimizer with a learning rate of 0.001, Dropout layers of 0.3 and a binary cross-entropy chosen as the loss function since the problem is one of binary classification (similar or not similar). Data augmentation was not used, as the aim was to work with the original data set, even with a few examples, and observe the behaviour of the Siamese network, which was proposed precisely for this reason. Training was carried out over 200 epochs on the Google Colab platform in conjunction with the Keras library.

3.2 Image Classification Results

The test stage was designed specifically for the context of this work. The Siamese network receives two images as input, but in a real scenario, the specialist treats each image individually, using the Neural Network to classify the patient's image as healthy or NAFLD. Given this, it was necessary to devise a test strategy to meet this context.

The strategy adopted was to find the two best training images, one of a healthy patient and the other of a patient with NAFLD. This choice was made by analyzing the output of the network that had already been trained. For all the training pairs, among the pairs of the same class, the healthy pair and the NAFLD pair with the best classification values were extracted, and at the end of the process, the first image from each pair was chosen. In this way, each image in the test set could be tested at the same time with the most representative images of the healthy and NAFLD classes.

The result with the highest network response, as a similarity function, will be the final classification of the image tested. The results of the evaluation metrics obtained are shown in Table 3.2, along with the results of the related works carried out by [10, 11].

Table 1. Results obtained by the proposed method and comparison with related works

Method	Accuracy	Precision	Recall
Proposed Method	71%	57%	**96%**
Pinto et al. (2021) [10]	**96%**	91%	91%
Farias et al. (2024) [11]	88%	94%	**100%**

Compared to the few studies already carried out in the area [10, 11], the Siamese network, despite being an early-stage study, shows promising results. Recall, as the most relevant metric, reached 96%, demonstrating that the model is effective at detecting sick people among all the sick people. This represents a great improvement compared to the work of Pinto et al. (2021), and very close to what was achieved by Farias et al. (2024).

The Siamese Network has an advantage in that it does not require the use of Data Augmentation and has a much simpler architecture than those used in the related works. This means less computing power for the training and testing phase, and less time is required for pre-processing the images. Accuracy was the lowest of the three metrics, but this is to be expected for a work still in progress. The lowest metric recorded was accuracy, meaning that the model still classifies many healthy people as sick, increasing the number of revisions to be carried out by the specialist assisted by the network.

Analyzing the architectures of the networks compared, it is clear that the convolutional part of the proposed network, used for feature extraction, is much simpler than the others, which may cause low accuracy and low precision. The network learned better the characteristics of the images of sick patients than the images of healthy patients, revealing a point of attention in the images of this group in the used dataset. Figure 5 shows an example of each network classification, where (a) is a true positive, (b) is a true negative, (c) is a false positive and (d) is a false negative.

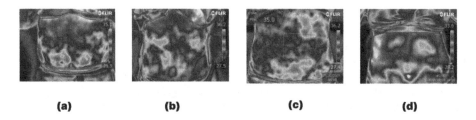

Fig. 5. Images classified by the proposed method

The results obtained show that the potential of the Siamese network is notorious. It represents a promising step in the classification of thermal images for non-alcoholic hepatic steatosis precisely because of the low quantity of images available for training the network, a situation in which Siamese networks acquire their greatest value.

4 Conclusion

This study presented an approach using Siamese Neural Networks (SNN) to classify non-alcoholic fatty liver disease (NAFLD). The use of thermal images has shown promise, highlighting the importance of adequate pre-processing to improve image quality and reduce noise. The proposed architecture achieved an accuracy of 71%, precision of 57% and recall of 96%, revealing that SNNs are a viable alternative for NAFLD classification, especially in scenarios similar to the context presented where there is a limited amount of training data. For future research, it is suggested to explore additional image enhancement techniques, as well as using the architecture already presented, but with modifications in the feature extraction stage, replacing the convolutional layers with a network specialized in extracting features such as, for example, different versions of EfficientNet.

Acknowledgments. The authors acknowledge the Coordenação de Aperfeiçoamento de Pessoal de Nível Superior (CAPES), Brazil - Finance Code 001, Conselho Nacional de Desenvolvimento Científico e Tecnológico (CNPq), Brazil, and Fundação de Amparo à Pesquisa Desenvolvimento Científico e Tecnológico do Maranhão (FAPEMA) (Brazil), Empresa Brasileira de Serviços Hospitalares (Ebserh) Brazil (Grant number 409593/2021-4) for the financial support.

References

1. Yalmaz, Y.: Review article: is non-alcoholic fatty liver disease a spectrum, or are steatosis and non-alcoholic steatohepatitis distinct conditions? Aliment. Pharmacol. Ther. **36**(9), 815–823 (2012). https://doi.org/10.1111/apt.12046
2. ANGULO, P.: GI Epidemiology: nonalcoholic fatty liver disease. Aliment. Pharmacol. Ther. **25**(8), 883–889 (2007). https://doi.org/10.1111/j.1365-2036.2007.03246.x
3. Schwabe, R.F., Tabas, I., Pajvani, U.B.: Mechanisms of fibrosis development in nonalcoholic steatohepatitis. Gastroenterology **158**(7), 1913–1928 (2020). https://doi.org/10.1053/j.gastro.2019.11.311
4. Santana, J.T., et al.: Perfil metabólico e antropométrico dos pacientes obesos e não obesos portadores de esteatose hepática não alcoólica. Revista Eletrônica Acervo Saúde **13**(2), e5525 (2021). https://doi.org/10.25248/reas.e5525.2021
5. Dowman, J.K., Tomlinson, J.W., Newsome, P.N.: Systematic review: the diagnosis and staging of non-alcoholicfatty liver disease and non-alcoholic steatohepatitis. Aliment. Pharmacol. Ther. **33**(5), 525–540 (2010)
6. Stern, C., Castera, L.: Non-invasive diagnosis of hepatic steatosis. Hepatol. Int. **11**(1), 70–78 (2017). https://doi.org/10.1007/s12072-016-9772-z
7. Olaru, A.: Infrared thermographic evaluation of patients with metastatic vertebral fractures after combined minimal invasive surgical treatment. Moldovan Med. J. **60**, 22–25 (2017)
8. Farooq, M.A., Corcoran, P.: Infrared imaging for human thermography and breast tumor classification using thermal images. In: 2020 31st Irish Signals and Systems Conference (ISSC), pp 1-6. IEEE (2020)
9. Abdul Wahab, A., Mohamad Salim, M.I., Yunus, J., Ramlee, M.H.: Comparative evaluation of medical thermal image enhancement techniques for breast cancer detection. J. Eng. Technol. Sci. **50**(1), 40–52 (2018). https://doi.org/10.5614/j.eng.technol.sci.2018.50.1.3
10. Pinto, D.M., Souza, J.C., Silva, A.C., de Araujo Martins Filho, H.M., de Paiva, A.C., Zangaro, R.A.: Classificação de esteatose hepática não alcoólica em imagens termicas da região do fígado utilizando redes neurais convolucionais. In: Anais do XXI Simposio Brasileiro de Computação Aplicada à Saúde, pp. 302–312. SBC (2021)
11. Farias, M., et al.: Method for detecting non-alcoholic fatty liver disease in abdominal thermography time series using temporal convolutional networks. In: Elsevir to be published
12. Li, Z., Liu, F., Yang, W., Peng, S., Zhou, J.: A survey of convolutional neural networks: analysis, applications, and prospects. In: IEEE Transactions on Neural Networks and Learning Systems (2021)
13. Mehmood, A., Maqsood, M., Bashir, M., Shuyuan, Y.: A depp siamese convolution neural network for mult-class classification of Alzheimer disease. Brain Sci. J. (2020)
14. Bromley, J., Guyon, I., LeCun, Y., et al.: Signature verification using a Siamese time delay neural network. In: Cowan, J.D., Tesauro, G., Alspector, J. (eds) Advances in Neural Information Processing Systems, voll 6, [7th NIPS Conference, Denver, Colorado, USA, 1993]. Morgan Kaufmann, pp. 737–744 (1993)
15. Ozougwu, J.C.: Physiology of the liver. Int. J. Res. Pharm. Biosci., 13–24 (2017)
16. Zuluaga-Gomez, J.., Al Masry, Z.., Benaggoune, K.., Meraghni, S.., Zerhouni, N..: A CNN-based methodology for breast cancer diagnosis using thermal images. Comput. Methods Biomech. Biomedical Eng. Imaging Vis. **9**(2), 131–145 (2021). https://doi.org/10.1080/21681163.2020.1824685

Classifying Graphs of Elementary Mathematical Functions Using Convolutional Neural Networks

Joaquim Viana[✉][iD], Helder Matos[iD], Marcelle Mota[iD], and Reginaldo Santos[iD]

Instituto de Ciências Exatas e Naturais, Universidade Federal do Pará, Belém-PA, Brazil
{joaquim.viana,helder.matos}@icen.ufpa.br, {mpmota,regicsf}@ufpa.br

Abstract. The classification of images of elementary mathematical function graphs presents a significant challenge in computer vision; this is due to the varied shapes and formats of each functions curves. This classification is crucial for identifying function graphs, which have important applications in text and mathematical symbol recognition technologies, aiding visually impaired individuals by providing access to printed content. In educational environments, this identification helps obtain the analytical expression of drawn graphs, facilitating the extraction of information from educational materials. This article investigates various convolutional neural network (CNN) architectures to identify the most suitable model for classifying images of elementary mathematical function graphs. We compare our model with other renowned architectures, such as ResNet, MobileNet, and EfficientNet, using a custom dataset of function graphs. Our experiments show that the proposed architecture significantly outperforms networks of general purpose, achieving an accuracy of 98.51% in classifying elementary mathematical function graphs.

Keywords: Graph of a function · Curve recognition · Image classification · Convolutional Neural Networks · Deep Learning

1 Introduction

Classifying images of elementary mathematical function graphs is challenging in computer vision because of the varied shapes and formats of these curves [7]. Identifying the type of graph of a function is a preliminary step toward identifying the analytic expression of the graph in question. Determining the analytic expression of graphs of elementary mathematical functions can be used in technologies for recognizing text and mathematical symbols, assisting visually impaired individuals by enabling them to access information contained in printed material [10]. In educational environments, such identification can aid in extracting information from educational materials, helping research, and the study of mathematical concepts.

© The Author(s), under exclusive license to Springer Nature Switzerland AG 2025
A. Paes and F. A. N. Verri (Eds.): BRACIS 2024, LNAI 15692, pp. 270–280, 2025.
https://doi.org/10.1007/978-3-031-79029-4_19

In this article, we propose a solution for classifying images of graphs of elementary mathematical functions using convolutional neural networks (CNNs). Our goal is to find a CNN architecture that best suits this problem.

This article is structured as follows: Sect. 2 discusses related work regarding the use of CNNs for image classification; Sect. 3 describes the proposed methodology; Sect. 4 presents results and discussions; and Sect. 5 lists the study's conclusions.

2 Related Work

In the literature, there are few works demonstrating techniques for extracting elements within images of graphs of elementary mathematical functions. In [9], a method is proposed that divides the graph image into regions: regions outside the axes (such as titles, axis labels, and scales) and contour regions of the line. Scale information is obtained from the regions outside the axes, and the coordinate values of each line graph are also extracted from these regions. Finally, this information is integrated to convert graph data into numerical information. However, the method in [9] was used only for graphs with a single configuration regarding the positioning of graph elements within the image.

In [21], an offline image segmentation approach based on the technique from [9] is proposed. Images are segmented into connected components and characters. The characters represent the analytical expressions typically accompanying the image, while the connected components encompass everything from the axes to the function curves. Subsequently, elements of connected components are grouped to identify the function curve. Thus, the author managed to separate 15 of a universe of 20 images and group 104 of a total of 112 components within the images. The study highlights the need to improve the accuracy of their results.

Another study, aimed at developing an automatic system for transforming function graphs into tactile graphics that can be used by visually impaired individuals, was conducted by [5]. The authors present a method for extracting graphical elements in mathematical function graphs. The method involves identifying all components of the images (curves, lines, axes, equations, dashed lines, points, etc.) and dividing them into primitive elements (x and y axes, main lines, and curves). After identifying the primitive elements, these elements are combined to form graphical elements. In an experiment with 33 scanned images from science textbooks, the authors were able to extract 502 primitive elements. Of these, 373 were successfully combined into 115 graphical elements, while another 129 were not.

Another approach suggested by [18] uses the *MathGraphReader*, a tool that extracts data from graphs and creates alternative textual descriptions. Using text processing, image processing, plotting, and mathematical concepts, the authors built a tool that allows visually impaired students to access graph information interactively. The technique involves first detecting the origin of the axes and then tracing the intersection points of the axes with the function curve to find inflection points. However, the study does not address cases where the dashed line of the graph curve is displayed without intersecting the axes.

The methods mentioned so far propose offline extraction techniques, where the extraction of elements is performed on previously provided images. A different method was suggested by [19], which proposes obtaining the analytic expression from a function constructed online, i.e., when the graph curve is drawn in real-time. This method captures the pixels of the curve on the screen and then uses polynomial regression to find a curve that best fits the pixel points. Although the method yields good results, it has not been evaluated on a large dataset to verify accuracy. Additionally, it is limited to polynomial expressions and is restricted to the range $x \in [-1, 1]$.

CNNs have revolutionized the field of computer vision and have been widely used in a broad range of applications [1,3], such as image recognition, object detection, image segmentation, among others [2,15,16]. However, the use of CNNs for classifying images of graphs of elementary mathematical functions has been limited due to the scarcity of datasets [4]. The proposed methodology encompasses not only the creation of a dataset that contains different types of functions and their variations but also conducts an investigation into CNN architectures applied to the recognition of mathematical function patterns in graphs. Additionally, it compares these results with a set of renowned architectures applied to more generalizable computer vision tasks.

3 Methodology

This section describes the generation of a dataset of images of elementary functions graphs, the rationale behind the chosen deep learning model architecture, and how the model's performance was assessed and benchmarked against other existing methods.

3.1 Process of Data Generation

The lack of open public datasets addressing the problem of classifying function graphs motivated the development of a systematic dataset generation process. Figure 1 illustrates the process of creating a dataset containing images of various elementary functions to be modeled.

To create the dataset, we first listed various types of elementary functions, as depicted in Fig. 2: linear (2a), quadratic (2b), cubic (2c), exponential (2d), logarithmic (2e), square root and cubic root (2f), sine (2g), cosine (2h), tangent (2i), and cotangent (2j). We randomly generated analytic expressions that could plot visible graphs within the range $y \in [-5, 5]$ and for the domain x the interval was set to $x \in [-2\pi, 2\pi]$ for trigonometric functions and $x \in [-5, 5]$ for other functions. Subsequently, a script was generated to automate the construction of these functions using Winplot [20], a free software for plotting two-dimensional and three-dimensional graphs of mathematical functions, including explicit, implicit, parametric, and polar function graphs.

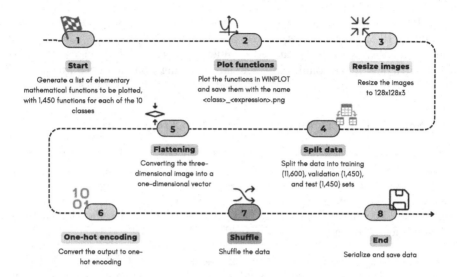

Fig. 1. Image processing in function graphs dataset.

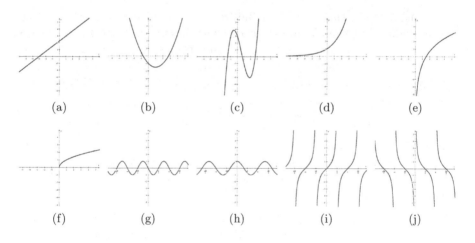

Fig. 2. Graphs of elementary mathematical functions.

Some functions, such as exponential and logarithmic functions, do not offer many variations within the specified range and therefore could generate a limited number of images. To avoid class imbalance, we limited the number of generated images to 1,450 per class. In total, 14,500 images were generated with a resolution of 128 × 128 pixels and saved with names like `<class>_<expression>.png`.

As for generated images, we divided them into training (80%, 11,600 images), validation (10%, 1,450 images), and test sets (10%, 1,450 images). Each group contains all classes with a balanced number of samples. The images were flattened into a one-dimensional vector and labeled using a one-hot encoding format. Finally, the dataset was shuffled and saved in serialized *Pickle* files.

3.2 Selection and Definition of the CNN Architecture

Due to the scarcity of studies focusing on constructing deep neural network architectures for classifying images in elementary function graphs data, we will compare the performance of the proposed architecture with architectures that have excelled on broader datasets. Three CNN architectures were selected for comparison, known for their strong performance in image classification tasks on datasets such as ImageNet [8] and CIFAR-10 [14]:

- **ResNet** [11]: Designed to ensure that learning is not hindered by adding more layers, using shortcuts or residual connections to facilitate optimization. Variants include ResNet-50 (50 layers), ResNet-101 (101 layers), ResNet-152 (152 layers). ResNet-50 was chosen for comparison as it achieved 93% to 94% accuracy on CIFAR-10 dataset.
- **MobileNet** [12]: Designed for mobile and low-power devices, known for its lightweight and low computational cost. Variants include V1 (depthwise separable convolutions), V2 (inverted residual blocks and linear bottlenecks), and V3 (automated neural architecture search and additional optimizations). MobileNetV3 was chosen for comparison as it achieved 91% to 92% accuracy on CIFAR-10 dataset.
- **EfficientNet** [22]: Designed to optimize the balance between accuracy and computational efficiency, using a compound scaling method that simultaneously scales depth, width, and resolution of the network. Variants include EfficientNet-B0 (more efficient) to B7 (more layers, filters, and resolution). EfficientNet-B0 was chosen for comparison as it achieved approximately 95% accuracy on CIFAR-10 dataset.

The CNN architecture proposed for classifying elementary function graph images was named *F-Graphs*. The *hyperband* algorithm [17] was used to perform an optimized search of neural network hyperparameters that enhance its performance in image classification tasks.

From the initializations made, the *hyperband* algorithm returned the CNN architecture illustrated in Fig. 3. The network consists of three convolutional layers activated with the Rectified Linear Unit (ReLU) function, combined with max pooling layers after the convolutional layers. In the fully connected layer, only one hidden layer with 128 neurons activated with the sigmoid function was included. At the output, a dense layer with 10 neurons corresponding to the 10 classes of the dataset and activated with the *softmax* function produces the probability distribution for the output classes.

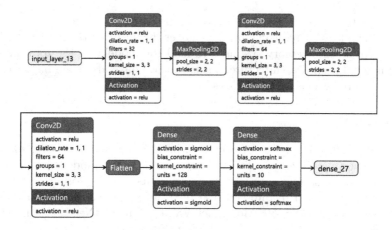

Fig. 3. F-Graphs' network architecture.

3.3 Model Training and Evaluation

F-Ggraphs was created using the Keras library for Python [6], which simplifies the construction and training of neural networks, often used to create and experiment with deep learning models. Figure 4 shows the loss behavior and accuracy evolution during the training of the proposed model. Training included five different initializations for each architecture and 40 epochs using *early stopping* based on monitoring loss on the validation data as the stopping criteria.

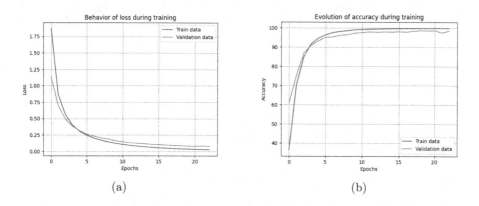

Fig. 4. Loss and accuracy values for training and validation of the proposed model.

To optimize training, the *Adaptive Moment Estimation* (Adam) optimizer [13] and a learning rate of 9×10^{-5} were used. Loss estimation during training for the 10 classes was performed using *categorical cross-entropy*. The metric used to evaluate performance during training was accuracy.

The results will be shown through comparisons between the accuracy, precision, recall, and F1-score metrics averaged over the 5 runs of the model fit on the data, as well as the memory size used to load the model and the total number of parameters for each architecture. Visualizations have also been built to show the model's predictions between classes.

4 Results

The average behavior results for accuracy, precision, recall, F1-score, size, and number of parameters for the five runs of the compared models are presented in Table 1. The proposed architecture achieved its best performance with the configuration described in Subsect. 3.2, reaching 98.51% accuracy on the test data. In contrast, ResNet-50, MobileNet-V3, and EfficientNet-B0 returned low accuracies of 10.57%, 13.02%, and 11.09%, respectively.

Table 1. Mean and standard deviation results for training on the function graphs dataset.

	ResNet-50		MobileNet-V3		EfficientNet-B0		F-Graphs	
	μ	σ	μ	σ	μ	σ	μ	σ
Accuracy	10.57%	0.01	13.02%	0.01	11.09%	0.04	98.51%	0.0
Precision	8.09%	0.09	25.87%	0.06	9.12%	0.01	98.43%	0.0
Recall	10.57%	0.01	13.02%	0.01	11.09%	0.04	98.41%	0.0
F1-Score	2.99%	0.01	6.72%	0.02	7.93%	0.03	98.41%	0.0
Size (MB)	138.00		38.63		45.46		74.16	
Parameters	36,174,880		10,126,560		11,918,147		19,440,800	

The proposed model *F-Graphs* also achieved better results in precision, recall, and F1-score, with values of 98.43%, 98.41%, and 98.41%, respectively. However, *F-Graphs* is surpassed in terms of memory size and number of parameters by MobileNet-V3, which has the smallest memory footprint at 38.63MB and the lowest number of parameters at 10,126,560. As expected, the latter presents a lower space complexity, due to the fact that it is specifically designed for mobile devices, which are constrained by limited computational resources.

ResNet-50 is a deep CNN with multiple convolutional layers designed to extract more features from images [11]. The number of convolutional layers influences the models learning process, leading to inferior performance in classifying images of function graphs. In terms of memory size, it occupies approximately 138MB, making it heavier due to the large number of network parameters.

Similarly, the number of convolutional layers heavily influenced the generalization power of MobileNet-V3 and EfficientNet-B0, although these architectures outperformed ResNet-50. To improve the performance of ResNet-50, MobileNet-V3, and EfficientNet-B0 on the dataset of elementary function images, parameter tuning would be necessary.

Although the proposed model *F-Graphs* achieved an accuracy of 98.51%, it struggled with misclassifying some functions, as illustrated in Fig. 5. For example, the misclassification of the cosine function with sine occurs quite often, since these functions are very similar, especially when translated horizontally.

Another important fact is that the *F-Graphs* model tends to generalize linear functions more frequently. Functions such as the exponential function (Eq. 1), shown in Fig. 6a, and the cubic function (Eq. 2), shown in Fig. 6b, were often misclassified as linear. This is because the coefficients of the functions were generated randomly, forming analytical expressions that generate functions of this type. To address this issue, we replaced some of these functions that tended to be confused with other less ambiguous functions.

$$f(x) = 1.25^x - 1 \tag{1}$$

$$g(x) = x^3 + 2x^2 + 5x + 0 \tag{2}$$

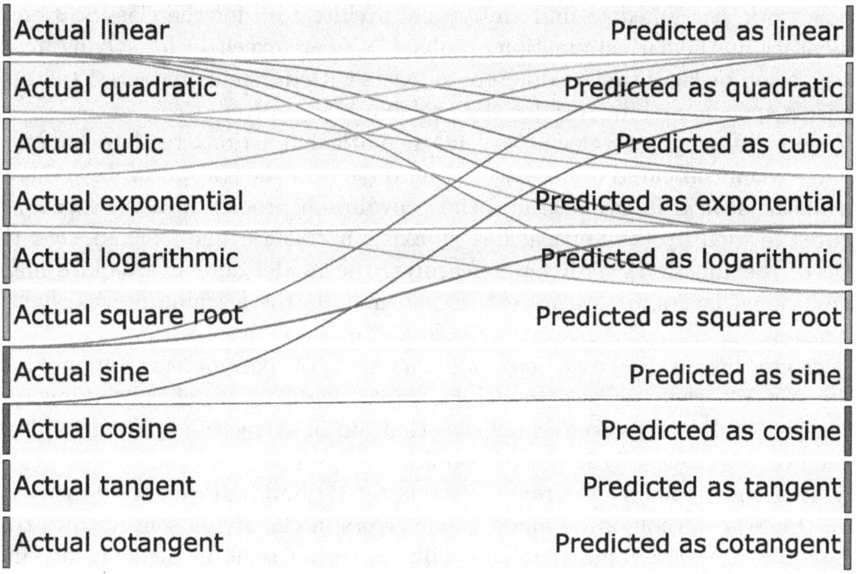

Fig. 5. Misclassifications among classes.

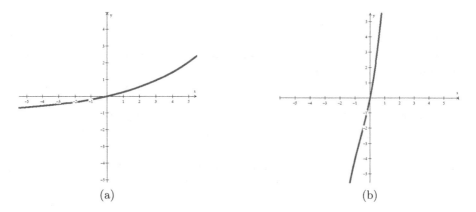

Fig. 6. Exponential and cubic functions that resemble linear ones.

5 Conclusion

In this work, we aimed to find an optimal architecture for the classification of elementary mathematical function graphs. CNNs are excellent for solving image classification tasks. However, different datasets with different characteristics may require a specific architecture for the problem being solved.

Images of graphs of elementary mathematical functions tend to have few features when compared to images of other objects in the real world. In an image of a mathematical function graph, the convolution process involves extracting features to identify the vertical line (y-axis), horizontal line (x-axis), and the curve of the function. With fewer features, the model does not require many convolutional layers for feature extraction, making the learning process lighter and faster.

Despite having a larger size and number of parameters compared to MobileNet-V3, the proposed model *F-Graphs* could be beneficial for incorporation into mobile assistive technologies that aid in extracting information from printed texts for visually impaired people.

The proposed model *F-Graphs* demonstrated good performance on the test data. However, although the model made errors in classifying some graphs that had ambiguities, we think that this will not be an issue in deployment environments, since scenarios of this kind are unlikely to occur. One limitation of this model arises when presented with two or more graphs sketched on the same axes. In some real-world scenarios, the model tends to misclassify functions when there are multiple analytical expressions in the image.

Future work should focus on extracting additional information from images, such as identifying the analytical expressions of functions, to provide more comprehensive details about graph behaviors.

Acknowledgments. This research was funded by the *Brazilian Federal Agency for Support and Evaluation of Graduate Education* (CAPES) and developed with the support of the *Center for High-Performance Distributed Computing of Pará* (CCAD) of the *Federal University of Pará* (UFPA).

Disclosure of Interests. The authors have no competing interests to declare that are relevant to the content of this article.

Dataset Availability. The dataset for this project is publicly available to facilitate transparency, reproducibility, and further research through the following link: https://github.com/jvianna07/fgraphs_dataset/.

The script used is available here: https://github.com/jvianna07/functions_graphs/

References

1. Algan, G., Ulusoy, I.: Image classification with deep learning in the presence of noisy labels: a survey. Knowl.-Based Syst. **215**, 106771 (2021). https://doi.org/10.48550/arXiv.1912.05170
2. Bakhtiarnia, A., Zhang, Q., Iosifidis, A.: Efficient high-resolution deep learning: a survey. ACM Comput. Surv. **56** (2024). https://doi.org/10.1145/3645107
3. Bhatt, D., et al.: CNN variants for computer vision: history, architecture, application, challenges and future scope. Electronics **10**(20) (2021). https://doi.org/10.3390/electronics10202470
4. Cao, P., Zhu, Z., Wang, Z., Zhu, Y., Niu, Q.: Applications of graph convolutional networks in computer vision. Neural Comput. Appl. **34**(16), 13387–13405 (2022). https://doi.org/10.1007/s00521-022-07368-1
5. Chen, J., Nagaya, R., Takagi, N.: Development of a method for extracting graph elements from mathematical graphs. In: 2012 IEEE International Conference on Systems, Man, and Cybernetics (SMC), pp. 2921–2926. IEEE (2012)
6. Chollet, F., et al.: Keras (2015). https://keras.io
7. Dai, J., Lin, S.: Image recognition: current challenges and emerging opportunities (2018). https://www.microsoft.com/en-us/research/lab/microsoft-research-asia/articles/image-recognition-current-challenges-and-emerging-opportunities/. Accessed 20 Jun 2024
8. Deng, J., Dong, W., Socher, R., Li, L.J., Li, K., Fei-Fei, L.: ImageNet: a large-scale hierarchical image database. In: 2009 IEEE Conference on Computer Vision and Pattern Recognition, pp. 248–255. IEEE (2009).https://doi.org/10.1109/cvpr.2009.5206848
9. Fuda, T., Omachi, S., Aso, H.: Recognition of line graph images in documents by tracing connected components. Syst. Comput. Japan **38**(14), 103–114 (2007). https://doi.org/10.1002/scj.10615
10. Goncu, C., Marriott, K.: Accessible graphics: graphics for vision impaired people. In: Cox, P., Plimmer, B., Rodgers, P. (eds.) Diagrams 2012. LNCS (LNAI), vol. 7352, pp. 6–6. Springer, Heidelberg (2012). https://doi.org/10.1007/978-3-642-31223-6_5
11. He, K., Zhang, X., Ren, S., Sun, J.: Deep residual learning for image recognition. In: Proceedings of the IEEE Conference on Computer Vision and Pattern Recognition, pp. 770–778 (2016). https://doi.org/10.1109/CVPR.2016.90
12. Howard, A.G., et al.: MobileNets: efficient convolutional neural networks for mobile vision applications. Comput. Res. Repository (CoRR) (2017). https://doi.org/10.48550/arXiv.1704.04861

13. Kingma, D.P., Ba, J.: Adam: a method for stochastic optimization. arXiv (2014). https://doi.org/10.48550/arXiv.1412.6980
14. Krizhevsky, A., Hinton, G.: Learning multiple layers of features from tiny images. Tech. Rep. 0, University of Toronto, Toronto, Ontario (2009). https://www.cs.toronto.edu/~kriz/learning-features-2009-TR.pdf
15. Krizhevsky, A., Sutskever, I., Hinton, G.E.: ImageNet classification with deep convolutional neural networks. Commun. ACM **60**(6), 84–90 (2017). https://doi.org/10.1145/3065386
16. LeCun, Y., Bengio, Y., Hinton, G.: Deep learning. Nature **521**(7553), 436–444 (2015). https://doi.org/10.1038/nature14539
17. Li, L., Jamieson, K., DeSalvo, G., Rostamizadeh, A., Talwalkar, A.: Hyperband: a novel bandit-based approach to hyperparameter optimization. J. Mach. Learn. Res. (2016). https://doi.org/10.48550/ARXIV.1603.06560
18. Nazemi, A., Fernando, C., Murray, I., A. McMeekin, D.: Accessible and navigable representation of mathematical function graphs to the vision-impaired. Comput. Inf. Sci. **9**(1), 31 (2015). https://doi.org/10.5539/cis.v9n1p31
19. Phyo, Y.K.: Graphical functions made from an effortless sketch. MS paint that returns equations. https://towardsdatascience.com/graphical-functions-made-from-an-effortless-sketch-266ccf95c46d (2020). Accessed 8 Apr 2024
20. Souza, S.D.A.: Usando o winplot. http://www.mat.ufpb.br/~sergio/winplot/winplot.html (2004). Accessed 08 Apr 2024
21. Takagi, N.: Mathematical figure recognition for automating production of tactile graphics. In: 2009 IEEE International Conference on Systems, Man and Cybernetics, pp. 4651–4656 (2009). https://doi.org/10.1109/ICSMC.2009.5346749
22. Tan, M., Le, Q.V.: EfficientNet: rethinking model scaling for convolutional neural networks. Computing Research Repository (CoRR) (2019). https://doi.org/10.48550/ARXIV.1905.11946

Comparing Neural Network Encodings for Logic-Based Explainability

Levi Cordeiro Carvalho, Saulo A. F. Oliveira, and Thiago Alves Rocha[✉]

Instituto Federal do Ceará (IFCE), Fortaleza, Brazil
{levi.carvalho,saulo.oliveira,thiago.alves}@ifce.edu.br

Abstract. Providing explanations for the outputs of artificial neural networks (ANNs) is crucial in many contexts, such as critical systems, data protection laws and handling adversarial examples. Logic-based methods can offer explanations with correctness guarantees, but face scalability challenges. Due to these issues, it is necessary to compare different encodings of ANNs into logical constraints, which are used in logic-based explainability. This work compares two encodings of ANNs: one has been used in the literature to provide explanations, while the other will be adapted for our context of explainability. Additionally, the second encoding uses fewer variables and constraints, thus, potentially enhancing efficiency. Experiments showed similar running times for computing explanations, but the adapted encoding performed up to 18% better in building logical constraints and up to 16% better in overall time.

Keywords: Artificial Neural Networks · Explainable Artificial Intelligence · Logic-based Explainable AI

1 Introduction

Artificial neural networks (ANNs) are widely applied in tasks like computer vision, speech recognition, and pattern recognition [13]. Despite their success, ANNs are often considered black-box algorithms. Such a lack of interpretability poses risks in critical domains such as medical and financial applications, where understanding model decisions is crucial. Additionally, the presence of adversarial examples highlights the need for explainability in machine learning algorithms, including neural networks. An adversarial example is an instance misclassified by a machine learning model and also slightly different from another correctly classified instance [6].

In this work, an explanation for a prediction made by an ANN is a subset of features and their values that alone suffice for the prediction. If an instance has the features in this subset, the ANN makes the same prediction, regardless of the values of other features. For example, given an instance $\{sneeze = True, weight = 70\ kg, headache = True, age = 40\ years\}$ and its ANN output flu, a possible explanation could be $\{sneeze = True, headache = True\}$. That is, if an instance has the features $sneeze = True$ and $headache = True$, the ANN prediction is *flu*, regardless of *weight* and *age* values. A minimal explanation avoids redundancy by including only essential information. An explanation is considered minimal when removing any feature results in the loss of assurance that every instance satisfying the explanation maintains the same output. Then, a minimal explanation avoids redundancy, providing only essential information.

Heuristic methods, such as ANCHOR [15] and LIME [14], have been used to provide explanations for machine learning models. However, these approaches explore the instance space locally, not resulting in explanations that have minimal sizes and formal guarantees of correctness. Correctness guarantees are provided when there are no instances with the values specified in the explanation such that the ANN makes a different prediction. Moreover, minimal explanations are desired since they do not contain redundancy, making them easier to understand and interpret.

Some approaches aim to provide explanations for machine learning models with formal guarantees of correctness [1, 4, 7, 8, 16, 18]. Ignatiev et al. [8] proposed a logic-based algorithm that gives minimal and correct explanations for ANNs, utilizing logical constraints originally designed for finding adversarial examples Fischetti and Jo [5]. These constraints include linear equations, inequalities, and logical implications, solved using a Mixed Integer Linear Programming (MILP) solver. However, scalability issues arise, particularly with large ANNs, necessitating further development before deployment in large-scale production environments.

This work explores two different encodings to improve the scalability of providing correct minimal explanations for ANNs, building upon [8]. In addition to the logical constraints of [5], we adopt the encoding proposed by Tjeng et al. [17], which uses fewer variables and constraints, and excludes logical implications. By reducing variables and constraints compared to [5], our approach aims to enhance explanation computation performance. To adapt the approach of [17] for explanations, we introduce new constraints to ensure correctness. In line with the encodings proposed by Fischetti and Jo [5] and Tjeng et al. [17], we also compute lower and upper bounds for each neuron. These bounds are found through optimization using a MILP solver. Moreover, these bounds can aid the solver in computing explanations more rapidly. In this manner, we compare the time required for constructing logical constraints with lower and upper bounds of each neuron, along with the time needed for computing explanations.

We conducted experiments to evaluate both encodings. Our adaptation of the encoding proposed in [17] exhibits a better running time in building encodings for ANNs with two layers and tens of neurons, showing an improvement of up to 18%. Surprisingly, both methods exhibit similar running times for computing explanations. Furthermore, our adaptation outperforms the other encoding in the overall time, encompassing both building logical constraints and computing explanations. In this case, the results indicate an improvement of up to 16%. In summary, our main contributions are described in the following:

- Adaptation of the encoding proposed in [17] to provide explanations for ANNs; additional constraints were incorporated to address the problem of computing explanations.
- Comparative analysis of the running time for building the logical constraints between the two approaches. Additionally, we analyze the time for generating explanations using both encodings.
- Publicly available implementations of both encodings for finding explanations for ANNs[1].

In the next section, we review some concepts and terminologies about Logic, MILP and ANNs. Sections 3 and 4 show how to compute explanations with and without implications, respectively. Section 4 describes our adaptation of the encoding proposed in [17]. Experiments and results are presented in Sect. 5. Finally, conclusions and future work are described in Sect. 6.

2 Background

In this section, we introduce some initial concepts and terminology to understand the rest of this work.

2.1 First-Order Logic over LRA

In this work, we use first-order logic (FOL) to give explanations with guarantees of correctness. We use quantifier-free first-order formulas over the theory of linear real arithmetic (LRA). Then, first-order variables are allowed to take values from the real numbers $\mathbb{R}$. For details, see [11]. Therefore, we consider formulas as defined below:

$$F, G := p \mid (F \wedge G) \mid (F \vee G) \mid (\neg F) \mid (F \rightarrow G),$$
$$p := \sum_{i=1}^{n} w_i x_i \leq b \mid \sum_{i=1}^{n} w_i x_i < b, \tag{1}$$

such that F and G are quantifier-free first-order formulas over the theory of linear real arithmetic. Moreover, p represents the atomic formulas such that $n \geq 1$, each w_i and b are fixed real numbers, and each x_i is a first-order variable. Observe that we allow the use of other letters for variables instead of x_i, such as s_i, z_i, q_i. For example, $(2.5x_1 + 3.1x_2 \geq 6) \wedge (x_1 = 1 \vee x_1 = 2) \wedge (x_1 = 2 \rightarrow x_2 \leq 1.1)$ is a formula by this definition. Observe that we allow standard abbreviations as $\neg(2.5x_1 + 3.1x_2 < 6)$ for $2.5x_1 + 3.1x_2 \geq 6$.

Since we are assuming the semantics of formulas over the domain of real numbers, an *assignment* $\mathcal{A}$ for a formula F is a mapping from the first-order variables of F to elements in the domain of real numbers. For instance, $\{x_1 \mapsto 2.3, x_2 \mapsto 1\}$ is an assignment for $(2.5x_1 + 3.1x_2 \geq 6) \wedge (x_1 = 1 \vee x_1 = 2) \wedge (x_1 =$

---

[1] https://github.com/LeviCC8/Explications-ANNs.

$2 \to x_2 \leq 1.1$). An assignment $\mathcal{A}$ *satisfies* a formula F if F is true under this assignment. For example, $\{x_1 \mapsto 2, x_2 \mapsto 1.05\}$ satisfies the formula in the above example, whereas $\{x_1 \mapsto 2.3, x_2 \mapsto 1\}$ does not satisfy it.

A formula F is *satisfiable* if there exists a satisfying assignment of F. To give an example, the formula in the above example is satisfiable since $\{x_1 \mapsto 2, x_2 \mapsto 1.05\}$ satisfies it. As another example, the formula $(x_1 \geq 2) \wedge (x_1 < 1)$ is unsatisfiable since no assignment satisfies it. Given formulas F and G, the notation $F \models G$ is used to denote *logical consequence* or *entailment*, i.e., each assignment that satisfies F also satisfies G. As an illustrative example, let $F = (x_1 = 2 \wedge x_2 \geq 1)$ and $G = (2.5x_1 + x_2 \geq 5) \wedge (x_1 = 1 \vee x_1 = 2)$. Then, $F \models G$. The essence of entailment lies in ensuring the correctness of the conclusion G based on the given premise F. In the context of computing explanations, as presented in [8], logical consequence serves as a fundamental tool for guaranteeing the correctness of predictions made by ANNs. Therefore, our adaptation of the encoding proposed by Tjeng et al. [17] also incorporates the principles of entailment for computing explanations.

The relationship between satisfiability and entailment is a fundamental aspect of logic. It is widely known that, for all formulas F and G, it holds that $F \models G$ iff $F \wedge \neg G$ is unsatisfiable. For instance, $(x_1 = 2 \wedge x_2 \geq 1) \wedge \neg((2.5x_1 + x_2 \geq 5) \wedge (x_1 = 1 \vee x_1 = 2))$ has no satisfying assignment since an assignment that satisfies $(x_1 = 2 \wedge x_2 \geq 1)$ also satisfies $(2.5x_1 + x_2 \geq 5) \wedge (x_1 = 1 \vee x_1 = 2)$ and, therefore, does not satisfy $\neg((2.5x_1 + x_2 \geq 5) \wedge (x_1 = 1 \vee x_1 = 2))$. Since our approach builds upon the concept of logical consequence, we can leverage this connection in the context of computing explanations for ANNs.

2.2 Mixed Integer Linear Programming

In Mixed Integer Linear Programming (MILP), the objective is to optimize a linear function subject to linear constraints, where some or all of the variables are required to be integers [2]. MILP is a crucial technique in our work for determining the lower and upper bounds of each neuron in the ANNs. For example, we utilize a minimization problem to determine the lower bound of neurons within ANNs. This process involves formulating an objective function that seeks to minimize the lower bound, subject to constraints that reflect the behaviour of ANNs. To illustrate the structure of a MILP, we provide an example below:

$$
\begin{aligned}
\min \quad & y_1 \\
\text{s.t.} \quad & 1 \leq x_1 \leq 3 \\
& 3x_1 + s_1 - 2 = y_1 \\
& 0 \leq y_1 \leq 3x_1 - 2 \\
& 0 \leq s_1 \leq 3x_1 - 2 \\
& z_1 = 1 \to y_1 \leq 0 \\
& z_1 = 0 \to s_1 \leq 0 \\
& z_1 \in \{0, 1\}
\end{aligned}
\tag{2}
$$

In the MILP in (2), we want to find values for variables x_1, y_1, s_1, z_1 minimizing the value of the objective function y_1 among all values that satisfy the constraints. Variable z_1 is binary since $z_1 \in \{0,1\}$ is a constraint in the MILP, while variables x_1, y_1, s_1 have the real numbers $\mathbb{R}$ as their domain. The constraints in a MILP may appear as linear equations, linear inequalities, and indicator constraints. Indicator constraints can be seen as logical implications of the form $z = v \to \sum_{i=1}^{n} w_i x_i \leq b$ such that z is a binary variable, v is a constant 0 or 1 [3].

An important observation is that a MILP problem without an objective function corresponds to a satisfiability problem, as discussed in Sect. 2.1. Given that the approach for computing explanations relies on logical consequence, and considering the connection between satisfiability and logical consequence, we employ a MILP solver to address explanation tasks. Additionally, throughout the construction of the MILP model, we utilize optimization, specifically employing a MILP solver, to determine tight lower and upper bounds for the neurons of ANNs.

2.3 Classification Problems and Artificial Neural Networks

In machine learning, classification problems are defined over a set of n features $\mathcal{F} = \{x_1, ..., x_n\}$ and a set of $\mathcal{N}$ classes $\mathcal{K} = \{c_1, c_2, ..., c_\mathcal{N}\}$. In this work, we consider that each feature $x_i \in \mathcal{F}$ takes its values v_i from the domain of real numbers. Moreover, each feature x_i has an upper bound u_i and a lower bound l_i such that $l_i \leq x_i \leq u_i$, and its domain is the closed interval $[l_i, u_i]$. This is represented as a set of domain constraints or feature space $D = \{l_1 \leq x_1 \leq u_1, l_2 \leq x_2 \leq u_2, ..., l_n \leq x_n \leq u_n\}$. For example, a feature for the height of a person belongs to the real numbers and may have lower and upper bounds of 0.5 and 2.1 meters, respectively. Furthermore, $\{x_1 = v_1, x_2 = v_2, ..., x_n = v_n\}$ represents a specific point or instance of the feature space such that each v_i is in the domain of x_i.

An ANN is a function that maps elements in the feature space into the set of classes $\mathcal{K}$. A feedforward ANN is composed of $L + 1$ layers of neurons. Each layer $l \in \{0, 1, ..., L\}$ is composed of n_l neurons, numbered from 1 to n_l. Layer 0 is fictitious and corresponds to the input of the ANN, while the last layer, K corresponds to its outputs. Layers 1 to $L - 1$ are typically referred to as hidden layers. Let x_i^l be the output of the ith neuron of the lth layer, with $i \in \{1, ..., n_l\}$. The inputs to the ANN can be represented as x_i^0 or simply x_i. Moreover, we represent the outputs as x_i^L or simply o_i.

The values x_i^l of the neurons in a given layer l are computed through the output values x_j^{l-1} of the previous layer, with $j \in \{1, ..., n_{l-1}\}$. Each neuron applies a linear combination of the output of the neurons in the previous layer. Then, the neuron applies a nonlinear function, also known as an activation function. The output of the linear part is represented as $\sum_{j=1}^{n_{l-1}} w_{i,j}^l x_j^{l-1} + b_i^l$ where $w_{i,j}^l$ and b_i^l denote the weights and bias, respectively, serving as parameters of the ith neuron of layer l. In this work, we consider only feedforward ANNs with the Rectified Linear Unit (ReLU) as activation function because it can be represented by linear constraints due to its piecewise-linear nature. This function is a widely used activation whose output is the maximum between its input value and zero. Then, $x_i^l = \text{ReLU}(\sum_{j=1}^{n_{l-1}} w_{i,j}^l x_j^{l-1} + b_i^l)$ is the output of the ReLU.

For classification tasks, the last layer L is composed of $n_L = \mathcal{N}$ neurons, one for each class. Moreover, it is common to normalize the output layer using a Softmax layer. Consequently, these values represent the probabilities associated with each class. The class with the highest probability is chosen as the predicted class. However, we do not need to consider this normalization transformation as it does not change the maximum value of the last layer. Thus, the predicted class is $c_i \in \mathcal{K}$ such that $i = \arg\max_{j \in \{1,...,\mathcal{N}\}} x_j^L$.

3 Explanations for ANNs with Logical Implications

Ignatiev et al. [8] proposed an algorithm that computes minimal explanations for ANNs, yielding a subset of the input features sufficient for the prediction. This approach is based on logic with guarantees on the correctness and minimality of explanations. A flowchart for computing explanations using such an algorithm is shown in Fig. 1.

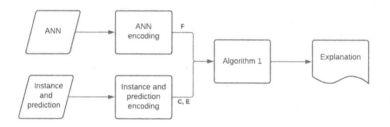

Fig. 1. Flowchart for calculating explanations.

First, the ANN and the feature space $\{l_1 \leq x_1 \leq u_1, l_2 \leq x_2 \leq u_2, ..., l_n \leq x_n \leq u_n\}$ are encoded as a formula F, an instance $\{x_1 = v_1, x_2 = v_2, ..., x_n = v_n\}$ of the feature space is encoded as a conjunction in a formula C, and the associated prediction by the ANN is encoded as a formula E. Then, it holds that $C \wedge F \models E$. The minimal explanation C_m of C is calculated removing feature by feature from C. For example, given a feature x_i with value v in C, if $C \setminus \{x_i = v\} \wedge F \models E$, feature x_i may be considered as irrelevant in the explanation and is removed from C. Otherwise, if $C \setminus \{x_i = v\} \wedge F \not\models E$, then x_i is kept in C since the same class cannot be guaranteed. This $C \setminus \{x_i = v\}$ notation represents the removal of $x_i = v$ from formula C. This process is described in Algorithm 1 and is performed for all features. Then, C_m is the result at the end of this procedure. This means that for the values of the features in C_m, the ANN makes the same classification, whatever the values of the remaining features. Since to check entailments $C \wedge F \models E$ is equivalent to test whether

$C \wedge F \wedge \neg E$ is unsatisfiable and F, C and $\neg E$ are encoded as linear constraints and indicator constraints, such a entailment can be addressed by a MILP solver.

Algorithm 1. Computing a minimal explanation

Input: ANN and domain constraints F, input data C, prediction E
Output: minimal explanation C_m
1: **for** $x_i = v$ in C **do**
2: **if** $C \setminus \{x_i = v\} \wedge F \models E$ **then**
3:   $C \leftarrow C \setminus \{x_i = v\}$
4: $C_m \leftarrow C$
5: **return** C_m

The encoding of ANNs used in [8] and originally proposed by Fischetti and Jo [5] uses implications to represent the behavior of the ReLU activation function. We encode an ANN with $L+1$ layers as in Eqs. (3)–(5). In the following, we explain the notation. The encoding uses variables x_i^l and o_n with the same meaning as in the notation for ANNs. Auxiliary variables s_i^l and z_i^l control the behaviour of ReLU activations. Variable z_i^l is binary and if z_i^l is equal to 1, the ReLU output x_i^l is 0 and $-s_i^l$ is equal to the linear part. Otherwise, the output x_i^l is equal to the linear part and s_i^l is equal to 0. The constant $ub_{s,i}^l$ is the upper bound of variable s_i^l, and the constant $ub_{x,i}^l$ is the upper bound of variable x_i^l. Each variable x_i^0 has also lower and upper bounds l_i, u_i, respectively, defined by the domain of the features.

$$\left. \begin{array}{l} \sum_{j=1}^{n_{l-1}} w_{i,j}^l x_i^{l-1} + b_i^l = x_i^l - s_i^l \\ z_i^l = 1 \rightarrow x_n^l \leq 0 \\ z_i^l = 0 \rightarrow s_n^l \leq 0 \\ z_i^l \in \{0, 1\} \\ 0 \leq x_i^l \leq ub_{x,i}^l \\ 0 \leq s_i^l \leq ub_{s,i}^l \end{array} \right\} l = 1, ..., L-1, \; i = 1, ..., n_l \quad (3)$$

$$l_i \leq x_i \leq u_i, \quad i = 1, ..., n_0 \quad (4)$$

$$o_i = \sum_{j=1}^{n_{L-1}} w_{i,j}^L x_i^{L-1} + b_i^L, \quad i = 1, ..., n_L \quad (5)$$

The constraints in (3)–(5) represent the formula F. The bounds $ub_{x,i}^l$ are defined by isolating variable x_i^l from other constraints in subsequent layers. Then, x_i^l is maximized to find its upper bound. A similar process is applied to find the bounds $ub_{s,i}^l$ for variables s_i^l. This optimization is possible due to the bounds of the features. Furthermore, these bounds can assist the solver in accelerating the computation of explanations. Therefore, the time required for this process must be considered when building F.

To check the unsatisfiability of the expression $C \wedge F \wedge \neg E$, we still need to take into account the formula $\neg E$, referring to the prediction of the ANN. Given an input C predicted as class c_i by the ANN, formula E must be equivalent to $\bigwedge_{j=1, j \neq i}^{\mathcal{N}} o_i > o_j$. This formula asserts that the maximum value of the last layer is in output o_i. Therefore, $\neg E$ must ensure that $\bigvee_{j=1, j \neq i}^{\mathcal{N}} o_i \leq o_j$. Since MILP solvers can not directly represent disjunctions, we use implications (6) and a linear constraint (7) over binary variables to define $\neg E$.

$$q_j = 1 \to o_i \leq o_j, \quad j \in \{1, ..., \mathcal{N}\} \setminus \{i\} \tag{6}$$

$$\sum_{j=1, j \neq i}^{\mathcal{N}} q_j \geq 1 \tag{7}$$

$$q_j \in \{0, 1\}, \quad j \in \{1, ..., \mathcal{N}\} \setminus \{i\} \tag{8}$$

If an assignment $\mathcal{A}$ satisfies $\bigvee_{j=1, j \neq i}^{\mathcal{N}} o_i \leq o_j$, then $o_i \leq o_j$ is true under $\mathcal{A}$ for some j. Therefore, the assignment $\mathcal{A} \cup \{q_j \mapsto 1\}$ satisfies the Eqs. (6)–(8). Conversely, if an assignment $\mathcal{A}$ satisfies Eqs. (6)–(8), it clearly also satisfies $\bigvee_{j=1, j \neq i}^{\mathcal{N}} o_i \leq o_j$. We conclude this section with a proposition regarding the number of variables and constraints of the encoding discussed above.

Proposition 1. *Let C be an instance predicted as a class $c \in \mathcal{K}$ by an ANN, the formula $C \wedge F \wedge \neg E$ has $n_0 + n_L + 2 \sum_{l=1}^{n_L-1} n_l$ real variables and $n_L - 1 + \sum_{l=1}^{n_L-1} n_l$ binary variables. Also, the formula has $n_0 + 2 n_L + 5 \sum_{l=1}^{n_L-1}$ constraints.*

4 Explanations for ANNs Without Implications

In this section, we present an adaptation of the encoding proposed by Tjeng et al. [17] for logic-based explainability. In such a work, the authors originally used the encoding to find adversarial examples without using logical implications. Even more importantly, such an encoding uses fewer variables and constraints compared to [5]. Then, we expect that our adaptation can lead to a better execution time for both building the logical constraints and computing explanations. Adapting the encoding in [17] to the context of computing explanations requires incorporating additional constraints that were not part of the original work. These new constraints represent the class predicted by the ANN as a formula E, as seen in Sect. 3. However, to maintain the concept of the original encoding without implications, we define these additional constraints without implications.

In the following, we apply the same algorithm from Fig. 1, but replacing the encoding of F as in [5] with the one in [17]. We encode an ANN with $L+1$ layers as in Eqs. (9), (4) and (5). The variables x_i^l and o_i have the same meaning as in Eqs. (3)–(5). Furthermore, auxiliary variables s_i^l are not required, as observed in the encoding by Fischetti and Jo [5]. Constants lb_i^l and ub_i^l are, respectively, the lower and upper bounds of $\sum_{j=1}^{n_{l-1}} w_{i,j}^l x_j^{l-1} + b_i^l$. Again, we find such bounds via a MILP solver. The behavior of ReLU is modeled using these bounds and binary variables z_i^l. If z_i^l is equal to 0, the ReLU output x_i^l is 0. Otherwise, x_i^l is equal to $\sum_{j=1}^{n_{l-1}} w_{i,j}^l x_j^{l-1} + b_i^l$. The bounds $lb^l i$ and $ub^l i$ are necessary to maintain the integrity of the set of constraints for the entire feature space. Regardless of the value of $z^l i$, the bounds ensure that the constraints remain valid for the entire feature space.

$$\left.\begin{aligned} x_i^l &\leq \sum_{j=1}^{n_{l-1}} w_{i,j}^l x_j^{l-1} + b_i^l - lb_i^l(1 - z_i^l) \\ x_i^l &\geq \sum_{j=1}^{n_{l-1}} w_{i,j}^l x_j^{l-1} + b_i^l \\ x_i^l &\leq ub_i^l z_i^l \\ z_i^l &\in \{0,1\} \\ x_i^l &\geq 0 \end{aligned}\right\} l = 1,...,L-1,\ i = 1,...,n_l \quad (9)$$

In our proposal for computing explanations, constraints in Eqs. (9), (4) and (5) represent the formula F. As in Sect. 3, an instance is a conjunction C, and the associated prediction by the ANN is a formula E. Given an input C predicted as class c_i by the ANN, again formula $\neg E$ must ensure that $\bigvee_{j=1, j\neq i}^{\mathcal{N}} o_i \leq o_j$. Therefore, we must add new constraints to represent $\neg E$. Maintaining the concept of the original encoding in [17] without implications, we define these additional constraints accordingly. We employ binary variables q_j and the upper and lower bounds ub_j and lb_j of variables o_j. As for lb_i^l and ub_i^l, we find the bounds ub_j and lb_j through a MILP solver. We recall such elements are not originally present in [17]. However, they are necessary for the context of computing explanations for ANNs. In Equations (10)-(12) we represent our proposal for encoding formula $\neg E$, where the prediction associated with an input C is class c_i.

$$o_i - o_j \leq (ub_i - lb_j)(1 - q_j), \quad j \in \{1,...,\mathcal{N}\} \setminus \{i\} \quad (10)$$

$$\sum_{j=1, j\neq i}^{\mathcal{N}} q_j \geq 1 \quad (11)$$

$$q_j \in \{0,1\}, \quad j \in \{1,...,\mathcal{N}\} \setminus \{i\} \quad (12)$$

In what follows, we prove that Equations (10)-(12) correctly ensure that $\bigvee_{j=1, j\neq i}^{\mathcal{N}} o_i \leq o_j$.

Proposition 2. *Let $\neg E$ be defined as in Equations (10)-(12). Let $i \in \{1,...,\mathcal{N}\}$ be fixed and ub_i be such that $o_i \leq ub_i$. Let lb_j be such that $lb_j \leq o_j$, for $j \in \{1,...,\mathcal{N}\} \setminus \{i\}$. Therefore,*

$$\neg E \text{ is satisfiable iff } \bigvee_{j=1, j\neq i}^{\mathcal{N}} o_i \leq o_j \text{ is satisfiable.}$$

Proof. If an assignment $\mathcal{A}$ satisfies $\bigvee_{j=1, j\neq i}^{\mathcal{N}} o_i \leq o_j$, then $o_i \leq o'_j$ is true under $\mathcal{A}$ for $j' \neq i$. Let $\mathcal{A}' = \mathcal{A} \cup \{q_j \mapsto v \mid v = 1 \text{ if } j = j', \text{ else } v = 0, \text{ for } j \in \{1, ..., \mathcal{N}\}\setminus\{i\}\}$ be an assignment. Then, $\mathcal{A}'$ imposes that $o_i \leq o'_j$ in Equation (10) for $j = j'$, which in clearly true under this assignment. For $j \neq j'$, it follows that $o_i - o_j \leq (ub_i - lb_j)$ must hold, which is also true under $\mathcal{A}'$ since $lb_j \leq o_j$ and $o_i \leq ub_i$.

Conversely, if an assignment $\mathcal{A}$ satisfies Equations (10)-(12), it satisfies some q'_j, for $j' \neq i$ by Equation (11). Moreover, $\mathcal{A}$ satisfies $o_i \leq o_{j'}$ by Equation (10), for $j = j'$. Therefore, $\mathcal{A}$ also satisfies $\bigvee_{j=1, j\neq i}^{\mathcal{N}} o_i \leq o_j$.

Finally, we give a proposition on the number of variables and constraints in our adaptation of the encoding in [17].

Proposition 3. *Let C be an instance predicted as a class $c \in \mathcal{K}$ by an ANN, then formula $C \wedge F \wedge \neg E$ has $n_0 + n_L + \sum_{l=1}^{n_L-1} n_l$ real variables and $n_L - 1 + \sum_{l=1}^{n_L-1} n_l$ binary variables. Moreover, the formula has $n_0 + 2n_L + 4\sum_{l=1}^{n_L-1} n_l$ constraints.*

Therefore, this encoding has $\sum_{l=1}^{n_L-1} n_l$ fewer variables than the one presented in Sect. 3. Additionally, this encoding has $\sum_{l=1}^{n_L-1} n_l$ fewer constraints. Consequently, one would expect a reduction in running time for both building the logical constraints and computing explanations.

5 Experiments

In this section, we detail the experiments conducted to compare our proposal against the encoding presented in [5]. Our evaluation consists of two main experiments. In the first one, we compare the two encodings using 12 datasets. In the second one, we conduct a detailed comparison using a single dataset. We vary the architecture of the trained ANNs to explore the effect of the number of layers and neurons. We evaluate the performance of each encoding in terms of time for building logical constraints and time for computing explanations. We explained all instances in a given dataset and calculated the average time and standard deviation to compare times for computing explanations. In the other case, given a trained ANN on a dataset, we built the logical constraints 10 times and calculated the average time and standard deviation.

Next, we present the experimental setup, describing technologies, datasets, the trained ANNs and hyperparameters. After that, we discuss the results providing a comparative analysis of the running times for building logical constraints and computing explanations. Finally, we highlight specific improvements observed in our proposal.

5.1 Experimental Setup

We used Python to implement the approaches and to run the experiments. TensorFlow was used to manipulate ANNs, including the training and testing steps. CPLEX was used as the MILP solver and accessed by the DOcplex library.

We used 12 datasets from the UCI Machine Learning Repository[2] and Penn Machine Learning Benchmarks[3], each ranging from 9 to 32 integer, continuous, categorical or binary features. The number of instances in the selected datasets ranges from 156 to 691. The types of classification problems related to these datasets are binary and multi-class classification. The preprocessing performed on the datasets included one-hot encoding of the categorical data and normalization of the continuous features to the range $[0, 1]$. This normalization was not applied to the integer features to avoid transforming their space into continuous, which could compromise formal guarantees on the correctness of the algorithm. As far as we know, such a methodology was not considered in earlier works.

The ANNs training was accomplished using a batch size of 4 and a maximum of 100 epochs, applying early stopping regularization with 10 epochs based on validation loss. The optimization algorithm used was Adam and the learning rate was 0.001. The datasets split was 80% for training and 20% for validation. The ANN architectures were limited to 2 layers to reduce the total running time, because many solver calls were performed in the experiments due to the large number of instances. Each solver call deals with an NP-complete problem, therefore, impacting the experiments running time.

The first experiment compared the two encodings presented on the 12 datasets. The architecture of the trained ANNs is two hidden layers with 20 neurons each. For each dataset and the associated ANN, the explanation of each instance was obtained using the Algorithm 1 and both presented encodings. The second experiment performed the comparison of the two encodings presented using the voting dataset of the first experiment. This experiment was conducted in two cases. In the first case, the trained ANNs consist of one hidden layer with the number of neurons ranging from 10 to 100. In the second case, the ANNs consist of two hidden layers such that both layers contains the same number of neurons, ranging from 10 to 40. In both cases, the number of neurons in the layers increases in increments of 5. Again, the explanations of each instance was obtained using Algorithm 1 and both presented encodings. The objective of this experiment is to verify the influence of the number of layers and the number of neurons in both encodings.

5.2 Results

The results of the first experiment are shown in Table 1. For each dataset, its number of features is indicated in parentheses. The column **Exp (s)** refers to the average running time for computing explanations in seconds, and the standard deviation is also presented. The column **Build (s)** refers to the average running time, in seconds, for building the logical constraints of the trained ANN. The running time for finding the bounds of variables is included in the time for building the encodings.

---

[2] https://archive.ics.uci.edu/ml/.
[3] https://github.com/EpistasisLab/penn-ml-benchmarks/.

Table 1. Comparison of both encodings.

Datasets	Fischetti and Jo [5]		Our Proposal	
	Exp (s)	Build (s)	Exp (s)	Build (s)
breast-cancer (9)	0.39 ±0.24	2.98 ±0.05	0.39 ±0.21	**2.67 ±0.17**
glass (9)	0.25 ±0.09	4.25 ±0.18	0.26 ±0.1	**3.86 ±0.09**
glass2 (9)	0.39 ±0.29	3.45 ±0.1	0.47 ±0.34	**3.02 ±0.02**
cleve (13)	0.51 ±0.23	4.26 ±0.14	0.54 ±0.25	**3.55 ±0.01**
cleveland (13)	0.69 ±0.8	5.28 ±0.11	0.82 ±1.03	**4.57 ±0.17**
heart-statlog (13)	0.41 ±0.18	3.76 ±0.05	0.41 ±0.18	**3.06 ±0.02**
australian (14)	0.76 ±0.43	3.2 ±0.08	0.84 ±0.42	3.14 ±0.72
voting (16)	1.02 ±0.52	3.62 ±0.07	1 ±0.47	**3.01 ±0.02**
hepatitis (19)	1.17 ±1.27	5.66 ±0.14	1.1 ±1.13	**5.28 ±0.07**
spect (22)	2.64 ±1.56	**4.81 ±0.13**	3.21 ±1.82	5.74 ±0.15
auto (25)	0.79 ±0.3	6.86 ±0.25	0.83 ±0.32	6.62 ±0.14
backache (32)	11.44 ±10.3	5.04 ±0.16	11.39 ±9.74	**4.78 ±0.14**

Despite the encoding by Fischetti and Jo [5] achieved a better average running time for computing explanations in 7 out of 12 cases, both encodings generally perform similarly when considering the variability of the time. For instance, in the spect dataset, the average execution time of the encoding by Fischetti and Jo [5] (2.64 seconds) falls within the range of the average minus the standard deviation of our approach (3.21 − 1.81). This pattern is observed across several datasets. In summary, the results indicate that both encodings generally perform similarly in terms of running time for computing explanations, with minor variations across different datasets.

With respect to the average running time for building the logical constraints, our adaptation generally outperformed the encoding by Fischetti and Jo [5]. However, there were exceptions noted, such as in the spect dataset. Overall, our adaptation achieved an improvement of up to 18% compared to the other one, as seen in the heart-statlog dataset. In summary, the results indicate that our adaptation is consistently more efficient than the other approach for building logical constraints. The variability in the results, as shown by the standard deviations, further supports this conclusion. For instance, the average running time of our proposal is less than the average minus the standard deviation of the other approach in 9 out of 12 datasets. These cases are highlighted in bold in Table 1. It is important to note that, in the spect dataset, the average running time of the encoding proposed by Fischetti and Jo [5] is less than the average minus the standard deviation of our adaptation. This indicates that, while our proposal appears generally more efficient, specific dataset characteristics can influence which encoding is more advantageous. Our adaptation yielded notably superior results not only in the average time for building logical constraints but also in the overall time, which includes both computing explanations and constructing logical constraints. For instance, it achieved an improvement of up to 16% compared to the other approach, as seen in the heart-statlog dataset.

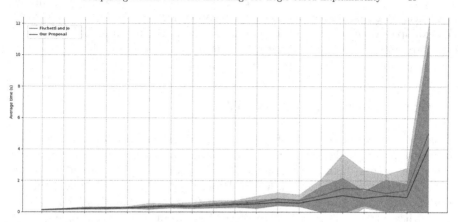

Fig. 2. Comparison of average running time for computing explanations, using ANNs with one hidden layer and the voting dataset.

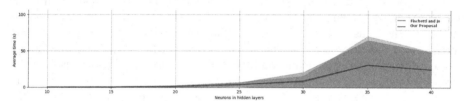

Fig. 3. Comparison of average running time for computing explanations, using ANNs with two hidden layers and the voting dataset.

The results of the second experiment are shown in Figs. 2 and 3. The x-axis shows the number of neurons in each hidden layer in both Figures. While the y-axis shows the average running time for computing explanations. Furthermore, the standard deviation is indicated by the shaded region in both Figures. Figure 2 refers to ANNs with one hidden layer, while Fig. 3 refers to ANNs with two hidden layers.

Figure 2 suggests that our proposal, for ANNs with one hidden layer, achieved a superior average running time for computing explanations in the voting dataset. This encoding outperforms the other approach with percentage improvements ranging from approximately 7.69% to 40.82%. The most significant improvements are observed with higher neuron counts. For instance, with 85 neurons per layer, our proposal shows an improvement of around 40.82%, and with 95 neurons, the improvement is about 31.65%. Moreover, our proposal generally exhibits similar or lower standard deviations, indicating more consistent performance across different number of neurons.

On the other hand, Fig. 3 depicts comparable results between both encodings for ANNs with two hidden layers. Moreover, similar standard deviations were achieved for both encodings, indicating comparable levels of consistency in performance. The variability increases significantly with more complex networks. For example, with 35 and 40 neurons per layer, both encodings exhibit significant variability. Furthermore, with 30 neurons per layer, our adaptation shows higher variability compared to the other approach.

6 Conclusions and Future Work

Explanations for the outputs of ANNs are fundamental in many scenarios, due to critical systems, data protection laws, adversarial examples, among others. Therefore, several heuristic methods have been developed to provide explanations for the decisions made by ANNs. However, these approaches lack guarantees of correctness and may also produce redundant explanations. Logic-based approaches address these issues but often suffer from scalability problems.

In this work, we compare two logical encodings of ANNs: one has been used in the literature to provide explanations [5,8], and another [17] that we have adapted for our context of explainability. Our experiments indicate that both encodings have similar running times for computing explanations, even as the number of neurons and layers increases. Furthermore, our experimental results suggest that our adaptation is generally more efficient for ANNs with one hidden layer, while the performance advantage diminishes for ANNs with two hidden layers. However, our proposal achieved a better running time for building the encoding for ANNs with two layers, showing an improvement of up to 18%. This can help to decrease the scalability issue for building the logical constraints given an ANN. Furthermore, this encoding obtained better results also in the overall time, i.e., the time for computing explanations plus the time for building the logical constraints, showing an improvement of up to 16%.

In the experiments of this work, we considered all instances of the datasets used, which considerably increased the experiments running time. As future work, we can change the design of experiments, using only a subset of the datasets, to allow the use of larger ANNs. More experiments are necessary, especially with additional layers and neurons, to further validate our findings and understand the performance of these encodings. Furthermore, others encodings [9,10] can be evaluated for computing logic-based minimal explanations for ANNs. Moreover, in order to improve the scalability of computing logic-based explanations, the ANNs can be simplified, before or during building their encodings, via pruning or slicing as proposed by [12]. This results in equivalent ANNs with smaller sizes.

References

1. Audemard, G., Lagniez, J.M., Marquis, P., Szczepanski, N.: Computing abductive explanations for boosted trees. In: 26th AISTATS (2023)
2. Bénichou, M., Gauthier, J.M., Girodet, P., Hentges, G., Ribière, G., Vincent, O.: Experiments in mixed-integer linear programming. Math. Program. **1**, 76–94 (1971)

3. Bonami, P., Lodi, A., Tramontani, A., Wiese, S.: On mathematical programming with indicator constraints. Math. Program. **151**(1), 191–223 (2015). https://doi.org/10.1007/s10107-015-0891-4
4. Choi, A., Shih, A., Goyanka, A., Darwiche, A.: On symbolically encoding the behavior of random forests. In: 3rd FoMLAS (2020)
5. Fischetti, M., Jo, J.: Deep neural networks and mixed integer linear optimization. Constraints **23**(3), 296–309 (2018). https://doi.org/10.1007/s10601-018-9285-6
6. Goodfellow, I.J., Shlens, J., Szegedy, C.: Explaining and harnessing adversarial examples. In: 3rd ICLR (2015)
7. Gorji, N., Rubin, S.: Sufficient reasons for classifier decisions in the presence of domain constraints. In: 36th AAAI (2022)
8. Ignatiev, A., Narodytska, N., Marques-Silva, J.: Abduction-based explanations for machine learning models. In: 33rd AAAI (2019)
9. Katz, G., Barrett, C., Dill, D.L., Julian, K., Kochenderfer, M.J.: Reluplex: an efficient SMT solver for verifying deep neural networks. In: Majumdar, R., Kunčak, V. (eds.) Computer Aided Verification: 29th International Conference, CAV 2017, Heidelberg, Germany, July 24-28, 2017, Proceedings, Part I, pp. 97–117. Springer International Publishing, Cham (2017). https://doi.org/10.1007/978-3-319-63387-9_5
10. Katz, G., et al.: The marabou framework for verification and analysis of deep neural networks. In: 31st CAV (2019)
11. Kroening, D., Strichman, O.: Decision Procedures: An Algorithmic Point of View. Springer, Berlin, Heidelberg (2016). https://doi.org/10.1007/978-3-662-50497-0
12. Lahav, O., Katz, G.: Pruning and slicing neural networks using formal verification. In: 21st FMCAD (2021)
13. Liu, W., Wang, Z., Liu, X., Zeng, N., Liu, Y., Alsaadi, F.E.: A survey of deep neural network architectures and their applications. Neurocomputing **234**, 11–26 (2017)
14. Ribeiro, M.T., Singh, S., Guestrin, C.: Why should I trust you?: explaining the predictions of any classifier. In: 22nd KDD (2016)
15. Ribeiro, M.T., Singh, S., Guestrin, C.: Anchors: high-precision model-agnostic explanations. In: 32th AAAI (2018)
16. Shih, A., Choi, A., Darwiche, A.: A symbolic approach to explaining Bayesian network classifiers. In: 27th IJCAI (2018)
17. Tjeng, V., Xiao, K.Y., Tedrake, R.: Evaluating robustness of neural networks with mixed integer programming. In: 7th ICLR (2019)
18. Wu, M., Wu, H., Barrett, C.: VeriX: towards verified explainability of deep neural networks. In: 37th NeurIPS (2023)

Deep Learning Approach to Temporal Dimensionality Reduction of Volumetric Computed Tomography

Lucas Almeida da Silva, Eulanda Miranda dos Santos, and Rafael Giusti[✉]

Instituto de Computação, Universidade Federal do Amazonas, Manaus, Brazil
{lucas.silva,emsantos,rgiusti}@icomp.ufam.edu.br

Abstract. A common approach for analyzing medical images on volumetric data is to employ deep 2D convolutional neural networks (2D CNN) on each individual slice of the volume. This is largely attributed to the challenges posed by the nature of three-dimensional data: variable volume size, high GPU and RAM requirements, costly parameter optimization, etc. However, handling the individual slices independently in 2D CNNs deliberately discards the temporal information that is contained within the depth of the volumes, which tends to result in poor performance. In order to maintain temporal information, current solutions reduce the temporal dimensionality of the data using non-adaptive sampling processes, such as taking slices at given intervals or by means of interpolation. However, although this allows to keep some temporal information, it may discard important slices. In this paper, we propose a method based on GradCam to select meaningful slices in computed tomography volumes by evaluating the activation map to reduce temporal data dimensionality. Extensive experiments demonstrate the effectiveness of our method when compared to the current state of the art.

Keywords: Pattern Recognition · Feature Selection · CT Classification · CT Slice Selection

1 Introduction

X-ray is a form of electromagnetic radiation, like visible light. It is less energetic than gamma rays and more energetic than ultraviolet light. While the human body is mostly opaque to visible light, X-rays easily pass through soft tissue, such as organs and muscles, but not as easily through hard tissue, such as bones and teeth. Consequently, X-ray imaging is well-suited for examining skeletal structures, but not so much for soft tissue, as it is the case of the brain. Computed tomography (CT) addresses this limitation, allowing thorough observation of internal structures, including the brain, the lungs, and other organs. The technique is non-invasive and provides good cross-sectional visualization.

Additionally, CT exams are usually quick, with a single imaging session often completed in less than a few minutes.

CT scans are obtained from multiple shots at different angles, and with complex geometric transformations they produce volumes of data called slices. A set of slices is comparable to a collection of images that provide information on body sections. In a sense, CT scans are considered a type of three-dimensional data, as each slice can be transformed into a gray-scale image in a two-dimensional space and a full CT exam can be transformed into a sequence of images.

They are used by medical professionals to diagnose a multitude of diseases and conditions, and may provide a rich source of information to Artificial Intelligence (AI) methods. However, many AI engines are unable to make the best use of CT data. The reason is the often large volume of data, which renders the computational cost involved in training and inference infeasible for many information systems [1]. Instead, researchers often resort to two-dimensional approaches, which naturally disregard the sequential aspect of the volumetric data. While this reduces the computational cost and allows models to be trained and employed without the requirement of supercomputers, performance is usually also reduced. Furthermore, even when approaches designed to cope with three-dimensional (3D) data are used, such as 3D convolutional neural networks (CNN3D), performance tends to be lower than expected, especially when compared to two-dimensional approaches. This is primarily due to the difficulty of adjusting hyper-parameters, which require very time-consuming validation steps, and the necessity of reducing the resolution of the scans in favor of temporal data.

State-of-the-art approaches attempt to reduce the data's temporal dimensionality while preserving image resolution as much as possible. However, most techniques rely on non-adaptive sampling processes, such as taking slices at given intervals or interpolating through slices. As a result, slices that are irrelevant to the application may be selected while important slices are discarded.

In this work, we address the problem of selecting slices that are more relevant for machine learning models, while simultaneously discarding slices containing less information about a given decision problem (e.g., detecting signs of a disease in volumetric CT). Our approach is based in the Grad-CAM technique [25]. Ideally, this should result in accurate models with shorter training times and lower hardware demands. As a case study, two datasets addressing two different decision problems are investigated in this paper: a set of lung CTs to determine if the patient is infected with the COVID-19 virus; and a dataset for intracranial hemorrhage detection. The highlights of our paper are:

- We employ Gradient-weighted Class Activation Mapping (Grad-CAM) to eliminate the manual task of selecting meaningful slices.
- We use depthwise convolutions [7] and Grad-CAM on volumetric CT data to preserve temporal order and facilitate relevant slice selection through adaptation of CNN [27].
- Our proposed method reduces the temporal dimensionality of volumetric CT data, providing an advantage when a low amount of processing power is available.

- The GSS method consistently shows the best results compared to other slice selection methods, both in terms of AUC and F1 Score. This is true for all configurations and for both deep learning models (C3D and 3DCNN-C) investigated in this paper. Therefore, the GSS method is the most promising for the task of selecting meaningful slices on CT volumes.

2 Related Work

Deep learning has been utilized in various medical domains [3,5,19,24,31]. Many of these studies concentrate on techniques developed for two-dimensional (2D) data, such as images. Since volumetric CTs are inherently 3D, one common approach is to analyze each CT slice individually using algorithms designed for 2D data, primarily state-of-the-art 2D CNNs [10,15,18,23,26,28]. However, there is evidence suggesting that utilizing 3D data from CT scans leads to improved results [2,14,16,17,21], as it maintains the depth properties of the CT scans.

Several challenges arise when processing CT scans with volumetric data. Generally, data complexity increases exponentially with each added dimension [20]. In machine learning, this often results in substantially larger demands for memory, computation, and training data [12] because the complexity of the model must grow to more properly represent the complexity of the date. In the case of deep neural networks (DNN), learning from 3D data typically demands more layers and neurons than learning from 2D data. In other words, working with 3D data is significantly more computationally expensive, and one might not have enough resources to train with 3D data.

To provide some insight into this computational cost, we summarize the requirements for three machine learning methods employed in this study when classifying examples of the Mosmed data set. The Mosmed data set contains volumetric CT scan data that we resize in both width (image resolution) and depth (number of slices per instance). Our analysis is presented in Table 1. In all instances, the CT scan slice numbers were standardized to either 12 or 30, and the resolution of each scan remained at 512×512 pixels or was scaled down to 224×244 pixels. We measured the model size in memory, the number of trainable parameters, the number of floating-point operations (FLOPs), and the GPU memory needed to perform tasks with each model. The number of FLOPs directly influences the model's execution time, and the required GPU memory determines whether the model can be used with the available computational resources. For example, SqueezeNet is a compact neural network designed for low-end devices, requiring minimal GPU memory. Conversely, the traditional 3D CNN architecture demands nearly 100 gigabytes of GPU memory to train and test 30 slices at full resolution. At the time of writing this article, this memory requirement exceeds the capacity of all consumer-grade devices on the market.

Typical CT exam equipments can produce volumes ranging from 2 to 640 slices. Selecting the most meaningful slices is the optimal approach for analyzing this data without discarding temporal information. Even if one possesses

Table 1. Computational cost of three 3D CNN models when classifying instances from the Mosmed data set. The input width and depth correspond, respectively, to the width of the input, and the number of slices of each instance. The total cost was measured empirically in terms of memory, number of trainable parameters, number of floating point operations (FLOPs), and the amount of GPU memory required to perform the classification.

Model	Input Width	Input Depth	Size (in MB)	Trainable Param. (M)	FLOPs (M)	GPU (in GB)
C3D	512 × 512	30	4.521,53	1.185,29	68,58	94,94
		12	2.345,53	614,87	19,02	38,37
	224 × 224	30	1.065,53	279,32	13,73	18,23
		12	617,53	161,88	4,08	7,39
3DCNN-C	512 × 512	30	4,24	1.112,00	44,46	45,92
		12	2,4138	0,63	7,03	17,62
	224 × 224	30	4,24	1,11	8,42	8,65
		12	2,41	0,63	1,33	3,32
SqueezeNet	512 × 512	30	6,98	1,83	5,64	18,81
		12	6,98	1,83	0,88	6,76
	224 × 224	30	6,98	1,83	1,08	3,55
		12	6,98	1,83	0,17	1,27

sufficient computational power to process all 640 slices, reducing data complexity is usually preferable, as less complex models tend to generalize better and decrease the risk of overfitting [4]. Slice selection can be achieved using adaptive or non-adaptive approaches. Adaptive approaches examine the slice content to determine which ones contain the most relevant information, while non-adaptive methods select subsets of slices regardless of their content. There are five primary techniques found in the literature, described in this section. The first three are examples of non-adaptive methods, the fourth is an adaptive approach, while the remaining one is also non-adaptive.

Discard by Slice Similarity (DSS) is a category of methods where a similarity-based algorithm (e.g., Structural Similarity Index Measure (SSIM), Mean Squared Error (MSE), Euclidean Distance) compares each slice to its successor. Based on a defined threshold, the most similar slices are eliminated [29]. This approach aims to reduce redundancy in the data while preserving unique and informative slices.

Subset Slice Selection (SSS) is a method proposed in [34]. In this approach, slices are sampled from three specific positions of a volumetric CT: start, middle, and end. To achieve this, the volume is first divided into three equal parts. Then, a desired number of slices are extracted from each part. This method aims to capture the representative information from different regions of the CT volume.

Even Slice Selection (ESS), also known as *Uniform Slice Selection*, is a technique commonly used in video processing. In ESS, a "spacing factor" is calculated to enable the selection of equidistant slices [33]. Given a volume with D slices and a desired number of $K << D$ slices, ESS works by dividing the volume into disjoint subsets of approximately $\frac{D}{K}$ slices. Then, only the first slice of each subset is retained. In contrast to SSS, this technique mitigates semantic losses of temporal data. Algorithm 1 summarizes the ESS method.

Algorithm 1. ESS Method

Require: A 3D volumetric image V_0 of size W(*width*) * H(*height*) * D(*depth*)
Ensure: I is a rank 3 tensor
1: Set constant target depth of size K
2: Compute depth factor by $DF = \frac{D}{K}$
3: Initialize an empty processed volume V_f of dimension W * H * K
4: **for** $i = 0$ to $K - 1$ **do**
5:     Compute the index of the slice to be sampled $idx = \lfloor i * DF \rfloor$
6:     Extract the slice S_i from V_0 at depth idx
7:     Add the extracted slice S_i to the processed volume V_f at depth i
8: **end for**
9: Output processed volume V_f of dimension W * H * K

Slice Selection by Object Detection (SSOD) consists of methods in which an algorithm scans each slice and determines whether an object of interest is present. For instance, an AI model may be employed on CT scans to detect which slices display segments of organs such as lungs, brain, or other targeted organs [8]. The slices displaying the desired organ are then selected, while the others are discarded. While this method adapts to the data, it overlooks the inherent temporal relationships between slices. Often, it discards the initial and final subsets of slices, as the largest portion of the target organ is typically located in the middle of the volume.

As previously mentioned, SSOD is an adaptive method, whereas DSS, SSS, and ESS are non-adaptive. In some applications, employing non-adaptive strategies to discard slices may lead to loss of essential information (e.g., a critical segment of an organ may not be present in any of the selected slices). *Spline Interpolation Zoom* (SIZ) is a non-adaptive strategy aimed at reducing these losses. In SIZ, the temporal dimension is reduced by interpolating all slices. The volume is enlarged or compressed by replicating the closest pixel of each slice along the depth/z-axis. Although each resulting slice may be less precisely "located" in time, this technique may be advantageous for various applications since it does not produce "gaps" in the temporal axis [13,32]. As summarized in Algorithm 2, given K as the desired number of slices and D as the total number of slices of a volume, *zoom* is performed along the z-axis by a factor of $\frac{1}{D/K}$ using interpolation by *splines* [9].

Our work primarily focuses on comparing non-adaptive techniques against our proposed method because, although our method is adaptive in nature, many

Algorithm 2. SIZ Method

Require: A 3D volumetric image V_0 of size W($width$) * H($height$) * D($depth$)
Ensure: I is a rank 3 tensor
1: Set constant target depth of size K
2: Compute depth factor by $DF = \frac{1}{\frac{D}{K}}$
3: Initialize an empty processed volume V_f of dimension W * H * K
4: **for** $i = 0$ to $K - 1$ **do**
5:     Compute the index of the slice to be interpolated $idx = i * DF$
6:     Interpolate the slices from V_0 at depth $idx - \lfloor idx \rfloor$ using splines
7:     Compute the interpolated slice S_i from the interpolation result
8:     Add the interpolated slice S_i to the processed volume V_f at depth i
9: **end for**
10: Output processed volume V_f of dimension W * H * K

of the available adaptive methods, including SSOD, do not leverage the temporal relationship present in the slices when selecting. This omission may result in a loss of critical information. Furthermore, most adaptive slice selection methods in the medical domain currently focus on object detection rather than preserving temporal information. We believe our adaptive approach that emphasizes temporal relationships offers a unique perspective in the slice selection domain, warranting its comparison with more traditional non-adaptive methods.

3 Materials and Methods

In this section we consider the definitions provided by Tran et al. [30], a primary reference for 3D learning, to help us to explain our proposed method. In that work, Tran et al. provide empirical evidence that 3D convolutions are effective feature extractors when modelling appearance and motion or depth simultaneously. This makes them well-suited for learning sequence and spatiotemporal data. For instance, in 3D CNNs, even the output of each layer is a volume.

In 2D convolution, the kernel is a 3-dimensional matrix and both the input layer and the filters have the same depth (channel number equals their kernel number). However, the 3D filter only moves in two directions, as in over the height and the width of the image. In other words, it deals with only the spatial dimensions of the input. Therefore, even when using 3D filters, the output is a 2D array.

Conversely, 3D convolution not only involves a 3D filter, but the filter can move in all three directions—height, width and time/sequence. At each position, multiplication and elementary addition provide a number. As the filter slides through the 3D space, the output numbers are also arranged in a 3D space, i.e. the output is 3D data. CNNs exploit the spatially-local correlation by enforcing a local connectivity pattern between neurons of adjacent layers. Such 3D convolutions also take temporal features into account, typically through a sliding window, which is a filter with trainable weights over the input, and producing as output a weighted sum of weights and input. The weighted sum is the feature space used as the input for the next layers.

In this work we employ depthwise convolution, which is a type of convolution where a single convolutional filter is applied for each input channel. In regular 2D convolution performed over multiple input channels, the filter is as deep as the input and mixes channels to generate each element of the output feature map. In contrast, depthwise convolution keeps each channel separated. In our case, it maintains each slice separated, preserving the temporal relationship at different timestamps. We summarize the steps involved in this process bellow:

1. The input tensor of 3 dimensions (volumetric CT, in our case) is divided into separate channels/slices.
2. Each slice is convolved with its respective filter.
3. The obtained convolved output are stacked together to provide the output as an entire 3D tensor.

Figure 1 exemplifies this process. At the top part of the image an instance containing three slices (a volumetric CT) is shown. Each slice is trated separated and convolution is only performed within each slice. It is necessary to maintain the depth size until the final convolution so that the feature maps of each slice is preserved to allow the extraction of Grad-Cam heatmap, which is performed in our proposed method, as detailed in next sections.

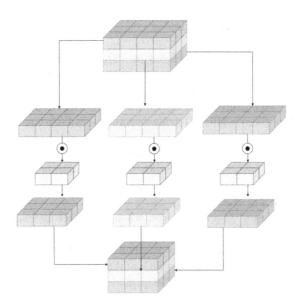

Fig. 1. Illustration of the Depthwise Convolution used in the Proposed Method on a volume with 3 slices.

3.1 Inference Method

The proposed method consists of a 3DCNN designed to capture the activation map of a volumetric CT. The objective is to obtain as output a 3D matrix representing the regions of interest in each slice of the volume. This is achieved by adapting the *Grad-Cam* technique, which uses the gradients obtained when evaluating the class of interest by flowing to the last convolutional layer in order to map the activation regions of an image. Grad-CAM requires no re-training and is broadly applicable to any CNN-based architectures. Additionally, the generated heatmaps may provide explanations, making the predictions understandable to users, consequently helping the users to trust the predictions made by the system, especially in medical applications.

As previously mentioned, we use a model that preserves the depth dimension up to the last convolutional layer, performing depthwise convolutions on the volume so as to maintain the order of each slice. In this way, a feature map is created for each slice. However, since CT are temporal high-dimensional data, a critical point here is the viability of training a 3DCNN model using all the slices of the volume, depending on the chosen hardware. Therefore, in order to allow this training, we focus on providing a condition of lower computational cost to train the model by applying image compression techniques to identify macro-regions of interest. Thus, the resizing of the width and height dimensions is a fundamental aspect.

The following steps were performed to obtain our proposed Grad-CAM Slice Selection method (GSS), in which Grad-Cam is applied to obtain the meaningful slices:

1. Normalize the volumetric data with one of the techniques described in Sect. 2. SIZ (Sect. 2) was used in this work.
2. Train a CNN3D in the classification problem represented by the volumetric data preserving the depth dimension, e.g. the binary classification of data as COVID-19 or not.
3. Remove the layers stacked after the last convolutional layer after finishing the network training, since they are no longer needed.
4. Extract the representative matrix of *heatmaps*.

These steps are also summarized in Algorithm 3 and are further visualized in Fig. 2 that shows that the last Conv Layer will be analyzed for the heatmap acquisition.

Algorithm 3. GSS Method

1: GSS−Model ←Custom CNN
2: **for** each epoch in EPOCHS **do**
3:    **for** each batch of Dataset **do**
4:       batch = normalize(batch)
5:       batch = SIZ(batch)
6:       train(GSS−Model, batch)
7:    **end for**
8:    Remove last layers after the last convolutional layer since they're no longer needed.
9: **end for**
10: Return GSS Model

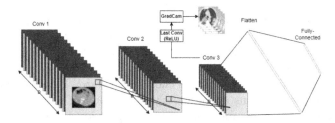

Fig. 2. Inference Pipeline of the GSS Model that will be used for heatmap acquisition on the last Conv Layer.

4 Experimental Results

In this section, we present the experimental results obtained using two 3DCNN architectures: C3D and 3DCNN-C. The experiments were conducted using the canonic houldout partitions established as training and test splits of the two datasets investigated in order to provide state-of-the-art comparison. The datasets employed are MosmedData [22] and CQ500 [6], detailed below. One notable aspect of our experimental design is the conscious decision to not employ any data augmentation techniques. Data augmentation is often employed to artificially expand the dataset size by introducing variations such as rotation, flipping, and scaling. While this can improve the model's generalization ability, it also introduces an additional layer of complexity that could potentially skew results or make direct comparisons with other methodologies more challenging. In our study, we aim to evaluate the efficacy of the deep learning architectures in recognizing patterns in unaltered, real-world medical imaging data. By excluding data augmentation, we seek to establish a baseline performance metric that reflects the model capabilities to generalize from the raw, original data, thereby ensuring that any observed performance differences are attributable solely to the architectures themselves, rather than to external manipulations of the data.

MosMedData contains anonymized human lung CT scans with COVID-19 related findings, as well as those without such findings. The CT scans were obtained between 1st March 2020 and 25th April 2020, provided by medical hospitals in Moscow, Russia.

CQ500 is a dataset that contains 491 scans with 193,317 slices, provided as anonymized DICOMs by the Centre for Advanced Research in Imaging, Neurosciences, and Genomics (CARING), New Delhi, India. The reads were performed by three radiologists with 8, 12, and 20 years of experience in cranial CT interpretation, respectively.

The 3D data from these datasets are in NIfTI format, which are volumetric (3D) images. After pre-processing, all images were set to a size of 512 with variable depth. Due to the nature of CT datasets, the images were imported as color Grayscale images, demanding modifications according to each network's input for data compatibility. In addition, the datasets were partitioned following the holdout procedure into 70-15-15 for training, test and validations subsets, respectively.

The two 3DCNN models investigated were trained end-to-end on both datasets for 100 epochs with early stopping method so as to better tackle potential overfitting. Moreover, overfitting was also addressed by incorporating dropout and L1/L2 regularization techniques. The Adam optimizer was used in both architectures with a learning rate of 10^{-4}, determined experimentally. Weights were initialized using Glorot Normal Initialization [11], with convolution layers featuring the activation function *Rectified Linear Activation Function (ReLU)* and the final layer using the activation function *sigmoid* for the binary problem. Training was performed in an environment with a Tesla V100-SXM2 video card with 16 Gb memory.

As previously mentioned, all four non-adaptive slice selection methods described in Sect. 2 are used in this work as baselines. The experiments were divided into two comparison scenarios according to the number of selected slices, precisely 30 and 12 slices were selected with each method and compared against the proposed GSS. Table 2 presents the results of this comparison on the two distinct datasets, Mosmed and CQ500, representing binary classification problems. Besides the number of selected frames (30 and 12), the four investigated slice selection methods (SSS, ESS, SIZ, and GSS) are also compared considering two different image resolutions: 512 × 512 and 224 × 224. Finally, two 3DCNN models are used: C3D and 3DCNN-C adapted from [27]. The evaluation metrics used are AUC (area under the ROC curve) and F1 Score.

Analyzing the results, the following conclusions are observed:

- GSS consistently presents the best results compared to other methods, both in terms of AUC and F1 Score. This is true across all configurations and for both deep learning models (C3D and 3DCNN-C). Therefore, GSS is the most promising method for selecting the most relevant slices from tomography volumes.
- The baseline SIZ outperforms SSS and ESS in general, but still cannot surpass the results from GSS.

- The results show that selecting 30 frames generally performs better than selecting 12 frames. This indicates that having more frames available for the model can lead to better model performance.
- The 512 × 512 resolution generally produces better results compared to the 224 × 224 resolution, suggesting that a higher resolution provides more valuable information for the model.
- The 3DCNN-C model adapted from [27] tends to achieve better results than the C3D model across all configurations and datasets.

Selecting slices without decreasing accuracy is a significant advancement towards using AI as an assistant in clinical management with lower computational power requirements. Therefore, the development of a clinical prognostic model based on our AI system, utilizing CT parameters and clinical data, marks an essential step forward in using AI to assist clinical management in this kind of scenario.

Table 2. Results comparing the proposed method to three non-adaptive slice selection methods when varying the number of selected slices (12 or 30) on two different CT datasets and using two different 3DCNN models.

Model	Width	Depth	Mosmed				CQ500			
			C3D		3DCNN-C		C3D		3DCNN-C	
			AUC	F1	AUC	F1	AUC	F1	AUC	F1
SSS	512 × 512	30	0.693	0.845	0.788	0.816	0.710	0.760	0.760	0.760
		12	0.617	0.845	0.733	0.851	0.680	0.756	0.730	0.743
	224 × 224	30	0.671	0.775	0.721	0.398	0.660	0.496	0.710	0.584
		12	0.522	0.781	0.707	0.742	0.640	0.687	0.690	0.689
ESS	512 × 512	30	0.715	0.825	0.763	0.855	0.720	0.782	0.770	0.776
		12	0.683	0.800	0.727	0.701	0.700	0.701	0.750	0.724
	224 × 224	30	0.671	0.800	0.737	0.834	0.690	0.755	0.740	0.748
		12	0.652	0.795	0.625	0.701	0.670	0.685	0.720	0.702
SIZ	512 × 512	30	0.674	0.846	0.754	0.857	0.760	0.806	0.810	0.808
		12	0.697	0.815	0.737	0.870	0.740	0.800	0.790	0.795
	224 × 224	30	0.703	0.787	0.736	0.838	0.730	0.780	0.780	0.780
		12	0.615	0.809	0.693	0.797	0.710	0.751	0.760	0.755
GSS	**512 × 512**	**30**	**0.753**	**0.865**	**0.804**	**0.838**	**0.800**	**0.819**	**0.850**	**0.834**
		12	**0.732**	**0.866**	**0.775**	**0.881**	**0.780**	**0.827**	**0.830**	**0.829**
	224 × 224	**30**	**0.690**	**0.769**	**0.737**	**0.853**	**0.770**	**0.810**	**0.820**	**0.815**
		12	**0.628**	**0.845**	**0.714**	**0.795**	**0.750**	**0.772**	**0.800**	**0.786**

5 Conclusions

In this paper we propose GSS (Grad-CAM Slice Selection), a technique to deal with the problem of classifying complex 3D spatiotemporal data in the form of volumetric computer tomography (CT) imagery. By selecting slices from each CT in a way that takes into account the classification problem at hand, we are able to dynamically select the slices that provide more information, and reduce the computational cost required to train and apply a model.

We compared different deep learning techniques under the same conditions when applied to classification problems with two well-known data sets. By applying the proposed GSS to each deep learning model before training and inference, we achieved accuracies on par with or surpassing the state-of-the-art and, in most cases, better than other agnostic sampling techniques. This was especially true for edge cases requiring a greater number of slices to be selected.

Our findings emphatically underline the superiority of the GSS approach combined with 30 frames at 512×512 resolution, leveraging the CNN model cited in [27]. As anticipated, this high-resolution configuration with an optimal number of frames yielded the most promising results in the domain of tomography volume anomaly detection. However, the strategic selection between resolution and the number of frames remains pivotal. For detecting nuanced or minute pathologies, it is imperative to lean towards higher spatial resolutions. Conversely, for challenges deeply rooted in the three-dimensional intricacies, a richer frame selection is recommended.

These findings support the hypothesis that using Grad-CAM in deep learning models for learning simple and macro-relevant features in a CT volume dataset is effective and warrants further investigation. Subsequently, transferring this knowledge to another model to learn complex patterns proves fruitful.

Acknowledgments. This study was financed in part by the Coordenação de Aperfeiçoamento de Pessoal de Nível Superior - Brasil (CAPES-PROEX) - Finance Code 001. This work was partially supported by Amazonas State Research Support Foundation - FAPEAM - through the project PDPG/CAPES.

References

1. Ahmed, E., et al.: A survey on deep learning advances on different 3D data representations. arXiv preprint arXiv:1808.01462 (2018)
2. Alebiosu, D.O., Dharmaratne, A., Lim, C.H.: Improving tuberculosis severity assessment in computed tomography images using novel DAvoU-net segmentation and deep learning framework. Expert Syst. Appl. **213**, 119287 (2023)
3. Becker, A.S., et al.: Detection of tuberculosis patterns in digital photographs of chest X-Ray images using deep learning: feasibility study. Int. J. Tuberculosis Lung Disease **22**(3), 328–335 (2018)
4. Bejani, M.M., Ghatee, M.: A systematic review on overfitting control in shallow and deep neural networks. Artif. Intell. Rev. **54**(8), 6391–6438 (2021). https://doi.org/10.1007/s10462-021-09975-1

5. Chen, W.W., et al.: A deep learning approach to classify fabry cardiomyopathy from hypertrophic cardiomyopathy using cine imaging on cardiac magnetic resonance. Int. J. Biomed. Imaging **2024**(1), 6114826 (2024)
6. Chilamkurthy, S., et al.: Development and validation of deep learning algorithms for detection of critical findings in head CT scans. arXiv preprint arXiv:1803.05854 (2018)
7. Chollet, F.: Xception: deep learning with depthwise separable convolutions. In: Proceedings of the IEEE Conference on Computer Vision and Pattern Recognition, pp. 1251–1258 (2017)
8. da Cruz, L.B., et al.: Kidney segmentation from computed tomography images using deep neural network. Comput. Biol. Med. **123**, 103906 (2020)
9. De Boor, C.: Bicubic spline interpolation. J. Math. Phys. **41**(1–4), 212–218 (1962)
10. Gao, X.W., Hui, R., Tian, Z.: Classification of CT brain images based on deep learning networks. Comput. Methods Programs Biomed. **138**, 49–56 (2017)
11. Glorot, X., Bengio, Y.: Understanding the difficulty of training deep feedforward neural networks. In: Proceedings of the Thirteenth International Conference on Artificial Intelligence and Statistics, pp. 249–256. JMLR Workshop and Conference Proceedings (2010)
12. Goodfellow, I., Bengio, Y., Courville, A.: Deep learning. MIT press (2016)
13. Gordaliza, P.M., Vaquero, J.J., Sharpe, S., Gleeson, F., Munoz-Barrutia, A.: A multi-task self-normalizing 3D-CNN to infer tuberculosis radiological manifestations. arXiv preprint arXiv:1907.12331 (2019)
14. Grewal, M., Srivastava, M.M., Kumar, P., Varadarajan, S.: RadNet: radiologist level accuracy using deep learning for hemorrhage detection in CT scans. In: 2018 IEEE 15th International Symposium on Biomedical Imaging (ISBI 2018), pp. 281–284. IEEE (2018)
15. Hamadi, A., Cheikh, N.B., Zouatine, Y., Menad, S.M.B., Djebbara, M.R.: ImageCLEF 2019: deep learning for tuberculosis CT image analysis. In: CLEF (Working Notes) (2019)
16. Huang, X., Shan, J., Vaidya, V.: Lung nodule detection in CT using 3D convolutional neural networks. In: 2017 IEEE 14th International Symposium on Biomedical Imaging (ISBI 2017), pp. 379–383. IEEE (2017)
17. Ji, S., Xu, W., Yang, M., Yu, K.: 3D convolutional neural networks for human action recognition. IEEE Trans. Pattern Anal. Mach. Intell. **35**(1), 221–231 (2012)
18. Kavitha, S., Nandhinee, P., Harshana, S., S, J.S., Harrinei, K.: ImageCLEF 2019: a 2D convolutional neural network approach for severity scoring of lung tuberculosis using CT images. In: CLEF (Working Notes) (2019)
19. Lakhani, P., Sundaram, B.: Deep learning at chest radiography: automated classification of pulmonary tuberculosis by using convolutional neural networks. Radiology **284**(2), 574–582 (2017)
20. LeCun, Y., Bengio, Y., Hinton, G.: Deep learning. Nature **521**(7553), 436–444 (2015)
21. Li, B., Zhang, T., Xia, T.: Vehicle detection from 3D lidar using fully convolutional network. arXiv preprint arXiv:1608.07916 (2016)
22. Morozov, S.P., et al.: MosMedData: data set of 1110 chest CT scans performed during the Covid-19 epidemic. Digit. Diagn. **1**(1), 49–59 (2020)
23. Ronneberger, O., Fischer, P., Brox, T.: U-Net: convolutional networks for biomedical image segmentation. In: International Conference on Medical Image Computing and Computer-Assisted Intervention, pp. 234–241. Springer (2015). https://doi.org/10.1007/978-3-319-24574-4_28

24. Santosh, K., Allu, S., Rajaraman, S., Antani, S.: Advances in deep learning for tuberculosis screening using chest X-rays: the last 5 years review. J. Med. Syst. **46**(11), 82 (2022)
25. Selvaraju, R.R., Cogswell, M., Das, A., Vedantam, R., Parikh, D., Batra, D.: Gradcam: visual explanations from deep networks via gradient-based localization. In: Proceedings of the IEEE International Conference on Computer Vision, pp. 618–626 (2017)
26. Shah, A.A., Malik, H.A.M., Muhammad, A., Alourani, A., Butt, Z.A.: Deep learning ensemble 2D CNN approach towards the detection of lung cancer. Sci. Rep. **13**(1), 2987 (2023)
27. da Silva, L.A., et al.: Spatio-temporal deep learning-based methods for defect detection: an industrial application study case. Appl. Sci. **11**(22), 10861 (2021)
28. Sultana, A., et al.: A real time method for distinguishing COVID-19 utilizing 2D-CNN and transfer learning. Sensors **23**(9), 4458 (2023)
29. Thasneem, A.H., Sathik, M.M., Mehaboobathunnisa, R.: A fast segmentation and efficient slice reconstruction technique for head CT images. J. Intell. Syst. **28**(4), 533–547 (2019)
30. Tran, D., Bourdev, L., Fergus, R., Torresani, L., Paluri, M.: Learning spatiotemporal features with 3D convolutional networks. In: Proceedings of the IEEE International Conference on Computer Vision, pp. 4489–4497 (2015)
31. Wajgi, R., et al.: Optimized tuberculosis classification system for chest X-ray images: fusing hyperparameter tuning with transfer learning approaches. Engineering Reports, e12906 (2024)
32. Yang, J., et al.: Reinventing 2D convolutions for 3d images. IEEE J. Biomedical Health Inform. (2021)
33. Zhu, H., Liu, X., Mao, X., Wong, T.T.: Real-time deep video deinterlacing. arXiv preprint arXiv:1708.00187 (2017)
34. Zunair, H., Rahman, A., Mohammed, N., Cohen, J.P.: Uniformizing techniques to process CT scans with 3D CNNs for tuberculosis prediction. In: Rekik, I., Adeli, E., Park, S.H., Valdés Hernández, M.C. (eds.) Predictive Intelligence in Medicine: Third International Workshop, PRIME 2020, Held in Conjunction with MICCAI 2020, Lima, Peru, October 8, 2020, Proceedings, pp. 156–168. Springer International Publishing, Cham (2020). https://doi.org/10.1007/978-3-030-59354-4_15

Deployment of IBM Federated Learning Platform and Aggregation Algorithm Comparison: A Case Study Using the MNIST Dataset

Hans Herbert Schulz[✉][iD] and Benjamin Grando Moreira[iD]

Joinville Technological Center (CTJ), Federal University of Santa Catarina (UFSC), R. Dona Francisca, 8300, CEP 89219-600 - Joinville, Santa Catarina, Brazil
`hansherbert99@gmail.com`, `benjamin.grando@ufsc.br`

Abstract. With the exponential growth in artificial intelligence, privacy and data acquisition concerns have surged, prompting stricter data protection laws. Federated Learning (FL) addresses these issues by enabling model training without accessing private data, allowing geographically dispersed clients to participate without sharing data. This study deployed an IBM FL platform using Docker containers across two clients and conducted training with the MNIST dataset using the two most common FL strategies: Federated Stochastic Gradient Descent (FedSGD) and Federated Averaging (FedAvg). The results validated the platform's deployment and assessed the performance of each strategy in terms of model accuracy and client hardware capabilities. Performance metrics, including CPU and RAM usage, network traffic, and model accuracy, were collected. Despite the higher resource demands, both strategies achieved satisfactory model accuracy, with FedAvg showing slightly better efficiency for the small-scale deployment. The results emphasize the potential of FL for predictive maintenance in industrial applications, enabling decentralized data utilization while ensuring data privacy and security.

Keywords: Federated Learning · Data Privacy · IBM Federated Learning Platform

1 Introduction

Traditional machine learning methods typically rely on centralized datasets to train models. However, this centralized approach carries significant risks, especially concerning data breaches that could expose all the training data [1,9]. In response to these concerns, Federated Learning (FL) has emerged as a promising solution. FL enables model training across geographically dispersed clients without the need to share their private data, thereby preserving data privacy and security.

This work is part of an industrial solution to enhance predictive maintenance within production engineering. The project focuses on developing a multi-sensory platform capable of acquiring and transmitting air parameters (e.g., saturation, temperature, and flow speed) from geographically apart industrial air exchange systems. The collected data will be used to train machine learning models to improve predictive maintenance routines.

The primary goal of this study is to deploy an FL platform using Docker container technology to train models and check its functionality by comparing the performance of machine learning models using different FL strategies. The deployment process involves several tasks aimed at installing a pre-developed application into its intended operational environment [5]. After a successful deployment, performance tests are conducted utilizing the usual FL techniques for machine-learning model algorithms and the results are discussed.

2 Federated Learning Overview

As industries grow, more technology is being required and developed. Among those technologies, Federated Learning is proposed as a pioneering approach to some common issues faced in machine learning, such as privacy and data storage. In [7] a review application in industrial engineering is presented.

With the advent of deep learning (DL), the need for data acquisition in several science instances has increased. This is mainly because DL models are usually interpolators, meaning they present the best results when the input is similar to the data used for training models. Therefore, a huge amount of heterogeneous data is required to achieve decent accuracy [16].

In light of this characteristic, Google presented in 2016 the concept of federated learning (FL), which can be described as a machine learning technique to provide decentralized collaborative learning across different nodes [10]. The FL architecture comprises multiple parties and an aggregator agent. Firstly, a global model is generated and sent to each party. Then, the model will be trained in each party with its own data set, meaning each model will have a different result by the end of the process [9].

An update that contains the difference between the base model and the new one is then generated by the parties and sent to the aggregator agent. The agent then merges the updates and creates a new global model sent to each party, and so on [10].

To understand Federated Learning comprehensively, we will explore its key components and methodologies. The following subsections will delve into data distribution methods, training models, and essential algorithms that form the backbone of federated learning.

2.1 Data Distribution

Data can be distributed and used in the FL context in different ways. The conventional way consists of having multiple parties training a model independently, and the data in all parties possess the same features. Such distribution results in the so-called Horizontal Federated Learning (HFL) approach [8].

In contrast to HFL, one can find the Vertical Federated Learning (VFL) which consists of training a model where each party's data contains different features of the same set of individuals [8]. Essentially, all parties have data related to the same entities but with different features. For example, in a healthcare scenario, one party might have medical records (features) for a set of patients (entities), while another party might have lifestyle information (different features) for the same set of patients.

To summarize, the main difference between HFL and VFL lies in how the training data is distributed among parties. The HFL will focus on training multiple data samples sharing the same features, whereas the VFL will use the same sample but different features [13].

2.2 FL Training Design

From a system perspective, the FL scenarios aim to train machine learning models using disparate data to maintain privacy and overall model performance. It is imperative that data is not moved across parties and not even visualized by the central server. To achieve that, there are mainly two system designs: cross-device and cross-silo [9].

The objective of cross-device FL is minimizing the function $F(w)$, which is defined by Eq. 1 [9]:

$$F(w) = \sum_{k=1}^{n} p_k F_k(w) \qquad (1)$$

In Eq. 1, $F_k(w)$ is the local objective function for the device k with model weights w. The term p_k represents the importance given to the contribution of the k device to the global model objective function.

Usually, in the cross-device setting, the parties are user-owned IoT (Internet of Things) instruments such as cell phones, tablets, edge devices, and others [8]. Figure 1 depicts the covered steps of the cross-device learning system whose main steps are [8]:

1. Initially, the model G_0 (represented as a neural network image in the upper rectangle) is untrained and initialized with random weights. A subset of the available devices is selected at the beginning of each round;
2. After selecting k devices, the initial weights w of the global model G_n are sent to the devices;
3. Each device trains the model on its dataset D_k, resulting in local models g_{nk};
4. Privacy mechanisms anonymize the local models, which are then sent to the aggregator;
5. The aggregator combines local models using an aggregation algorithm to create a new global model G_{n+1}. The process repeats from Step 1 until a stopping condition, like the maximum number of rounds or model convergence, is met.

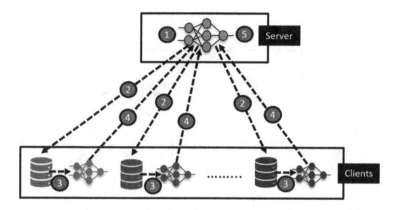

Fig. 1. Cross-device federated learning system overview. Source: [8]

On the other hand, the main objective of Cross-Silo Federated Learning can be formally described as minimizing the $L(W)$ function described by Eq. 2.

$$L(W) = \sum_{j=1}^{m} \sum_{i=1}^{n} L(w_j, x_j^i) \qquad (2)$$

where $L(W)$ is the global loss function for the global model W, i represents the client, j is the data provider (i.e., silos) with the local model w_j trained from data feature x_j^i [8]. In the context of cross-silo, the data silos used as parties tend to be of commercial grade, providing a less heterogeneous environment and generally having better computational resources than IoT devices [8]. Figure 2 depicts the chain of events for the cross-silo system whose main steps are [8]:

0. Participants align datasets using techniques like secure multi-party communication or key-sharing, ensuring data alignment by ID. Unmatched data is discarded;
1. A third-party aggregator distributes encryption key pairs to each party and initializes partial models. The last party usually holds the labels, while others have training features;
2. The first party trains on a mini-batch of local data, encrypts the output with homomorphic encryption, and sends it to the next party;
3. The second party performs a forward pass, calculates the loss (since it holds the labels), and sends intermediate outputs to the first party. The loss is sent to the aggregator;
4. Both parties compute partial gradients, add encryption, and send them to the aggregator to prevent data leakage;
5. The aggregator decrypts the gradients, calculates exact gradients using the loss, and sends them back to the parties for model updates. The process repeats from Step 1.

The training procedure for cross-silo FL also happens in rounds but with one significant difference: all silos take part in every round of the training.

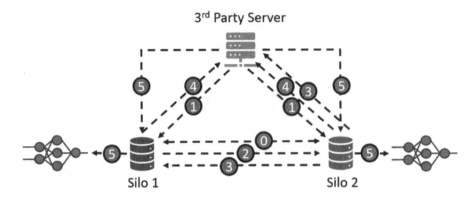

Fig. 2. Cross-silo federated learning system overview. Source: [8]

Therefore, the choice of targeted devices significantly influences the system selection. In cross-device FL, user-owned IoT devices like cellphones, tablets, and edge devices are common, leading to a diverse range of specifications. Conversely, cross-silo FL typically involves commercial-grade data silos, offering a more uniform environment with superior computational resources compared to IoT devices [8]. Thus, the type of devices used as parties in training will determine the most suitable FL system.

2.3 Federated Learning Algorithms

After establishing the main concepts of FL and its benefits and challenges, it becomes clear that the next step is optimization to address common issues, such as unbalanced data sets and limited communication.

Two algorithms used to contour these challenges are the Federated Stochastic Gradient Descent (FedSDG) and the Federated Averaging (FedAvg) as they are the most common in the FL context [8].

To apply the FedSDG in FL, a C fraction of clients is chosen in each round and the gradient of the loss over all the data held by the clients is computed. Therefore, C refers to the global batch size with $C = 1$ representing the full batch setup [11].

A typical implementation is with $C = 1$ and a fixed learning rate η [11]. In this scenario, each client k will compute $g_k = \nabla F_k(w_t)$ (the average gradient on its local data for model w_t). Then, the central server aggregates these gradients and applies the update $w_{t+1} \leftarrow w_t - \eta \sum_{k=1}^{K} \frac{n_k}{n} g_k$ since that $\sum_{k=1}^{K} \frac{n_k}{n} g_k = \nabla f(w_t)$. An equivalent update is found by $\forall k, w_{t+1}^k \leftarrow w_t - \eta g_k$ followed by $w_{t+1} \leftarrow \sum_{k=1}^{K} \frac{n_k}{n} w_{t+1}^k$. This means that each client will locally take one step of gradient descent on the current model using its local data. The server then takes a weighted average of the resulting models [11].

Conversely, the Federated Averaging algorithm, FedAvg, is an extension of the FedSDG. The main concept is following the same steps motioned above, but also iterating the local update $w_k \leftarrow w^k - \eta \nabla F_k(w^k)$ in each client multiple times before the averaging step [11].

Besides the previous hyperparameters C and η, the FedAVG will count with two more. B represents the local mini-batch size and E is the number of iterations through the local data before the global model is updated [14].

It is worth noting that both algorithms work with cross-silo and cross-device training design. For this work, cross-silo was used since it is an industry-oriented project.

3 Methodology

This section outlines the sequential steps taken in this study, including the selection of the Federated Learning (FL) library, deployment mechanism, hardware setup, dataset, neural network model, FL parameters, network configuration, and evaluation criteria.

3.1 FL Library

Given the enterprise-oriented nature of the application, which prioritizes privacy, security, and rapid model specification, the IBM FL library was selected for this project. This library was chosen for its robust cryptographic methods, extensive range of FL strategies, and support for various machine learning models. According to [15], from the most notable libraries, only Flower and IBM FL are production-ready libraries.

A comparison between FL platforms (Pysyft, Flower, IBM FL, TFF and FedML) was performed and is available at [17].

3.2 Hardware

The FL platform's main components include a central server (aggregator) and client devices. Two devices were chosen as silos for this study. Table 1 presents the hardware configurations for each device (silo).

Table 1. Hardware Configurations of Federated Learning Client Devices

Hardware Description	IPT-N-3011	IPT-N-0007
HDD [TB]	1	1
RAM Memory [GB]	4	8
Processing Unit	4-core @ 2.9 GHz	4-core @ 2.9 GHz
Linux Distribution	Ubuntu 18.04	Ubuntu 18.04

3.3 Dataset

To test the platform, the Modified National Institute of Standards and Technology (MNIST) dataset [3] was used. This dataset contains 28 × 28 pixel grayscale images of handwritten digits (0 through 9) and corresponding labels. It is a standard benchmark for machine learning algorithms.

3.4 Neural Network Overview

For pattern recognition tasks involving images, Convolutional Neural Networks (CNN) are commonly used [4]. The training model used in the experiment is a Keras CNN classifier, a built-in feature in the FL library. Since this model aims to test the platform, the suggested model in the IBM guide was used. It was configured to possess two convolutional layers with activated ReLu, one pooling layer, one dropout, flatten, and two dense layers to compose the fully-connected layers segment.

3.5 FL Parameters

Two categories of tests were conducted to explore the platform's capabilities, each repeated five times. The main variable between the tests was the FL strategy used: FedSGD and FedAvg. Table 2 summarizes the training parameters.

Table 2. Federated Learning Training Parameters

Parameters	Values
Training data points	2000
Testing data points	5000
Parties	2
Learning Rate	0.01
Rounds	20
Epochs	20
Termination Accuracy	0.9

3.6 Network Configuration

The platform's deployment involved creating a private VPN connection to ensure secure data transmission. A WebSocket connection was established between the FL Client App and the Web App backend, along with a Flask connection between the aggregation service and local training services.

3.7 Evaluation Criteria

In terms of model training, threshold metrics were used. A threshold metric quantifies classification prediction errors, meaning that it is a fraction or ratio of a class classification and its expected classification. They are usually used when it is expected to minimize the number of errors [2].

The platform's performance was evaluated based on hardware metrics (CPU and RAM usage), network traffic, and model training metrics. Threshold metrics, which measure classification prediction errors, were used to assess accuracy, loss, precision, recall, F1 score, and the confusion matrix for each global model.

4 Results

This section addresses the achieved results and discusses how they compare to the expected output from the platform setup. Data was collected from each FL strategy (FedAvg and FedSDG) through a Python script using the psutil and time libraries.

4.1 Resource Usage for FedAvg Strategy

The project's initial proposition was to use edge devices such as Raspberry Pi devices to perform the role of parties. Therefore, the main concern with the deployment was the resources required to execute a full FL round, given that edge devices are known for their limited hardware capabilities.

Figure 3(a) and Fig. 3(b) show the CPU usage while using the FedAvg strategy from clients IPT-N-0007 and IPT-N-0311 respectively. Each CPU presents the same 20-peak behavior, indicating that model training occurred as expected, with each peak representing one training round.

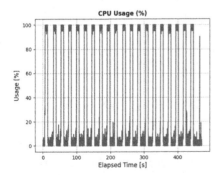
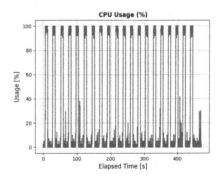

Fig. 3. (a) CPU usage from device IPT-N-0007 in FedAvg. (b) CPU usage from device IPT-N-0311 in FedAvg.

Figures 4(a) and 4(b) show the RAM usage for both devices during the experiment. The scale indicates that device IPT-N-0311 uses almost 600 MB/s more

RAM than IPT-N-0007, likely due to having more available memory. This may jeopardize using edge devices as parties due to their limited RAM resources.

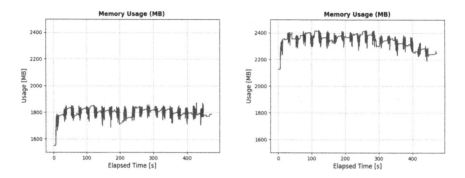

Fig. 4. (a) RAM usage from device IPT-N-0007 in FedAvg. (b) RAM usage from device IPT-N-0311 in FedAvg.

The next criterion to analyze is the network traffic from both parties. Figure 5(a) and 5(b) depicts both devices network traffic. Since the training procedure encloses receiving and sending the model parameters, peak patterns like CPU usage are also expected.

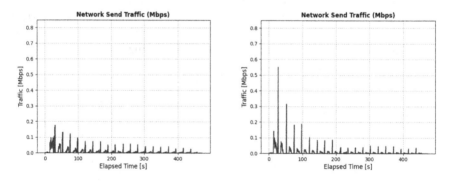

Fig. 5. (a) Network traffic from device IPT-N-0007 in FedAvg. (b) Network traffic from device IPT-N-0311 in FedAvg.

Both figures show a very distinctive exponential-shaped curve from the network traffic. That can happen due to a phenomenon entitled *communication overhead*.[1] After every round, the model gets more accurate, which means that some parameters will no longer be altered. Therefore, the size of the exchanged communication keeps getting smaller after every round.

---

[1] Communication overhead is defined as the total number of packets to be transferred or transmitted from one node to another [6].

4.2 Resource Usage for FedSDG Strategy

The same analysis was conducted for the FedSDG strategy. Figure 6 show the CPU usage for both parties.

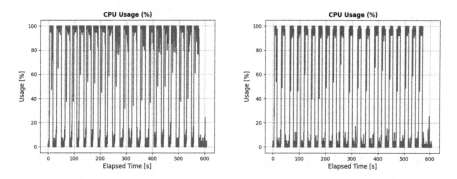

Fig. 6. (a) CPU usage from device IPT-N-0007 in FedSDG. (b) CPU usage from device IPT-N-0311 in FedSDG.

The presented behavior is almost identical to the one observed while using the FedAvg strategy. However, since the FedAvg algorithm is a variation of the FedSDG, a similar behavior was expected.

RAM usage in the experiment has also been tracked and is shown in Fig. 7 from devices IPT-N-0311 and IPT-N-0007 respectively. Once again, the observed behavior is similar to the one observed within the FedAvg strategy.

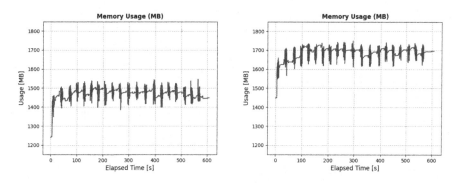

Fig. 7. (a) RAM usage from device IPT-N-0007 in FedSDG. (b) RAM usage from device IPT-N-0311 in FedSDG.

Finally, the network traffic is analyzed and the results are shown in Fig. 8(a) and 8(b) for both IPT-N-0311 and IPT-N-0007 parties respectively.

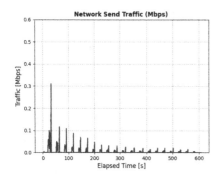 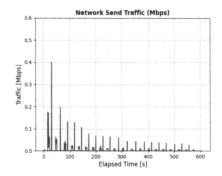

Fig. 8. (a) Network traffic from device IPT-N-0007 in FedSDG. (b) Network traffic from device IPT-N-0311 in FedSDG.

4.3 Training Results for FedAvg

The training results of the 5 experiments for FedAvg is presented in Table 3 and a confusion matrix is depicted in Fig. 9.

Table 3. Global model metrics achieved with FedAvg strategy

Metrics	Experiment 1	Experiment 2	Experiment 3	Experiment 4	Experiment 5
Accuracy	0.8840	0.8782	0.8828	0.8806	0.8774
Loss	0.7848	0.8463	0.8284	0.8835	0.8283
Precision	0.8817	0.8728	0.8763	0.8771	0.8713
Recall	0.8734	0.8696	0.8706	0.8698	0.8662
F1 Score	0.8820	0.8755	0.8808	0.8781	0.8747

The F1 Score for all five experiments was above 0.87, indicating a decent classification model. Since the main objective of the experiments was testing the platform, the enhancement of the model performance by using extra layers, for instance, was neglected. The confusion matrix in Fig. 9 shows clearly defined diagonal squares.

4.4 Training Results for FedSDG

Table 4 shows the training metrics for the FedSDG strategy, with the corresponding confusion matrix depicted in Fig. 10.

Similar to the FedAvg experiments, the metrics for the FedSDG strategy were within the established thresholds. The confusion matrix in Fig. 10 shows well-defined diagonal squares.

The results evaluated the performance and resource usage of FedAvg and FedSDG Federated Learning strategies using multi-sensory device data. Both

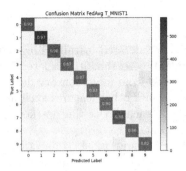

Fig. 9. Confusion matrix for Experiment 1 using FedAvg

Table 4. Global model metrics achieved with FedSDG strategy

Metrics	Experiment 1	Experiment 2	Experiment 3	Experiment 4	Experiment 5
Accuracy	0.8802	0.8814	0.8772	0.8768	0.8838
Loss	0.7647	0.9040	1.0133	1.0645	1.0040
Precision	0.8813	0.8814	0.8793	0.8796	0.8838
Recall	0.8781	0.8792	0.8751	0.8750	0.8819
F1 Score	0.8781	0.8788	0.8746	0.8744	0.8813

strategies achieved high accuracy and F1 scores, indicating effective classification models, with FedAvg showing slightly better resource efficiency. CPU and RAM usage were significant, particularly for devices with lower RAM capacity, and network traffic increased with model accuracy, with FedSDG consuming more bandwidth and taking longer. Scalability issues and data heterogeneity challenges were noted, underscoring the need for careful consideration in practical implementations.

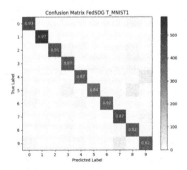

Fig. 10. Confusion matrix for Experiment 1 using FedSDG

5 Conclusion

This work aimed to deploy a federated learning library into an industrial platform and test its functionalities so that it can later be adapted to train with data acquired and transmitted from multi-sensory devices. To achieve the main objective, the library and several other features such as the dataset itself, were containerized and deployed in each device that worked as a client in the training process, through a Docker container.

Afterwards, the Websocket and Flask connections were established, and minor configurations such as export model function buttons were added. With the deployment process completed, two sets of five different global model training were executed using two different FL strategies, the FedAvg and FedSDG.

Nevertheless, when comparing the strategies hardware-wise, the results have shown that the FedSDG strategy consumes higher network bandwidth, ranging from 30% to almost 70% more, depending on the analyzed experiment. Moreover, the FedSDG took an average of 1,8 more minutes to finish the training procedure than the FedAvg. For a test training with only two parties, two minutes does not impact the final results, however, the full concept of FL is to be able to train on dozens or hundreds of devices, meaning, this network consumption can exceed the available network traffic and impact on several more minutes of delay.

Although the testing was a proof of concept made with only two parties, when extending to dozens of clients, some other challenges may appear such as the heterogeneity of data. According to [12], since each company is physically apart, the way data is sampled can vary for each party, and achieving a good performance model requires similar data sets in all parties.

As the project unfolded, the federated learning approach seemed even more fitting since one company could train a model with data from multiple branches, enhancing the predictive maintenance schedule and mitigating production costs. Moreover, this approach affords the company the ability to retain all data within its storage infrastructure, thereby ensuring the privacy and governance of each client's data.

Future work could explore scalability testing with larger networks, incorporating more diverse datasets beyond MNIST, and investigating advanced aggregation techniques to improve performance and privacy. Additionally, comparing the IBM FL platform with other frameworks, examining the impact of communication delays, and customizing federated learning algorithms for specific use cases could further refine the platform's effectiveness and applicability.

References

1. Anthem pays OCR $16 million in record HIPAA settlement following largest U.S. health data breach in history (2020). https://www.hhs.gov/guidance/document/anthem-pays-ocr-16-million-record-hipaa-settlement-following-largest-us-health-data-breach
2. Brownlee, J.: Imbalanced classification with Python: better metrics, balance skewed classes, cost-sensitive learning. Mach. Learn. Mast. (2020)

3. Cohen, G., Afshar, S., Tapson, J., van Schaik, A.: EMNIST: an extension of MNIST to handwritten letters (2017)
4. Herman, R.L.: Introduction to Partial Differential Equations. R.L. Herman (2015). https://doi.org/10.1007/978-3-030-96896-0, https://math.libretexts.org/Bookshelves/Differential_Equations/Introduction_to_Partial_Differential_Equations_(Herman)/09%3A_Transform_Techniques_in_Physics/9.06%3A_The_Convolution_Operation
5. Heydarnoori, A., Mavaddat, F.: Reliable deployment of component-based applications into distributed environments. Third International Conference on Information Technology: New Generations (ITNG'06), pp. 52–57 (2006). https://doi.org/10.1109/ITNG.2006.112, https://ieeexplore.ieee.org/document/1611570
6. Kumar, N., Singh, Y.: Routing Protocols in Wireless Sensor Networks, p. 86–128. IGI Global (2017). https://doi.org/10.4018/978-1-5225-0486-3.ch004, http://dx.doi.org/10.4018/978-1-5225-0486-3.ch004
7. Li, L., Fan, Y., Tse, M., Lin, K.Y.: A review of applications in federated learning. Comput. Ind. Eng. **149**, 106854 (2020). https://doi.org/10.1016/j.cie.2020.106854, https://www.sciencedirect.com/science/article/pii/S0360835220305532
8. Ludwig, H., Baracaldo, N. (eds.): Federated Learning. Springer (2022). https://doi.org/10.1007/978-3-030-96896-0
9. Ludwig, H., et al.: IBM federated learning (2020). https://doi.org/10.48550/arXiv.2007.10987
10. Manias, D.M., Shami, A.: Making a case for federated learning in the internet of vehicles and intelligent transportation systems. IEEE Network **35**(3), 88–94 (2021). https://doi.org/10.1109/MNET.011.2000552
11. McMahan, H.B., Moore, E., Ramage, D., Hampson, S., y Arcas, B.A.: Communication-efficient learning of deep networks from decentralized data (2023). https://doi.org/10.48550/arXiv.1602.05629
12. Laydner de Melo Rosa, G., Mohanram, P., Gilerson, A., Schmitt, R.H.: Architecture for edge-based predictive maintenance of machines using federated learning and multi sensor platforms (2023). https://doi.org/10.20944/preprints202305.1563.v1
13. Nguyen, D.C., Ding, M., Pathirana, P.N., Seneviratne, A., Li, J., Vincent Poor, H.: Federated learning for internet of things. IEEE Commun. Surv. Tutor. **23**(3), 1622–1658 (2021). https://doi.org/10.1109/COMST.2021.3075439
14. Nilsson, A., Smith, S., Ulm, G., Gustavsson, E., Jirstrand, M.: A performance evaluation of federated learning algorithms. In: Proceedings of the Second Workshop on Distributed Infrastructures for Deep Learning (2018). https://doi.org/10.1145/3286490.3286559, https://dl.acm.org/doi/pdf/10.1145/3286490.3286559
15. Saidani, A.: A Systematic Comparison of Federated Machine Learning Libraries. Master's thesis, Technische Universität München (2023). https://wwwmatthes.in.tum.de/pages/1giyhi2qf7es2/Master-s-Thesis-Ahmed-Saidani
16. Sarma, K.V., et al.: Federated learning improves site performance in multicenter deep learning without data sharing. J. Am. Med. Inform. Association: JAMIA **28**, 1259–1264 (2021)
17. Schulz, H.H.: Decentralized brilliance: deploying a federated learning platform and evaluating aggregation algorithms (2024). https://repositorio.ufsc.br/handle/123456789/255868

Detection of Pathological Regions of the Gastrointestinal Tract in Capsule Images Using EfficientNetV2 and YOLOv8

Anderson Lopes Silva[✉], Hellen Guterres França,
Carlos Mendes dos Santos Neto, Alexandre César Pinto Pessoa,
Darlan Bruno Pontes Quintanilha, Aristófanes Corrêa Silva,
and Anselmo Cardoso de Paiva

Núcleo de Computação Aplicada, Universidade Federal do Maranhão (UFMA),
Caixa Postal, São Luís, MA 65.085-580, Brazil
{anderson.silva,hellen.guterres,carlos.mendes,alexandre.pessoa,
dquintanilha,ari,anselmo.paiva}@nca.ufma.br

Abstract. Diseases of the gastrointestinal (GI) tract are among the most common pathologies in the world population and are responsible for thousands of deaths every year. This work proposes an automatic method for detecting regions with GI Tract abnormalities, intending to reduce the number of lesions missed in Wireless Capsule Endoscopy (WCE) video exams by expert endoscopists. By taking advantage of convolutional neural networks (CNNs) and YOLO detection models, the proposed method not only increases the reliability of pathological detection in WCE images, but also sets a new benchmark in this field. Our results for binary classification between healthy and pathological images are promising, with an accuracy of 87.8%, precision of 91.6%, recall of 89.1% and F1-Score of 90.3%. In addition, the detection model showed an Intersection over Union (IoU) of 31.33% among all the images classified as pathological. The impact of this research is significant, as it provides a method capable of detecting GI Tract diseases in WCE images and contributing to better clinical decision-making and patient care.

Keywords: Wireless Capsule Endoscopy · EfficientNet · YOLO · Detection of gastrointestinal diseases

1 Introduction

The gastrointestinal (GI) tract is a tubular system whose main task is digestion. Its structure is made up of the mouth, pharynx, esophagus, stomach, small intestine, large intestine, rectum and anal canal. Inflammatory diseases associated with the GI Tract have a high incidence rate in the world population, especially in developing countries, where colon, liver, and stomach cancer are among the most recurrent diseases [15]. According to the World Health Organization (WHO), in 2022, there were approximately 5.1 million cases of GI Tract cancers worldwide, and, in the Americas region alone, 431,116 deaths were recorded [27].

Based on the need to combat the harmful effects of GI Tract pathologies, endoscopy is one of the most widely used techniques for analyzing GI Tract pathologies, but it is characterized by being a painful and invasive process [12]. In contrast, videos can be taken using Wireless Capsule Endoscopy (WCE), a less aggressive and uncomfortable method. During this procedure, the patient swallows a capsule with a light source, a micro-camera, and a signal emitter that travels painlessly throughout the GI Tract until it is naturally expelled by the patient's body. However, as they scan the entire GI Tract, the videos acquired by the capsule are around 8 h long. Therefore, analyzing this video is an arduous and tedious task, making it possible to miss abnormalities during the examination due to its dependency on the analyst's constant concentration on the images [4]. For this reason, it is important to use automation to help the specialist analyze the video.

From this perspective, machine learning techniques, which consist of using a dataset to develop a computer-aided diagnosis (CAD) system, have been used in the analysis of medical images since the 1990s [13]. These techniques are widely used in medical tasks due to their ability to automatically learn discriminative features from raw data and to reduce the need for pre-processing and feature extraction [14]. An example of this type of method is Convolutional Neural Networks (CNNs), a computational resource inspired by the workings of the human brain commonly used in the processing and analysis of digital images [26]. Another well-known method is YOLO (You Only Look Once), which stands out as a real-time object detection algorithm and a single-stage detector that can predict all objects in a single pass through the image. In this way, the adoption of YOLO can help in the early detection and diagnosis of GI Tract diseases [19].

Furthermore, the importance of identifying the region of the pathology lies in enabling a more precise diagnosis, potentially avoiding additional invasive diagnostic procedures [12]. It allows for the comparison of images from subsequent exams to evaluate disease progression or treatment response and informing patients about the exact location of their condition can improve their understanding of the disease and increase treatment adherence. Thus, it becomes crucial to enhance the capability of tools to automatically detect and characterize pathologies, increasing the efficiency of CADs.

Therefore, this work aims to introduce detection methodologies for WCE images of the GI Tract, providing a solid foundation for subsequent research efforts to build upon the automated detection of diseases in WCE images. Thus, this study developed the application of a method to classify and detect the pathologies using the best CNN architecture experimented and YOLO. Next, Sect. 2 highlights the related work, Sect. 3 describes the proposed method, and Sect. 4 presents the results, experiments, comparisons, and discussions of the proposed method.

2 Related Works

In recent years, neural networks based on deep learning methods have proven to be a powerful tool for image analysis, including medical imaging tasks [3]. This has led to several advances in the field, including computer-aided classification and localization of pathologies from the GI Tract. This section presents some of the work that has been done on this subject.

Xu et al. (2019) introduced a multi-task anatomy detection convolutional neural network (MT-AD-CNN) designed to evaluate esophagogastroduodenoscopy inspection quality. The model classifies images and distinguishes between informative and non-informative frames, thus displaying detection boxes only on informative frames and reducing false positives. The MT-AD-CNN achieves a mean average precision (mAP) of 93.74% for the detection task and an accuracy of 98.77% for the classification task.

The work by Oukdach et al. (2024) presented a new framework that synergizes the strengths of CNNs and Vision Transformers (ViTs). This approach incorporates an attention mechanism within a CNN to extract local features. Experiments performed on the Kvasir Capsule, a large-scale dataset of WCE images [21], showed a promising result of 97% in accuracy, precision, recall, and F1-Score.

Muruganantham and Balakrishnan (2022) proposed a method for classifying images with ulcers, bleeding, polyps, or no abnormality using ResNet-50 associated with a self-attention mechanism aggregate spatial characteristics globally. The method was also tested on the Kvasir-Capsule dataset and achieved a classification accuracy of 94.7%.

In addition, Srivastava et al. (2022) proposed a focal modulation network called FocalConvNet. FocalConvNet makes use of focal modulation and an added convolutional block to extract features in order to improve detection. The study used 13 classes from anatomy and luminal findings from the Kvasir-Capsule dataset, obtaining F1-Score, recall, and MCC (Matthews Correlation Coefficient) of 67.34%, 63.73%, and 29.74, respectively.

As videos are sequences of captured images and the mentioned methods are applied per image (individual video frame), some work [5,7,17,18,22] do not clearly specify how sampling is done to split the training and test sets. Ideally, all video frames for a patient should be in the same set. Applying a random sampling of video frames can result in "data leakage" [9], as similar frames (neighbors) from the same patient can end up in both the training and test sets, compromising the results' reliability.

In addition, although there are many published studies to assist in the diagnosis of GI pathologies in WCE images, there is a notable lack of studies focused on the detection of the region in which these anomalies are found. To this end, a pipeline was developed to classify abnormal images using the best-performing CNN architecture in the conducted experiments, followed by the detection of the region with the presence of the disease using YOLOv8 [8], a recent model for detecting objects from images in real-time. This work's contribution lies in the proposal of an automated method that not only identifies the presence of different types of anomalies in WCE images but also indicates the region affected by the pathology, reducing the time and work involved in analyzing videos of WCE exams.

3 Materials and Method

This section describes the proposed method. Firstly, the image dataset used to train and evaluate the proposed method is determined, and then the best performing CNN architecture is selected to classify the images in the dataset. Then, the model that has shown the best performance during the experiments sends the images classified as abnormal for detection. Finally, the detection models trained for a specific class by YOLOv8, make predictions based on the images received. The predicted region is the one with the highest confidence between the model's predictions. A representation of the process can be seen in Fig. 1.

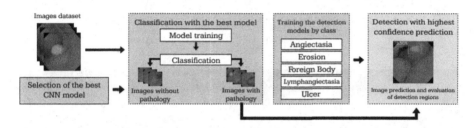

Fig. 1. Proposed method

3.1 Dataset

In this work, the experiments were carried out using the Kvasir-Capsule dataset, which was built by collecting WCE scans from a Norwegian hospital [21]. Currently, Kvasir-Capsule is the largest and most diverse WCE dataset with expert-labeled data and publicly available. It comprises a total of 117 videos collected from WCE exams, totaling around 4.7 million frames extracted from these videos.

In addition, the dataset features 47.238 frames labeled by specialists, 336×336 in size and divided into two categories: anatomical findings, consisting of Pylorus, Ileocecal Valve and Ampulla of Vater, and luminal findings, composed of Angiectasia, Blood - fresh, Blood - hematin, Erosion, Erythema, Foreign Body, Polyp, Lymphangiectasia, Reduced Mucosal View and Ulcer, in addition to the Normal Clear Mucosa class, which represents normal images. Figure 2 shows the sample image for each class. The division of the dataset, the number of patients, and the availability of the bounding boxes per class can be seen in Table 1.

Table 1. Kvasir-Capsule images by class

Class	Image Quantity	Patients	Bounding Boxes
Ampulla of Vater	10	1	Available
Angiectasia	866	6	Available
Blood - fresh	446	2	Available
Blood - hematin	12	1	Available
Erosion	506	9	Available
Erythema	159	3	Unavailable
Foreign Body	776	3	Available
Ileocecal Valve	4189	35	Unavailable
Lymphangiectasia	592	3	Available
Normal Clear Mucosa	34.338	37	–
Polyp	55	1	Available
Pylorus	1.538	32	Available
Reduced Mucosal Vision	2.906	7	Unavailable
Ulcer	854	3	Available

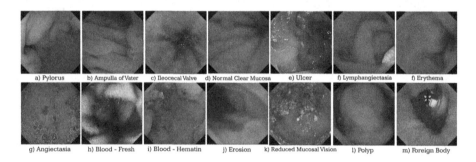

a) Pylorus b) Ampulla of Vater c) Ileocecal Valve d) Normal Clear Mucosa e) Ulcer f) Lymphangiectasia f) Erythema
g) Angiectasia h) Blood - Fresh i) Blood - Hematin j) Erosion k) Reduced Mucosal Vision l) Polyp m) Foreign Body

Fig. 2. Sample of labeled images in Kvasir-Capsule

3.2 Classification Stage

In this stage, we previously implemented and evaluated CNN architectures to choose the model to be used in the method, defined the metrics to be used to evaluate the binary classification models, and trained and tested the dataset with the architecture that obtained the best performance.

Selection of CNN Model. After defining the dataset, the best-performing CNN is selected to classify the images. For this task, CNNs widely used in the literature in the field of image classification were chosen. The architectures evaluated in this stage were VGG [20], Resnet [6], Inception [23], Xception [1], EfficientNet [24] and EfficientNetV2 [25]. The architectures mentioned were used according to their respective original authors and the implementation provided by the Keras library [2].

Binary Classification Training and Evaluation. After choosing the best architecture to perform the binary classification, the model is evaluated during and after the training process using the following metrics:

$$\text{Accuracy } (ACC) : (TP+TN)/(TP+TN+FP+FN) \tag{1}$$

$$\text{Precision } (PRE) : TP/(TP+FP) \tag{2}$$

$$\text{Recall } (REC) : TP/(TP+FN) \tag{3}$$

$$\text{F1-Score } (F1) : 2*(PRE*REC)/(PRE+REC) \tag{4}$$

where TP and TN represent correctly classified images with pathology and healthy images. In addition, FP represents the healthy images wrongly classified as having abnormalities, and FN means the classifications of pathological images as healthy images.

In the context of the proposed method's objective, accuracy is useful for giving an overview of the model's performance, but it is misleading in unbalanced datasets, as in the case of Kvasir-Capsule, where one class is more frequent than the other. Thus, precision is also used to ensure that images classified as containing pathologies actually do have pathologies, thus reducing false positives that can overload the region detection stage.

In addition, recall is crucial to ensure that as many images with pathologies as possible are identified, minimizing false negatives that could result in undetected pathologies. The F1-Score will be the most important metric to evaluate due to its balance between precision and recall, providing a single performance measure that considers both false positives and false negatives. Furthermore, although specificity is a useful metric in many contexts, in this work, the main focus is ensuring that all images with pathologies are identified (maximizing recall) and that the identified images contain pathologies (maximizing precision).

Thus, using the metrics presented, the model selects the images with pathologies and sends them to the detection phase to determine the region where the classified pathology appears.

3.3 Detection Stage

During this stage, the detection models based on YOLOv8 were trained and the pathological area was predicted based on the model with the highest prediction confidence.

YOLOv8 Architecture. YOLOv8 was unveiled in January 2023 by Ultralytics, providing five versions: YOLOv8n (nano), YOLOv8s (small), YOLOv8m (medium), YOLOv8l (large), and YOLOv8x (extra-large) [8]. YOLOv8 provides support for classification, segmentation, and object detection tasks. In this work,

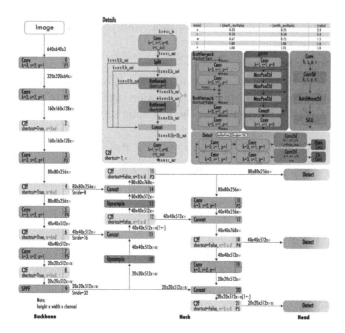

Fig. 3. YOLOv8 architecture [8]

YOLOv8s was used as a model to detect the regions where anomalies appear in the dataset images. The YOLOv8 architecture can be seen in Fig. 3.

Among the differences in YOLOv8 compared to previous versions is the C2f (partial inter-stage bottleneck with two convolutions) module, which combines high-level features to improve detection accuracy based on contextual information. In addition, YOLOv8 has an approach that allows each branch to be focused on its task in order to improve the overall accuracy of the model. In the output layer of YOLOv8, the sigmoid function is used as the activation function to represent the probability of a bounding box containing an object. In addition, the softmax function is used for class probabilities.

YOLOv8 uses the CIoU (Complete Intersection over Union) and DFL (Distribution Focal Loss) loss functions for bounding box loss and binary cross-entropy for classification loss. These losses improved object detection performance, especially when dealing with smaller objects [11]. YOLOv8's architecture uses a modified backbone CSPDarknet53 combined with an SPPF layer (Spatial Pyramid Pooling Fast), which speeds up the calculation by grouping features into a fixed-size map. Furthermore, each convolution has batch normalization and SiLU activation (Sigmoid Linear Unit), and the head of the architecture is decoupled, so the objectivity, classification, and regression tasks are processed independently [16].

Detection Model Training. After classifying the images with pathologies, the next step is to train the detection models with YOLOv8. The aim was to train a dedicated model to detect each class so that each model could predict based on the abnormal images received from the previous stage. YOLOv8s was therefore used to train the detection models. Along with accuracy and sensitivity, the following metrics were also used to measure the assertiveness of the predicted regions in relation to the true region with the pathology:

- Average Precision (AP): $\int_0^1 P(R)\, dR$
- mean Average Precision (mAP): $(\frac{1}{n} \sum_{i=1}^{n} AP_i)$
- Intersection over Union (IoU): $\frac{\text{Intersection Area}}{\text{Union area}}$

mAP is a widely used metric in object detection tasks, as it helps calculate average accuracy at various IoU thresholds. Thus, it provides a comprehensive assessment of the algorithm's accuracy in object detection. The mAP calculation considers the AP, the average of the accuracies calculated at various recovery points. The recovery point is where the rate of true positives is calculated in relation to the total number of positive examples.

Additionally, the IoU metric was used to measure how close the predicted bounding boxes are to the real box described in the dataset. This metric is calculated as the area of intersection over the area of the union of the prediction with the true area. It is commonly used to assess the accuracy of object detection. Thus, after training the dedicated models for each class, all these metrics were calculated and applied to a test set to evaluate the performance of each model.

Pathology Region Detection. After training the dedicated model for each class, the images classified as abnormal are passed through all the detection models to assess the confidence in the region predicted in each image. From the results stored in text files in the YOLO format, we will have access to the following prediction information: detected class, x and y coordinates of the center of the bounding box, width, and height of the predicted bounding box, and the prediction confidence. So this step consists of identifying the prediction with the highest confidence and then collecting the coordinates of the predicted bounding box.

Then, the given coordinates are translated into Cartesian format on top of the original image size and the region of intersection between the prediction and the true location of the pathology is stipulated. Once all the images have been passed through and the region of interest has been identified, the IoU corresponding to all the intersection regions is calculated in order to obtain an overall evaluation of the proposed method.

4 Results and Discussion

This section presents the results achieved using the proposed method for detecting regions containing GI tract pathologies. The experiment carried out, and the impact of the results on the advancement of detection models in GI Tract images are also described.

4.1 Dataset Preparation

Initially, the dataset was divided by patient, given that the images were taken from WCE videos and depict frames of the same patient. This division was then designated to prevent data leakage, the influence of extremely similar images of the same patient being present in the same set (training, validation, or test). As a result, only classes with at least three patients were used, since each set must contain at least one different patient with each pathology. In addition, only classes of luminal findings in which the bounding boxes were also available in the dataset were considered. Therefore, based on Table 1, the classes used were Angiectasia, Erosion, Foreign body, Lymphangiectasia, and Ulcer.

As a result, there was an imbalance between the number of pathological images belonging to each patient. So, in order to increase the difference in the number of samples in each set, the patient with the most images was sent to the training set, the second to the validation set, the third to the test set, and so on, in that order, until all the patients in the class were distributed. The number of images with pathology belonging to each of the sets can be seen in Table 2.

Table 2. Distribution of images by class for each set

Class	Train	Validation	Test
Angiectasia	687	129	50
Erosion	230	152	124
Foreign Body	579	180	17
Lymphangiectasia	368	150	74
Ulcer	582	145	127
Total	**2446**	**756**	**392**

In the healthy images used for classification, it was observed that many images were from the same patient. Therefore, efforts were made to diversify the number of normal images obtained from each patient, ensuring a balance with the set of abnormal images. Thus, 200 images from 13 patients were used for the training set, 80 images from 10 patients for validation, and 30 images from 14 patients for testing, totaling 2600 images for training, 800 images for validation, and 420 images for testing.

4.2 Abnormal Image Classification

This section describes the results achieved in the stage of selecting the best-performing CNN architecture. All CNN architectures were trained with 100 epochs and the following hyperparameters: Adam [10] optimizer with a learning rate of 0.001, batch size of 32 and early stopping with patience of 25 epochs.

Moreover, the images were normalized and resized to 224 × 224. In order to evaluate the models to determine the best performing one, experiments were carried out on the following architectures: VGG16, ResNet50, InceptionV3, Xception, EfficientNetB0, EfficientNetB1, EfficientNetB2, EfficientNetV2B0, EfficientNetV2B1 and EfficientNetV2B2. All the architectures were trained with pre-trained weights from imagenet. Table 3 shows the results achieved with each architecture and the number of parameters in each model.

Table 3. Comparison of CNN architectures

Model	ACC	PRE	REC	F1	Parameters
VGG16	51.7%	100%	51.7%	68.1%	138.4M
ResNet50	88.5%	75.4%	82.2%	78.7%	25.6M
InceptionV3	90.9%	64.7%	78.4%	70.9%	23.9M
Xception	77.2%	76.6%	76.2%	76.4%	22.9M
EfficientNetB0	72.4%	90.7%	77.3%	83.4%	5.3M
EfficientNetB1	**91.3%**	80.0%	85.7%	82.7%	7.9M
EfficientNetB2	82.9%	85.7%	83.4%	84.5%	9.2M
EfficientNetV2B0	87.8%	**91.6%**	**89.1%**	**90.3%**	7.2M
EfficientNetV2B1	87.5%	70,4%	79.5%	74.7%	8.2M
EfficientNetV2B2	67.9%	65.2%	66.1%	65.6%	10.2M

Although VGG16 has perfect precision, its accuracy, recall and F1-score are low, which indicates that the model makes few positive predictions and is not ideal for applications that require balanced detection. Something similar occurs in EfficientNetB0, as the model has high accuracy but fails to detect part of the true positives.

Xception and EfficientNetV2B2 presented modest results in all metrics, demonstrating that neither model was well-suited to the proposed task. Therefore, ResNet50, InceptionV3 and EfficientNetV2B1 have good accuracy, but their precisions are relatively low, resulting in moderate F1-score values, 78.7% and 70.9%, respectively.

EfficientNetB1 and EfficientNetB2 presented relatively good and balanced precision and recall values, although EfficientNetB2 presented low accuracy compared to the other models. As a result we have 82.7% and 84.5% F1-Score, respectively. For these two architectures, the model does not usually miss a positive instance and also does not make many wrong predictions.

However, EfficientNetV2B0 is the model that presents the best combination of accuracy (87.8%), precision (91.6%), recall (89.1%) and F1-score (90.3%). This indicates that it makes many correct predictions and detects most true positives. In other words, it performs the best correct classification of pathologies among all detections made by the system, thus being the most suitable model to undertake

Table 4. EfficientNetV2B0 confusion matrix

		Prediction	
		Anormal	Normal
Real	Anormal	329	63
	Normal	40	380

the proposed task of classifying pathological images. Table 4 shows the confusion matrix for the classification with EfficientNetV2B0.

4.3 Detection of Regions with Anomalies

To train the detection models, data augmentation was performed on the images of each class. The operations applied were: Vertical and horizontal flip and blur variations with filter size varying from 1×1 to 7×7 pixels. The results of training the dedicated detection models for each class are shown in Table 5.

Table 5. Metrics of the models trained with YOLOv8

Class	PRE	REC	mAP50	mAP50-95	IoU
Angiectasia	94.6%	69.9%	79.3%	32.8%	67.3%
Erosion	46,6%	18.6%	16.9%	4.8%	52.7%
Foreign Body	93.6%	94.1%	97.6%	76.0%	87.2%
Lymphangiectasia	79.4%	41.7%	66.7%	37.8%	49.7%
Ulcer	69.9%	50.4%	52.7%	14.7%	73.1%

The mAP50 returns the average precision of the detections with an IoU threshold of 0.5, while the mAP50-95 returns the results of the mAP with IoU thresholds of 0.5 to 0.95. Finally, the IoU indicates the extent to which the prediction box actually intersected the region with the anomaly. Of particular note were the Angiectasia, Foreign Body and Ulcer classes, which achieved 67.3%, 87.2% and 73.1% IoU, respectively, demonstrating the models' ability to detect the majority of pathological regions in their respective class.

At the end of training the detection models, the next step is to determine the coordinates of the predicted box with the highest confidence prediction in the images classified as abnormal. Thus, in order to evaluate the model's prediction in relation to the true region with the pathology, the IoU was used. The results obtained from the predictions per model and the overall evaluation of all the models can be seen in Table 6.

To evaluate overall performance, it should be noted that for normal images, which have passed through the classification filter for detection, a 100% true positive value will be considered if there are no predictions for these images

Table 6. Evaluation of detection regions

	Angiectasia	Erosion	Foreign Body	Lymphangiectasia	Ulcer	General
IoU	33.59%	39.80%	43.38%	65.85%	32.00%	31.33%

since there are no abnormalities to be found. Otherwise, if there is any kind of prediction in an image without pathology, a 0% true positive value will be considered, as there are no pathologies to be found in such images. In addition, abnormal images that have been classified as normal automatically count as a 0% hit in the method's overall performance since the detection models won't have the chance to predict the pathologies present in these images. Some examples of the acquired results can be seen in Fig. 4. The red boxes represent the model's prediction, while the green boxes show the true anomaly region. In addition, the IoU value of each prediction is highlighted in the top left corner of the images. Although the overall result was not satisfactory, it should be noted that this work represents an initial step in the detection of pathological regions in WCE images, especially with regard to the criteria for dividing up the dataset.

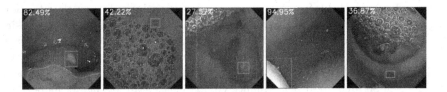

Fig. 4. Sample results of pathology region prediction

4.4 Discussion

The results shown in Tables 5 and 7 demonstrate the superiority of the dedicated models for each class compared to the single model for all classes in detecting pathologies. For the angiectasia classification, the dedicated model obtained a precision of 94.6% and a mAP50 of 79.3%, in contrast to the single model, which showed a precision of only 21.1% and a mAP50 of 29.9%. Although the recall of the single model is slightly higher, the combination of high precision and mAP50 of the dedicated model indicates a more accurate and reliable identification of angiectasia images. In erosion detection, the dedicated model was also superior, with a precision of 46.6% and a mAP50 of 16.9%, while the single model showed substantially lower values.

For the foreign body class, although both models showed high performances, the dedicated model was still superior in its performance. In the lymphangiectasia and ulcer classes, the dedicated models significantly outperformed the single model in all evaluated metrics. As for the IoU, the dedicated models were

Table 7. Metrics of the model trained for all classes with YOLOv8

Class	PRE	REC	mAP50	mAP50-95	IoU
Angiectasia	21.1%	72.0%	29.9%	10.4%	55.5%
Erosion	1,34%	0.08%	0.23%	0.06%	0.06%
Foreign Body	68.4%	88.2%	92.8%	79.8%	85.39%
Lymphangiectasia	65.3%	30.5%	53.2%	33.2%	21.21%
Ulcer	18.7%	2.36%	3.59%	1.94%	0.01%

superior in detecting regions containing anomalies in all classes. Especially in the erosion and ulcer classes, where the model trained for all classes obtained insignificant results, while the dedicated models could detect more than half of the pathological regions. These results highlight that the approach of training dedicated models for each class not only improves efficiency in detecting specific pathologies, but also reduces the false positive rate and increases confidence in the detections made. Therefore, using dedicated models for pathology detection in WCE is a more effective solution, providing more reliable results than a single model trained for all classes.

In addition, in Table 8 a comparison was made with the binary classification methods of works related to the study carried out in this paper, all using Kvasir-Capsule. Each study contributed different methodologies and made significant advances in the field. However, based on the division criteria adopted by the works that relate to the classification stage carried out in this study, the proposed method still showed good performance compared to state-of-the-art models. In comparison, Muruganantham et al. [17] achieved 94.70% accuracy, Fonseca et al. [5] obtained 97.00% precision, Oukdach et al. [18] achieved F1-Score of 97.00% and Jain et al. [7] achieved 98.00% across all metrics with a hybrid network approach and residual attention, all using the same random division of data.

Table 8. Comparison of binary classification works using Kvasir-Capsule

Author	Approach	Dataset division criteria	ACC	PRE	REC	F1
Muruganantham [17]	ResNet-50 + self-attention mechanism	Random split	94.7%	–	–	–
Fonseca el al. [5]	ResNet50 + tranfer learning	Random split	–	97.0%	69.0%	81.0%
Srivastava et al. [22]	FocalConvNet	Random split	63.7%	75.5%	63.7%	67.3%
Jain [7]	Hybrid CNN + residual attention network	Random split	98.0%	98.0%	99.0%	98.0%
Oukdach et al. [18]	ViT + attention mechanism	Random split	97.0%	97.0%	97.0%	97.0%
Proposed Model	EfficientNetV2B0	Split by patient	87.8%	91.6%	89.1%	90.3%

It is emphasized that the patient splitting criterion adopted in the proposed method is crucial to avoid data leakage between training and testing, ensuring more reliable results when evaluating the model's performance. This method ensures the integrity of the results by mitigating potential biases introduced by less rigorous data-splitting practices. It is worth noting that during the search for research related to this work, no other work was identified that utilized the Kvasir-Capsule dataset and proposed or evaluated a method to classify abnormalities and detect the region in which the pathology is found. As such, this work aims to propose advances in the area of pathology detection in WCE images of the GI Tract, presenting some baseline results for future work in the area.

5 Conclusion

In conclusion, early diagnosis is essential so that the necessary treatment can be carried out, but the detection of GI Tract pathologies in WCE images remains a challenging task due to the complex nature of the gastrointestinal system and the low quality of the images generated by this exam. The proposed method proved to be a promising tool for detecting GI Tract pathologies present in WCE images. The F1-Score of 90% obtained in the classification stage indicates that the EfficientNetV2B0 model is able to help in the early identification of pathological images, allowing the detection models dedicated to each class to work on the images that present the diseases. Thus, the detection showed the ability to find some of the pathological regions based on the strategy of treating the prediction region as the one with the highest confidence among the trained detection models.

As future work, in order to build on the results obtained in this study, an expansion of the proposed method is intended, with pre-processing methods to improve the models' ability to identify pathologies. In addition, it is proposed to investigate multi-class classification methods to serve as a step between the first classification and detection, to use the best detection model for the frame analyzed. It is also planned to test the method on other datasets in order to validate the robustness of the proposed method further. Therefore, it is understood that this work can help health specialists in the task of identifying pathologies in WCE images, potentially reducing the workload on doctors and improving patient outcomes through early and accurate diagnosis.

Acknowledgments. The authors acknowledge the Coordenação de Aperfeiçoamento de Pessoal de Nível Superior (CAPES), Brazil - Finance Code 001, Conselho Nacional de Desenvolvimento Científico e Tecnológico (CNPq), Brazil, and Fundação de Amparo à Pesquisa Desenvolvimento Científico e Tecnológico do Maranhão (FAPEMA) (Brazil), Empresa Brasileira de Serviços Hospitalares (Ebserh) Brazil (Grant number 409593/2021-4) for the financial support.

References

1. Chollet, F.: Xception: deep learning with depthwise separable convolutions. In: Proceedings of the IEEE Conference on Computer Vision and Pattern Recognition, pp. 1251–1258 (2017)
2. Chollet, F.: Keras (2015). https://keras.io

3. Costa, C.L., Lima, D.A., Zorzo Barcelos, C.A., Travençolo, B.A.: Ensemble architectures and efficient fusion techniques for convolutional neural networks: an analysis on resource optimization strategies. In: Brazilian Conference on Intelligent Systems, pp. 107–121. Springer (2023)
4. Delagah, B., Hassanpour, H., et al.: Feature extraction for polyp detection in wireless capsule endoscopy video frames. J. Healthc. Eng **2023**, 6076514 (2023)
5. Fonseca, F., Nunes, B., Salgado, M., Cunha, A.: Abnormality classification in small datasets of capsule endoscopy images. Procedia Comput. Sci. **196**, 469–476 (2022)
6. He, K., Zhang, X., Ren, S., Sun, J.: Deep residual learning for image recognition. In: Proceedings of the IEEE Conference on Computer Vision and Pattern Recognition, pp. 770–778 (2016)
7. Jain, S., Seal, A., Ojha, A.: A hybrid convolutional neural network with meta feature learning for abnormality detection in wireless capsule endoscopy images. arXiv preprint arXiv:2207.09769 (2022)
8. Jocher, G., Chaurasia, A., Qiu, J.: YOLO by Ultralytics (2023). https://github.com/ultralytics/ultralytics. Accessed 28 Feb 2023
9. Kaufman, S., Rosset, S., Perlich, C., Stitelman, O.: Leakage in data mining: formulation, detection, and avoidance. ACM Trans. Knowl. Discov. Data **6**(4) (2012). https://doi.org/10.1145/2382577.2382579
10. Kingma, D.P., Ba, J.: Adam: a method for stochastic optimization. arXiv preprint arXiv:1412.6980 (2014)
11. Li, X., et al.: Generalized focal loss: learning qualified and distributed bounding boxes for dense object detection. Adv. Neural. Inf. Process. Syst. **33**, 21002–21012 (2020)
12. Lima, D.L.S., Pessoa, A.C.P., De Paiva, A.C., da Silva Cunha, A.M.T., Júnior, G.B., De Almeida, J.D.S.: Classification of video capsule endoscopy images using visual transformers. In: 2022 IEEE-EMBS International Conference on Biomedical and Health Informatics (BHI), pp. 1–4. IEEE (2022)
13. Litjens, G., et al.: A survey on deep learning in medical image analysis. Med. Image Anal. **42**, 60–88 (2017)
14. Mendes, A.C., Pessoa, A.C.P., de Paiva, A.C.: Multi-label classification of pathologies in chest radiograph images using densenet. In: Brazilian Conference on Intelligent Systems, pp. 167–180. Springer (2023)
15. Milivojevic, V., Milosavljevic, T.: Burden of gastroduodenal diseases from the global perspective. Curr. Treatment Opt. Gastroenterol. **18**, 148–157 (2020)
16. MMYOLO: YOLOv8 by MMYOLO (2023). https://github.com/open-mmlab/mmyolo/tree/main/configs/yolov8. Accessed 15 Mar 2024
17. Muruganantham, P., Balakrishnan, S.M.: Attention aware deep learning model for wireless capsule endoscopy lesion classification and localization. J. Med. Biol. Eng. **42**(2), 157–168 (2022)
18. Oukdach, Y., Kerkaou, Z., El Ansari, M., Koutti, L., Fouad El Ouafdi, A., De Lange, T.: Vitca-NDet: a framework for disease detection in video capsule endoscopy images using a vision transformer and convolutional neural network with a specific attention mechanism. Multimedia Tools Appl. 1–20 (2024)
19. Qureshi, R., et al.: A comprehensive systematic review of yolo for medical object detection (2018 to 2023). Authorea Preprints (2023)
20. Simonyan, K., Zisserman, A.: Very deep convolutional networks for large-scale image recognition. arXiv preprint arXiv:1409.1556 (2014)
21. Smedsrud, P.H., et al.: Kvasir-capsule, a video capsule endoscopy dataset. Sci. Data **8**(1), 142 (2021)

22. Srivastava, A., Tomar, N.K., Bagci, U., Jha, D.: Video capsule endoscopy classification using focal modulation guided convolutional neural network. In: 2022 IEEE 35th International Symposium on Computer-Based Medical Systems (CBMS), pp. 323–328. IEEE (2022)
23. Szegedy, C., Vanhoucke, V., Ioffe, S., Shlens, J., Wojna, Z.: Rethinking the inception architecture for computer vision. In: Proceedings of the IEEE Conference on Computer Vision and Pattern Recognition, pp. 2818–2826 (2016)
24. Tan, M., Le, Q.: EfficientNet: rethinking model scaling for convolutional neural networks. In: International Conference on Machine Learning, pp. 6105–6114. PMLR (2019)
25. Tan, M., Le, Q.: Efficientnetv2: smaller models and faster training. In: International Conference on Machine Learning, pp. 10096–10106. PMLR (2021)
26. Valentim, N.A., Dorça, F.A., Asnis, V.P., Elias, N.C.: The artificial intelligence as a technological resource in the application of tasks for the development of joint attention in children with autism. In: Brazilian Conference on Intelligent Systems, pp. 306–320. Springer (2023)
27. World Health Organization: The Global Cancer Observatory (2024). https://gco.iarc.fr/en. Accessed 15 Mar 2024
28. Xu, Z., et al.: Upper gastrointestinal anatomy detection with multi-task convolutional neural networks. Healthcare Technol. Lett. **6**(6), 176–180 (2019)

Dual-Bandwidth Spectrogram Analysis for Speaker Verification

Rafaello Virgilli[1]($^{\boxtimes}$), Arnaldo Candido Junior[2], Augusto Seben da Rosa[3], Frederico S. Oliveira[1], and Anderson da Silva Soares[1]

[1] Universidade Federal de Goiás, Goiânia, Brazil
rafaello.virgilli@discente.ufg.br
[2] Universidade Estadual Paulista, São José do Rio Preto, Brazil
[3] Universidade Tecnológica Federal do Paraná, Medianeira, Brazil

Abstract. The variability of the human voice is a challenge for speaker verification systems, influenced by individual traits and environmental conditions. This research introduces a novel approach that uses dual-bandwidth spectrograms with the Fast ResNet-34 neural network architecture for speaker verification. Dual-bandwidth spectrograms are data structures similar to multi-channel images, generated by stacking spectrograms derived from the same audio segment using two different window sizes. In this study, we employed window sizes of 5 ms and 30 ms. This approach captures a wider range of voice features across multiple temporal and spectral resolutions. Our findings demonstrate a statistically significant improvement in system performance, achieving an Equal Error Rate (EER) of 1.64% ±0.13%. This represents a 26% enhancement over the previously reported benchmark EER of 2.22% ±0.05%, validating our hypothesis that dual-bandwidth spectrograms offer a more detailed and comprehensive representation of voice features for accurate speaker verification. Analysis of individual bandwidth contributions reveals that narrowband spectrograms carry more relevant features for speaker verification, while the combination with broadband spectrograms provides complementary information.

Keywords: speaker verification · feature fusion · dual-bandwidth spectrogram · broadband · narrowband

1 Introduction

Speaker verification (SV) is the task of determining whether two voice samples belong to the same speaker. Unlike speaker identification [1], which attempts to identify a speaker from a closed set of individuals, SV compares the features extracted from any pair of voices to assess their similarity. This open-set approach offers greater flexibility and applicability in biometric systems [2].

The variability of the human voice poses a significant challenge for SV systems. Individual traits such as age, accent, emotional state, intonation, and language, as well as environmental factors like background noise, reverberation, and recording devices, can influence the voice signal [3,4]. Despite these variations, the voice possesses unique attributes that remain relatively stable, primarily due to the anatomical characteristics of the speaker's vocal tract.

In the context of machine learning, SV is typically addressed by training a model to extract individualized voice characteristics and compare them to determine a similarity score between two voice samples. This score is then compared to a decision threshold to classify the samples as originating from the same speaker or different speakers. The Equal Error Rate (EER) is a commonly used metric for calibrating the decision threshold. It balances the rate of false positives and false negatives.

In this work, our primary contribution is the exploration of dual-bandwidth spectrograms for speaker verification. We propose adapting an existing deep neural network architecture for SV to process a combination of two spectrograms generated from the same audio segment. Our hypothesis is that the integration of narrowband and broadband spectrograms will significantly enhance the accuracy of SV tasks by leveraging the complementary nature of these two spectrogram types.

Our dual-bandwidth approach combines narrowband spectrograms, generated with longer analysis windows (approximately 30 ms), and broadband spectrograms, generated with shorter windows (around 5 ms). Narrowband spectrograms offer higher frequency resolution, allowing better visualization of harmonic structures related to the speaker's fundamental frequency. Conversely, broadband spectrograms provide better temporal resolution, capturing rapid acoustic events and broader spectral patterns like formants.

By exploring this novel combination of representations, we aim to capture a more comprehensive set of speaker-specific features across different temporal and spectral resolutions. This dual-bandwidth approach potentially allows the model to leverage both fine-grained frequency information and broader spectral patterns, leading to more robust speaker verification performance. While combined spectrograms have shown success in related tasks such as keyword detection, voice activity detection, and emotion recognition, our work is the first to thoroughly investigate their potential in speaker verification.

We chose the model presented in [5], which achieved state-of-the-art SV performance in 2020, as the reference to validate our hypothesis. Although more recent models with better performance exist, the simplicity of this model's construction facilitates the isolation of the input's influence during experiments. Moreover, its relatively low computational cost enables training on available hardware resources.

The remainder of this paper is organized as follows. Section 2 presents related work, positioning our research within the current landscape of speaker verification techniques. Section 3 describes our proposed method in detail, elaborating on the dual-bandwidth spectrogram approach and the neural network model created to validate our hypothesis. Section 4 outlines the experimental setup, including the dataset, hardware configuration, and training procedure. Section 5 presents and discusses the results, covering the evaluation protocol, performance metrics, and a comparison of the proposed model with the baseline. This section also addresses the limitations of our current approach and outlines potential directions for future work. Finally, Sect. 6 concludes the paper, summarizing the main findings and their implications for the field of speaker verification.

2 Related Work

Speaker verification has seen significant advancements through various techniques including data augmentation, novel neural network architectures, and multi-bandwidth spectrogram approaches.

In terms of data augmentation for speaker recognition, [6] proposed using a voice conversion model to generate additional data. They also applied a bandwidth extension to augment narrowband speech, generating missing frequency bands from lowband information.

Regarding neural network architectures, [7] and [8] introduced different attention layers on top of the feature extractor to better encode variable-length vector representations compared to global average pooling. Further advancing this concept, [9] adapted Selective Kernel Attention (SKA) modules with a multi-scale frequency and channel module, modifying feature extractor blocks to use an attention mechanism that selects from multiple 1D convolutional kernel sizes.

The use of multiple spectrograms as input to neural network models has been explored by [10]. They combined Mel spectrograms, Gammatone spectrograms, and spectrograms extracted with continuous wavelet transform into a single 3D-channel spectrogram for a Convolutional Neural Network (CNN).

Multi-bandwidth spectrograms, which combine multiple frequency bandwidths to capture a broader range of audio features, have shown promise in various speech and audio processing tasks. [11] analyzed combined spectrograms and proposed a new representation obtained by computing the pixel-wise geometric mean of narrowband and broadband spectrograms from the same audio signal.

In the work of [12], multi-bandwidth spectrograms were applied to voiced and unvoiced sound detection. The authors used different window analyses depending on the frequency range, resulting in performance improvements compared to classical spectrogram generation approaches.

While multi-bandwidth spectrograms focus on combining different frequency representations, other researchers have explored combining different types of features to improve speaker recognition performance. Feature fusion techniques have been explored in speaker recognition. [13] proposed fusing Mel Frequency Cepstral Coefficients (MFCCs) with new features based on temporal domain statistical indicators such as mean, median, and standard deviation. Their model, trained and evaluated on different subsets of LibriSpeech [14], outperformed the baseline models they evaluated.

These studies demonstrate the potential of multi-bandwidth spectrograms and feature fusion techniques in various speech and audio processing tasks, motivating further exploration in speaker verification applications.

Our research indicates that the use of dual-bandwidth spectrograms, as defined in this work, has not been previously applied to speaker verification tasks. While studies such as [10] and [11] have explored various combinations of spectral representations in speech processing, they have primarily focused on speech recognition or general audio processing tasks. The specific application of dual-bandwidth spectrograms to speaker verification appears to be novel.

This approach, combining narrowband and broadband spectrograms, extends the existing work on spectral analysis in speaker recognition systems. By leveraging complementary information from different spectral resolutions, our method aims to enhance the accuracy of speaker verification tasks. Our approach addresses limitations in previous works by providing a simple yet effective means of improving speaker verification performance using readily available spectral information.

3 Method

This study introduces a novel approach to SV analysis, leveraging a deep neural network model that processes combined narrowband and broadband spectrograms. Narrowband spectrograms are generated with longer analysis windows of 30ms. They capture finer harmonic details. Whereas broadband spectrograms are generated with shorter windows of 5ms. They better represent rapid temporal variations and spectral envelopes. The window lengths of 30ms for narrowband and 5ms for broadband spectrograms are defined in [15] and [16].

The baseline for this study is the Fast ResNet-34 architecture proposed by [5]. We have implemented it using the original code provided by the authors. This model is a streamlined version of the traditional ResNet-34 [17], characterized by a significant reduction in complexity. Fast ResNet-34 has been optimized to use only 1.4 million parameters, while the classical ResNet-34 comprises 63.5 million parameters. This efficient design greatly enhances computational efficiency and simplifies the training process, enabling us to explore our hypothesis without requiring high-end hardware resources.

The baseline for this study is the Fast ResNet-34 architecture proposed by [5]. We have implemented it using the original code provided by the authors. This model is a streamlined version of the traditional ResNet-34 [17], characterized by a significant reduction in complexity. Fast ResNet-34 has been optimized to use only 1.4 million parameters, while the classical ResNet-34 comprises 63.5 million parameters. This efficient design allows us to explore our dual-bandwidth hypothesis while maintaining reasonable computational requirements, even with the additional preprocessing step of generating two spectrograms per audio sample.

3.1 Adaptation for Dual-Bandwidth Spectrograms

To accommodate our dual-bandwidth spectrogram approach, we modified the baseline Fast ResNet-34 model. This adaptation primarily involved changing the initial convolutional layer to accept two input channels instead of one, corresponding to the narrowband and broadband spectrograms. Despite this modification, the number of output channels in this layer remained the same as in the original model. Consequently, the dimensions of the tensor after the first layer, and throughout the rest of the network, remain unchanged. This allows the subsequent layers to process the combined information from both spectrograms seamlessly, without requiring further architectural changes.

Our modified model processes and combines features from both narrowband and broadband spectrograms from the very first layer, enabling it to simultaneously capture fine spectral details and broader temporal patterns. Additionally, we incorporated Self-Attentive Pooling (SAP) [7] for temporal aggregation, utilizing attention mechanisms to emphasize critical segments in the SV process. Despite these adaptations, the modified model maintains the parameter efficiency of the original Fast ResNet-34, retaining a total of 1.4 million parameters. This efficient design allows us to investigate the potential of dual-bandwidth spectrogram input for improved SV performance without significantly increasing computational requirements.

3.2 Angular Prototypical Loss Function

In this study, we employ the Angular Prototypical loss function for metric learning, as introduced by [5]. This method involves using M audio clips from each speaker per training batch[1]. These M clips are divided into two sets: S clips for the support set and Q clips for the query set, with Q set to 1, following the approach in [5].

As described in [5], a class prototype, or centroid, is calculated from the support set to represent each speaker class using the formula:

$$c_j = \frac{1}{M-1} \sum_{m=1}^{M-1} \mathbf{x}_{j,m}$$

Here, $\mathbf{x}_{j,m}$ is the feature from the m-th audio of the j-th speaker, and $M-1$ represents the number of clips in the support set.

The similarity between each class centroid, c_j, and the query audio feature, $x_{j,M}$, is then measured using:

$$S_{j,k} = w \cdot \cos(x_{j,M}, c_k) + b \quad (1)$$

where w and b are learnable parameters.

Finally, the Angular Prototypical loss is calculated by comparing these similarity scores across all classes within a batch, as defined in [5]:

$$L_p = -\frac{1}{N} \sum_{j=1}^{N} \log \frac{e^{S_{j,j}}}{\sum_{k=1}^{N} e^{S_{j,k}}}$$

---

[1] Here, a "batch" refers to a subset of data for one training iteration, sometimes called a "mini-batch" in other studies.

Here, $S_{j,j}$ is the similarity between the centroid and the query vector of the same class, as defined in Eq. 1. The remaining terms $S_{j,k}$ represent comparisons with centroids of other classes.

3.3 Dual-Bandwidth Spectrogram Representation

We represent the input audio as a dual-bandwidth spectrogram, which is a data structure similar to a multi-channel image with dimensions of width, height, and number of channels, i.e., $X \in \mathbb{R}^{C \times H \times W}$.

The way we do so is the following, we combine narrowband and broadband spectrograms extracted from the same audio segment. We apply the Hamming window function [18] with a window width of 30 ms for the narrowband spectrogram and 5 ms for the broadband spectrogram. The step size between consecutive windows is set to 6.25 ms.

Considering the audio data sampling rate of 16,000 Hz, the window lengths correspond to 480 samples for the narrowband spectrogram, 80 samples for the broadband spectrogram, and a step size of 100 samples.

We then apply the mel-function to both narrowband and broadband spectrograms, resulting in 40 mel channels each. Hence, let $X_n \in \mathbb{R}^{M \times T}$ and $X_b \in \mathbb{R}^{M \times T}$ denote the narrowband and broadband mel-spectrograms, respectively, where M is the number of mel channels, and T is the variable length dimension determined by the audio segment duration. To construct the dual-bandwidth spectrogram, X_n and X_b are concatenated along a new channel dimension C, yielding $X \in \mathbb{R}^{C \times M \times T}$ with $C = 2$. Figure 1 illustrates this construction process for a 2-s audio segment, resulting in a dual-bandwidth spectrogram.

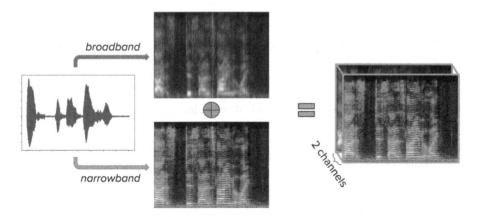

Fig. 1. Construction of the dual-bandwidth spectrogram by concatenating narrowband and broadband spectrograms along a new channel dimension.

4 Experiments

We conducted three distinct experiments to evaluate SV models using the Voxceleb2 [19] dataset's development subset. This dataset features 5,994 unique speakers and over one million audio clips from YouTube videos, varying in quality, presenting a gender imbalance, and predominantly in English. For each experiment, the models were initialized with random weights.

The training was conducted on a setup with a single V100 GPU with 64GB of memory and 8 CPU cores, allowing for batch sizes of 240 audio samples. We used the Adam optimizer [20] with an initial learning rate of 0.001, which was decreased by a factor of 0.95 every 10 epochs over 200 epochs. We observed a significant reduction in training loss early on, leading to stabilized training progress. The average training duration was 1.8 h per epoch for single spectrogram models and 2.21 h for dual-bandwidth spectrogram models. This sums to 360 h for single spectrogram models and 442 h for dual-bandwidth models.

We performed three experiments with the models detailed in [5] 1. We have explored broadband mel-spectrograms with an 80-sample window 2. Narrowband mel-spectrograms with a 480-sample window 3. Dual-bandwidth mel-spectrograms. The core of our study is the exploration of a dual-bandwidth mel-spectrogram approach that combines both 80 and 480 sample windows, highlighting our innovative proposal.

In the context of metric learning, accuracy is defined by calculating class centroids (prototypes) from the support set and classifying feature vectors inferred by the model from the query set based on the most similar centroid for each batch. At the end of the training, the model based on broadband spectrograms achieved an accuracy of 81.94%, the narrowband-based model obtained 85.19% accuracy, and the model using both spectrograms resulted in an accuracy of 88.22%.

5 Results and Discussion

This section presents the main findings of our study, including the evaluation protocol, performance metrics, and a comparison of the proposed models with the baseline. We also discuss the significance of the results and their implications for SV.

5.1 Evaluation Protocol

The trained models were evaluated using the test set and following the protocol proposed in [5], which utilizes a list of test cases developed by the authors of Voxceleb1 [21]. Each test case specifies two audio files and a label indicating if they belong to the same or different speakers. The test files are from the Voxceleb1 test set, comprising 40 classes and 4,874 files. The list encompasses 8 tests for each file, with an equal division between intra-class and inter-class comparisons.

For each audio file in the list of test cases, 10 two-second segments were extracted, evenly distributed (overlap may occur). A feature vector was extracted from each segment using the trained model, resulting in 10 vectors per audio, which were then normalized. The similarity between two audios was calculated using the average Euclidean distance among all 100 possible vector pairs.

The decision threshold L, set between 0 and 2, classifies audios as same or different classes based on similarity values being less than/greater than L.

5.2 Performance Metrics

We evaluated the models using two main performance metrics: Equal Error Rate (EER) and Minimum Detection Cost Function (MinDCF) [22].

Table 1. *Equal Error Rate* and *MinDCF* from the trained models and from the baseline [5]. To calculate the 95% confidence intervals for the EERs of our trained models, we considered the EER as a random variable following a Bernoulli distribution, approximated by a normal distribution for our large test set (37,720 instances) [23]. In [5], the authors repeated all experiments three times and reported the mean and standard deviation to account for random initialization.

Model	EER (%)	MinDCF
Pre-trained from [5]	2.22 ± 0.05	0,17
Broadband Mel-Spectrogram (window of 80 samples)	2.65 ± 0.16	0.20
Narrowband Mel-Spectrogram (window of 480 samples)	2.01 ± 0.14	0.15
Dual-Bandwidth Mel-Spectrogram (windows of 80 and 480 samples)	**1.64 ± 0.13**	0.14

The system is calibrated to equate the costs of both errors, setting the prior probability of the target speaker at 5% for defining MinDCF.

Figure 2 presents the probability distributions of similarity scores obtained for the audio test cases using the model trained with dual-bandwidth spectrograms. It demonstrates the overlap between same-speaker and different-speaker cases.

Figure 3 displays the variation of True Positive Rate (TPR), True Negative Rate (TNR), and Detection Cost Function (DCF) with respect to the decision threshold for the model trained with dual-bandwidth spectrograms, as observed in our experiments.

The EER and MinDCF values monitored during the training of each model are shown in Fig. 4.

5.3 Results and Comparison

Table 1 presents the EER and MinDCF values obtained for the trained models, along with the results from the baseline [5]. Among the single spectrogram models, the narrowband mel-spectrogram model (EER 2.01% ±0.14%) significantly outperformed the broadband mel-spectrogram model (EER 2.65%

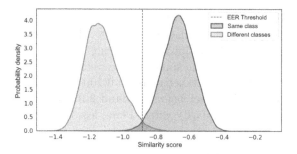

Fig. 2. Distribution of similarity scores obtained for the test cases from the list provided in [21] using the model trained with dual-bandwidth spectrograms.

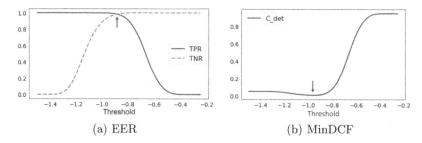

Fig. 3. Success rate curves for positive cases (TPR) and for negative cases (TNR) and Equal Error Rate (red arrow); (3(a)) and detection cost [22] with minimum MinDCF (red arrow) for the model trained with dual-bandwidth spectrograms. (Color figure online)

±0.16%). The minimal overlap in their confidence intervals suggests that this difference is statistically significant. This indicates that the narrowband representation likely carries more relevant features for SV, aligning with the baseline model [5] which used a 25 ms window (EER 2.22%). The superior performance of narrowband spectrograms suggests that fine-grained frequency resolution may be more critical for speaker verification than high temporal resolution, as it better captures certain speaker-specific characteristics in speech signals. The Dual-Bandwidth Mel-Spectrogram model achieved the lowest EER of 1.64% ±0.13%, representing a 26% relative improvement compared to the baseline. This further improvement demonstrates that while narrowband information is more relevant, the temporal resolution provided by broadband spectrograms contributes valuable complementary information. The statistically significant differences between these models, evidenced by the minimal overlap in confidence intervals, validate the potential of multi-resolution approaches in capturing the complex nature of speaker-specific information in speech signals. These confidence intervals were calculated considering the EER as a random variable following a Bernoulli distribution, approximated by a normal distribution given our large test set (37,720 instances). This statistical approach allows us to confidently state that

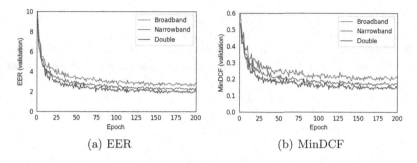

Fig. 4. EER (4(a)) and MinDCF (4(b)) evolution during the training.

the observed improvements, particularly with the Dual-Bandwidth model, are statistically significant and unlikely due to random chance.

5.4 Limitations and Future Work

While our study demonstrates the effectiveness of dual-bandwidth spectrograms in enhancing SV performance, there are areas that could benefit from further exploration to optimize and expand upon the current findings. Our investigation primarily focused on two specific bandwidth configurations-5ms and 30ms windows-which were selected based on their promising results. However, exploring a wider range of window sizes could reveal more effective combinations for various SV tasks. Additionally, integrating our dual-bandwidth approach with more recent SV architectures might yield further performance enhancements, as the current comparisons were primarily against a single baseline model from 2020.

Moreover, the potential of multi-bandwidth spectrograms, extending beyond two bandwidths, offers a promising direction for future research. This could involve investigating spectrograms with three or more bandwidths to assess potential performance gains and identify points of diminishing returns. Alternative techniques for combining spectrograms may also present opportunities for more optimal representations of multi-bandwidth information.

Finally, while our experiments were conducted on the widely used Voxceleb2 dataset, extending the evaluation to include more diverse datasets and real-world scenarios could help to establishing the robustness and broader applicability of the approach. By addressing these considerations, future research can continue to advance the understanding and application of multi-bandwidth spectrogram approaches in speaker verification.

6 Conclusion

This study investigated the impact of using dual-bandwidth spectrograms on the performance of Speaker Verification (SV) models. Our research validated the hypothesis that combining narrowband and broadband spectrograms provides complementary information, leading to improved SV accuracy.

The proposed dual-bandwidth spectrograms model achieved an Equal Error Rate (EER) of 1.64% ±0.13%, outperforming the reference model from [5] which had an EER of 2.22% ±0.05%. This represents a 26% relative improvement in performance. The statistical analysis, based on 95% confidence intervals, supports the robustness of these findings, indicating that the performance differences observed are unlikely to be due to random variation.

Our analysis revealed that narrowband spectrograms (EER 2.01% ±0.14%) carry more relevant features for SV compared to broadband spectrograms (EER 2.65% ±0.16%). This suggests that when using conventional spectrograms, opting for analysis windows around 30 ms, which produce narrowband spectrograms, is preferable. However, the superior performance of the dual-bandwidth approach demonstrates that while narrowband information is more relevant, the temporal resolution provided by broadband spectrograms contributes valuable complementary information.

The limitations of our study, including the exploration of only two bandwidth configurations and the use of a single dataset, have been acknowledged. These limitations, along with the promising results obtained, pave the way for future research. These include exploring additional bandwidth configurations, integrating our approach with more recent state-of-the-art models, and investigating the potential of multi-bandwidth spectrograms beyond dual-bandwidth.

In summary, this study demonstrates the effectiveness of dual-bandwidth spectrograms in enhancing SV performance. The statistically significant improvements achieved by our approach highlight its potential to advance the state-of-the-art in SV systems.

References

1. Reynolds, D.A.: An overview of automatic speaker recognition technology. In: 2002 IEEE International Conference on Acoustics, Speech, and Signal Processing, vol. 4, pp. IV–4072. IEEE (2002)
2. Jain, A.K., Flynn, P., Ross, A.A.: Handbook of Biometrics. Springer (2007)
3. Spazzapan, E.A., Cardoso, V.M., Fabron, E.M.G., Berti, L.C., Brasolotto, A.G., Marino, V.C.D.C.: Acoustic characteristics of healthy voices of adults: from young to middle age. In: CoDAS, vol. 30. SciELO Brasil (2018)
4. Geoffrey, S.M., Ewald, E., Ramos, D., González-Rodríguez, J., Lozano-Díez, A.: Statistical models in forensic voice comparison. In: Handbook of Forensic Statistics, pp. 451–497. Chapman and Hall/CRC (2020)
5. Chung, J.S., et al.: In defence of metric learning for speaker recognition. In: Interspeech (2020)
6. Yamamoto, H., Lee, K.A., Okabe, K., Koshinaka, T.: Speaker augmentation and bandwidth extension for deep speaker embedding. In: Interspeech, pp. 406–410 (2019)
7. Cai, W., Chen, J., Li, M.: Exploring the encoding layer and loss function in end-to-end speaker and language recognition system (2018)

8. Kye, S.M., Kwon, Y., Chung, J.S.: Cross attentive pooling for speaker verification [c]‖ 2021 IEEE Spoken Language Technology workshop (SLT). 19–22 January 2021, Shenzhen, China (2021)
9. Mun, S.H., Jung, J.-W., Han, M.H., Kim, N.S.: Frequency and multi-scale selective kernel attention for speaker verification. In: IEEE Spoken Language Technology Workshop (SLT). IEEE 2023, pp. 548–554 (2022)
10. Arias-Vergara, T., Klumpp, P., Vasquez-Correa, J.C., Nöth, E., Orozco-Arroyave, J.R., Schuster, M.: Multi-channel spectrograms for speech processing applications using deep learning methods. Pattern Anal. Appl. **24**, 423–431 (2021)
11. Cheung, S., Lim, J.: Combined multiresolution (wide-band/narrow-band) spectrogram. IEEE Trans. Signal Process. **40**(4), 975–977 (1992)
12. Annabi-Elkadri, N., Hamouda, A.: Automatic silence/sonorant/non-sonorant detection based on multi-resolution spectral analysis and ANOVA method. In: International Workshop on Future Communication and Networking. Szczecin, Poland (2011)
13. Jahangir, R., et al.: Text-independent speaker identification through feature fusion and deep neural network. IEEE Access **8**, 32 187-32 202 (2020)
14. Panayotov, V., Chen, G., Povey, D., Khudanpur, S.: Librispeech: an ASR corpus based on public domain audio books. In: 2015 IEEE International Conference on Acoustics, Speech and Signal Processing (ICASSP), pp. 5206–5210 (2015)
15. Boersma, P., Weenink, D.: PRAAT: doing phonetics by computer (version 6.1.48) (2021). http://www.praat.org
16. Styler, W.: Using PRAAT for linguistic research. University of Colorado at Boulder Phonetics Lab (2013)
17. He, K., Zhang, X., Ren, S., Sun, J.: Deep residual learning for image recognition (2015)
18. Smith, J.O.: Spectral audio signal processing. W3K (2011)
19. Chung, J.S., Nagrani, A., Zisserman, A.: Voxceleb2: deep speaker recognition. In: Interspeech 2018 (2018). http://dx.doi.org/10.21437/Interspeech.2018-1929
20. Kingma, D., Ba, J.: Adam: a method for stochastic optimization. In: International Conference on Learning Representations (2014)
21. Nagrani, A., Chung, J.S., Zisserman, A.: Voxceleb: a large-scale speaker identification dataset. Interspeech 2017 (2017). http://dx.doi.org/10.21437/Interspeech.2017-950
22. Omid.sadjadi@nist.gov. Nist 2018 speaker recognition evaluation (2018). https://www.nist.gov/itl/iad/mig/nist-2018-speaker-recognition-evaluation
23. Agresti, A., Coull, B.A.: Approximate is better than "exact" for interval estimation of binomial proportions. Am. Stat. **52**(2), 119–126 (1998). https://doi.org/10.1080/00031305.1998.10480550

Dynamicity Analysis in the Selection of Classifier Ensembles Parameters

Jesaías Carvalho Pereira Silva[1]($\boxtimes$), Anne Magaly de Paula Canuto[1], and Araken de Medeiros Santos[2]

[1] Federal University of Rio Grande do Norte, Natal RN, 59.078-970, Brazil
jesayassilva@gmail.com, anne@dimap.ufrn.br
[2] Federal University of Rural do Semi-Árido, Angicos, RN 59.515-000, Brazil
araken@ufersa.edu.br

Abstract. Over the years, significant progress has been made in the realm of classifier ensembles research. Several methods to enhance their efficiency have been proposed, applicable to both homogeneous and heterogeneous ensemble structures. A key challenge in employing classifier ensembles lies in determining their structure (hyper-parameters). Basically, the ensemble structure selection can be done in two different ways, static and dynamic selection. Unlike static selection, dynamic selection defines the ensemble structure for each testing instance. Different dynamic selection methods have been proposed in the literature, mainly for ensemble members and features, but very little effort has been done to propose dynamic selection methods for combination methods. Therefore, it is important to evaluate the impact of a dynamic selection of combination methods or both (methods and members) in the creation of robust classifier ensembles. In this paper, an exploratory analysis of dynamic selection of the main ensemble structure parameters will be performed. In order to do this, three different scenarios will be assessed: Full Static ensemble, Partial Dynamic ensemble and Full Dynamic ensemble. Finally, an empirical analysis of these scenarios will be carried out. Our findings show that the use of a full dynamic selection provides more robust classifier ensembles, in most of the analyzed cases.

Keywords: Classifier ensembles · Dynamic ensemble selection · Individual classifiers · Combination methods

1 Introduction

In a classifier ensemble, the individual classifiers work together to create a more robust and accurate system for classifying patterns. These individual classifiers, known as ensemble members, operate in parallel, each one receiving the same input pattern and independently producing its own output. A combination method receives the members outputs and provides the global output of the system [1].

Classifier ensemble is a two-step classification structure in which the main parameters are the individual classifiers (first step) and the combination method (second step). In these systems, one important aspect is the definition of the ensemble structure.

Several studies have proposed different ways to define the ensemble structure such as: Optimization techniques, meta-learning, among others [2,3]. However, most of these studies are related to the selection of individual classifiers. We can find some studies that investigate efficient methods for combining classifiers in ensemble systems, such as in [4–7].

The selection of ensemble parameters (classifiers and/or combination methods) can be static or dynamic. In static selection, the ensemble structure is defined during the training phase and remains fixed throughout. In contrast, dynamic selection adapts the ensemble structure for each test instance, often enhancing predictive performance [4,8].

Dynamic selection can be applied to both classifiers and combination methods, with most studies focusing on the dynamic selection of classifiers [6,7,9]. This is often based on techniques such as Region of Competence, Hyper-boxes, or Meta-learning. However, very little has been done to define a fully dynamic selection, making the selection of the ensemble parameters an automatic process.

To advance the design of efficient classifier ensembles, this paper conducts an exploratory analysis of integrating dynamic selection into the main ensemble parameters. In this investigation, the dynamic selection will be applied in one ensemble parameter (classifier or combination) and in both parameters. The main aim of this analysis is to assess the impact of the dynamic selection in the two most important parameters of a classifier ensemble. In other words, to analyze whether the dynamic selection leads to more efficient ensembles when applied to individual classifiers, combination methods or in both parameters at the same time.

In this analysis, three different scenarios will be evaluated. In the first scenario, the individual classifiers will be selected dynamically while the combination method will be selected statically. For this scenario, three well-known DES (dynamic ensemble system) methods are used, which are: KNORA-Eliminate (KNORA-E [4]) FH-DES [10] and META-DES [5]. In the second scenario, the combination method is selected dynamically while the individual classifiers will be selected statically. For this scenario, a dynamic fusion method is presented.

Finally, in the last scenario, both individual classifiers and combination method will be selected dynamically. As baseline, a full static ensemble structure will also be investigated in order to assess the impact of the dynamic selection in the performance of the classifier ensembles. All the ensemble structures will be evaluated using 20 classification datasets.

This paper is divided into 6 sections and its organization is defined as followed. Section 2 describes the theoretical concepts and related work of this paper, while Sect. 3 describes in more detail the ensemble structures to be used in the exploratory analysis of this paper. Section 4 presents the experimental methodology of the empirical analysis, while its results are presented in Sect. 5. Finally, Sect. 6 describes the final remarks of this paper.

2 Theoretical Concepts and Related Work

2.1 State of the Art

There are several studies that investigate the dynamic selection of ensemble structure, mainly for ensemble members [6,7] and features [11] and both of them [9,12].

Regarding ensemble members, in [6], for instance, a new method for dynamic ensemble member selection is presented. In this method, the confidence of the base classifiers during the classification and its general credibility is used as selection criterion.

Another interesting way is to use region of competence as selection criterion, making it possible to improve the combination of classifiers, in which the most competent ones in a certain region are selected. The use of region of competence as selection criterion helps to maximize results by focusing only on the most competent classifiers, and examples can be found in KNORA-E [4] and META-DES [5].

In terms of dynamic feature selection, in [11], a dynamic feature selection approach was proposed. The main aim of this approach is to select a different subset of features for one instance or a group of instances. The main goal of this approach is to explore the full potential of all instances in a classification problem.

In [9], an initial study on how to combine these two dynamic selection techniques was performed. According to the authors, an improvement in performance was detected with the use of this integrated dynamic selection technique. Already in [13], the authors presented an initial method of classifier fusion using K-nearest neighbors (KNN). Although the results are promising, there is no general comparison with static ensembles.

Although there are several studies to propose dynamic selection of ensemble members and feature selection, very little has been done in order to propose efficient dynamic selection of combination methods. This paper tries to bridge this gap and it proposes a dynamic selection method based on region of competence.

2.2 Classifier Ensembles

It is well-known that there is not a single classifier which can be considered optimal for all problem domains [1]. Therefore, it is difficult to select a good single classifier which provides the best performance in practical pattern classification tasks [14].

In this context, classifier ensembles have emerged as an efficient classification structure since it combines the advantages and overcomes the limitations of the individual classifiers. Thus, studies have shown that classifier ensemble provides better generalization and performance ability, when compared to the individual classifiers [14,15].

In a classifier ensemble, an input pattern is presented to all individual classifiers [16,17], and a combination method combines their outputs to produce the overall output of the system [1]. The Machine Learning literature has ensured that diversity plays an important role in the design of ensembles, contributing to their accuracy and generalization [1].

One important issue regarding the design of classifier ensembles involves the appropriate selection of its structure (individual classifiers and combination methods) [18]. As previously mentioned, there are basically two main selection approaches, static and dynamic. In this paper, we will focus on the dynamic approach. The next subsection will describe some existing dynamic selection methods that will be used in this paper.

2.3 Dynamic Ensemble Member Selection

The Dynamic Ensemble Selection (DES) methods perform the dynamic selection of a subset of classifiers to classify each test instance. The selection of the classifier subset is done through the use of a selection procedure and each DES method has its own procedure. There are several DES methods proposed in the literature. In this paper, we will use two well-known DES methods, KNORA-E and META-DES.

KNORA-E. Knora [4] is a well-known DES method and it seeks to find the best subset of classifiers for a given test instance. It applies a k-Nearest Neighbors method. The neighbors of a testing instance are selected from the validation set and the competence of each classifier is calculated. Based on a certain selection criterion, the classifier subset is selected.

KNORA-E is a Knora-based method, and the selection criterion is to select a set of classifiers formed only by the classifiers that correctly classify all k neighbors of a testing instance. In the case where no classifier can correctly classify all k neighbors, the k value is decremented by one and this is done until at least one classifier can be selected [4].

META-DES. The META-DES [5] is a DES method that uses the idea of selection using meta-learning. In this method, a meta-problem is created to determine whether a classifier is competent for a given test instance. According to [12], the META-DES method uses five criteria for extracting meta-features in order to establish the new region of a meta-problem.

After that, a meta-classifier is trained, based on the defined meta-features. This meta-classifier is then used to identify whether a classifier is competent or not to classify a testing instance. Classifiers that are labeled as competent will be selected to compose the ensemble to classify the test instance.

FH-DES. The FH-DES is also a DES method, but based on fuzzy hyperboxes [19] to solve the local sensitivity problem using KNN. Hyperboxes represent a group of samples using maximum and minimum corners. They are formed based on regions that classifiers work well or areas of competence, but can also be applied to regions that present poor classifications or areas of incompetence.

In the latter case, the classifiers whose hyperboxes have a lower degree of relevance will be further away from the query sample, therefore, they will be more competent to classify the query sample [10]. The tool can be applied in three ways, using a sum rule based on hyperbox weights, performing only the selection of competent classifiers, and hybrid, in which it uses the weights and selection of classifiers.

3 Proposal

3.1 Dynamic Selection Scenarios

In order to carry out the exploratory analysis of the dynamic selection in classifier ensemble, three scenarios are defined, which are described as follows.

1. Full static ensemble selection (FSES): In the first scenario, we will have full static ensembles. In other words, all ensemble parameters are selected statically.
2. Partially dynamic ensemble selection (PDES): In the second scenario, we will have partial dynamic ensembles, having the dynamic selection in only one parameter: ensemble members or fusion, but not both of them. Therefore, this leads to two sub-scenarios:
   (a) Partial - Dynamic Member Selection (P-DMS): In this case, only the ensemble members (individual classifiers) will be dynamically selected.
   (b) Partial - Dynamic Fusion Selection (P-DFS): In this cases, only the combination method will be dynamically selected. The Dynamic Fusion Selection method is presented in Sect. 3.2.
3. Full dynamic ensemble selection (FDES): in this scenario, full dynamic ensembles, both parameters (ensemble member and combination methods) will be chosen dynamically for each new instance of teste, as well as which fusion.

These three scenarios were selected because they gradually increase the dynamicity of the selection of the ensemble parameters. Therefore, we aim at evaluating the impact of the dynamic selection in the design of robust classifier ensembles.

3.2 The Dynamic Fusion Selection Method

In this paper, in order to achieve dynamicity in the selection of combination methods, we present the DFS (Dynamic Fusion Selection) method, which is an algorithm that dynamically selects the combination methods from a set of methods.

In other words, for each test instance, the most appropriate combination methods is selected. The selection is carried out in the testing phase. DFS calculates the competence of each combination method with respect to the presented test instance. Algorithm 1 presents the main steps of the DFS processing.

As it can be observed, DFS has two main parameters: the number of neighbors and the set of combination methods, which comes from the algorithm input. The number of neighbors determines the size of the neighborhood used to calculate the selection criterion. The set of combination methods define the methods that will be used in the dynamic selection method.

Algorithm 1 : Algorithm for Dynamic Fusion Selection

1: **procedure** DYNAMIC FUSION($ti, V1, P, C$) ▷ Testing instance (ti), Validation 1 set (V1), Pool of classifiers (P), Combination methods (C)
2:   **for** $i = k$ **until** $i = size(V1/2)$ **do**
3:     $N \leftarrow NearestNeighbors(i, ti, V1)$ ▷ Find the i neighbors of ti.
4:     **for** $j = 1$ **until** $j = size(C)$ **do**
5:       $Acc(C_j) \leftarrow Accuracy(P, C_j, N)$ ▷ Calculate the accuracy of C_j in N.
6:     **for** $j = 1$ **until** $j = size(C)$ **do**
7:       **if** $Acc(C_j) = Max(Acc(C_j), j = 1, 2, ..., size(C))$ **then**
8:         **if** $CountMax(Acc(C_j), j = 1, 2, ..., size(C)) = 1$ **then**
9:           **return** C_j ▷ C_j has the best accuracy in N.
10:      **else**
11:        $Remove(C_j)$ ▷ Remove C_j from C.
12: **return** C ▷ Returns the combination methods that still a tie in precision and have not been removed.

Competence is calculated based on the accuracy of the combination methods using the neighbors of the test instance (line 5). If there is a tie in the local accuracy, the number of neighbors is increased by 1 ($k = k + 1$) (line 2) until one combination method is selected. There is still a tie, the combination methods with accuracy equal to the maximum accuracy are selected (line 12).

Finally, DFS can be applied to a pool of classifiers selected statically or dynamically. In this paper, this method will be used in two dynamic selection scenarios, partially dynamic selection and the full dynamic selection one.

4 The Experimental Methodology

In this section, the main aspects of the empirical analysis will be described, mainly the used datasets, as well as its methods and materials.

4.1 Datasets

This paper uses datasets extracted from the UCI Machine Learning repository. Table 1 presents some characteristics of these datasets, including the number of instances (Inst), the number of attributes (Att) and the number of classes (Class).

Each dataset is divided into training, Validation1, Validation2 and Testing sets, in a proportion of 50%, 16.7%, 16.7%, and 16.6%, respectively. The training set is used to train the pool of classifiers (ensemble members). The testing set is used to assess the performance of the classifier ensembles. The Validation2 set is used to train the trainable combination methods (Neural Networks and Naive Bayes) while the Validation1 set is used to obtain the selection criteria of the presented dynamic member selection methods.

Table 1. Description of the used datasets.

Dataset	Name	Inst	Att	Class
D1	Cardiac insufficiency	368	53	2
D2	Car	1728	6	4
D3	Seismic-bumps	2584	18	2
D4	Zoo	101	16	7
D5	Ionosphere	351	34	2
D6	Prognostic	198	33	2
D7	Wine	178	13	3
D8	Dermatology	366	34	6
D9	Heart	303	13	2
D10	Bone marrow	187	36	2
D11	Algerian Forest Fires	244	13	2
D12	Congres Voting Records	435	16	2
D13	Maternal Health Risk	1014	6	3
D14	Risk Factors Cervical Cancer	855	28	2
D15	Phishing Website	2456	30	2
D16	Blood	748	4	2
D17	Monk2	432	6	2
D18	Banknote	1372	4	2
D19	Mammographicq	850	5	2
D20	Banana	1000	2	2

This division is performed 30 times and the presented results of each ensemble configuration represent the average values over these 30 values.

4.2 Methods and Materials

In this paper, all classifier ensembles used decision trees as individual classifiers, generated through the Bagging method. In all analyzed scenarios, 6 different pool sizes are used, which are: 5, 10, 15, 20, 25, and 30 individual classifiers. The remaining parameter values were defined through extensive grid search experimental evaluation.

For the FSES scenario, the classifier ensembles will be evaluated using 12 different combination methods, which are: Majority Vote, Sum, Max, Min, Geometric Mean, Naive Bayes, Weigthed Sum, Weigthed Vote, Edge and three Multilayer Perceptron (MLP) versions: Hard, Soft, and Soft-Class. The three Neural Networks (NN) versions differ in the input information received by the ensemble members. In the Hard version, the ensemble member provides only the winner class for the testing instance. In other words, this MLP version is trained and tested using only the winner class of each ensemble member.

In the other two MLP versions, the prediction probability for each class is used. In this sense, the prediction probability for each class is provided for both MLP versions. Additionally, the Weighted sum and Weighted vote methods use weights in their functioning. The used weight is 1/(distance-of-classes), and it is applied to the voting procedure in the Weighted Vote as well as to the outputs of the classifiers in the Weighted sum method.

For the first case of the PDES scenario (P-DMS), three well-known methods are used Knora-E [4], META-DES [5] and FH-DES [19]. In the second case of the PDES scenario (P-DFS), the dynamic fusion method presented in Sect. 4.2 is used to dynamically select the most suitable combination method for each testing instance. For the FDES scenario, all three methods used in the F-DMS will be combined with the method for the P-DFS case, leading to 3 PDES variations (Knora-E with dynamic fusion selection, META-DES with dynamic fusion selection, and FH-DES with dynamic fusion selection in fusion).

It is important to highlight that the DFS method, and some of the DMS methods (KNORA-E and META-DES) use the idea of region of competence. In this sense, the same number of neighbors are used for all cases, in the selection of the combination method (DFS) and in the dynamic member selection (KNORA-E and META-DES), being defined in each iteration by means of a grid search between the values of 3, 7 and 11, using KNN in the dataset Validation 1. For the FH-DES, there is no idea of a region of competence, but rather the use of hyper-boxes, which were created based on the areas of incompetence as presented by the authors in [10].

The results of all analyzed methods will be evaluated using the Friedman statistical test [20]. The Friedman test is used to be able to state the hypothesis that the k-related observations derive from the same population (similar performance) or not (superiority in performance). In this test, the significance level used was set to 0.05.

Hence, if the p-value is less than the established value, the null hypothesis is rejected, with a confidence level greater than 95%. In cases where a statistically significant difference is detected, the Nemenyi post-hoc test is applied [21]. In order to present the obtained results by the post-hoc test, the critical difference diagram (CD) [21] is used. This diagram was selected in order to have a visual illustration of the statistical test, making it easier to interpret the obtained results.

5 The Obtained Results

This section presents the results obtained by the empirical analysis, in terms of accuracy levels. First, all Full Static configurations is evaluated. Then we evaluate separately the two cases of the PDES scenario, along with the best full static configuration. Finally the FDES scenario is evaluated, along with the best full static configuration, the best dynamic member fusion configuration, and the best dynamic fusion configuration. In order to define the best configuration, the critical difference diagram is used. If there is no statistical difference between the best techniques, then the one with the highest average is selected.

5.1 Full Static Ensembles

Tables 2, 3, 4, 5 and 6 present the accuracy results of all analyzed methods. As mentioned previously, 6 configurations were made for pool sizes; each configuration was executed 30 times. Therefore, the values in Tables 2, 3, 4, 5 and 6 represent the average of all 180 results. Additionally, the last row of all tables represent the overall accuracies for all 20 datasets. Finally, the numbers in bold represent the highest accuracy for each data set.

For the FSES scenario (Table 2), it can be seen that all methods (except Edge Fusion) provided the highest accuracy levels in at least three datasets. In a general perspective, it can be observed that the ensembles combined by Sum (second column) presented the highest overall accuracy (89.09%) and highest accuracy in 8 out of 20 datasets. They were followed by all three NNs (6 data sets) and Majority vote (5 data sets).

In order to evaluate the obtained results from a statistical point of view, the Friedman test [20] was applied to verify if there are statistical differences among all ensemble classifiers. The Friedman test was applied to all 12 FSES configurations. Therefore, the statistical test detected statistical differences among all analyzed methods, with a p-value < 0.05. In this sense, the post-hoc test was applied, and the results are presented in the Critical Difference Diagram [21], depicted in Fig. 1.

As it can be seen in this figure, the CD diagram detected no statistically significant difference in the accuracy of Sum, MLP Soft, and MLP Soft Class. For the remaining configuration, they provided superior performance, detected by the statistical test. Since no method stood out, SUM-combined ensemble is selected the remaining tests since it achieved the best overall highest accuracy level.

Table 2. Score of fusion methods in full static ensemble.

Dataset	FSES Majority Vote	FSES SUM	FSES MAX	FSES MIN	FSES Geometric Mean	FSES Weighted Sum	FSES Weighted Vote	FSES MLP HARD	FSES MLP SOFT	FSES MLP SOFT CLASS	FSES Edge	FSES Naive Bayes
D1	93.42	93.42	**94.33**	**94.33**	**94.33**	93.39	93.39	93.20	93.83	93.83	93.42	89.61
D2	96.46	**96.49**	96.39	96.05	96.05	**96.49**	96.48	96.26	96.45	96.45	96.25	87.10
D3	89.75	89.74	90.93	90.93	90.94	89.64	89.73	**93.55**	93.01	93.01	89.75	65.05
D4	**97.03**	**97.03**	96.93	95.26	95.26	51.21	51.21	68.79	96.18	96.18	95.69	01.54
D5	**88.56**	**88.56**	87.72	87.72	87.72	88.53	88.53	87.94	88.53	88.53	**88.56**	68.29
D6	66.43	66.43	60.74	60.74	60.74	66.60	66.60	75.03	**75.94**	**75.94**	66.43	57.90
D7	90.86	90.86	89.04	87.66	87.66	90.84	90.84	86.67	**91.25**	**91.25**	90.61	10.54
D8	**97.94**	**97.94**	97.46	97.02	97.02	94.64	94.64	87.42	96.31	96.31	**97.94**	60.95
D9	75.22	75.22	73.68	73.68	73.68	**75.33**	**75.33**	74.54	74.06	74.06	75.22	60.91
D10	92.28	92.28	**92.63**	**92.63**	**92.63**	92.11	92.11	91.99	92.54	92.54	92.28	82.96
D11	96.42	96.42	96.03	96.03	96.03	96.43	96.43	**96.54**	96.38	96.38	96.42	83.38
D12	93.54	93.54	92.86	92.86	92.86	93.53	93.53	93.48	**93.63**	**93.63**	93.54	75.25
D13	74.81	**74.86**	74.75	74.68	74.69	74.74	74.76	70.32	73.83	73.83	74.85	56.61
D14	89.85	89.86	90.94	90.94	90.94	89.57	89.70	**93.25**	88.10	88.10	89.85	76.06
D15	94.66	**94.79**	94.72	94.72	94.73	94.77	94.66	94.67	93.65	93.65	94.66	76.06
D16	72.37	72.39	72.78	72.78	72.85	71.87	72.31	**74.58**	68.70	68.70	72.37	61.05
D17	99.79	99.79	99.79	99.79	99.79	99.79	99.79	99.79	99.79	99.79	99.79	83.70
D18	97.90	97.90	97.88	97.88	97.88	97.83	97.83	97.81	**97.95**	**97.95**	97.90	76.75
D19	77.75	77.82	77.67	77.67	77.66	77.81	77.73	77.93	**77.99**	**77.99**	77.75	61.26
D20	96.39	96.39	96.31	96.31	96.31	96.39	96.39	96.39	93.21	93.21	96.39	74.32
Acc Ave	89.07	**89.09**	88.68	88.48	88.49	86.58	86.60	87.21	89.07	89.07	88.98	65.46

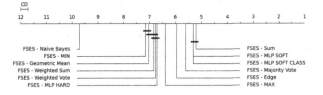

Fig. 1. Critical Difference Diagram for Full Static Ensemble

5.2 The PDES Scenario

In this subsection, the results for the PDES scenario are presented. For a comparative analysis, in each table, the accuracy of the best FSES configuration (FSES-SUM) is also presented. As mentioned previously, three well dynamic member selection are used. In this paper, Table 3, 4 and 5 represent the values of KNORA-E, FH-DES and META-DES, respectively. Thus, for each method, the result of the P-DFS case is also presented, leading to a total of 14 analyzed methods.

KNORA-E. For KNORA-E (Table 3), it can be seen that the P-DFS method presented the best overall accuracy level (89.88%) and obtained the best results in 4 out of 20 datasets. However, the KNORA-E method combined by the Majority Vote method achieved the best results in 9 datasets, presenting the same average accuracy as KNORA-E combined by Sum (89.85%) with 8 best results.

Weight fusion methods showed the worst results. The Friedman test was then applied, which identified statistical differences with a p-value < 0.05.

Table 3. Score of fusion methods for partial dynamic ensembles in KNORA-E

Dataset	KNORA-E Majority Vote	KNORA-E SUM	KNORA-E MAX	KNORA-E MIN	KNORA-E Geometric Mean	KNORA-E Weighted Sum	KNORA-E Weighted Vote	KNORA-E MLP HARD	KNORA-E MLP SOFT	KNORA-E MLP SOFT CLASS	KNORA-E Edge	Naive Bayes	P-DFS	FSES SUM
D1	94.77	94.77	95.16	95.16	95.16	87.77	87.77	94.48	94.67	94.85	94.77	92.61	94.25	93.42
D2	96.94	96.94	93.48	91.63	91.63	83.74	83.74	93.69	95.50	96.59	96.00	93.90	96.50	96.49
D3	91.95	91.94	92.97	92.97	92.97	91.43	91.44	92.33	91.39	92.03	91.95	86.95	93.18	89.74
D4	93.95	93.95	93.24	89.84	89.84	75.20	75.20	90.07	83.79	91.18	91.11	74.97	96.18	97.03
D5	90.80	90.80	80.89	80.89	80.89	77.45	77.45	88.64	88.87	88.97	90.80	85.31	88.74	88.56
D6	69.66	69.66	49.19	49.19	49.19	67.10	67.10	70.30	70.37	69.95	69.66	74.19	71.26	66.43
D7	93.75	93.75	80.13	74.18	74.18	63.93	63.93	85.90	91.42	93.49	92.41	56.76	90.84	90.86
D8	95.94	95.94	90.29	83.95	83.95	78.56	78.56	89.38	93.59	95.72	93.95	77.83	97.66	97.94
D9	77.72	77.72	65.03	65.03	65.03	68.58	68.58	74.92	73.38	72.93	77.72	74.86	75.91	75.22
D10	91.59	91.59	89.37	89.37	89.37	91.16	91.16	90.63	88.19	89.68	91.59	90.23	93.26	92.28
D11	97.38	97.38	96.10	96.10	96.10	97.10	97.10	96.18	96.31	96.47	97.38	96.57	96.26	96.42
D12	94.65	94.65	91.80	91.80	91.80	92.37	92.37	93.45	93.53	93.46	94.65	93.00	93.38	93.54
D13	77.65	77.55	78.25	78.17	78.17	63.50	63.49	75.25	75.01	75.97	77.71	72.57	74.80	74.86
D14	91.18	91.18	93.14	93.14	93.14	90.81	90.83	92.64	91.95	92.32	91.18	90.70	92.69	89.86
D15	95.59	95.58	92.61	92.61	92.61	92.70	92.70	95.11	95.00	95.17	95.59	91.09	94.83	94.79
D16	70.31	70.29	72.30	72.30	72.30	70.87	70.84	72.31	71.87	71.19	70.31	71.50	75.23	72.39
D17	99.94	99.94	98.26	98.26	98.26	99.18	99.18	99.48	99.46	99.48	99.94	98.86	99.79	99.79
D18	98.96	98.96	95.42	95.42	95.42	92.88	92.88	98.35	99.08	98.84	98.96	96.04	98.17	97.90
D19	77.15	77.25	75.46	75.46	75.46	75.61	75.62	77.01	76.39	75.21	77.15	74.81	78.23	77.82
D20	97.07	97.07	93.77	93.77	93.77	92.78	92.78	96.90	96.66	96.66	97.07	95.13	96.53	96.39
AccAve	89.85	89.85	85.84	84.96	84.96	82.64	82.64	88.35	88.32	89.01	89.50	84.39	89.88	89.09

When applying the post-hoc test, the CD Diagram, in Fig. 2, results showed that KNORA-E Sum and KNORA-E Vote provided the best performance. However, these methods provide similar performance from a statistical point of view. For the other methods, the statistical test detected a superiority, in terms of accuracy, of these methods.

The P-DFS and KNORA-E Edge methods also showed good results, being superior to the remaining methods, from a statistical point of view. As there was not just one method that stood out, the KNORA-E Vote method is selected for the best KNORA-E configuration. When comparing the P-DFS method and the best KNORA-E method (P-DMS) method, although the P-DFS method provided the best overall accuracy, the statistical test showed that the use of dynamic selection in the ensemble members provided more robust ensembles.

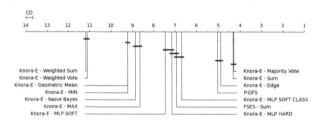

Fig. 2. Critical Difference Diagram for Partial Dynamic Ensembles in KNORA-E.

FH-DES. For FH-DES, in Table 4, it can be seen that FH-DES - Sum presented the best overall result (90.51%) and obtained the best results in 4 of the 20

Table 4. Score of fusion methods for partial dynamic ensembles in FH-DES.

Dataset	FH-DES Majority Vote	FH-DES SUM	FH-DES MAX	FH-DES MIN	FH-DES Geometric Mean	FH-DES Weighted Sum	FH-DES Weighted Vote	FH-DES MLP HARD	FH-DES MLP SOFT	FH-DES MLP SOFT CLASS	FH-DES Edge	Naive Bayes	P-DFS	FSES SUM
D1	**96.97**	**96.97**	87.97	81.27	81.27	78.09	78.09	87.04	93.36	94.84	90.90	86.16	94.25	93.42
D2	**96.81**	**96.81**	87.39	84.09	84.09	96.80	96.79	93.26	96.68	96.69	95.19	90.28	96.50	96.49
D3	92.52	92.42	93.42	93.42	93.42	92.31	92.35	**93.50**	92.33	93.22	92.52	82.91	93.18	89.74
D4	96.84	96.84	87.60	81.13	81.13	77.88	77.88	86.02	93.35	95.43	91.38	86.39	96.18	**97.03**
D5	**91.27**	**91.27**	76.11	76.11	76.11	91.18	91.18	90.01	90.81	91.25	**91.27**	83.53	88.74	88.56
D6	72.02	72.02	34.43	34.43	34.43	72.95	72.95	**77.86**	71.95	76.55	72.02	68.15	71.26	66.43
D7	93.41	93.41	71.61	64.96	64.96	93.30	93.30	89.73	92.24	**93.79**	92.05	77.43	90.84	90.86
D8	95.88	95.88	86.85	79.10	79.10	95.00	95.09	88.15	95.68	95.49	92.87	85.47	97.66	**97.94**
D9	77.64	77.64	58.42	58.42	58.42	**77.83**	77.83	76.76	75.08	76.61	77.64	68.00	75.91	75.22
D10	92.81	92.81	87.67	87.67	87.67	92.62	92.62	92.74	92.76	92.96	92.81	89.71	**93.26**	92.28
D11	96.88	96.88	93.94	93.94	93.94	96.86	96.86	96.86	96.88	**96.89**	96.88	95.13	96.26	96.42
D12	94.52	94.52	90.97	90.97	90.97	**94.54**	**94.54**	94.24	94.02	93.11	94.52	92.61	93.38	93.54
D13	76.65	**77.38**	68.51	64.75	64.86	77.31	76.51	74.08	75.90	77.23	76.39	65.61	74.80	74.86
D14	91.71	91.86	93.12	93.12	93.12	90.85	90.91	**93.23**	88.79	92.06	91.75	90.75	92.69	89.86
D15	95.44	95.48	89.11	89.11	89.11	**95.50**	95.46	95.38	94.84	95.24	95.44	90.11	94.83	94.79
D16	75.27	74.73	75.66	75.66	75.69	74.43	74.84	**75.94**	71.16	73.92	75.27	71.42	75.23	72.39
D17	99.62	99.62	97.34	97.34	97.34	99.68	99.68	99.77	**99.88**	99.83	99.62	98.92	99.79	99.79
D18	98.42	98.42	92.56	92.56	92.56	98.46	98.46	98.46	**98.58**	98.48	98.42	95.57	98.17	97.90
D19	78.16	78.12	73.05	73.05	73.04	78.12	78.04	78.19	76.26	77.62	78.16	73.90	**78.23**	77.82
D20	97.05	97.05	91.63	91.63	91.63	**97.09**	**97.09**	97.05	97.01	96.73	97.05	94.20	96.53	96.39
AccAve	90,49	**90,51**	81,87	80,14	80,14	88,54	88,52	88,91	89,38	90,40	89,61	84,31	89,88	89,09

datasets. Followed by the FH-DES - Majority Vote method with 90.49%, which obtained the best results in 3 datasets.

The FH-DES - MLP Hard and Weighted Sum methods also presented the best results in 4 datasets; however, the results were slightly lower than the best FH-DES methods. Fusion by Min and Geometric Mean provided the worst results. In order to check whether the accuracy levels derive from the same population, the Friedman test identified statistical differences with a p-value < 0.05.

In the post-hoc test, the Critical Difference Diagram (Fig. 3) in accuracy of FH-DES - Sum, Majority Vote, and MLP Soft Class did not present statistically significant differences. The FH-DES - Min and Geometric Mean presented the worst results, detected by the statistical test. Once again, as there was not just one FH-DES method that stood out in the statistical test, the FH-DES - Sum method was selected to be the best FH-DES method.

When comparing the P-DFS method and the best FH-DES method (P-DMS) method, we can observe a superiority of the P-DMS case, showing that the use of dynamic selection in the ensemble members provided more robust ensembles.

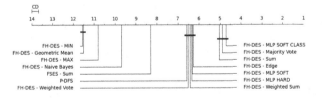

Fig. 3. Critical Difference Diagram for Partial Dynamic Ensembles in FH-DES.

META-DES. In Table 5, for Meta-DES, we can observe that META-DES - Sum and Majority Vote presented the best overall results (90.33%). META-DES - Sum obtained the best results in 8 of the 20 datasets. The META-DES - Vote method obtained the best results in 7 datasets, presenting a slightly lower accuracy than Sum. The Friedman test was then applied, which identified statistical differences, with a p-value < 0.05 in the tests.

When applying the post-hoc test, the results of the CD Diagram (Fig. 4) did not show any statistically significant difference in the accuracy of META-DES - Sum and Majority Vote again, but providing higher levels of accuracy, when compared to the other methods, from a statistical point of view.

Table 5. Score of fusion methods for partial dynamic ensembles in META-DES.

Dataset	META-DES Majority Vote	META-DES SUM	META-DES MAX	META-DES MIN	META-DES Geometric Mean	META-DES Weighted Sum	META-DES Weighted Vote	META-DES MLP HARD	META-DES MLP SOFT	META-DES MLP SOFT CLASS	META-DES Edge	Naive Bayes	P-DFS	FSES SUM
D1	94.48	94.48	94.48	94.48	94.48	89.64	89.64	93.83	94.85	**95.12**	94.48	91.69	94.25	93.42
D2	96.80	96.80	90.29	87.43	87.43	84.25	84.25	93.77	96.83	**96.96**	95.87	91.15	96.50	96.49
D3	92.05	92.05	93.13	93.13	93.13	92.28	92.29	92.54	92.11	92.42	92.05	81.81	**93.18**	89.74
D4	94.31	94.31	90.42	86.05	86.05	75.65	75.65	90.00	92.45	91.76	89.54	74.54	96.18	**97.03**
D5	**90.45**	**90.45**	81.08	81.08	81.08	76.58	76.58	87.82	88.97	88.49	**90.45**	84.13	88.74	88.56
D6	**74.68**	**74.68**	44.75	44.75	44.75	71.14	71.14	70.35	69.71	70.57	**74.68**	64.01	71.26	66.43
D7	93.43	93.43	74.25	67.80	67.80	64.00	64.00	86.05	93.26	**93.66**	92.43	59.39	90.84	90.86
D8	96.07	96.07	87.74	80.54	80.54	78.79	78.79	89.25	95.96	96.21	93.66	78.21	**97.66**	**97.94**
D9	**78.93**	**78.93**	64.93	64.93	64.93	68.37	68.37	75.77	76.09	75.60	**78.93**	69.47	75.91	75.22
D10	92.03	92.03	89.53	89.53	89.53	91.99	91.99	90.54	90.91	91.02	92.03	89.12	**93.26**	92.28
D11	**97.25**	**97.25**	95.06	95.06	95.06	96.58	96.58	95.83	96.31	96.33	**97.25**	95.58	96.26	96.42
D12	**94.72**	**94.72**	91.80	91.80	91.80	92.73	92.73	93.55	93.88	93.87	**94.72**	92.94	93.38	93.54
D13	77.82	77.80	77.03	75.95	76.04	63.18	63.23	75.42	76.75	77.69	**77.94**	68.81	74.80	74.86
D14	92.00	91.97	**93.70**	**93.70**	**93.70**	92.13	92.14	92.76	92.54	92.33	92.00	89.05	92.69	89.86
D15	95.45	**95.46**	92.32	92.32	92.32	92.39	92.39	94.96	95.14	95.01	95.45	90.48	94.83	94.79
D16	72.50	72.28	74.50	74.50	74.55	72.97	73.08	72.66	76.23	71.94	72.50	70.55	**75.23**	72.39
D17	**100.00**	**100.00**	97.99	97.99	97.99	99.38	99.38	99.54	99.54	99.54	**100.00**	99.84	99.79	99.79
D18	98.56	98.56	94.09	94.09	94.09	92.66	92.66	98.42	**98.74**	98.65	98.56	95.41	98.17	97.90
D19	77.76	78.06	75.59	75.58	75.58	75.74	75.73	77.85	77.20	76.73	77.76	72.48	**78.23**	77.82
D20	**97.22**	**97.22**	92.72	92.72	92.72	92.34	92.34	96.95	97.01	97.06	**97.22**	94.40	96.53	96.39
AccAve	**90.33**	**90.33**	84.77	83.67	83.68	83.14	83.15	88.39	89.52	89.55	89.88	82.60	89.88	89.09

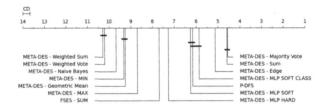

Fig. 4. Critical Difference Diagram for Partial Dynamic Ensembles in META-DES.

When comparing the P-DFS method and the best META-DES method (P-DMS) method, once again, we can observe a superiority of the P-DMS case, showing that the use of dynamic selection in the ensemble members provided more robust ensembles.

5.3 The FDES Scenario

For the FDES scenario, the Dynamic Fusion methods were combined with META-DES, KNORA-E and FH-DES, leading to 3 FDES configurations. For comparison purposes, the best 3 PDES configurations (one for META-DES, one for KNORA-E and one for FH-DES) are also presented, along with P-DFS and the best FSES configuration. Table 6 presents the results of all evaluated methods.

Table 6. Score of fusion methods in full dynamic ensembles.

Dataset	FDES (FH-DES)	FDES (KNORA-E)	FDES (META-DES)	FH-DES SUM	KNORA-E Majority Vote	META-DES Majority Vote	P-DFS	FSES - SUM
D1	**97.48**	95.97	95.56	96.97	94.77	94.48	94.25	93.42
D2	96.70	96.74	**96.97**	96.81	96.94	96.80	96.50	96.49
D3	**93.19**	92.66	92.82	92.42	91.95	92.05	93.18	89.74
D4	**97.06**	94.64	96.14	96.84	93.95	94.31	96.18	97.03
D5	91.24	90.14	89.97	**91.27**	90.80	90.45	88.74	88.56
D6	74.24	72.59	73.82	72.02	69.66	**74.68**	71.26	66.43
D7	94.20	**94.46**	94.16	93.41	93.75	93.43	90.84	90.86
D8	95.93	96.14	96.43	95.88	95.94	96.07	97.66	**97.94**
D9	76.88	76.80	77.87	77.64	77.72	**78.93**	75.91	75.22
D10	92.90	91.90	92.20	92.81	91.59	92.03	**93.26**	92.28
D11	96.79	**97.64**	97.00	96.88	97.38	97.25	96.26	96.42
D12	94.59	94.34	94.59	94.52	94.65	**94.72**	93.38	93.54
D13	76.93	79.37	**79.56**	77.38	77.65	77.82	74.80	74.86
D14	93.01	93.25	**93.44**	91.86	91.18	92.00	92.69	89.86
D15	95.00	**95.76**	95.48	95.48	95.59	95.45	94.83	94.79
D16	**76.32**	73.58	74.27	74.73	70.31	72.50	75.23	72.39
D17	99.97	**100.00**	**100.00**	99.62	99.94	**100.00**	99.79	99.79
D18	98.77	**99.18**	99.02	98.42	98.96	98.56	98.17	97.90
D19	**78.58**	77.94	78.20	78.12	77.15	77.76	78.23	77.82
D20	97.38	97.32	**97.52**	97.05	97.07	97.22	96.53	96.39
Acc Ave	**90.86**	90.52	90.75	90.51	89.85	90.33	89.88	89.09

From Table 6, we can see that FDES (FH-DES) achieved the best overall accuracy levels (90.86%), closely followed by FDES (META-DES) (90.75%), and then FDES (KNORA-E) (90.52). These methos delivered the best result in 5 out of 20 datasets each. Then, the FH-DES SUM method is the best PDES case (90.51%), followed by META-DES VOTE and KNORA-E VOTE and then the P-DFS method. Finally, the worst result was obtained by the FSES - SUM method. As it can be seen, the use of dynamic selection on both ensemble members and combination methods provides the most robust classifier ensembles.

Figure 5 presents the CD Diagram of the post-hoc test on the results of Table 6. From this figure, it was possible to detect that the accuracy of all three FDES configurations showed the most accurate classifier ensembles. Additionally, the statistical test detected superiority, in terms of accuracy of the FDES configurations, when compared to the remaining analyzed methods.

The results obtained in Fig. 5 corroborates the results of Table 6, in which the use of dynamic selection on both ensemble members and combination methods provides the most robust classifier ensembles. When using dynamicity in the selection of one parameter, the dynamic selection of ensemble members provided the best results. Finally, the static selection delivered the worst results, detected by the statistical test.

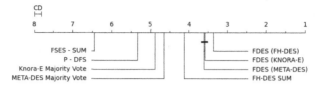

Fig. 5. CD Diagram for Full Dynamic Ensemble

6 Final Remarks

This paper proposed an exploratory analysis of the dynamic selection of the most important parameters of a classifier ensemble. In order to perform this analysis, three dynamic selection scenarios are defined, which are: FSES (Full Static Ensemble Selection); PDES (Partial Dynamic Ensemble Selection) with two cases: P-DFS (Partial - Dynamic Fusion Selection) and P-DMS (Partial - Dynamic Member Selection); and FDES (Full Dynamic Ensemble Selection). The main aim is to assess the impact of dynamic selection in the performance of classifier ensembles.

Through this exploratory analysis, it can be observed that the use of dynamic selection on both ensemble members and combination methods provides the most robust classifier ensembles. When using dynamicity in the selection of one parameter, the dynamic selection of ensemble members provided the best results. Finally, the static selection delivered the worst results, detected by the statistical test.

As future analysis, it is necessary to expand this analysis using different dynamic selection approach that are not based on region of competence. A statistical analysis comparing methods across different strategies is also needed. Finally, this empirical study was limited to 20 classification datasets. It is important to perform an analysis with more robust datasets.

Acknowledgement. The authors would like to thank the financial support provided by the Coordenação de Aperfeiçoamento de Pessoal de Nıvel Superior do Brasil (CAPES), during the development of this work.

References

1. Kuncheva, L.I.: Combining Pattern Classifiers: Methods and Algorithms. [S.l.]: Wiley (2004)
2. Feitosa, A.A., Canuto, A.M.P.: An exploratory study of mono and multi-objective metaheuristics to ensemble of classifiers. Appl. Intell. **48**(2), 416–431 (2018)
3. Kordík, P., Cerný, J., Frýda, T.: Discovering predictive ensembles for transfer learning and meta-learning. Mach. Learn. **107**(1), 177–207 (2018)
4. Ko, A.H.R., Sabourin, R., Britto, A.S., Jr.: From dynamic classifier selection to dynamic ensemble selection. Pattern Recogn. **41**(5), 1718–1731 (2008)
5. Cruz, R.M.O., et al.: META-DES: a dynamic ensemble selection framework using meta-learning. Pattern Recogn. **48**(5), 1925–1935 (2015)

6. Nguyen, T.T., et al.: Ensemble selection based on classifier prediction confidence. Pattern Recogn. **100**, 107104 (2020)
7. Cruz, R.M.O., et al.: Deslib: a dynamic ensemble selection library in python. J. Mach. Learn. Res. **21**(1), 283–287 (2020)
8. Woods, K., Kegelmeyer, W. P., Bowyer, K.: Combination of multiple classifiers using local accuracy estimates. [S.l.]: IEEE (1997). 405–410 p.
9. Dantas, C., et al.: Instance hardness as a decision criterion on dynamic ensemble structure. In: 2019 8th Brazilian Conference on Intelligent Systems (BRACIS). IEEE (2019)
10. Davtalab, R., Cruz, R.M.O., Sabourin, R.: Dynamic ensemble selection using fuzzy hyperboxes. In: 2022 International Joint Conference on Neural Networks (IJCNN), pp. 1–9. IEEE (2022)
11. Nunes, R.O., et al.: An unsupervised-based dynamic feature selection for classification tasks. In: 2016 International Joint Conference on Neural Networks (IJCNN). IEEE (2016)
12. Dantas, C. A.: An integration analysis of dynamic selection techniques for a classification system building. (Ph.D. thesis) Federal University of Rio Grande do Norte (2021)
13. Silva, J.C.P., de Paula Canuto, A.M., de Medeiros Santos, A.: The dynamic selection of combination methods in classifier ensembles by region of competence. In: International Conference on Artificial Neural Networks. Springer, Cham (2023)
14. Dietterich, T.G.: Ensemble methods in machine learning. In: Kittler, J., Roli, F. (eds.) MCS 2000. LNCS, vol. 1857, pp. 1–15. Springer, Heidelberg (2000). https://doi.org/10.1007/3-540-45014-9_1
15. Silva, J. C. .P, et al.: Ensemble classifiers in a serious game for medical students in clinical cases. In: Proceedings of the 10th Euro-American Conference on Telematics and Information Systems. Association for Computing Machinery, pp. 1–5 (2020)
16. Kuncheva, L.I.: A theoretical study on six classifier fusion strategies. IEEE Trans. Pattern Anal. Mach. Intell. **24**(2), 281–286 (2002)
17. Brown, G., et al.: Diversity creation methods: a survey and categorisation. Inf. Fus. **6**(1), 5–20 (2005)
18. Canuto, A.M.P.: Investigating the influence of the choice of the ensemble members in accuracy and diversity of selection-based and fusion-based methods for ensembles. Pattern Recogn. Lett. **28**(4), 472–486 (2007)
19. Simpson, P.K.: Fuzzy min-max neural networks-part 1: classification. IEEE Trans. Neural Networks **3**(5), 776–786 (1993)
20. Friedman, M.: A comparison of alternative tests of significance for the problem of m rankings. Ann. Math. Stat. **11**(1), 86–92 (1940)
21. Demšar, J.: Statistical comparisons of classifiers over multiple data sets. J. Mach. Learn. Res. **7**, 1–30 (2006)

Embedding Representations for AutoML Pipelines

Camila Santana Braz, Matheus Cândido Teixeira(✉), and Gisele Lobo Pappa

Computational Science Department, Federal University of Minas Gerais, Belo Horizonte, Brazil
`camilabraz@ufmg.br`, `{matheus.candido,glpappa}@dcc.ufmg.br`

Abstract. The area of Automated Machine Learning (AutoML) emerged to automate the tedious process of manually testing different sets of algorithms hyperparameters and other data engineering tasks involved in the process of solving a machine learning (ML) problem. While many researchers focus on developing and refining techniques, there have been few advances in understanding the models and how the optimization works. One way to tackle this problem is to investigate the fitness landscape and analyze the distance between solutions to describe this environment. These techniques require calculating distances between solutions in the search space. This is a problem, as in a diverse range of AutoML methods, machine learning pipelines are represented by a tree structure, which has limitations in computational time for calculating distances and does not account for the semantics for the solutions. In this direction, this paper proposes a new way to represent ML pipelines using embeddings. We use a Transformer model to generate embeddings of machine learning pipelines, and then evaluated the embeddings using the correlation between the distances calculated when using the two representations. We also perform a qualitative and a visual analyses to compare both representations. Developing this representation allows researchers to improve current AutoML methods by providing a better understanding of how difficult it is to search for them.

Keywords: Deep Learning · AutoML · Embedding Space · machine learning pipelines

1 Introduction

The area of Automated Machine Learning (AutoML) emerged to automate the tedious process of manually testing different sets of algorithms hyperparameters and other data engineering tasks involved in the process of solving a machine learning (ML) problem. AutoML methods reduce or eliminate the need of specialized human intervention in building and tuning models. Given its importance, nowadays all major tech companies like Google, Microsoft, and Amazon offer AutoML techniques on their platforms.

An ML pipeline is defined as a set of ML operations that include data preprocessing, data classification and post-processing steps, which can be executed sequentially or in parallel. Most AutoML tasks are tackled as an optimization problem, with a search space that includes a range of algorithms and hyperparameters used to build machine learning task pipelines. The search method optimizes a metric of learning quality, such as the accuracy or f-measure [11].

Different optimization methods have been used in AutoML, including Bayesian optimization, evolutionary algorithms [5] and hybrids techniques, but few have looked at how difficult this optimization problem is. One way to measure problem difficulty is to look at the fitness landscape of a problem [12]. The fitness landscape is defined by the set of viable solutions within the search space associated with the quality metric being optimized. Investigating the fitness landscape is important because it contributes to understanding the space from an optimization perspective, such as whether it has multiple global/local optima, saddle points, plateaus, or regions of low variability. Understanding these characteristics of the space leads to the construction of more effective AutoML methods, allowing for a more informed exploration of the space.

The analysis of fitness landscapes requires calculating distances between solutions to determine neighbourhoods of solutions with, for example, the same quality (plateaus) or peaks that may contain global or local optima, in the case of maximization problems. Calculating these distances will depend on how solutions are represented.

The current canonical representation of an AutoML pipeline is a tree. This representation is used because it suits the complex and hierarchical search spaces of AutoML [9]. However, this model presents two important limitations: (i) high computational complexity to calculate distance between solutions and (ii) difficulty in considering the semantic aspects of the solutions. With these drawbacks in mind, this work proposes to represent these pipelines as embeddings.

An embedding is essentially a numerical vector, generated, in our case, from training a neural network with data from a range of ML pipelines. The network is able to capture latent features of the original space and map it to the weights of the network, which can be later used to present these pipelines. Calculating the distance between embeddings is simple and has low computational complexity. In addition, embeddings have the capability to preserve the semantic aspects of the data they represent, which can be observed when they are plotted in spaces of a lower number of dimensions.

Being able to plot ML pipelines in 2D spaces also allows for the visualization of both the path taken during optimization and the distance between different model configurations, facilitating the analysis of these aspects and, consequently, promoting a more efficient exploration of the search space. Finally, many state-of-art works in machine learning have been using embeddings representations of spaces, suggesting them as a promising representation solution [15] [6].

Inspired by the work of [3] – which developed an embedding representation for symbolic regression trees and showed them to be effective to measure the distance between trees and its potential to capture semantics – this paper proposes to generate embeddings to represent AutoML pipelines. The embeddings generated are evaluated by comparing distance metrics of different representations. Moreover, visual and qualitative analysis of the search space and pipelines distance are performed to better asses the proposed representation.

The main contributions of this study are:

- Comparison of two models to represent AutoML pipelines representations: tree and embeddings;
- Analysis and evaluation of the use of this linear representation in the context of AutoML;
- Investigation of semantic preservation in the representation through embeddings;
- Development of a method for visualizing the search spaces.

In the long term, this research may contribute to the development of more robust, effective and efficient AutoML models, starting from the analysis of fitness landscapes.

This work is organized as follows. Section 2 reviews relevant literature, positioning the research within the state-of-the-art . Section 3 outlines the methodology, detailing the techniques, metrics, and tools used to achieve the objectives. Section 4 discusses the results, analyzing the effectiveness of embeddings as representations of AutoML pipelines and visualizations of their search spaces. Finally, Sect. 5 presents the conclusions from the experiments and future work directions.

2 Related Work

A few studies have proposed new representations for AutoML pipelines and different visualization tools to better understand how small changes affect the fitness landscape.

Concerning pipeline representation, research in [10] employs deep learning techniques in embeddings to optimize AutoML pipelines. The method involves creating latent representations for pipeline configurations through DeepPipe, which uses Bayesian optimization to find ideal configurations. The system improves its generalization and prediction capabilities through meta-learning. Experiments on three meta-datasets show state-of-the-art results, surpassing existing methods like OBOE, SMAC and AutoPrognosis.

With a different objective, the approach in [4] integrates NLP techniques with AutoML methods. The authors use embeddings to represent metadata from datasets and algorithm documentation, allowing AutoML systems to recommend machine learning pipelines based on textual descriptions. Their system increases pipeline optimization efficiency and shows significant improvements over frameworks like OBOE, AutoSklearn, AlphaD3M and TPOT, recommending solutions in under a second without prior runs. They also made their data, models and code publicly available.

Turning to visualization techniques, study [8] presents PipelineProfiler, an interactive visualization tool for exploring and comparing machine learning pipelines generated by different AutoML systems. Integrated into Jupyter Notebook, it facilitates pipeline analysis, including hyperparameters and evaluation metrics, and helps identify patterns that lead to better results.

In [16], AutoAIViz is presented as an interactive visualization tool that uses Conditional Parallel Coordinates (CPC) to explore model and pipeline generation steps. CPC visualization helps users understand how AutoML decisions impact fitness. A usability study showed the tool's effectiveness in increasing the comprehensibility of AutoML processes, highlighting the importance of transparency in AutoML systems.

The authors in [14] introduce Atmseer, a visualization tool that enhances transparency and controllability in AutoML processes. It allows users to refine the search space and analyze results interactively. Atmseer offers multi-level visualizations, enabling real-time monitoring and modifications.

The methodology proposed in this paper can be coupled to any of the visualization tools aforementioned to represent ML pipelines in a 2D space. It can also help improve search by using informed decisions based on the shape of the search space.

3 Methodology

The main objective of this paper is, given a set of machine learning pipelines represented as trees, map these trees to a new representation space, where the pipelines are represented as embeddings. The embeddings are generated by training a neural network, and then extracting the weights of the final encoder layer.

The proposed methodology can be divided into 3 steps: (i) generate a training set, which involves preparing machine learning pipelines to be given as input to neural networks; (ii) configure and train the neural network; (iii) given a new tree, extract the embedding from the network (Fig. 1).

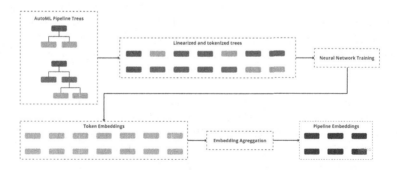

Fig. 1. Methodology steps to generate the pipeline represented as an embedding.

3.1 Generating the Training Set

As previously explained, an ML pipeline includes preprocessing steps (data cleaning, feature selection), a machine learning model (classification or regression) and post-processing steps. The search space used in this paper was borrowed from [9], where pipelines were generated by following a Context-Free Grammar with 38 production rules, 92 terminals, and 45 non-terminals. These include three dimensionality reduction algorithms (PCA, Select K-Best, Standard Scaler) and five classification methods (Logistic Regression, MLP, KNN, Random Forest, AdaBoost), with parameter counts ranging from two to seven. Default values from Scikit-learn were used for some continuous parameters while relative values were employed for parameters that depend on the number of attributes. This search space was chosen because it has size 69,960, allowing all solutions to be enumerated. This facilitates testing a new representation methodology, as we know which solutions should be more similar to each other.

These generated pipelines are initially structured as a tree. They must be converted into a linear, non-hierarchical format of fixed size to be used as input for the neural network. A Depth-First Search (DFS) is performed on the graph structure to investigate the connected nodes and leaves [3], and a grammar-based parser generates a token sequence representing the original pipeline in the appropriate format.

3.2 Neural Network Configuration

The neural network purpose is to learn a function to encode the linearized pipeline string into an embedding by mapping each token of this string into an embedding space.

The trained neural network is a Transformer [13], an architecture developed for both supervised and unsupervised machine learning tasks, primarily applied to natural language processing (NLP) to capture semantics in large datasets. Transformers revolutionized NLP by implementing Self-Attention mechanisms, allowing the encoder to extract information from the entire input sequence and enabling the decoder to give more importance to specific elements of the input, depending on the token being processed at the moment. With multiple Self-Attention heads, it is possible to map various pieces of information to a single word, allowing each position in the input sequence to attribute importance to all others, improving the understanding of context and semantic relationships. Table 1 presents the Transformer parameters and tested values.

Two tree representation strategies will be examined: a detailed one with tags indicating algorithm types (i.e., PCA points to <features_dim>, <whiten> and <svd_solver> and these nodes points to the values of each parameter) and a straightforward one with only algorithm names and parameters (i.e., PCA points directly to the values of each parameter). For the first one, the grammar vocabulary is comprised of a total of 137 tokens. In the second, 69 tokens are included. For both of them three of the tokens are for specific scenarios: <pad_token> (used for padding sequences to the same length), <unk_token>

Table 1. Transformer parameters.

Parameter Name	Tested Values	Description
n_epochs	50	Number of training epochs.
batch_size	128	Number of samples processed before updating the model.
loader_workers	2	Number of subprocesses to load the data.
d_model	16, 32, 64, 128, 256	Dimension of the output vectors
test_proportion	0.3	Proportion of the data used for testing.
lr	1e-3, 1e-4	Learning rate.
N	1	Number of layers in the model.
d_ff	64	Dimension of the internal feed-forward layers.
h	8	Number of multi-attention heads.
dropout	0.3	Proportion of neurons randomly deactivated during training.

(represents unknown words or parameters) and <sos_token> (marks the start of a sequence).

Finally, the loss function measures the difference between the model's predictions and the actual tokens and the LabelSmoothing strategy helps make the model less confident and more general. The loss is normalized by the number of non-padding tokens to ensure fair evaluation across sequences of varying lengths.

3.3 Extracting Embeddings from Network Layers

The encoding function learned by the Transformer in the previous step generates embeddings for each token in the input sequence. These embeddings are extracted from the final encoder layer, after passing through the attention and feed-forward sublayers and undergoing normalization. To create the complete tree embedding, it is necessary to aggregate the individual mapped tokens to obtain the full tree representation. This can be done using three aggregation functions:

- **Sum**: Summing the value of each dimension of the tokens across all tokens and dimensions.
- **Mean**: Calculating the mean value of each dimension of the tokens across all tokens and dimensions.
- **Concat**: Concatenating all dimensions of the tokens across all tokens and dimensions.

The three aggregation functions proposed above will be tested to determine the most appropriate one.

4 Generating Tree Embeddings

Given the proposed methodology, the next step was to train the proposed transformer to generate tree representations. These datasets were generated from 3 datasets, listed in Table 2. They are available on Kaggle and the UCI Machine Learning Repository, and were selected considering their number of attributes and number of classes.

Table 2. Datasets characteristics.

Dataset Name	Instances	Attributes	Classes
Ml-prove	6118	51	6
Mushrooms	8124	22	7
Raisin	900	7	2

For each dataset, the trees previously defined in Sect. 3.1 are used as input to train the transformer. The only differences in tree representation are in the parameters that depend on the input variables, such as the number of features in the SelectKBest algorithm. For this type of parameter, relative values were set.

Having the datasets, we tuned the number of embedding dimensions (16, 32, 64, 128 and 256) and learning rate (0.001 and 0.0001), while other parameters were kept with their default values presented in Table 3.

The network was trained for up to 50 epochs, with an early stopping mechanism. If the loss did not decrease more than a predefined threshold of $1e-8$ for five consecutive epochs, it was considered that the model had converged, and training was stopped.

Table 3 presents the configuration and test set loss for each trained neural network. We ran each model five times to generate the variance and confidence interval.

Table 3. Transformer configuration and test loss for each model.

Dataset	Strategy	LearnR.	Dimen.	Test Loss	Var.	CI
ML-Prove	W/o tag	0.0001	64	1.102e−08	3.231e−17	[1.193e−08, 2.605e−08]
	With tag	0.001	256	0.000	1.820e−15	[−2.836e−08, 7.758e−08]
Mushrooms	W/o tag	0.0001	64	3.885e−09	1.237e−16	[1.848e−08, 4.609e−08]
	With tag	0.001	128	0.000	1.940e−16	[−3.478e−09, 3.111e−08]
Raisin	W/o tag	0.001	128	2.007e−09	9.155e−17	[1.645e−08, 4.021e−08]
	With tag	0.001	128	0.000	7.354e−16	[−5.179e−09, 6.216e−08]

Figure 2 shows the evolution per epoch of the loss function of each model that resulted from the best configuration. The training loss and the test loss decrease rapidly within the first few epochs, which indicates that the model is learning quickly and effectively adjusting its parameters early in the training process.

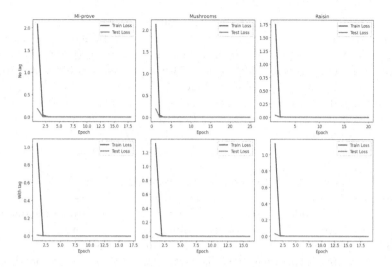

Fig. 2. Learning curve.

5 Contrasting Tree-Based and Embedded Representations

One of the difficulties of the proposed methodology is how to evaluate the embedded representations when compared to the tree representation. We start by looking at the correlations between the distances considering trees and embedding representations.

The traditional way to calculate distances between tree representations is to use the tree edit distance, defined in Equation (1):

$$D(T_1, T_2) = \begin{cases} 0 & \text{if } T_1 = T_2 \\ \min \begin{cases} D(T_1 - l_1, T_2 - l_2) + \text{cost}(l_1, l_2) \\ D(T_1 - l_1, T_2) \\ D(T_1, T_2 - l_2) \end{cases} & \text{otherwise.} \end{cases} \quad (1)$$

The distances between the embeddings will be measured using two measures: cosine distance (1 - the cosine similarity) and Euclidean distance. The goal is to compare whether the embeddings are suitable representations for the pipelines, evaluating the Pearson and Spearman correlation between the two representation distances. By calculating the correlation between these distances, evidence of the suitability of using embeddings for AutoML can be obtained.

Additionally, as the embeddings are encoded in high-dimensional spaces, we use UMAP [7] to reduce their dimensionality and plot the solutions in a two-dimensional space, allowing the visualization of the search space and a qualitative analysis of the relationships between pipelines. Both are reported in the next section.

To calculate the distances between solutions using tree edit distance metrics, Euclidean distance and cosine distance, five samples with 5000 trees each were collected through random and stratified sampling to ensure diversity and representativeness in the sampled population. The use of samples is justified by the complexity of calculating tree edit distances [1,2].

Figures 3 and 4 present the results of Spearman correlations between edit tree distance and embedding distances. The Pearson correlations are omitted due to space restrictions, but Spearman presented better results. This indicates that the relation between the tree edition and the other metrics might be monotonic but not necessarily linear.

Results showed a strong correlation between embedding distances and tree edit distances, especially in models without tags and using "sum" as the aggregation function. Euclidean distances consistently outperformed cosine distances.

The ML-Prove and Raisin datasets exhibited the highest correlations, ranging from 0.73 to 0.79, with the ML-Prove model (64 dimensions, learning rate 0.0001) performing better. The Mushrooms dataset showed lower correlations but still reached values between 0.66 and 0.71 in certain configurations. In summary:

- **Spearman Correlation with Euclidean Distance:** Spearman correlations were slightly higher, with no-tag models using mean and sum functions achieving correlations up to 0.78. The Mushrooms dataset showed significantly lower correlations.
- **Spearman Correlation with Cosine Distance:** Results were slightly inferior to Euclidean distance, with maximum correlations of 0.79 for no-tag models in ML-Prove. The Mushrooms dataset also showed lower correlations.

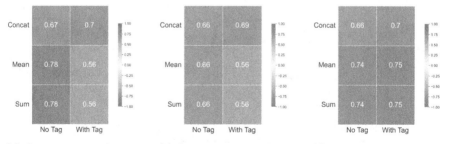

(a) Spearman correlation for ML-Prove. (b) Spearman correlation for Mushrooms. (c) Spearman correlation for Raisin.

Fig. 3. Average across samples of Spearman correlations for Euclidean distance.

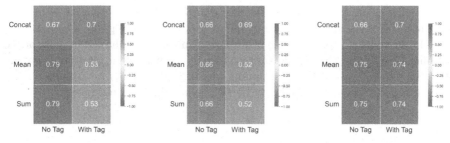

(a) Spearman correlation for ML-Prove. (b) Spearman correlation for Mushrooms. (c) Spearman correlation for Raisin.

Fig. 4. Average across samples of Spearman correlations for cosine distance.

6 Visual Analysis of the Search Space

As previously mentioned, the embeddings also underwent a dimensionality reduction process to enable plotting them on a Cartesian plane and generating a coordinate chart. UMAP was trained using a greedy algorithm that optimized the number of neighbours, minimum distance and distance metric.

Based on the correlation results, plots generated with the aggregation functions mean and sum in the no-tag strategy are plotted. Additionally, the datasets ML-Prove and Mushrooms were selected for visualizations as they demonstrated the best and worst performance, respectively. The objective is to investigate how visualizations occur in datasets with different performance levels, understanding the potential of semantic preservation in scenarios of high correlations (0.78) and lower correlations (0.66).

In the coordinate plot in Fig. 5, each point represents a pipeline, and the set of points is the search space. Since embeddings have the potential to preserve the semantic aspects of the data they represent, similar pipelines should be closer together in this space, while different pipelines should be farther apart. The following analysis highlights this potential for semantics preservation, as pipelines with similar algorithms and hyperparameters are grouped closely together.

Figure 5a shows the visualization of the ML-Prove search space. Note that similar algorithms were clustered together. Random Forest and K-Nearest Neighbors display a clear boundary of separation in the search space, while the other algorithms - MLP, Logistic Regression and Ada Boost - are concentrated in the lower right region of the search space. Furthermore, a cluster for each of these algorithms is identifiable, although Logistic Regression is slightly more spread out than the other two. Since the search space is predominantly composed of Random Forest and K-Nearest Neighbors, it is consistent with the results that the Transformer has acquired more information about these algorithms to classify them with higher accuracy.

Figure 5b presents the search space of the Mushrooms dataset. Although this model showed the worst correlation results, an analysis of the images reveals an acceptable algorithm clustering, with the locations of each algorithm in the space being easily visualized. Again, Random Forest and K-Nearest Neighbors occupy most of the space and are in the most identifiable regions. However, the model also managed to group the other classification algorithms into perceptible regions. Comparing this visualization with the previous one, it is noteworthy that this model was better at grouping identical algorithms into smaller more concise regions, revealing similarity by both algorithm and parameters.

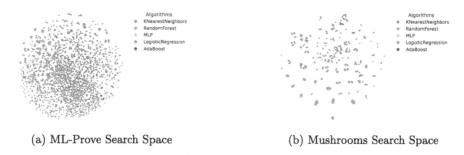

(a) ML-Prove Search Space   (b) Mushrooms Search Space

Fig. 5. Search space by classification algorithm.

The search space divided by preprocessing algorithms is presented in Fig. 6. It is worth highlighting that this type of algorithm was not the best choice to primarily cluster the pipelines but can still be useful if analyzed together with the plot showing classification algorithm 5. As observed with the classification algorithms, the Mushrooms model better separated the pipelines into groups.

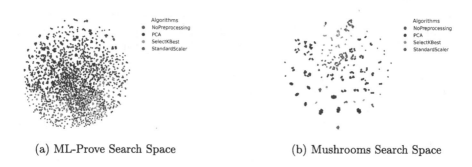

(a) ML-Prove Search Space   (b) Mushrooms Search Space

Fig. 6. Search space grouped by preprocessing algorithm.

6.1 Investigating Clustering

Two types of clustering are worth investigating: regions where the model incorrectly grouped the pipelines, such as the lower right region in the ML-Prove dataset, and regions well-separated, such as the clustering observed in the Mushrooms search space.

When analyzing Fig. 5a, observe that the Logistic Regression, MLP, and AdaBoost algorithms appear in locations where they should not be, mixed with the KNN region in the lower right part of the search space. By analyzing Fig. 6a, notice that the preprocessing algorithm also fails to differentiate the pipelines. Although the clustering is not pure, the model still managed to separate the groups to some extent. One possible reason for the poor grouping of these three classification algorithms could be the number of pipelines that use them. With fewer training data on these three, the model has less information and does not learn as well how to classify them.

Another important investigation is to analyze the composition of clusters. It is expected that subgroups are composed of pipelines with similar algorithms and hyperparameters. Figure 7a provides a closer look at a region within one of the Random Forest clusters. In this figure, we observe sub-clusters with similar preprocessing parameters and algorithms:

- **Preprocessing algorithm:** PCA with 8, 15 and 21 principal components;
- **Classification algorithm:** Random Forest with entropy as criterion and sqrt as max_features hyperparameters.

Figure 7b presents another group with similar hyperparameters:

- **Preprocessing algorithm:** Standar Scaler;
- **Classification algorithm:** Random Forest with gini as criterion and sqrt as max_features hyperparameters

(a) ML-Prove Search Space (b) Mushrooms Search Space

Fig. 7. Search space by preprocessing algorithm.

6.2 Qualitative Analysis

Finally, to provide a qualitative analysis of the embedding distance, a set of four pipelines from the best and worst-performing models - as done in the previous subsection - were selected for this experiment. The goal is to choose a pair of pipelines that appear similar and a pair of very different pipelines and then investigate their distances and locations in the search space. Table 4 presents these results for the ML-Prove dataset.

Note that the similar pairs are closer than the different ones, which is expected. The similar pairs use the same preprocessing algorithm with different parameters and the same classification algorithm with different parameters.

Table 4. Qualitative analysis for ML-Prove

	Pipeline	Cosine	Euclidean
Similar	SelectKBest 18 KNearestNeighbors 5 uniform ball_tree 100 minkowski 3	0.0587	23.2038
	SelectKBest 1 KNearestNeighbors 5 distance kd_tree 100 manhattan		
Different	NoPreprocessing KNearestNeighbors 1 distance kd_tree 60 minkowski 3	0.3601	64.2433
	SelectKBest 50 RandomForest gini False False balanced_subsample 50 True log2 10		

For the Mushrooms dataset, results are shown in Table 5. As expected, the similar pair is closer than the different one. The comparison between the sum without the tag model and the mean with the tag model indicates that the former better captures the structures and semantics of the pipelines, as evidenced by the greater distance between the pipelines. Nevertheless, the second model can satisfactorily identify distance relationships.

Table 5. Qualitative analysis for Mushrooms.

	Pipeline	Cosine	Euclidean
Similar	PCA 1 False arpack MLP 300 logistic	0.0448	1.7502
	PCA 1 False arpack MLP 500 logistic		
Different	PCA 15 True full AdaBoost SAMME.R	0.1460	3.2360
	StandardScaler False False RandomForest entropy False False balanced_subsample 30 True sqrt 10		

Figure 8 illustrates the positions of each pair of pipelines. The points were jittered to disperse overlapping regions. The visual analysis suggests that UMAP successfully preserved the high-dimensional structure and mapped it into a two-dimensional space. However, it is worth noting that the Euclidean distance between the worst and best pipelines is greater in the ML-Prove dataset than in the Mushrooms dataset, yet, in the plot, it appears to be smaller.

(a) Location of pipelines pair for ML-Prove.

(b) Location of pipelines pair for Mushrooms.

Fig. 8. Location of each pair of pipeline. Stars represent similar pipelines and circles mark different ones.

7 Conclusion

This paper developed and evaluated a novel linear representation for AutoML pipelines. By transforming traditional tree-based representations into embeddings, we addressed the challenges of computational complexity to calculate distances between solutions and semantic preservation inherent in the canonical methods. The experimental results indicate that different embedding models exhibit strong correlations with traditional tree edit distances, suggesting that embeddings can effectively capture the underlying relationships between different pipeline configurations.

The visualizations of the search space validate the potential of embeddings as a suitable representation for pipelines. Analysis of the plots reveals distinct algorithm regions and clusters and qualitative analysis shows that the embeddings' distances align with the apparent pipeline distances.

Future work will focus on further studies regarding embeddings, such as investigating the preservation of local optima networks (LONs) and using neighborhood plots to explore these relationships more deeply.

Overall, this research contributes to the advancement of AutoML by proposing a robust, efficient, and semantically meaningful representation for pipelines, enhancing the explicability of the optimization process and paving the way for more effective and transparent AutoML solutions.

Acknowledgments. A bold run-in heading in small font size at the end of the paper is used for general acknowledgements, for example: This study was funded by X (grant number Y). The authors have no competing interests to declare that are relevant to the content of this article.

References

1. Bille, P.: A survey on tree edit distance and related problems. Theoret. Comput. Sci. **337**(1–3), 217–239 (2005)
2. Bringmann, K., Gawrychowski, P., Mozes, S., Weimann, O.: Tree edit distance cannot be computed in strongly subcubic time (unless apsp can). ACM Trans. Algorithms (TALG) **16**(4), 1–22 (2020)
3. Caetano, V., Teixeira, M.C., Pappa, G.L.: Symbolic regression trees as embedded representations. In: Proceedings of the Genetic and Evolutionary Computation Conference, pp. 411–419 (2023)
4. Drori, I., Liu, L., Nian, Y., Koorathota, S.C., Li, J.S., Moretti, A.K., Freire, J., Udell, M.: Automl using metadata language embeddings. arXiv preprint arXiv:1910.03698 (2019)
5. Hutter, F., Kotthoff, L., Vanschoren, J.: Automated machine learning: methods, systems, challenges. Springer Nature (2019)
6. Kim, H., Park, M., Cho, S.: Learning representations for medical images with deep embeddings. Med. Image Anal. **78**, 102343 (2023)
7. McInnes, L., Healy, J., Melville, J.: Umap: Uniform manifold approximation and projection for dimension reduction. arXiv preprint arXiv:1802.03426 (2018)
8. Piazentin Ono, J., Castelo, S., Lopez, R., Bertini, E., Freire, J., Silva, C.: Pipeline-profiler: a visual analytics tool for the exploration of automl pipelines. arXiv e-prints pp. arXiv–2005 (2020)
9. Pimenta, C.G., de Sá, A.G., Ochoa, G., Pappa, G.L.: Fitness landscape analysis of automated machine learning search spaces. Threshold **10**(6), 103
10. Pineda Arango, S., Grabocka, J.: Deep pipeline embeddings for automl. In: Proceedings of the 29th ACM SIGKDD Conference on Knowledge Discovery and Data Mining, pp. 1907–1919 (2023)
11. Smith, J., Brown, K., Williams, D.: Automated machine learning: Hyperparameter optimization in practice. J. Artif. Intell. Res. **65**, 101–120 (2023)
12. Stadler, P.F.: Fitness landscapes. In: Biological Evolution and Statistical Physics, pp. 183–204. Springer (2002)
13. Vaswani, A., et al.: Attention is all you need. Advances in neural information processing systems **30** (2017)
14. Wang, Q., Ming, Y., Jin, Z., Shen, Q., Liu, D., Smith, M.J., Veeramachaneni, K., Qu, H.: Atmseer: Increasing transparency and controllability in automated machine learning. In: Proceedings of the 2019 CHI Conference on Human Factors in Computing Systems, pp. 1–12 (2019)
15. Wang, Y., Liu, S., Yang, J.: Embedding-based approaches for recommender systems: a survey. IEEE Trans. Knowl. Data Eng. **35**(4), 1234–1248 (2023)
16. Weidele, D.K.I., et al.: Autoaiviz: opening the blackbox of automated artificial intelligence with conditional parallel coordinates. In: Proceedings of the 25th International Conference on Intelligent User Interfaces, pp. 308–312 (2020)

Enhancing Graph Data Quality by Leveraging Heterogeneous Node Features and Embeddings

Silvio Fernando Angonese(✉)[iD] and Renata Galante[iD]

Institute of Informatics, Federal University of Rio Grande do Sul (UFRGS), Ave. Bento Gonçalves, 9500 Porto Alegre, Rio Grande do Sul, Brazil
sfangonese@inf.ufrgs.br
https://www.inf.ufrgs.br

Abstract. Heterogeneous Graphs are important data sources due to their rich representation of knowledge, primarily based on node features and relationships. It is common for these graphs to have significant data gaps, particularly in the nodes. Graph Neural Networks are state-of-the-art solutions that achieve excellent results by extracting information based on node relationships. However, they suffer from severe limitations when there is no available information in the graph elements, weakening their representation. This paper proposes the specifications and an algorithm to process different types of node features, such as text, images, and subgraphs, generating both single and composition embeddings. To evaluate the effectiveness of the proposed algorithm, experiments were conducted to generate the features and their respective node embeddings in a Heterogeneous Graph. The achieved performance was measured using the average of Accuracy, F1-Score, and their Standard Deviations based on the Recommender System tasks applied to the embeddings generated in the experiments. We can highlight the performance achievement in the experiments as the Node Classification task, using the composition of Aggregated Features with Metapaths embedding, which achieved an F1-Score of 83.66% overcoming the 60.70% achieved by the approach without embeddings.

Keywords: Heterogeneous Embedding · Graph Embedding · Heterogeneous Graph · Representation Knowledge · Graph Neural Networks

1 Introduction

With the exponential growth of data, the challenge lies in acquiring raw data and transforming it into valuable information. Although the data exists, extracting and utilizing it is not a simple task. Complex business cases, for example, can be effectively modeled as graphs, which may contain rich data about the entities they represent. Enhancing learning from graph representations signals a viable path to uncovering the hidden knowledge in raw data [6]. Deep Learning and Machine Learning excel at discovering hidden knowledge and characteristics, enabling the use of specific techniques to generate rich data from each graph node. The semantically rich data generated can then be used in various downstream applications, such as Recommender Systems (RecSys) [1,17].

However, an important restriction is associated with the common low level of information in the graphs. If graph elements, especially nodes, have more data attached, the Representation of Knowledge will increase and enhance the subjacent performance of applications. Node embedding is a technique that maps graph nodes to low-dimensional vectors, preserving the graph structure and node features, representing the nodes [16]. Thus, node embeddings are an excellent option for collecting and aggregating data from nodes, resulting in high-quality node representation.

Some work [10,18] introduces the generation of embeddings based on the neighboring nodes, increasing their expressiveness and recommendation performance. Meanwhile, the works of [17,19] evaluate the heterogeneity of data types embeddings, opening up a new perspective. Researches [7,16] propose a new approach for generating embeddings from metapaths, capturing semantics based on node relationships. Vision GNN [11] and Superpixel Image Classification [3] are examples of the few works where the graph is an image and the nodes are parts of it, demonstrating the feasibility of having images as nodes. None of the related work addresses the use of Heterogeneous Graphs with heterogeneous data types, such as texts, images, and subgraphs.

This paper proposes the specifications and a new algorithm that leverages Deep Learning techniques to extract information from images. Also, the feature generation is provided by Specialized Autoencoders, which are Machine Learning models designed to generate embeddings by mapping high-dimensional data to a compact and meaningful latent representation, preserving the essential data from the node features. The experiment results clearly demonstrate that the performance of embeddings, particularly when a composition of different types of embeddings is created, can yield superior results compared to evaluations conducted without them. In general, this paper contributes to providing an unexplored way of generating information and embeddings on nodes in Heterogeneous Graphs, through the definition of a modeling approach using texts, images, and subgraphs as representations of the graph nodes.

In previous work AGHE [2], we proposed an approach for generating and processing heterogeneous embeddings. In this paper, we specify each step of the approach, with the main contributions as follows: a) Specification of procedures for processing and generating different types of embedding; b) Proposes an algorithm based on procedures for processing and generating heterogeneous embeddings; c) Introduce composition embeddings as an alternative to generate high semantic nodes and present the evaluation of their performances achieved by the proposed algorithm; d) Share the public datasets, including the final recommended graph in JSON and CSV formats, with the research community, particularly those focused on Heterogeneous Graphs and RecSys.

The remainder of this paper is organized as follows: Sect. 2 conceptualizes the background techniques applied in this paper. Section 3 describes the related work. Section 4 presents the approach AGHE. Section 5 defines the specification of the procedures that serve as the foundation for the proposed algorithms shown in this section. Section 6 conducts some experiments and evaluates the results achieved while Sect. 7 exposes the conclusions and future works.

2 Background

The aim of this section is to present the Machine Learning and Deep Learning concepts and techniques adopted in this paper.

Heterogeneous Graph and Embeddings. In Heterogeneous Graphs the nodes and edges can be of different types, e.g., a person group represented by a Heterogeneous Graph can have nodes with different types. The bipartite graph is a special, commonly used type of Heterogeneous Graph, where edges exist between nodes of two different types. Thus, multiple types of nodes and different relationships contain comprehensive information and rich semantics [8]. A recurring problem in Representation Learning in Heterogeneous Graphs is Over-Smoothing issue. This implies that, after too many aggregations of the features in a graph, the node embeddings start to converge to nearly the same or even the very same value. Losing the capacity to discriminate between the different types of nodes and their various characteristics. During the aggregation process, if the neighboring nodes have very similar features or if the aggregation process is repeated many times, the local differences between the nodes are smoothed out, leading to a homogenization of the features. Two simple techniques used to minimize Over-Smoothing are discarding similar features from nodes during the embedding generation step and removing random edges during each training epoch [12].

Embedding captures the graph topology, node features, node-to-node relationships, and other relevant information about graphs, subgraphs, and nodes. Hence, embedding represents the nodes, and the similarity between node embeddings indicates their similarity in the graph [10,14].

MetaPath2Vec. It is a Heterogeneous Graph embedding model that formalizes metapath based on random walk to construct the heterogeneous neighborhood of a node and then leverages a heterogeneous Skip-Gram model to perform node embedding. Maximizing the probability of preserving both the structures and semantics of a given heterogeneous network, being able to learn desirable node representations in heterogeneous networks [7].

Autoencoders. They are useful for incorporating structural graph information from nodes, providing specific implementations. Autoencoder (AE) is a kind of neural network architecture that imposes a bottleneck on the network that forces a compressed knowledge representation of the original input. If the input resources were independent of each other, the compression and subsequent reconstruction would be a very difficult task. Thus, AEs are neural networks that aim to copy their input to their output, compressing the input in a latent space representation, called Encoder $h = f(x)$, and after that, rebuilding the output through this representation, called Decoder $r = g(h)$ [5]. Autoencoder may be modified or combined to form new models for various applications such as generative models, classification, clustering, anomaly detection, recommendation, dimensionality reduction, and capture information [5].

3 Related Work

Heterogeneous Graph can be traced back to generate data embedding from node features based on random walk approach citing Representation Learning on Graphs [10,18] improving the node expressivity. More close to the aims of our proposal is [19] which defines of Heterogeneous Graph Neural Network with the processing of embedding. The survey Graph Neural Networks in Recommender Systems [17] shows GNNs have been widely used in downstream applications essentially because graph structure and GNN have superiority in graph Representation Learning, citing GraphSAGE [10] as an important work regarding generating node embedding from node feature information.

MetaPath2Vec [7] captures the structure of Heterogeneous Graph, guiding random walks to generate sequences of heterogeneous nodes with rich semantics. Hence, metapath plays an important role in this paper capturing vital information by leveraging the relationships among heterogeneous nodes, transforming it into a form of node embedding. Vision GNN [11] and Superpixel Image Classification [3] works are other sources of inspiration that illustrate image representation in the form of a graph. In this context, each node corresponds to a distinct part of the same image, implying that every node encapsulates an image. Adopting a similar conceptualization, we can extend this idea to employ an image for representing node content.

Heterogeneous Graph can be integrated with some applications. However, one usually needs to carefully consider two factors: the first is how to construct Heterogeneous Graph for a specific application, and the second is what information or domain knowledge should be incorporated into a Heterogeneous Graph to ultimately benefit the application [16]. In the RecSys, the interaction between the user and items can be naturally modeled as a Heterogeneous Graph with two types of nodes. Thus, the application of Heterogeneous Graph embedding to RecSys constitutes an important research area, as highlighted in the survey on Heterogeneous Graph Embedding [16]. Hence, our proposal uses recommendations as an assessment to validate the value of performance by employing various types of node embeddings from the Heterogeneous Graph.

4 AGHE Approach for Generating Enhanced Heterogeneous Embeddings from Heterogeneous Graphs

This section presents the *Approach for Generating Enhanced Heterogeneous Embeddings from Heterogeneous Graphs* (AGHE) [2] shown in Fig. 1. AGHE generates heterogeneous embeddings through the processing of texts, images, and subgraphs represented in the nodes of Heterogeneous Graphs, such as the following steps:

1. *Graph Creation* - generates the Heterogeneous Graph along with all its components, such as nodes, edges, and node features. This step is critical because all other steps and results depend on it;
2. *Generating Text Node Embeddings* - is the process of creating node embeddings from their corresponding node features or extracted from images embedded in the nodes;
3. *Metapath and Aggregated Node Embeddings* - generates of aggregated feature embeddings using the random walks approach. This includes defining metapaths that represent the business rules through the relationships among the nodes, followed by the generation of their embeddings using the MetaPath2Vec algorithm;
4. *Graph Enhancement with RecSys tasks* - represents the experiments conducted in this paper, aiming to predict the type of nodes, predict some links, and cluster the nodes based on the Heterogeneous Graph generated in the first step;
5. *Rebuilding the Graph* - involves incorporating the generated embeddings and predictions saved into the graph nodes.

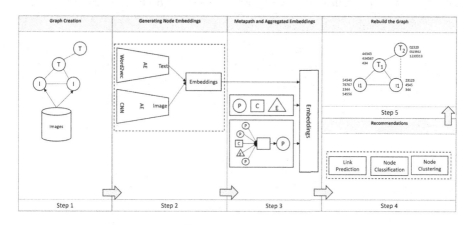

Fig. 1. Steps of AGHE - Approach for Generating Enhanced Heterogeneous Embeddings from Heterogeneous Graphs.

In the following sections, we describe the main contributions of this paper, specifying the key procedures to be performed at each step, serving as the foundation for building the proposed algorithm.

5 Specification for Enhancing Graph Data Quality

This section builds the AGHE approach [2] defining the procedures to be used in the proposed Algorithms 1 and 2 as highlighted in Fig. 1, where each of them is described as follows.

5.1 Graph Creation

Graph creation is the first step and there are two parts, the first one regards adding nodes and defining their types, characterizing a Heterogeneous Graph. Subsequently, node features should be attached using a set of short descriptions or simple words. In this context, edge features are not considered as part of the Heterogeneous Graph in this paper. The second part involves mapping relationships between nodes by adding edges to the Heterogeneous Graph, which enables navigation between the nodes. If the node type is Image, then another important operation is needed that involves uploading the node image from an image database provided by the application that understands the graph. As a result, the Heterogeneous Graph is created and ready to be used in subsequent steps. Figure 2(a) shows a Heterogeneous Graph $\mathcal{HG}(V, E)$ where V is a set of nodes, and E is a set of edges, which the node $v \in V$ can have different types of content with various data type features, such as text, image, and subgraph. Figure 2(b) shows the same graph simplified with heterogeneous data features embedded into the nodes, after the graph creation is done.

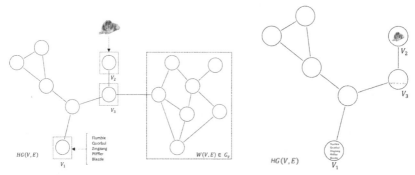

(a) Heterogeneous Graph supporting heterogeneous node data types.

(b) Heterogeneous Graphs with heterogeneous features embedded.

Fig. 2. Heterogeneous Graph within heterogeneous data types.

5.2 Embeddings Creation and Rebuilding the Graph

After the graph is ready for use, the following steps describe the creation of embeddings from the processing of heterogeneous features, detailing the techniques used in the proposed algorithm:

1. *Specialized Autoencoder for Processing Image Feature Data Type* - nodes of the type image are processed in two steps. First, the Convolutional Neural Network (CNN) classifies the image, and second, extracts the characteristics from the image. Both results, class, and characteristics are merged and saved into the nodes as data text features. CNNs can be represented like the approximation function $f^* : \mathbb{R}^{\alpha \times \beta \times \gamma} \to \{1, ..., c\}$ that takes as input an image from unknow distribution $\mathbb{R}^{\alpha \times \beta \times \gamma}$ where $\alpha \times \beta$ are pixels of image and γ is the number of color bands and determines which of class the image is from a set of classes $\{1, ..., c\}$.

    Our proposed algorithm uses the ResNet50 model, a CNN with 50 layers and ImageNet weights, to implement the Specialized Autoencoder for extracting classes and characteristics from nodes where the types are images. The extracted data was saved into the respective node feature as a text vector;

2. *Specialized Autoencoder for Generating Aggregated Embedding* - the text features of nodes are captured on k interaction over their neighbors and combined with the information embedding on the before $k-1$ interaction. The approach has two phases, Message-Passing and Readout [9,13]. The Message-Passing phase or Propagation step runs for T time steps and contains two subfunctions: message function M_t and a node update function U_t. During the message passing phase, hidden states h_v^t at each node in the graph are updated based on messages m_v^{t+1} according to:

$$m_v^{t+1} = \sum_{w \in N_{(v)}} M_t(h_v^t, h_w^t), \quad h_v^{t+1} = U_t(h_v^t, m_v^{t+1}), \tag{1}$$

where the sum $N_{(v)}$ denotes the neighbors of node v in graph G, h_w^t node embedding from node v to w, and function U_t updates the hidden states h_v^t. The Readout phase uses function R to compute an embedding vector representation for the entire graph as follows:

$$\hat{y} = R(\{h_v^T \mid v \in G\}), \tag{2}$$

where T denotes the total time steps.

    The Message-Passing phase in our algorithm was implemented through random walks, which accessed k neighboring nodes, extracted their features, and aggregated them with the local node. In an attempt to avoid the Over-Smoothing issue in the first layer, the node features were saved into a data structure of type set, removing duplicate features. The Readout phase in our proposed algorithm used a Word2Vec model to generate the vector node embeddings from the node features produced by the Message-Passing phase;

3. *Specialized Autoencoder for Generating Metapaths Embedding* - the first step of the process is to read the original graph as input, and through the Meta-Path2Vec algorithm using the pre-defined metapaths, within a sequence of nodes dependent on their node type, walking transversely in the graph captures the semantics between the nodes. The information of nodes is saved into the latent vector space represented by the function $f : v \to \mathbb{R}^d$ where v is the node embedding, and $\mathbb{R}$ is the vector space. The specific nodes of

the data type graph already have an adjacency matrix linked, like a subgraph embedded, where the MetaPath2Vec exploits it and generates the corresponding metapath embedding [15]. MetaPath2Vec uses random walk to generate the metapaths embeddings, where Skip-Gram or Node2Vec model maintains the proximity of node v and its neighbors in the random walk sequences, and Heterogeneous Graph $\mathcal{HG} = (V, E, T)$ is considered, where T_V represents the set of node types and $|T_V| > 1$ maximizing the probability of having heterogeneous context $N_t(v), t \in T_V$ given a node v:

$$arg \max_{\theta} \sum_{v \in V} \sum_{t \in T_V} \sum_{c_t \in N_t(v)} \log p(c_t \mid v; \theta), \qquad (3)$$

where $N_t(v)$ denotes $v's$ neighbourhood with the tth type of nodes and $p(c_t \mid v; \theta)$ is defined as a Softmax function, and

$$p(c_t \mid v; \theta) = \frac{e^{X_{c_t} \cdot X_v}}{\sum_{u \in V} e^{X_u \cdot X_v}}, \qquad (4)$$

where X_v is the vth row of X, representing the embedding vector for node v [7].

The proposed algorithm implements the MetaPath2Vec algorithm by performing walks on the Heterogeneous Graph based on a set of predefined metapaths. The hyperparameters "length" and "walks" define the length of each random walk and the number of random walks to generate from each node, respectively. By tracking valid walks based on node types, a Node2Vec model is built to generate the metapath vector embeddings for each node;

4. *Consolidating Data Embedding and Rebuilding the Graph within Node Features and Embeddings* - each Autoencoder $A : \mathbb{R}^n \to \mathbb{R}^p$ generates its data vectors output that is used as input per the Decoder $B : \mathbb{R}^p \to \mathbb{R}^n$ to rebuild the entire graph, associating the nodes with their data embedding generated, that satisfy

$$arg \min_{A,B} E[\Delta(x, B \circ A(x)], \qquad (5)$$

where E is the expectation over the distribution of x, Δ is the reconstruction loss function, which measures the distance between the output of the Decoder and the input, and A and B are neural networks [4,5];

Based on the data saved during the entire process, the proposed algorithm rebuilds the Heterogeneous Graph with the original nodes and edges, adding the features generated, vector embeddings, and prediction information to the respective nodes and edges.

5.3 Algorithms

The Algorithm 1 aims to process Heterogeneous Graphs with features of different data types as input, according to the procedures described in Sect. 5.2. It produces node representations in latent vector spaces, consolidating various data type features into a unified format and generating the respective node embeddings. The algorithm takes a Heterogeneous Graph $\mathcal{HG}$ as input data, and it consists of several main blocks. Lines 1 to 3 locate the subgraphs in the $\mathcal{HG}$ and call itself to solve the generation of features and embeddings. Lines 4 to 7 iterate over the nodes, uploading the images and extracting their features using a specialized image Autoencoder based on a ResNet50 CNN model with ImageNet weights. Line 8 generates node embeddings from node features using specialized text Autoencoder with Word2Vec algorithm. Lines 9 and 10 iterate over the set of hyperparameter metapaths M, generating metapath embeddings into the local node from its neighbors according to the MetaPath2Vec algorithm, using Node2Vec model to create the respective embeddings. Lines 11 to 15 generate aggregated embeddings from direct neighbors and create compositions of embeddings, such as features with metapaths and aggregated features with metapaths. Thus, $\mathcal{HG}$ has all the features and embeddings returned in line 17, upon completion of the execution of Algorithm 1.

Algorithm 1 delivers a Heterogeneous Graph $\mathcal{HG}$ with vector embeddings v_h, where $v_h \in \{Features, Aggregated, Metapaths, Features + Metapaths, Aggregated + Metapaths\}$. Thus, the next step is to evaluate the performance of the downstream application using $\mathcal{HG}$ with embeddings v_h to determine if it achieves better results compared to without node embeddings.

Algorithm 2 implements RecSys tasks as an example of downstream application, where lines 1 to 5 compute the RecSys tasks based on the set H of embeddings. In line 3, the Link Prediction task can define a target node V_v to predict the new edges using Cosine-Similarity as the applied technique. In Line 4, the node classification task has hyperparameters Iterations = 5 and K-fold cross-validation = 10 to calculate performance based on the average metrics and evaluate their standard deviations. Line 5 calls the node clustering task, defining 3 clusters using the KMeans algorithm. The generation of JSON and CSV files aims to provide a new public dataset with the raw data for the research community.

The time complexity of the algorithms is determined by the loops that iterate over the graph nodes, which is $O(n)$ where n is the number of nodes in the graph in the worst case, $O(m \times n)$ where m is the number of metapaths, and $O(n)^2$ generated in the nested loops for generating embeddings composed of features neighbors of each node, which dominate over the other complexities. Therefore, the total time complexity of the proposed algorithms are polynomial $O(n)^2$ in the worst case.

6 Experiments

This section presents the experiments aimed at applying RecSys tasks to the graph with and without embeddings, and evaluating the resulting metrics in both cases. Special focus is given to the composition of embeddings, such as Features+Metapaths and Aggregated+Metapaths, as introduced in this paper. The experiments were guided by the methodology based on the approach for generating enhanced heterogeneous features and embeddings using the Algorithms

1 and 2. Figure 3(a) shows the main pipeline of the methodology used to execute the entire set of experiments, where they are guided by the follows steps:

Algorithm 1: GenProcHetEmbedding: Generating and processing heterogeneous data type features and embeddings.

Data: Graph $\mathcal{HG}(\mathcal{V}, \mathcal{E}, \mathcal{M})$ where $\mathcal{V}$, $\mathcal{E}$, and $\mathcal{M}$ are sets of Nodes, Edges, and Metapaths respectively.
Result: The graph $\mathcal{HG}$ with latent vector embedding
$v.embedding \in \{features, aggregated, metapath, compositions\} \in \mathcal{V}$.

1 **foreach** $node\ v \in \mathcal{V}$ **do**
2   **if** $v.dataType = Subgraph$ **then**
3   GenProcHetEmbedding(v, $\mathcal{HG}$)
4   **if** $v.dataType = Image$ **then**
5   **if** $v.image\ is\ empty$ **then**
6          $v.imagePath, v.image \leftarrow upload(imageDB)$
7       $v.features \leftarrow AE.Image\_CNN(algo = \text{``resnet50''}, weights = \text{``imagenet''})$
8    $v.embedding_{features} \leftarrow AE.Text\_word2vec(k = \text{``all''})$
9 **foreach** $metapath\ m \in \mathcal{M}$ **do**
10    $v.embedding_{metapath} \leftarrow metapath2vec(algo = \text{``node2vec''}, length = 10, walks = 5)$
11 **foreach** $node\ v \in \mathcal{V}$ **do**
12   **foreach** $node\ neighbors(v)$ **do**
13       $v.embedding_{aggregated} \leftarrow random\_walk(v)$
14    $v.embedding_{features\_metapath} \leftarrow v.embedding_{features\_metapath}$
15    $v.embedding_{aggregated\_metapath} \leftarrow v.embedding_{aggregated\_metapath}$
16 Decoder $D(\mathcal{HG})$ gets the entire graph data including latent vector spaces and rebuilds the graph
17 return $\mathcal{HG}$

---

1. *Graph Generation* - responsible for creating Heterogeneous Graphs, including nodes, edges, and node features when available;
2. *Embeddings Generation* - process of creating embeddings, where Features are generated from text node features; Aggregated embeddings are generated from the text node features of neighboring nodes; Metapath embeddings are generated from the defined set of metapaths; Features+Metapaths and Aggregated+Metapaths are compositions of those embeddings.
3. *RecSys and Performance Metrics* - iterates over the set of each embedding, applying RecSys tasks to each type of embedding and generating respective performance metrics.

In essence, the experiments involve the generation of features and embeddings from nodes, and the validation of RecSys performance metrics based on the

Algorithm 2: RecSys: Running RecSys tasks over the Heterogeneous Graph.

Data: Graph $\mathcal{HG}(\mathcal{V}, \mathcal{E})$ where $\mathcal{V}$, and $\mathcal{E}$ are sets of Nodes and Edges.
Embeddings $\mathcal{H} \in \{Features, Aggregated, Metapaths, Features + Metapaths, Aggregated + Metapaths\}$
Result: The recommended graph $\mathcal{HG}$, JSON, and CSV files.

1 **foreach** *type of embedding* $h \in \mathcal{H}$ **do**
2     $target \leftarrow specific\ V_v\ or\ None$
3     $v.predictedLink \leftarrow link\_prediction(h, target = target, algo =$ "$Cosine\_Similarity$")
4     $v.predictedType \leftarrow node\_classification(h, iter = 5, kfolds = 10, algo =$ "$XGBoost$")
5     $v.predictedCluster \leftarrow node\_clustering(h, k = 3, algo =$ "$KMeans$")
6 Generating JSON, CSV files of the Recommended Heterogeneous Graph
7 **return** $\mathcal{HG}$, graph JSON and CSV files

embeddings generated. To evaluate the generalization capability and robustness of XGBoost models used in the experiments, performance Accuracy and F1-Score metrics should be calculated using 10 stratified K-folds and the average of 5 iterations, along with their respective Standard Deviations.

6.1 Heterogeneous Graph Data Model

Figure 3(b) defines the Heterogeneous Graph employed in the experiments, containing features of different data types, such as Person, Car, and Pet. Each node has its own heterogeneous features, which may include texts, images (e.g., cars and pets), or subgraphs, such as the family of Mary embedded into the Mary node, each with its corresponding embeddings.

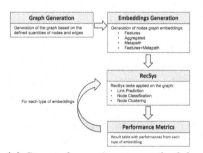
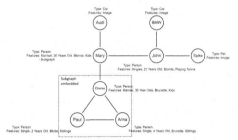
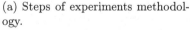

(a) Steps of experiments methodology.
(b) Heterogeneous Graph use case model used in the experiments.

Fig. 3. Methodology and Heterogeneous Graph within heterogeneous data types model.

Based on the scope of experiments and the use case shown in Fig. 3(b), the graph data model was defined according to Table 1. The entire Heterogeneous Graph used in the experiments is available at https://github.com/silviofernandoangonese/datasets/blob/main/experiments_initial_het_graph.json.

Table 1. Heterogeneous Graph data model used in the experiments.

Field	Graph	Depth	Data	Description Domain
Nodes	Nodes	1	Empty	JSON root node
ID	Nodes	2	String	Node identification
Attributes	Nodes	2	Empty	Set of node attributes
Type	Nodes	3	String	Node types. Define the heterogeneity
Data_Type	Nodes	3	String	Data types: Text, Image, and Subgraph
Image	Nodes	3	String	Pillow (PIL) image representation
Image_Path	Nodes	3	String	Path of the image uploaded
Features	Nodes	3	String	Vector of text features
Aggregated	Nodes	3	String	Vector of text features from neighbors
Predicted_Type	Nodes	3	String	Node type predicted
Predicted_Cluster	Nodes	3	String	Node cluster predicted
Embedding	Nodes	3	Empty	Set of vector node embeddings
Features	Nodes	4	Number	Text Features
Aggregated	Nodes	4	Number	Aggregated Features from neighbors
Metapaths	Nodes	4	Number	Metapaths
Features_Metapaths	Nodes	4	Number	Composition Features and Metapaths
Aggregated_Metapaths	Nodes	4	Number	Composition Aggregated and Metapaths
Edges	Edges	1	Empty	Set of graph edges
Source	Edges	2	String	Node ID from the source node
Target	Edges	2	String	Node ID from the target node
Predicted	Edges	2	Bool	True indicates the edge was predicted

6.2 Execution of Experiments

The experiments were guided according to the methodology shown in Fig. 3(a) and the pre-definition of the metapaths used in the experiments, which were $\mathcal{M} \leftarrow \{(Car, Person), (Pet, Person), (Car, Person, Person), (Pet, Person, Pet),$ $(Person, Person, Car), (Person, Person, Pet), (Person, Person, Person)\}$. These 13 metapaths represent the interactions between different types of nodes, providing the business semantics for the respective node embeddings.

The second step was to generate single and composed node embeddings, which were defined as Features, Aggregated Features, Metapaths, a composition of Features and Metapaths, and a composition of Aggregated Features and Metapaths, which guided the entire experiments and analysis shown in Table 2.

The third and fourth methodological steps aim to validate the graph data quality by applying Link Prediction, Node Classification, and Node Clustering tasks to all the previously defined types of embeddings. Link Prediction was calculated using Cosine

Similarity with a threshold of 80% based on the target node. The algorithm supports selecting a specific node or all the nodes as a target. For Node Classification, the node type was used as the class. Link prediction from "No Embeddings" was calculated using the Jaccard algorithm based on the intersection and union set operation $J(A,B) = \frac{|A \cap B|}{|A \cup B|}$. For Node Clustering, we pre-defined three clusters $C0, C1, C2$. "No Embeddings" clusters were calculated using the Louvain algorithm based on the nodes community where nodes without community identified have no cluster assigned.

The experiments results are collected and presented in Table 2, where each column represents the following: "Type Node Embeddings" - indicating the type of generated embeddings, using an ablation perspective, where the experiments aim to identify which embedding has the greatest impact on the performance of the models; "Prediction Links" - the count of links predicted by RecSys Link Prediction task; "Classification Avg Acc and F1-Score" - are average of Accuracy and F1-Score metrics, and "STDs" - are the Standard Deviation from each average of metrics respectively, illustrating the overall accuracy of the entire node graph classification; "Cor" - is the count of correctly predicted links applied to the final model over the entire graph; and "Inc" - is the count of incorrectly predicted links. Columns "Clusters C0 C1 C2" - display the count of nodes clustered by the RecSys task.

Table 2. Performance metrics achieved from different types of node embeddings.

Nodes: 872, Edges: 799, Metapaths: 13										
	Prediction	Classification						Clusters		
Type Node Embeddings	Links	Avg Acc	STD	F1-Score	STD	Cor	Inc	C0	C1	C2
No Embeddings	6	43.58%	2.61%	60.70%	2.13%	380	492	63	54	48
Features	866	43.46%	1.50%	60.59%	1.22%	379	493	1	447	424
Aggregated	72	61.01%	0.11%	75.78%	0.13%	532	340	26	681	165
Metapaths	6	63.76%	1.96%	77.87%	2.05%	556	316	462	184	226
Features+Metapaths	6	67.09%	1.14%	80.30%	1.08%	585	287	462	184	226
Aggregated+Metapaths	1	**71.90%**	1.98%	**83.66%**	1.95%	627	245	214	191	467

6.3 Evaluation of Results

The assumption that enriching the Heterogeneous Graph within heterogeneous embeddings, generated from processing the available data within the graph, could enhance the performance of downstream applications, was validated. Table 2 illustrates the evolution of the performance starting with Features embeddings and progressing to the best performance achieved by the composition of Aggregated+Metapaths node embeddings, as indicated by the average of Accuracy and F1-Score metrics. Link Prediction count using Features embedding is so high, it can indicate data homogeneity with a lack of distinctive features. Although it varies based on certain factors, predictions from Metapath and Features+Metapaths may be deemed more reliable. Node Clustering already reveals similar cluster distributions independent of the embedding used, except Aggregated embedding.

The best results were achieved with the composition Aggregated+Metapaths embeddings, where the results indicate that combining aggregated features from neighbors and metapaths embeddings led to better performance in all tested RecSys tasks. The experiments reveal that the best average Accuracy was 71.90% with a 1.98% Standard Deviation, which means the Accuracy could range from 69.92% to 73.88%. The average F1-Score is already at 83.66% with a 1.95% of Standard Deviation, thus the F1-Score can vary between 81.68% and 85.61%. A higher Standard Deviation indicates a greater spread of results, suggesting that the model is more sensitive to variations in input data or other training conditions. On the other hand, a lower Standard Deviation indicates greater consistency in results, which may indicate that the model is more stable and robust. This suggests that combining information from node content with information from semantics defined by features and structural relationships defined by metapaths can lead to more powerful graph node representation. The final recommended heterogeneous graph JSON file is available at https://github.com/silviofernandoangonese/datasets/blob/main/experiments_final_recommended_het_graph.json and CSV file at https://github.com/silviofernandoangonese/datasets/blob/main/experiments_final_recommended_het_graph.csv.

The failure cases were due to a high number of links predicted from the Features embedding, specifically 866 out of 872 nodes. This result is not acceptable, indicating the need for a deeper analysis of the model to understand the reason for the discrepancy in values. Additionally, another topic related to the Features embedding is the smallest cluster, C0, which contains only 1 node. This raises an important question: why is there only 1 node in this cluster, and what does it mean? Future investigations should explore this anomaly to enhance our understanding of the clustering behavior and improve the accuracy model.

7 Conclusion

This paper proposed an algorithm based on the AGHE - Approach for Generating Enhanced Heterogeneous Embeddings from Heterogeneous Graphs, enhancing the graph as a dataset for downstream applications. The performance achieved by the experiments conducted, especially when we compare "No Embeddings" with "Aggregated+Metapaths" embeddings demonstrates how the proposed algorithm effectively collaborates with the data enhancements in Heterogeneous Graph. Represented by the RecSys, which was used as a downstream application reference in this paper. Developing effective and efficient graph analytics from information embedding, can greatly help to better understand complex graphs, and provide innovative solutions for data models. Based on the obtained results, we believe that the conducted studies can open the doors for its use in different downstream applications as demonstrated. Some specific evaluations can be achieved, where the choice of appropriate embeddings plays a crucial role in the performance of downstream tasks. The results indicate that a one-size-fits-all embedding approach is not necessarily the best for all tasks and datasets.

An important lesson learned from the experiments is the significance of exploring a variety of embedding generation techniques and considering the unique characteristics of the data and tasks at hand. Combining information from different sources, such as node features and structural relationships defined by metapaths, can lead to more comprehensive and informative node representations in the graph. This highlights the importance of exploring hybrid approaches.

Future works include: aggregating edges data features to enhance the nodes data embedding; evaluating the performance of recommendations using a tabulated dataset and the same dataset modeled as a Heterogeneous Graph with heterogeneous embeddings; automatic detection of subgraphs and embedded it into the interconnected node; evaluation of the impact of the embeddings vectors elements normalization;

References

1. Alslaity, A., Tran, T.: Towards persuasive recommender systems. In: 2019 IEEE 2nd International Conference on Information and Computer Technologies (ICICT) on Proceedings, pp. 143–148. Publisher (2019)
2. Angonese, S. F., Galante, R.: AGHE: approach for generating enhanced heterogeneous embeddings from heterogeneous graphs. In: 2024: Proceedings of the 51st Integrated Software and Hardware Seminar (SEMISH) on Proceedings, pp. 252–263. Publisher (2024)
3. Avelar, P.H.C., Tavares, A.R., da Silveira, T.L.T., Jung, C.R., Lamb, L.C.: Superpixel image classification with graph attention networks. In: 2020 33rd SIBGRAPI Conference on Graphics, Patterns and Images (SIBGRAPI), pp. 203-209. Publisher (2020)
4. Baldi, P.: Autoencoders, unsupervised learning and deep architectures. In: Proceedings of the 2011 International Conference on Unsupervised and Transfer Learning Workshop - Volume 27 on Proceedings, pp. 37-50. Publisher (2011)
5. Bank, D., Koenigstein, N., Giryes, R.: Autoencoders. Publisher (2021)
6. Barret, N., Gauquier, A., Law, J.J., Manolescu, I.: PathWays: entity-focused exploration of heterogeneous data graphs. In: The Semantic Web: ESWC 2023 Satellite Events on Proceedings, pp. 91–95. Publisher (2023)
7. Dong, Y., Chawla, N.V., Swami, A.: MetaPath2Vec: Scalable Representation Learning for Heterogeneous Networks. In: Proceedings of the 23rd ACM SIGKDD International Conference on Knowledge Discovery and Data Mining on Proceedings, pp. 135-144. Publisher (2017)
8. Fu, X., Zhang, J, Meng, Z., King, I.: MAGNN: metapath aggregated graph neural network for heterogeneous graph embedding. In: Proceedings of The Web Conference 2020 on Proceedings, pp. 2331–2341. Publisher (2020)
9. Gilmer, J., Schoenholz, S.S., Riley, P.F., Vinyals, O., Dahl, G.E.: Neural Message Passing for Quantum Chemistry. Publisher (2017)
10. Hamilton, W.L., Ying, R., Leskovec, J.: Inductive representation learning on large graphs. In: Proceedings of the 31st International Conference on Neural Information Processing Systems on Proceedings, pp. 1025–1035. Publisher (2017)
11. Han, K., Wang, Y., Guo, J., Tang, Y., Wu, E.: Vision GNN: an image is worth graph of nodes. In: Advances in Neural Information Processing Systems, pp. 8291–8303. Publisher (2022)
12. Li, J., Zhang, Q., Liu, W., Chan, A. B., Fu, Y-G. Koishekenov, Y.: Another perspective of over-smoothing: alleviating semantic over-smoothing in deep GNNs. IEEE Trans. Neural Networks Learn. Syst. Proc., 1–14. Publisher (2024)
13. Liu, Z., Zhou, J.: Introduction to Graph Neural Networks. Publisher (2020)
14. Rozemberczki, B., Davies, R., Sarkar, R., Sutton, C.: GEMSEC: graph embedding with self clustering. In: GEMSEC: Graph Embedding with Self Clustering on Proceedings, pp. 65–72. Publisher (2020)
15. Sun, Y., Han, J.: Mining Heterogeneous Information Networks: Principles and Methodologies. Publisher (2012)

16. Wang, X., Bo, D., Shi, C., Fan, S., Ye, Y., Yu, Philip S.: A survey on heterogeneous graph embedding: methods, techniques, applications and sources. In: IEEE Transactions on Big Data on Proceedings, pp. 415–436. Publisher (2023)
17. Wu, S., Fei, S., Wentao, Z., Xie, X, Cui, B.: Graph Neural Networks in Recommender Systems: A Survey. Publisher (2023)
18. Ying, R., He, R., Chen, K., Eksombatchai, Hamilton, P.W., Leskovec, J.: Graph convolutional neural networks for web-scale recommender systems. In: Proceedings of the 24th ACM SIGKDD; Data Mining on Proceedings, pp. 974–983. Publisher (2018)
19. Zhang, C., Song, D., Huang, C., Swami, A., Chawla, N.: Heterogeneous graph neural network. In: Proceedings of the 25th ACM SIGKDD; Data Mining on Proceedings, pp. 793–803. Publisher (2019)

Ensemble of CNNs for Enhanced Leukocyte Classification in Acute Myeloid Leukemia Diagnosis

Leonardo P. Sousa[1]($\boxtimes$), Romuere R. V. Silva[1], Maíla L. Claro[2], Flávio H. D. Araújo[1], Rodrigo N. Borges[1], Vinicius P. Machado[1], and Rodrigo M. S. Veras[1]

[1] Federal University of Piauí, Teresina, Brazil
{leonardosousa,romuere,flavio86}@ufpi.edu.br,
{vinicius,rveras}@ufpi.edu.br
[2] Federal Institute of Piauí, Paulistana, Brazil

Abstract. Acute Myeloid Leukemia (AML) is one of the most lethal and aggressive forms of hematological cancer, characterized by the rapid proliferation of immature leukocytes. This disease is diagnosed by highly trained specialists who meticulously analyze microscopic images of blood smears. This work explores the feasibility and effectiveness of using eight Convolutional Neural Network (CNN) architectures to form specialized ensembles capable of accurately differentiating between mature and immature leukocytes. We used voting ensemble techniques and the bagging method to integrate CNNs that achieved the best individual performances. The bagging strategy, explicitly using the EfficientNet B3 CNN, stood out by achieving an accuracy of 96.62%, a precision of 98.11%, and a Kappa index of 92.27% on a dataset of 48,100 blood cell images. This performance enhancement highlights the superior diagnostic capabilities of this approach compared to the individual architectures of CNNs in identifying cell types in the context of AML diagnosis.

Keywords: Convolutional Neural Networks (CNNs) · Leukocyte Classification · Acute Myeloid Leukemia Diagnosis

1 Introduction

Blood plays a crucial role in transporting oxygen and nutrients to all organs and comprises three main types of cells: erythrocytes, platelets, and leukocytes. Erythrocytes are responsible for the transport of gases such as oxygen and carbon dioxide; platelets are essential in the blood clotting process [16]; and leukocytes, or white cells, are tasked with defending the organism. Additionally, leukocytes can be classified as mature (fully developed cells) or immature (cells in the development process).

All these cells originate in the bone marrow and are regularly released into the bloodstream. The healthy development of blood cells can be compromised by the disproportionate increase in abnormal blood cells, becoming the primary cause of blood cancer. Leukemia originates when the malignancy of blood precursor cells manifests through their uncontrolled proliferation, primarily affecting leukocytes [27].

Leukemias can be categorized as either acute or chronic. The acute form manifests rapidly, requiring immediate therapeutic intervention, while chronic leukemias progress more slowly, often not requiring immediate treatment after diagnosis. They can also be grouped based on the types of white blood cells they affect: lymphoid or myeloid [15]. Thus, the main types of leukemia are Acute Lymphoblastic Leukemia (ALL), Acute Myeloid Leukemia (AML), Chronic Myeloid Leukemia (CML), and Chronic Lymphocytic Leukemia (CLL).

AML stands out as the most lethal variant among the four leukemia subtypes, with approximately 20,380 new cases and about 11,300 deaths in the United States in 2023, according to data from the National Cancer Institute. The five-year average survival rate is 31.7%[1]. From 2009 to 2019, it was the most prevalent leukemia in Brazil, with 10,554 cases[2]. The disease carries an unfavorable prognosis, and any delay in accurate diagnosis has a severe and negative impact on the patient's survival capacity.

Currently, no specific screening methods are available for early detection of leukemia before the onset of symptoms. Physicians advise that vulnerable individuals undergo regular medical examinations, including physical assessments and routine blood tests. The detection of an increase in leukocyte count during these tests may indicate the presence of various factors, such as infection, stress, inflammation, or, in some cases, bone marrow disorders, including the possibility of leukemia.

Subsequently, the blood smear examination is conducted by trained technical personnel to carry out morphological analyses, count, and identify anomalies in blood cells. This process prioritizes the detection of specific characteristics associated with each type of leukemia. According to Sadek et al. [19], the diagnostic criterion for AML is established when the count of blasts (immature leukocytes) reaches or exceeds 20%. Figure 1 displays examples of mature and immature leukocytes. Biermann et al. [3] described that mature leukocytes typically exhibit more segmented and ring-shaped nuclei, while immature ones have larger, more rounded, and less segmented nuclei.

Considering the low survival rate, especially among adults diagnosed with AML, and the inherent challenges of manual leukocyte classification for diagnosing AML, it becomes crucial and indispensable to develop tools that utilize computational technology and advances in machine learning methods. These tools aim to assist specialist technicians in validating and supporting the diagnostic process.

Convolutional Neural Networks (CNNs) have demonstrated exceptional performance in image classification. In this study, in addition to analyzing individual networks, we explored various ensemble techniques, including majority voting,

---

[1] https://seer.cancer.gov/statfacts/html/amyl.html.
[2] https://www.htct.com.br/pt-leucemia-mieloide-aguda-perfil-clinico-epidemiologico-articulo-resumen-S253113792101018X.

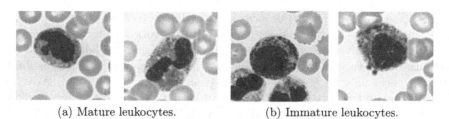

(a) Mature leukocytes.   (b) Immature leukocytes.

Fig. 1. Examples of (a) Mature and (b) Imature Leukocytes.

weighted voting and bagging. We chose to use distinct architectures, selected for their topological innovations and methods, to optimize image classification. Specifically, we propose an ensemble using the bagging technique with the EfficientNet B3 architecture to classify mature and immature leukocytes in blood smear images aimed at detecting AML. For analysis, we constructed a dataset by combining information from three public datasets, totaling 48,100 images, enhancing our model's diversity and robustness.

This paper is organized as follows: Sect. 2 presents related work, citing the most relevant studies. Section 3 introduces the image dataset, the techniques applied, and the evaluation metrics adopted in the study. Section 4 presents the results and discussions, and Sect. 5 discusses the conclusions and future work.

2 Related Work

In recent years, numerous studies in the literature have explored image datasets to classify various types of leukemia, employing Machine Learning (ML) and Deep Learning (DL) techniques. The principal methodologies identified are discussed below.

An example of leukemia classification is found in works such as Ahmed et al. [2] and Aftab et al. [1], which address the four basic types of leukemia: AML, CML, CLL, ALL, in addition to a class representing images without leukemia. On the other hand, there are papers like Claro et al. [5] that focus exclusively on classifying the two types of acute leukemias (AML and ALL), along with a class representing images without leukemia.

Another approach for diagnosing AML involves classifying and counting different types of leukocytes, focusing on immature leukocytes in blood samples. However, this approach still needs to be explored, with few publications addressing binary classification. Some of the works using this approach are discussed below.

Khanam et al. [10] presents an advanced approach for the diagnosis of AML using the ANCOM convolutional neural network, aiming to identify all AML subtypes and healthy leukocytes automatically. The study utilized an image set from Johns Hopkins Hospital, totaling 18,365 images. Random sampling and data augmentation techniques were employed to balance the dataset, achieving an accuracy of 95% for binary classification.

Rahman and Ahmad [18] propose a methodology to precisely detect immature leukocytes using the convolutional neural network architectures AlexNet, ResNet50, DenseNet161, and VGG-16. After training and validation with the optimization of various parameters, the best-performing model was the modified AlexNet, achieving an accuracy of 96.52%, an area under the curve (AUC) of 94.94%, and an F1-score of 97.00%. They utilized a dataset from Johns Hopkins Hospital with 18,365 images, consisting of 3,532 immature leukocytes and 14,833 mature leukocytes.

Dasariraju et al. [7] began their research with binary classification distinguishing between mature and immature leukocytes. Positive results for immature leukocytes were further subdivided into four classes: Erythroblasts, Monoblasts, Promyelocytes, and Myeloblasts. The LMU-DB dataset (University Hospital of Munich), consisting of 15,192 images, was used. The procedure started with the segmentation of the nucleus using Multi-Otsu thresholding. Subsequently, features were extracted, and an SVM-based classifier was applied. The achieved results show an accuracy of 92.99% for binary classification.

The three studies, as presented in Table 1, address the binary classification of leukocytes using different deep-learning techniques and methodologies. Khanam et al. [10] utilize the ANCOM neural network, achieving an accuracy of 95%. Rahman and Ahmad [18] implement transfer learning with a modified AlexNet, achieving an accuracy of 96.52%. Meanwhile, Dasariraju et al. [7] apply an SVM-based classifier, obtaining an accuracy of 92.99%. The main deficiencies include data imbalance, model complexity, lack of comprehensive comparisons with other methodologies, and the use of only a single fold in testing, which may limit the robustness and generalization of the results.

Table 1. Comparative Summary of Related Works.

Work	Image Dataset	Number of Images	Models Used	Acc(%)
Khanam et al. [10]	Johns Hopkins Hospital	18,365	ANCOM, VGG16, ResNet50, Inception V3	95
Rahman and Ahmad [18]	Johns Hopkins Hospital	18,365	AlexNet, ResNet50, DenseNet161, VGG-16	96.52
Dasariraju et al. [7]	University Hospital of Munich	15,192	SVM	92.99

3 Materials and Methods

To conduct a comparative analysis of acute myeloid leukemia (AML) image classifications, we evaluated pre-trained Convolutional Neural Networks. We selected eight distinct architectures: DenseNet201, EfficientNet B3, InceptionV3, ResNet50, ResNet101, VGG16, VGG19, and Xception. This choice was based on a meticulous review of the literature, in which each of these architectures demonstrated remarkable capabilities in image classification tasks. The diversity of the architectures allows for the exploration of a wide range of characteristics of leukocyte images, essential for enhancing the precision of classification between mature and immature cells.

3.1 Proposed Method

After analyzing the results obtained from both individual CNNs and committees, the proposed approach was developed as illustrated in Fig. 2, where we adopted a bagging ensemble technique to improve image classification using the EfficientNet B3 architecture, which achieved the best results in leukocyte classification. Initially, all images are standardized to 300 × 300 pixels, ensuring consistency in the input data. We used the bootstrap method to train ten distinct models (N = 10), each on randomly selected data subsets with replacement, processed over 100 epochs in batches of 64 images.

We also implemented an early stopping mechanism that monitors validation loss to prevent overfitting. If loss does not improve after a specific number of epochs, training is stopped, and the weights of the best-performing model are saved. This strategy saves training time and ensures that the models are optimally trained without overfitting.

During the testing phase, each model in the ensemble makes independent predictions, and the final classification is determined by averaging these predictions using a decision threshold of 0.5. The final classification distinguishes between mature and immature leukocytes. This approach leverages the diversity of the models in the committee and enhances the robustness of the classification system, ensuring an accurate analysis of leukocyte maturation.

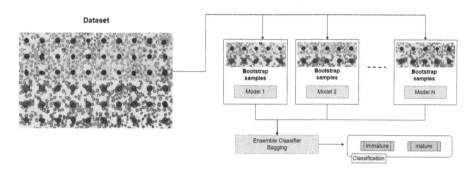

Fig. 2. Proposed Method using a Bagging Ensemble.

3.2 Image Dataset

This study combined three datasets, resulting in 48,100 images categorized into mature and immature classes. Table 2 illustrates the combination of images used.

The Johns Hopkins Hospital dataset, as described in Sidhom et al. [20], contains images of blood smears featuring different types of individual leukocytes (.jpg format, 360 × 360 pixels) from 106 patients diagnosed with AML and Acute Lymphoblastic Leukemia (ALL) across various age groups and genders. Initially, the images were organized by the patient, but they were reorganized by type of leukocyte since the leukocyte identification is present in each patient's folder. This reorganization process resulted in a total of 18,365 images.

In the dataset from the University Hospital of Munich, as referenced in Matek et al. [14], there are 15,192 images of blood smears with individual leukocytes (.tiff format, 400 × 400 pixels) from 100 patients diagnosed with AML. These data were collected between 2014 and 2017, covering 13 classes of images.

The dataset from the Central Laboratory of the Clinical Hospital of Barcelona, described by Boldu et al. [4], consists of 14,543 images of individual cells. The dataset is organized into eight distinct groups: neutrophils, eosinophils, basophils, lymphocytes, monocytes, immature granulocytes (promyelocytes, myelocytes, and metamyelocytes), erythroblasts, and platelets or thrombocytes. The images are 360×363 pixels in size, in .jpg format, and have been annotated by specialized clinical pathologists.

Table 2. Summary of the Image Datasets Used.

Leukocytes		Johns Hopkins	Munich	Barcelona	Total
Mature	Neutrophil (Segmented)	8484	2022	1646	**32624**
	Neutrophil (Band)	109	170	1633	
	Lymphocyte (Typical)	3937	3412	1214	
	Lymphocyte (Atypical)	11	470		
	Monocyte	1789	1311	1420	
	Eosinophil	424	107	3117	
	Basophil	79	51	1218	
Immature	Promyelocyte	70	667	592	**15476**
	Promyelocyte Biobled	18			
	Myelocyte	42	181	1137	
	Metamyelocyte	15	75	1015	
	Myeloblast	3268	3286	0	
	Monoblast	26		0	
	Erythroblast	78	520	1551	
	Smudge cells	15	2920	0	
Total		**18365**	**15192**	**14543**	**48100**

3.3 Evaluated CNN Architectures

CNNs have been widely employed in the field of machine learning. Thanks to their deep architectures, CNNs can map image features at various levels of abstraction, which has significantly contributed to the development of more accurate medical diagnostic tools. In many cases, CNNs surpass the accuracy of conventional feature extraction methods [21], thus demonstrating their potential and effectiveness in analyzing and interpreting medical images.

CNNs designed for the ImageNet Large Scale Visual Recognition Challenge (ILSVRC) were evaluated. In Kornblith et al. [11], the authors concluded that the better a CNN architecture adapts to the ImageNet dataset, the better it will transfer to other datasets. The architectures evaluated for ensemble formation are indicated in Table 7, which refers to the topological depth of the network, number of parameters, input resolution, and year of publication for each (Table 3).

Table 3. Characteristics of the Evaluated Architectures.

Architecture	Depth	Nº Parameters	Resolution	Year	Top-5 acc (ImageNet)
VGG-16	16	138.357.544	224 × 224	2014	90.1%
VGG-19	19	143.667.240	224 × 224	2014	90.0%
ResNet 50	50	25.636.712	224 × 224	2015	92.9%
ResNet 101	101	44.707.176	224 × 224	2015	93.7%
Inception V3	48	23.851.784	299 × 299	2015	94.4%
DenseNet 201	201	20.242.984	224 × 224	2017	93.7%
Xception	71	22.910.480	299 × 299	2017	94.5%
EfficientNet B3	112	12.467.032	300 × 300	2019	95.7%

3.4 Transfer Learning

Transfer Learning is a method where knowledge gained by a neural network on one task is used to improve performance on another related task. This technique reduces the need to re-adjust all the parameters of the CNN from scratch [22].

The proposed method utilizes pre-trained convolutional neural networks to identify mature and immature leukocytes. During development, eight neural networks were evaluated using Shallow Fine-Tuning (SFT), which involves freezing layers from the beginning of the CNN to extract features from the input images and classify them.

The network architecture was first loaded to implement this, excluding the dense layers at the top. Subsequently, the pre-trained convolutional layers were frozen. Custom dense layers were then added to the convolutional layers to tailor the model to the specific needs of binary leukocyte classification. Finally, the model was compiled using the Adam optimizer, the binary cross entropy loss function, and the accuracy metric, preparing it for training.

3.5 Ensembles of Classifiers

The ensemble approach, also known as the ensemble technique, is a method that existed long before the advent of the deep learning paradigm [6]. The theory behind it is pretty simple and is based on the well-known notion of the "wisdom of crowds": instead of relying on just one model for prediction, a set of multiple (pre-trained) models is created. The results of these models are combined into a final classification through a voting mechanism. The original idea was developed to reduce the variance of classifiers, aiming to achieve better overall performance [8].

Constructing an ensemble of classifiers involves three main phases: generating base classifiers, selecting ensemble members, and defining the decision mechanism [26]. In addition to the individual analysis of Convolutional Neural Networks (CNNs), this study created ensembles using the Voting method, which incorporates both majority and weighted voting. The bagging technique was also employed to enhance the diversity and robustness of the classifiers in the tests.

Voting is a fundamental ensemble technique that aggregates the predictions of several models to improve classification performance. In this approach, each model independently predicts the output, and the final decision is made based on either the majority vote or the weighted vote of all models. In the majority voting scheme, each classifier votes for a specific class, and the class receiving the most votes is selected as the final output. This technique is particularly effective when the individual classifiers are diverse, and their errors are uncorrelated [8]. In the weighted voting scheme, different weights are assigned to each classifier based on their performance. This ensures that more reliable classifiers have a greater influence on the final decision, potentially improving overall accuracy [12].

Bagging is an algorithm framework that trains several different models respectively and then lets all models vote to test the output of samples [17]. As shown in Fig. 3 of Zhang et al. [23], Bagging adopts a sampling with replacement to generate multiple training subsets, which are employed to train classifiers [25]. Each training process is independent, so parallel computing could accelerate the process [9]. Particularly, the training subset in Bagging is selected randomly, meaning different subsets can contain the same data. Additionally, Bagging introduces randomization in the training process of each classifier. After training, all classifiers are combined to reduce the variance in prediction results.

3.6 Evaluation Metrics

The technique of stratified k-fold cross-validation was used. This approach randomly distributes instances from the dataset into k subsets (or folds), which are mutually exclusive and of approximately equal size while maintaining the same proportion observed in the original dataset. Thus, the CNN model is fitted and tested k times, and in each iteration, a different subset is reserved for evaluation. In contrast, the remaining k-1 subsets are used for fine-tuning the network parameters. This technique is valuable for assessing the model's performance across different datasets and can provide a more reliable estimate of its generalization ability.

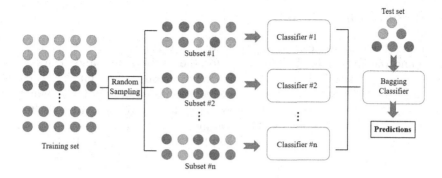

Fig. 3. Structure of Bagging classifier. Adapted from Zhang et al. [23].

The image dataset was divided into five folds (k = 5), resulting in 20% of the instances comprising the test subset. This subset is not used during the network training but is employed in the final evaluation of the classifiers' performance. The images in the remaining folds were divided into two subsets: the training subset, used to adjust the network weights based on the loss observed during training, and the validation subset, which comprises 20% of the total training subset. The latter monitors the training process and identifies potential overfitting issues.

Values from the confusion matrix were used to evaluate the methodology's performance. From these values, five metrics widely recognized in the literature were calculated: accuracy (A), precision (P), recall (R), F1-score (F), and Kappa index (K). The selection of these metrics is justified by their importance in providing a holistic view of the model's performance, as highlighted by Zhang et al. [24]. These metrics are crucial for comprehensively testing machine learning models, enabling not only the assessment of overall effectiveness (accuracy) but also the precision and the model's ability to identify all relevant positive instances (recall), balance these measures (F1-score), and measure agreement adjusted for chance (Kappa).

The Kappa index is recommended as an appropriate measure of accuracy because it can adequately represent the confusion matrix. This index considers all elements of the confusion matrix, not just those on the main diagonal, unlike overall classification accuracy, which considers only the main diagonal elements. The Kappa coefficient is calculated using the formula 1:

$$K = \frac{(observed - expected)}{1 - expected} \times 100 \qquad (1)$$

According to [13], the value of K can be interpreted in the following way: K $\leq$ 20%: Poor; 20% $<$ K $\leq$ 40%: Fair; 40% $<$ K $\leq$ 60%: Good; 60% $<$ K $\leq$ 80%: Very Good; and K $>$ 80%: Excellent.

4 Results and Discussions

From the fine-tuning techniques presented, experiments were conducted to determine the most effective approach for the problem at hand. During the experiments, the selected evaluation metrics were considered. The k-fold cross-validation method was adopted with K equal to 5 to ensure a robust evaluation of the models. Through empirical evaluation, different architectures were tested to optimize the obtained results. All experiments were performed on a computer with an Intel(R) Core(TM) i5 processor at 2.90 GHz, 16 GB of RAM, and an Nvidia Zotac GeForce RTX 3060 12 GB graphics card with 3584 cores.

4.1 Individual Classification

This section presents the individual results obtained with the eight evaluated architectures. The networks were trained for 100 epochs using a batch size of 128 images. A parameter was used to monitor the network's learning progress, setting a patience value of 15 epochs. If accuracy did not improve after this period, the training was terminated.

Additionally, during the CNNs' training, the binary cross-entropy cost function and the Adam optimizer were employed. The dense layers at the network's top were not included, and the pre-trained convolutional layers were frozen. Two custom dense layers were added for the binary classification step: the first composed of 256 neurons with ReLU activation, followed by an output layer with one neuron and sigmoid activation.

The results obtained are described in Table 4, highlighting EfficientNet B3 as the architecture with the best performance, achieving an accuracy of 94.21%, precision of 95.74%, recall of 95.76%, F1-Score of 95.75%, and a Kappa index of 86.69%, which is considered excellent. Including the Squeeze-and-Excitation (SE) attention mechanism in EfficientNet B3 likely contributed significantly to its superior performance. This mechanism enhances the network's ability to focus on the most relevant features within an image by adaptively recalibrating channel-wise feature responses, leading to more effective feature representation and extraction. The standard deviation analysis showed values close to 0, indicating little variation in results across the folds. These results surpassed those of the other individual architectures, demonstrating the effectiveness of the SE module in improving the network's accuracy and robustness.

4.2 Ensemble of CNNs

We conducted 120 experiments to evaluate the robustness of various convolutional neural network (CNN) architectures in an ensemble context. We explored combinations of three networks, selected from eight possibilities, using majority and weighted voting methods. Additionally, we implemented the bagging method on all eight networks evaluated to enhance the robustness of the classifications. The consolidated results indicated that the ensemble approach significantly improved performance metrics compared to individual classifiers. How-

Table 4. Results of the Evaluated Architectures.

Architecture	A(%)	P(%)	R(%)	F1-Score(%)	K(%)
VGG-16	90.79 ± 0.002	92.63 ± 0.005	93.95 ± 0.006	93.28 ± 0.001	78.63 ± 0.004
VGG-19	90.36 ± 0.002	92.57 ± 0.007	93.33 ± 0.007	92.94 ± 0.001	77.71 ± 0.005
ResNet 50	93.11 ± 0.003	94.95 ± 0.005	94.94 ± 0.006	94.94 ± 0.002	84.15 ± 0.006
ResNet 101	92.76 ± 0.004	94.43 ± 0.004	94.97 ± 0.003	94.70 ± 0.003	83.29 ± 0.009
Inception V3	91.36 ± 0.003	93.54 ± 0.007	93.80 ± 0.008	93.66 ± 0.002	80.10 ± 0.006
DenseNet 201	93.07 ± 0.003	94.74 ± 0.004	95.10 ± 0.005	94.92 ± 0.002	84.02 ± 0.006
Xception	91.27 ± 0.003	93.08 ± 0.004	94.18 ± 0.004	93.63 ± 0.002	79.79 ± 0.007
EfficientNet B3	**94.21 ± 0.003**	**95.74 ± 0.003**	**95.76 ± 0.005**	**95.75 ± 0.002**	**86.69 ± 0.006**

ever, the bagging strategy stood out, surpassing the ensembles that employed voting methods and achieving superior results.

The analysis of Table 5 reveals that forming ensembles by combining different convolutional neural network (CNN) architectures and using voting methods significantly increased classification metrics. Notably, the ensemble composed of DenseNet 201, EfficientNet B3, and ResNet 50, using majority voting, achieved the best results: an accuracy of 94.93%, a precision rate of 96.21%, a recall of 96.30%, an F1-score of 96.27%, and a Cohen's Kappa index of 88.35%. These results substantially surpass the metrics achieved by the individual networks, highlighting the superior efficacy of the ensemble approach.

Table 5. Best Results for the Evaluated CNN Ensemble.

Ensemble	Votting	A (%)	P (%)	R (%)	F1-Score (%)	K (%)
DenseNet 201EfficientNet B3ResNet 50	Majority	**94.93 ± 0.001**	**96.21 ± 0.001**	**96.30 ± 0.002**	**96.27 ± 0.001**	**88.35 ± 0.002**
DenseNet 201EfficientNet B3ResNet 50	Weighted	94.87 ± 0.002	94.86 ± 0.003	0.9487 ± 0.002	94.86 ± 0.003	88.18 ± 0.006
DenseNet 201ResNet 50Xception	Weighted	94.61 ± 0.001	94.61 ± 0.001	94.61 ± 0.001	94.61 ± 0.001	87.58 ± 0.003
DenseNet 201ResNet 50Inception V3	Majority	94.59 ± 0.002	95.53 ± 0.004	96.41 ± 0.002	96.03 ± 0.001	87.25 ± 0.005
DenseNet 201ResNet 50Xception	Majority	94.59 ± 0.002	95.53 ± 0.004	96.41 ± 0.003	96.03 ± 0.002	87.50 ± 0.005

Table 6 illustrates the best results achieved with bagging ensemble, applied to various individual neural networks. Among them, the set that used the EfficientNet B3 network stood out notably, achieving an accuracy of 96.62%, precision of 98.11%, recall of 96.89%, F1-Score of 97.50%, and a Kappa coefficient of 92.27%. The superiority of EfficientNet B3 can be largely attributed to its attention layer, which allows the model to focus on the most relevant features within complex images, significantly enhancing classification precision.

Furthermore, a low standard deviation in the results obtained through bagging must be considered, as it indicates high consistency and reliability of the classification metrics. By combining multiple instances of the same network, the bagging method forms a more stable model that better resists data variability and minimizes the chances of overfitting. This stability is enhanced by the attention layer of EfficientNet B3, which enables each model in the ensemble to concentrate on different nuances of the data efficiently. This combination of bagging and focused attention contributes to stronger generalization and more accurate classification.

Table 6. Best Results for the Bagging Ensemble.

Bagging	A (%)	P (%)	R (%)	F1-Score (%)	K (%)
DenseNet 201	95.07 ± 0.001	96.52 ± 0.004	96.22 ± 0.004	96.37 ± 0.001	88.67 ± 0.004
EfficientNet B3	**96.62 ± 0.003**	**98.11 ± 0.002**	**96.89 ± 0.002**	**97.50 ± 0.003**	**92.27 ± 0.002**
Inception V3	94.58 ± 0.007	95.85 ± 0.009	96.22 ± 0.006	96.03 ± 0.005	87.51 ± 0.017
ResNet 50	93.60 ± 0.006	95.07 ± 0.015	95.60 ± 0.011	95.31 ± 0.004	85.21 ± 0.015
Xception	94.58 ± 0.006	96.44 ± 0.013	95.61 ± 0.016	96.00 ± 0.004	87.60 ± 0.012

Figure 4 illustrates a comparative that synthesizes the top performances from individual, ensemble, and bagging approaches, providing a clear visual representation of how different methodologies enhance the effectiveness of convolutional neural network (CNN) architectures. The chart highlights that the bagging approach using EfficientNet B3 stands out across all metrics, achieving superior accuracy, precision, recall, F1-score, and Kappa coefficient. This method exhibits exceptional stability and minimal variability, underscoring its robustness against data inconsistencies and ability to maintain high performance.

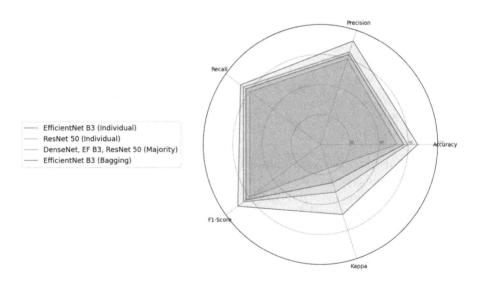

Fig. 4. Performance Comparison of CNN Models Across Different Methods.

The data presented in Table 7 provide a clear comparative view of the advancements achieved in the methodologies for classifying leukocyte images.

Our research, which implemented a bagging strategy using a combination of three public datasets, achieved an impressive accuracy of 96.62%. This not only surpasses the accuracy levels reported in other studies but also marks a significant milestone by utilizing a considerably larger dataset of 48,100 images.

Unlike previous studies, such as Khanam et al. [10], who used the ANCOM architecture on 18,365 images to achieve a 95% accuracy, or Rahman and Ahmad [18] who modified AlexNet to achieve a slightly higher accuracy of 96.52%, our approach significantly expands the size and diversity of the dataset. The use of machine learning by Dasariraju et al. [7] with Random Forest on the same number of images as the first two studies, but achieving only 92.99% accuracy, further underscores the effectiveness of our method. The substantial increase in the number of images in our study not only contributes to the robustness and generalization of the classification model but also enhances the heterogeneity and diversity of the data.

Table 7. Comparison of state-of-the-art and the proposed method: architectures used, number of images, and accuracy achieved.

Work	Number of Images	Used Architecture	Accuracy (%)
Khanam et al. [10]	18,365	ANCOM	95
Rahman and Ahmad [18]	18,365	AlexNet (Modified)	96.52
Dasariraju et al. [7]	18,365	Random Forest	92.99
Proposed Method	**48,100**	Bagging Ensemble	**96.62**

When considering computational efficiency, it's clear that despite its higher accuracy, the CNN ensemble naturally requires more computational resources for training and inference. In comparison, EfficientNet B3, the top-performing individual network, provides a more resource-efficient and faster option. The decision between these approaches should weigh the need for diagnostic accuracy against the available resources, especially in clinical environments where speed and efficiency are critical.

5 Conclusion

The implementation of ensemble methods using bagging with the EfficientNet B3 architecture has shown substantial promise in improving the accuracy of leukocyte classification for the early detection of AML. Our study demonstrated significant improvements in performance metrics compared to individual models.

Using a diverse dataset comprising over 48,100 images from three different public image collections provided robustness in the training and validation of the models. This extensive dataset allowed a thorough evaluation of the ensemble's performance under various image conditions, ensuring the generalization and reliability of our findings.

The promising results obtained in this study suggest that ensemble methods can be a significant step in developing more accurate diagnostic tools in medical imaging. Integrating multiple networks into a bagging ensemble reduced variance and enhanced predictive reliability, which is crucial for clinical applications.

While our research utilizing a bagging strategy with the EfficientNet B3 architecture achieved impressive precision, there are several opportunities for enhancement. Despite the substantial size of our dataset, which includes 48,100 images, further diversification of data sources is essential to improve the model's robustness and its ability to generalize across different clinical scenarios. Integrating generative networks or Large Vision Models (LVMs) during the transfer learning stage could significantly enhance feature robustness. Additionally, employing advanced ensemble techniques such as boosting and stacking may further refine model accuracy and adaptability. These techniques focus on improving training on challenging instances and synthesizing predictions from diverse models, which could lead to more precise and reliable diagnostic tools in medical imaging.

References

1. Aftab, M.O., Awan, M.J., Khalid, S., Javed, R., Shabir, H.: Executing spark bigdl for leukemia detection from microscopic images using transfer learning. In: 2021 1st International Conference on Artificial Intelligence and Data Analytics (CAIDA), pp. 216–220. IEEE (2021)
2. Ahmed, N., Yigit, A., Isik, Z., Alpkocak, A.: Identification of leukemia subtypes from microscopic images using convolutional neural network. Diagnostics **9**(3), 104 (2019)
3. Biermann, H., Pietz, B., Dreier, R., Schmid, K.W., Sorg, C., Sunderkötter, C.: Murine leukocytes with ring-shaped nuclei include granulocytes, monocytes, and their precursors. J. Leukoc. Biol. **65**(2), 217–231 (1999)
4. Boldú, L., Merino, A., Acevedo, A., Molina, A., Rodellar, J.: A deep learning model (alnet) for the diagnosis of acute leukaemia lineage using peripheral blood cell images. Comput. Methods Programs Biomed. **202**, 105999 (2021)
5. Claro, M., et al.: Convolution neural network models for acute leukemia diagnosis. In: 2020 International Conference on Systems, Signals and Image Processing (IWSSIP), pp. 63–68. IEEE (2020)
6. Dasarathy, B.V., Sheela, B.V.: A composite classifier system design: concepts and methodology. Proc. IEEE **67**(5), 708–713 (1979)
7. Dasariraju, S., Huo, M., McCalla, S.: Detection and classification of immature leukocytes for diagnosis of acute myeloid leukemia using random forest algorithm. Bioengineering **7**(4), 120 (2020)
8. Dietterich, T.G.: Ensemble methods in machine learning. In: International Workshop on Multiple Classifier Systems, pp. 1–15. Springer (2000)
9. Khan, F.H., Saadeh, W.: An eeg-based hypnotic state monitor for patients during general anesthesia. IEEE Trans. Very Large Scale Integr. (VLSI) Syst. **29**(5), 950–961 (2021)
10. Khanam, A.S., et al.: Classification of acute myeloid leukemia using convolutional neural network (2023)

11. Kornblith, S., Shlens, J., Le, Q.V.: Do better imagenet models transfer better? In: Proceedings of the IEEE/CVF Conference on Computer Vision and Pattern Recognition, pp. 2661–2671 (2019)
12. Kuncheva, L.I.: Combining pattern classifiers: methods and algorithms. John Wiley & Sons (2014)
13. Landis, J.R., Koch, G.G.: The measurement of observer agreement for categorical data. Biometrics **33**(1), 159–174 (1977)
14. Matek, C., Schwarz, S., Spiekermann, K., Marr, C.: Human-level recognition of blast cells in acute myeloid leukaemia with convolutional neural networks. Nature Mach. Intell. **1**(11), 538–544 (2019)
15. Mrozek, K., Heerema, N.A., Bloomfield, C.D.: Cytogenetics in acute leukemia. Blood Rev. **18**(2), 115–136 (2004)
16. Nascimento, M.d.L.P.: A interação entre as células hematológicas durante as suas atividades. Concluído em (2017)
17. Nguyen, T.T., Pham, X.C., Liew, A.W.C., Pedrycz, W.: Aggregation of classifiers: a justifiable information granularity approach. IEEE Trans. Cybern. **49**(6), 2168–2177 (2019)
18. Rahman, J.F., Ahmad, M.: Detection of acute myeloid leukemia from peripheral blood smear images using transfer learning in modified cnn architectures. In: Proceedings of International Conference on Information and Communication Technology for Development: ICICTD 2022, pp. 447–459. Springer (2023)
19. Sadek, N.A., et al.: Prognostic value of absolute lymphocyte count, lymphocyte percentage, serum albumin, aberrant expression of cd7, cd19 and the tumor suppressors (pten and p53) in patients with acute myeloid leukemia. Asian Pacific J. Cancer Biol. **5**(4), 131–140 (2020)
20. Sidhom, J.W., et al.: Deep learning for diagnosis of acute promyelocytic leukemia via recognition of genomically imprinted morphologic features. NPJ Precision Oncol. **5**(1), 38 (2021)
21. Tajbakhsh, N., et al.: Convolutional neural networks for medical image analysis: full training or fine tuning? IEEE Trans. Med. Imaging **35**(5), 1299–1312 (2016)
22. Yosinski, J., Clune, J., Bengio, Y., Lipson, H.: How transferable are features in deep neural networks? Advances in neural information processing systems **27** (2014)
23. Zhang, H., Zhou, T., Xu, T., Hu, H.: Remote interference discrimination testbed employing ai ensemble algorithms for 6g tdd networks. Sensors **23**, 2264 (02 2023)
24. Zhang, J.M., Harman, M., Ma, L., Liu, Y.: Machine learning testing: survey, landscapes and horizons. IEEE Trans. Software Eng. **48**(1), 1–36 (2020)
25. Zhang, L., Suganthan, P.N.: Benchmarking ensemble classifiers with novel co-trained kernel ridge regression and random vector functional link ensembles [research frontier]. IEEE Comput. Intell. Mag. **12**(4), 61–72 (2017)
26. Zhou, Z.H.: Ensemble methods: foundations and algorithms (2012)
27. Zolfaghari, M., Sajedi, H.: A survey on automated detection and classification of acute leukemia and wbcs in microscopic blood cells. Multimed. Tools Appl. **81**(5), 6723–6753 (2022)

ERASMO: Leveraging Large Language Models for Enhanced Clustering Segmentation

Fillipe dos Santos Silva[1,2,3]([✉]), Gabriel Kenzo Kakimoto[1,2,4], Julio Cesar dos Reis[1,3], and Marcelo S. Reis[1,2,3]

[1] Hub de Inteligência Artificial e Arquiteturas Cognitivas (H.IAAC), Campinas, Brazil
{fillipesantos,g234878,jreis,msreis}@ic.unicamp.br
[2] Artificial Intelligence Laboratory (Recod.ai), Campinas, Brazil
[3] Instituto de Computação, Universidade Estadual de Campinas (UNICAMP), Campinas, Brazil
[4] Faculdade de Engenharia Mecânica, Universidade Estadual de Campinas (UNICAMP), Campinas, Brazil

Abstract. Cluster analysis plays a crucial role in various domains and applications, such as customer segmentation in marketing. These contexts often involve multimodal data, including both tabular and textual datasets, making it challenging to represent hidden patterns for obtaining meaningful clusters. This study introduces ERASMO, a framework designed to fine-tune a pretrained language model on textually encoded tabular data and generate embeddings from the fine-tuned model. ERASMO employs a textual converter to transform tabular data into a textual format, enabling the language model to process and understand the data more effectively. Additionally, ERASMO produces contextually rich and structurally representative embeddings through techniques such as random feature sequence shuffling and number verbalization. Extensive experimental evaluations were conducted using multiple datasets and baseline approaches. Our results demonstrate that ERASMO fully leverages the specific context of each tabular dataset, leading to more precise and nuanced embeddings for accurate clustering. This approach enhances clustering performance by capturing complex relationship patterns within diverse tabular data.

Keywords: Clustering Segmentation · Transformer-based Models · Tabular Data Embeddings

1 Introduction

Tabular data is ubiquitous in various fields such as finance, healthcare, and marketing, where it serves as a primary source of information for Machine Learning

(ML) tasks [15]. Despite its widespread use, extracting meaningful insights from tabular data remains a complex challenge, particularly in clustering tasks [6]. These challenges include handling heterogeneous feature types, dealing with high-dimensional spaces, and ensuring meaningful distance metrics. These can significantly impact the effectiveness of clustering algorithms in identifying natural groupings within the data [14].

Recent researchers have explored traditional statistical methods and modern deep learning approaches [14,19,21] to address these challenges. Existing studies have utilized Large Language Models (LLMs) like OpenAI's GPT and LLaMA to create embeddings from textual datasets, enhancing data representation and analysis [6,14]. In addition, a method combining LLMs and Deterministic,Independent-of-Corpus Embeddings (DICE) has been proposed to generate consistent embeddings across datasets, improving segmentation accuracy [19]. However, these approaches often fail to fully leverage the specific context of each tabular dataset, resulting in less nuanced embeddings for precise clustering.

This study originally introduces ERASMO, our proposed framework designed to gen**ERA**te high-quality embeddings from tabular data using tran**S**former-based language **MO**dels. These embeddings excel in clustering analysis, revealing hidden patterns and groupings. Our solution can also be used in Retrieval-Augmented Generation (RAG) systems and other tasks to gain deeper insights from context information [7]. ERASMO operates through two stages: **(1)** fine-tuning a pretrained language model on textually encoded tabular data; and **(2)** generating embeddings from the fine-tuned model. Using techniques like random feature sequence shuffling and number verbalization, ERASMO produces contextually rich and structurally representative embeddings, outperforming all clustering strategies from the literature based on internal metrics.

Our experimental evaluation used three clustering quality metrics to compare ERASMO with state-of-the-art methods: Silhouette Score (SS), Calinski-Harabasz Index (CHI), and Davies-Bouldin Index (DBI). These metrics comprehensively assess clustering effectiveness results by measuring cohesion, separation, and overall cluster structure.

We extensively evaluated ERASMO on real-world datasets without true labels, including Banking Marketing Targets, E-Commerce Public Dataset by Olist, Yelp reviews, PetFinder.my, and Women's Clothing Reviews. These datasets encompass diversified information and present various challenges, rigorously testing ERASMO's clustering capabilities.

This article provides the main contributions as follows:

- We introduce ERASMO, a novel framework that leverages transformer-based language models to generate high-quality embeddings from tabular data, enhancing clustering analysis.
- We demonstrate that our framework significantly improves clustering performance by capturing the complex relationships within tabular data through random feature sequence shuffling and number verbalization techniques.

- We experimentally achieve state-of-the-art clustering results with ERASMO, showcasing its effectiveness in identifying patterns and groupings within diverse tabular datasets.
- To the best of our knowledge, ERASMO is the first framework to fully integrate and fine-tune transformer-based language models specifically for generating embeddings from tabular data, resulting in superior clustering outcomes.

The remainder of this article is organized as follows. Section 2 presents a synthesis of related work. Section 3 details the ERASMO framework. Section 4 outlines our experimental methodology. Section 5 presents the results of our evaluations. Section 6 discusses the implications of our findings. Finally, Sect. 7 provides conclusions and directions for future work.

2 Related Work

Several studies have explored the application of LLMs to transform tabular data for clustering tasks, demonstrating the potential to enhance user segmentation and data analysis [6,14,19,21,23,24]. Zhu et al. [24] proposed a novel method named Word Embedding of Dimensionality Reduction (WERD) for document clustering. Their approach integrates pre-trained word embeddings with dimensionality reduction techniques. In their work, Sentence-BERT embeds them into high-dimensional vectors after preprocessing documents, which PaCMAP then reduces. Spectral clustering is applied, followed by Non-Negative Matrix Factorization to extract keywords.

CLUSTERLLM [23], a novel text clustering framework, leverages feedback from LLMs such as ChatGPT. This method enhances clustering by utilizing LLMs to refine clustering perspectives and granularity through two stages: a triplet task for fine-tuning embedders based on user preferences and a pairwise task for determining cluster granularity. Extensive experiments on fourteen datasets demonstrated that CLUSTERLLM consistently improves clustering quality and is cost-effective, outperforming traditional clustering methods. Both WERD [24] and CLUSTERLLM [23] presented limitations compared to the ERASMO framework (our proposal). WERD might not fully capture the contextual nuances of each dataset due to its focus on dimensionality reduction techniques. At the same time, CLUSTERLLM's reliance on general-purpose LLMs for guidance may overlook specific dataset characteristics.

A method demonstrating that LLMs enables few-short learning applied to clustering tasks was proposed in [21]. Their study showed how LLMs can perform clustering tasks with minimal labeled data by leveraging their extensive pretraining, significantly reducing the need for large annotated datasets and achieving reasonable clustering performance with few-shot learning. Similarly, Tipirneni et al. [18] explored context-aware clustering using LLMs, highlighting how these models can utilize contextual information to enhance clustering accuracy. Both methods, however, may not fully leverage the dataset-specific nuances as effectively as ERASMO because we employ a fine-tuning step, allowing the model to capture better and utilize dataset-specific details, leading to more accurate and reliable clustering results.

Tissera, Asanka, & Rajapakse developed [19] an approach to enhancing customer segmentation using LLMs and DICE. Their method combined LLMs with DICE to generate consistent and deterministic embeddings across different datasets, improving segmentation accuracy and robustness. Their approach may not fully leverage the context-specific nuances of each dataset as effectively as ERASMO. Our fine-tuning process proposal allows it to adapt to the unique characteristics of the input data, providing more contextually rich and detailed embeddings that can result in more precise and meaningful clusters.

A comparative analysis of LLM embeddings for effective clustering was explored in [6]. As an extension, the study on text clustering with LLM embeddings [14] delves deeper, exploring additional models and datasets to demonstrate improvements in text data clustering. While existing approaches effectively capture complex semantic relationships and handle categorical, numerical, and textual data, they lack the fine-tuning specificity of ERASMO, our key originality aspect. ERASMO's tailored embeddings for tabular datasets and integration of feature order permutation provide more precise and contextually relevant clusters, offering superior versatility and robustness in various clustering applications.

3 ERASMO

This section introduces ERASMO, our framework that leverages transformer-based language models to generate high-quality embeddings from tabular data. These embeddings are particularly effective for clustering analysis, allowing for identifying patterns and groupings within the data that might not be immediately apparent.

The process involves two main stages: **(1)** fine-tuning a pretrained LLM on a textually encoded tabular dataset; and **(2)** utilizing the fine-tuned model to generate embeddings, which are used by a clustering algorithm. These designed stages were inspired by [2]. Subsection 3.1 details the fine-tuning phase, whereas Subsect. 3.2 reports on the embedding generation processes.

3.1 Phase 1: Fine-Tuning

Standard pretrained generative LLMs expect sequences of words as inputs. Hence, we convert each row of our dataset into a textual representation to apply an LLM to tabular data, which can contain categorical, numerical, and textual information.

Definition 1 (Textual Converter). *Given a tabular dataset with m columns with feature names $f_1, f_2, \ldots, f_m$ and n rows of samples $s_1, \ldots, s_n$, let the entry $v_{i,j}$, $i \in \{1, \ldots, n\}$, $j \in \{1, \ldots, m\}$ represent the value of the j-th feature of the i-th data point. Taking the feature name and value into account, each sample s_i*

of the table is transformed into a textual representation t_i using the following subject-predicate-object transformation:

$$t_{i,j} = [f_j, \text{``is''}, v_{i,j}, \text{``,''}], \quad \forall i \in \{1,\ldots,n\}, j \in \{1,\ldots,m\} \tag{1a}$$

$$t_i = [t_{i,1}, t_{i,2}, \ldots, t_{i,m}], \quad \forall i \in \{1,\ldots,n\}, \tag{1b}$$

where $t_{i,j}$, the textually encoded feature, is a clause with information about a single value and its corresponding feature name, and $[\cdot]$ denotes the concatenation operator.

By transforming a tabular feature vector into a sequence using the textual subject-predicate-object encoding scheme, pseudo-positional information is artificially introduced into the transformed tabular data sample. However, there is no spatial ordering relationship between features in tabular datasets. We randomly permute the encoded short sentences $t_{i,j}$ of the full textual representation t_i to reconstruct the feature order independence.

Definition 2 (Random Feature Sequence Shuffle). *Let $t_i, i \in \{1,\ldots,n\}$, be a textual representation. Consider a sequence $k = (k_1,\ldots,k_m)$ that is a permutation of the sequence of indices $(1,\ldots,m)$. A random feature sequence shuffle is defined as $t_i(k) = [t_{i,k_1}, t_{i,k_2}, \ldots, t_{i,k_m}]$.*

We fine-tune our generative language model on samples without order dependencies when using shuffled orders of the textually encoded features. Moreover, such permutations are highly beneficial as they allow for arbitrary conditioning in tabular data generation. In our experiments, we refer to ERASMO$_{base}$ as the baseline model, utilizing only the Textual Converter and Random Feature Sequence Shuffle. In addition, there is evidence that verbalizing numerical tokens can enhance effectiveness in specific scenarios [8]. In this sense, we explore this approach, naming it ERASMO$_{NV}$, as follows.

Definition 3 (Number Verbalizer). *Let $t_i, i \in \{1,\ldots,n\}$, be a textual representation, and $t_{i,j}$ be the set of words of the j-th feature of t_i. A number verbalizer is a function v that receives as input a word w of $t_{i,j}$ and is defined as:*

$$v(w) = \begin{cases} w, & \text{if } w \text{ is not numerical,} \\ \text{verbalized } w \text{ otherwise.} \end{cases} \tag{2}$$

By applying this transformation on every token of every textual representation, we ensure that any numerical information in the text is verbalized. In some NLP tasks, such as clustering with embeddings, sentiment analysis, and text classification, verbalizing numbers can improve the model's understanding of the context and meaning of numerical values, leading to more accurate and meaningful results. This transformation might not be beneficial in some cases, depending on the specific nature of the data and the task at hand [8,9].

Fine-Tuning a Pretrained Auto-Regressive Language Model : We describe the fine-tuning procedure of a pretrained LLM on the encoded tabular data for generation tasks. We suppose a textually encoded tabular dataset $T = \{t_i(k)\}_{i=1,\ldots,n}$ that was transformed into text by the proposed encoding scheme. Let k be a randomly drawn permutation, and n denote the number of rows. Based on user choice, the pipeline can proceed directly to fine-tuning the LLM to generate ERASMO$_{base}$, or it can first apply a number verbalizer to convert numerical tokens into their verbal representations before fine-tuning the LLM to generate ERASMO$_{NV}$.

To be processed with an LLM, the input sentences $t \in T$ must be encoded into a sequence of tokens from a discrete and finite vocabulary W. These tokens can be character, word, or subword encodings such as the Byte-Pair-Encodings (BPE). Thus, $t \in T$ is represented by a sequence of tokens $(w_1, \ldots, w_j) =$ TOKENIZE(t) with tokens $w_1, \ldots, w_j \in W$, where j denotes the number of tokens required to describe the character sequence t. Commonly, the probability of natural-language sequences is factorized in an auto-regressive manner in LLMs. It is represented as a product of output probabilities conditioned on previously observed tokens:

$$p(t) = p(w_1, \ldots, w_j) = \prod_{k=1}^{j} p(w_k \mid w_1, \ldots, w_{k-1}). \tag{3}$$

As a result, an end-user can choose any existing generative language model for tabular data modeling and exploit the vast amount of contextual knowledge presented in these models. Fine-tuning enables the model to leverage this contextual information with the feature and category names to enhance the model's capabilities. Figure 1 presents the pipeline for ERASMO's fine-tuning step.

3.2 Phase 2: Embedding Generation and Clustering Analysis

We generate embeddings from the model after fine-tuning the LLM on the textually encoded tabular dataset. These embeddings capture the contextual relationships and features encoded during the training phase.

We start by feeding the test dataset, transformed into its textual representation, into the fine-tuned LLM. The model generates embeddings for each input sequence, providing a high-dimensional representation for each sample. This process ensures that the embeddings preserve the contextual and feature relationships learned during fine-tuning. Depending on the user's choice in the pipeline, the embeddings are generated from either ERASMO$_{base}$ or ERASMO$_{NV}$ models, reflecting whether the number verbalizer step was applied.

To generate these embeddings, the input sentences $t \in T_{test}$ are encoded into sequences of tokens and processed by the fine-tuned LLM. The embeddings are obtained from the final hidden states of the model, resulting in rich and informative representations of the data. These embeddings can then be utilized for various downstream tasks, including clustering analysis, to gain deeper insights into the data structure (cf. Figure 2).

4 Experimental Methodology

Our experiments evaluated the quality assessment for the best-performing clustering algorithms for each dataset and approach (model) combination (cf. Table 1 for the obtained results). Subsection 4.1 describes the datasets used for training and testing. Subsection 4.2 presents an overview of the clustering algorithms. Subsection 4.3 describes the approaches used for comparison as baselines in our experiments. Subsection 4.4 reports on the evaluation metrics. Subsection 4.5 presents the implementation details. Each experimental setup for a given dataset assessed the different language models fine-tuned and pretrained. Each model considers the several clustering algorithms and their configuration.

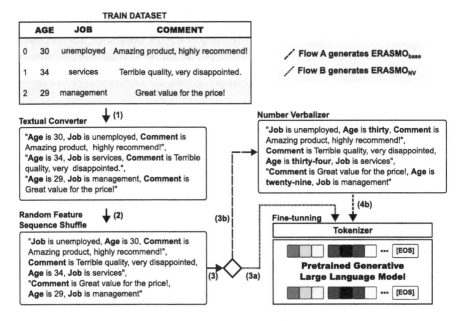

Fig. 1. The ERASMO data pipeline for the fine-tuning phase. First, a textual converter step transforms tabular data into meaningful text (1). Next, a random feature order permutation step is applied (2). Then, based on user choice, the pipeline diverges: it can proceed directly to fine-tuning a LLM (3a) to generate ERASMO$_{base}$, or apply a number verbalizer (3b) before fine-tuning the LLM (4b) to generate ERASMO$_{NV}$.

4.1 Datasets

We selected a diversified set of datasets to encompass a variety of challenges related to text categorization and clustering, and we used them to evaluate text clustering algorithms.

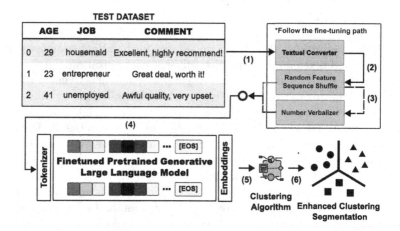

Fig. 2. The ERASMO pipeline for generating embeddings and cluster analysis. The input test tabular data is first transformed into text sequences (1). Next, a random feature order permutation step is applied (2). For ERASMO$_{NV}$, a number verbalizer step follows (3) before processing by the fine-tuned LLM to generate embeddings (4). For ERASMO$_{base}$, the pipeline goes directly from step (2) to step (4). The embeddings are subsequently used for clustering analysis.

- **Banking Marketing Targets:** Composed of data from direct marketing campaigns of a banking institution, which includes client attributes like age and job, along with the response to the campaign [10].
- **E-Commerce Public Dataset by Olist:** A Brazilian e-commerce dataset with over 100,000 orders from 2016 to 2018 across multiple marketplaces [11]. It includes 72,794 training and 18,199 testing samples. The Recency, Frequency, Monetary (RFM) model was used for customer segmentation, as described in [19].
- **Yelp:** Comprises reviews from Yelp businesses, including text reviews, star ratings, and business attributes, offering a rich resource for sentiment analysis and review classification tasks [4].
- **PetFinder.my:** Features adoption records from the PetFinder.my website, encompassing various pet attributes, descriptions, and adoption status, valuable for text classification and clustering related to animal welfare [5].
- **Women Clothing Reviews:** Contains reviews of women's clothing, with each review detailing text feedback, ratings, and customer information, suitable for sentiment analysis and recommendation system research [3].

Each unlabeled dataset was processed through the proposed pipeline, which involves training a pretrained LLM. This approach ensures that the clustering algorithms can perform optimally across diverse textual inputs, enhancing their ability to effectively identify and group related data points.

4.2 Clustering Algorithms

The clustering algorithms chosen are well-suited for handling complex patterns in structured and textual data, ensuring efficient categorization.

We used the k-means algorithm for its simplicity and effectiveness with large datasets and k-means++ for its strategic centroid initialization to enhance clustering efficiency and quality [12]. Unlike k-means, which assigns each data point to a single cluster, Fuzzy C-Means (FuzzyCM) employs a probabilistic membership approach, effectively capturing the nuances and polysemy typical of textual data. We used Agglomerative Hierarchical Clustering (AHC) to uncover hierarchical structures and spectral clustering for its proficiency in recognizing clusters based on the data's graph structure, effectively identifying non-convex shapes.

For k-means, the parameters were: initialization method set to random, number of initializations (*ninit*) set to 10, and the random seed set to 0. The k-means++ algorithm utilized k-means++ for initialization, *ninit* was 1, and the seed was 0. Agglomerative Hierarchical Clustering (AHC) employed the Euclidean metric with Ward linkage. For Fuzzy C-means (FuzzyCM), no initialization method was specified, the fuzziness parameter (m) was 2, error tolerance was set to 0.005, and the maximum number of iterations (*maxiter*) was 1000. Spectral clustering used 'discretize' for label assignment and a random seed of 10.

Implementations for these algorithms were sourced from the *scikit-learn library* [13], except for FuzzyCM, which used the *scikit-fuzzy* package [22]. For k-means and k-means++, *init* specifies the initial cluster centroid method, *ninit* indicates the number of algorithm runs with different seeds, and *seed* sets the random number for centroid initialization. In AHC, *metric* is the metric used for linkage computation, and *linkage* is the criterion measuring the distance between observation sets. We used Euclidean distance to measure point similarities and a nearest centroid approach to associate clusters.

For FuzzyCM, *init* is the initial fuzzy c-partitioned matrix (random if *None*), m is the fuzziness degree, *error* is the stopping criterion, and *maxiter* is the iteration limit. In Spectral clustering, *assign_labels* specifies the labeling strategy in the embedding space, and *seed* is the pseudorandom number for initializing the eigenvector decomposition. For all datasets, the number of clusters (k) was determined using the silhouette score to optimize cluster cohesion and separation.

4.3 Approaches (Baselines)

We utilized various embedding techniques from state-of-the-art LLMs, including OpenAI, Falcon, Llama 2, GPT-2 Medium, and an MPNet-based model, each enhancing text representation by capturing contextual nuances.

For the MPNet-based model, we used **sentence-transformers/all-mpnet-base-v2** (MPNet-v2) [17]. For the OpenAI model, we utilized **text-embedding-3-large**, and for the Falcon model, we used **tiiuae/falcon-7b** [1]. Additionally, we employed **Llama-2-7b-chat-hf** [20] for chat applications and

gpt2-medium [16] for faster textual representations. We used the embeddings from the last layer of all models for the most contextually rich text representations.

We integrated an additional baseline from a recent study that explores customer segmentation using LLMs combined with DICE [19]. They used the **paraphrase-multilingual-mpnet-base-v2** model from Sentence Transformers to generate 768-dimensional sentence embeddings. This model, based on the MPNet architecture, has about 278 million parameters and is designed for clustering and semantic search.

4.4 Evaluation Metrics

We use a set of metrics to evaluate our proposed framework and baselines thoroughly. Specifically, we employed the SS, CHI, and DBI metrics to assess the cohesion, compactness, and separation of clusters, ensuring a robust analysis of their structural integrity.

The SS metric, which assesses the separation and cohesion of clusters, is calculated for each data point i as:

$$s(i) = \frac{b(i)-a(i)}{\max\{a(i),b(i)\}},$$

where $a(i)$ measures the average intra-cluster distance, and $b(i)$ is the minimum inter-cluster distance for the point i.

CHI measures the ratio of between-cluster dispersion to within-cluster dispersion, providing insights into the overall clustering structure. The index is formulated as:

$$CHI = \frac{\text{Tr}(B_k)/(k-1)}{\text{Tr}(W_k)/(N-k)},$$

where $\text{Tr}(B_k)$ is the trace of the between-group dispersion matrix, and $\text{Tr}(W_k)$ the trace of the within-group dispersion matrix, thus evaluating both the separation and compactness of the clusters.

The DBI evaluates the average similarity ratio of each cluster with its most similar one, offering a measure of cluster separation. The index is calculated as:

$$DBI = \frac{1}{k}\sum_{i=1}^{k} \max_{j \neq i} \left(\frac{\sigma_i + \sigma_j}{d_{ij}}\right),$$

where σ_i and σ_j represent the average distance of all elements in clusters i and j to their respective centroids, and d_{ij} is the distance between centroids of clusters i and j. Lower values of DBI indicate better cluster separation.

4.5 Our Implemented Setup

We compare the described baselines (cf. Subsection 4.3) using a pretrained transformer-decoder LLM model, $GPT-2$ medium [16], which has 355 million trainable parameters, 24 layers, 16 attention heads, an embedding size of 1024, and a context size of 1024. To facilitate this comparison, we convert the tabular dataset into text for all baselines, applying the random feature sequence shuffling function and comparing the results with ERASMO$_{base}$ and ERASMO$_{NV}$. Both models were trained with a batch size of 8 over 60 epochs. We applied a dropout rate of 0.1 and utilized 500 warmup steps. The models incorporated a weight decay of 0.01 and used the Adam optimizer with ϵ set to 1e–8 and β values of [0.7, 0.9]. The initial learning rate was set to 5e–5, with a schedule starting at 1e–8, ranging from a minimum of 1e–5 to a maximum of 4e-5. The model was developed using PyTorch[1] and is made available at a GitHub repository[2]. It ran on a system equipped with five NVIDIA RTX A6000, each having 48 GB of Random Access Memory (RAM).

5 Experimental Results

We present key results obtained organized by datasets. Table 1 presents the SS, CHI, and DBI metrics results for the test dataset for all evaluated approaches; values in bold indicate the best outcomes. The best algorithm was determined by choosing the algorithm with the highest SS.

Banking. In the Banking dataset, the ERASMO$_{base}$ strategy outperformed all other strategies, achieving the highest SS of 0.75, the highest CHI of 12,038.44, and the lowest DBI of 0.37. This indicates well-defined and compact clusters. ERASMO$_{NV}$ also performed strongly with an SS of 0.71 and CHI of 7,570.38, though it had a slightly higher DBI of 0.43 compared to ERASMO$_{base}$. Other strategies like MPNet-v2 and Falcon showed moderate performance with SS values of 0.27 and 0.23, respectively, while OpenAI had the lowest SS at 0.11 and the highest DBI at 2.73, indicating poor clustering.

Olist. For the Olist dataset, ERASMO$_{NV}$ achieved the highest SS of 0.77, indicating the best clustering quality. It reached the highest CHI of 62,036.87 and a low DBI of 0.32. ERASMO$_{base}$ followed closely with an SS of 0.75 and a CHI of 54236.31, along with the lowest DBI of 0.30. Other strategies like LLaMA-2 and Falcon performed reasonably well, with SS values of 0.71 and 0.66, respectively. However, OpenAI and MPNet-v2 showed lower SS values, with OpenAI achieving an SS of 0.19 and MPNet-v2 an SS of 0.24.

Yelp. In the Yelp dataset, ERASMO$_{NV}$ and ERASMO$_{base}$ both demonstrated superior performance, with SS values of 0.79 and 0.78, respectively. ERASMO$_{NV}$ also achieved the highest CHI of 8,410.94 and tied with ERASMO$_{base}$ for the lowest DBI of 0.28. Other strategies, such as GPT2 Medium and LLaMA-2, showed moderate clustering performance with SS values of 0.39 and 0.29, respectively. OpenAI, with an SS of 0.07, and MPNet-v2, with an SS of 0.23, indicated less effective clustering.

---
[1] pytorch.org.
[2] ERASMO - GitHub.

Table 1. This table shows the clustering quality assessment results for the top-performing algorithms across each dataset and approach combination. The best algorithm was chosen based on the highest SS. We provide the optimal number of clusters (k) and the results for SS, CHI, and DBI. Bold values indicate the best results for each metric.

Dataset	Approach	Best alg.	Best k	SS	CHI	DBI
Banking	MPNet-v2	k-means	2	0.27	1,981.53	1.46
	OpenAI	k-means	9	0.11	212.96	2.73
	LLaMA-2	k-means++	8	0.22	593.66	1.66
	Falcon	k-means	2	0.23	1,776.67	1.56
	GPT2 Medium	k-means	2	0.40	4,764.08	0.90
	PMV2 + DICE	k-means	2	0.31	2,389.25	1.33
	ERASMO$_{base}$	k-means	2	**0.75**	**12,038.44**	**0.37**
	ERASMO$_{NV}$	AHC	2	0.71	7,570.38	0.43
Olist	MPNet-v2	k-means	2	0.24	5,927.77	1.59
	OpenAI	k-means	3	0.19	3,946.14	1.83
	LLaMA-2	k-means	4	0.71	43,306.34	0.45
	Falcon	k-means	6	0.66	45,512.63	0.55
	GPT2 Medium	k-means	2	0.48	26,471.54	0.75
	PMV2 + DICE	SC	2	0.61	27,578.16	0.67
	ERASMO$_{base}$	SC	2	0.75	54,236.31	**0.30**
	ERASMO$_{NV}$	k-means	2	**0.77**	**62,036.87**	0.32
Yelp	MPNet-v2	AHC	2	0.23	36.61	2.25
	OpenAI	AHC	2	0.07	38.78	3.86
	LLaMA-2	k-means	10	0.29	445.87	1.42
	Falcon	k-means++	14	0.32	442.83	1.22
	GPT2 Medium	k-means	2	0.39	1,898.25	1.00
	PMV2 + DICE	AHC	2	0.53	32.86	1.08
	ERASMO$_{base}$	SC	2	0.78	7,702.89	**0.28**
	ERASMO$_{NV}$	AHC	2	**0.79**	**8,410.94**	**0.28**
PetFinder.my	MPNet-v2	k-means	2	0.14	236.58	2.28
	OpenAI	AHC	2	0.16	3.29	1.75
	LLaMA-2	k-means++	17	0.35	179.00	1.38
	Falcon	k-means++	2	0.20	397.65	1.86
	GPT2 Medium	AHC	2	0.55	636.2	0.70
	PMV2 + DICE	k-means	5	0.18	242.83	1.85
	ERASMO$_{base}$	k-means	2	0.72	**3,351.95**	0.40
	ERASMO$_{NV}$	AHC	2	**0.73**	3,063.55	**0.34**
Clothings	MPNet-v2	k-means	3	0.12	195.01	2.47
	OpenAI	SC	5	0.07	70.17	2.90
	LLaMA-2	k-means++	9	0.37	435.45	1.53
	Falcon	k-means	12	0.24	326.64	1.54
	GPT2 Medium	k-means	2	0.52	3,541.87	0.68
	PMV2 + DICE	AHC	2	0.17	72.74	1.97
	ERASMO$_{base}$	k-means	2	**0.72**	**6,208.52**	**0.39**
	ERASMO$_{NV}$	k-means++	2	0.71	5,916.57	**0.39**

PetFinder.my. In the PetFinder.my dataset, $ERASMO_{NV}$ slightly outperformed $ERASMO_{base}$ with an SS of 0.73 compared to 0.72. $ERASMO_{base}$ had the highest CHI of $3,351.95$ and a low DBI of 0.40, while $ERASMO_{NV}$ had a CHI of $3,063.55$ and the lowest DBI of 0.34. Other strategies, such as GPT2 Medium and Falcon, showed moderate results with SS values of 0.55 and 0.20, respectively. MPNet-v2 had the lowest SS of 0.14, indicating poor clustering performance.

Clothings. For the Clothings dataset, $ERASMO_{base}$ achieved the highest SS of 0.72 and the highest CHI of $6,208.52$, along with the lowest DBI of 0.39. $ERASMO_{NV}$ also performed well with an SS of 0.71 and a CHI of $5,916.57$, matching the lowest DBI of 0.39. Other strategies like GPT2 Medium and LLaMA-2 showed moderate performance, with SS values of 0.52 and 0.37, respectively. OpenAI and MPNet-v2 had the lowest SS values of 0.07 and 0.12, respectively, indicating poor clustering quality.

Figure 3 shows a 2D t-Distributed Stochastic Neighbor Embedding (t-SNE) visualization for the Yelp dataset across all strategies. The Yelp dataset was chosen for its rich and diverse user reviews, making it ideal for clustering evaluation. Our models, $ERASMO_{base}$ and $ERASMO_{NV}$, display distinct and well-separated clusters, reflecting their lowest DBI scores. This highlights the superior clustering performance of our models.

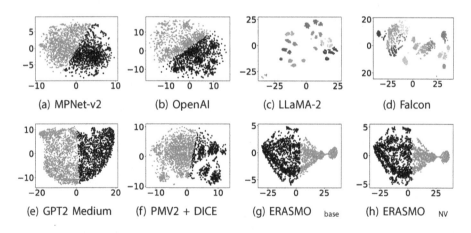

Fig. 3. t-SNE visualization of embedding representations on the Yelp dataset for different models: (a) MPNet-v2, (b) OpenAI, (c) LLaMA-2, (d) Falcon, (e) GPT2 Medium, (f) PMV2 + DICE, (g) $ERASMO_{base}$, and (h) $ERASMO_{NV}$.

6 Discussion

The SS results consistently highlight the superior clustering effectiveness of the $ERASMO_{base}$ and $ERASMO_{NV}$ strategies across all datasets. $ERASMO_{base}$

achieved the highest SS in the Banking and Clothing datasets, while ERASMO$_{NV}$ led in the Olist, Yelp, and PetFinder.my datasets. This indicates that both strategies help to form well-defined clusters, with ERASMO$_{base}$ having a slight edge in some datasets. The higher SS values for these models suggest they are better at creating distinct clusters than other strategies like MPNet-v2, OpenAI, and Falcon, which showed significantly lower SS values, indicating poorer representation relevance for clustering quality.

The DBI results further support the effectiveness of ERASMO$_{base}$ and ERASMO$_{NV}$. Both strategies consistently achieved the lowest DBI values, particularly excelling in the Banking, Olist, and Yelp datasets. A lower DBI indicates more compact and well-separated clusters, affirming that our proposed models form tight and distinct clusters. ERASMO$_{base}$ generally showed slightly better DBI scores, suggesting it produces more compact clusters than ERASMO$_{NV}$. In contrast, approaches like OpenAI and MPNet-v2 had significantly higher DBI values, reflecting less adequate cluster compactness and separation.

The CHI results corroborate the trends observed in SS and DBI. ERASMO$_{base}$ and ERASMO$_{NV}$ achieved the highest CHI scores across most datasets, with ERASMO$_{base}$ leading in the Banking and Clothing datasets, and ERASMO$_{NV}$ in the Olist and Yelp datasets. High CHI scores indicate that our proposed models create clusters that are not only well-separated but also highly dense. These superior results across multiple metrics underscores the robustness and effectiveness of the ERASMO approach. Other models like MPNet-v2, OpenAI, and Falcon demonstrated lower CHI scores, indicating less effective clustering effectiveness.

ERASMO has limitations, including reliance on high-quality data, challenges with ambiguous datasets, and significant computational demands for fine-tuning. Random sequence shuffling can affect reproducibility, and verbalizing numerical tokens does not consistently improve outcomes, warranting further investigation. Furthermore, applying traditional metrics such as SS, CHI, and DBI to analyze clusters from various representation spaces appears as an open research challenge yet to be resolved in the literature. While these metrics are designed for uniform representation spaces, they are frequently applied to different representation spaces, as evidenced by previous studies [6,14]. We recognize these limitations and stress the relevance of investigating and developing further refined metrics capable of accurately assessing embeddings from varied dimensional spaces.

Despite these challenges, ERASMO significantly advances clustering techniques by enhancing data representation. This innovation sets the stage for the potential standardization of metrics and improves clustering quality. ERASMO stands as a crucial development in pursuing more robust clustering strategies.

7 Conclusion

Exploring structured and textual data simultaneously for clustering analysis is a challenging problem. This study presented and evaluated the ERASMO$_{base}$ and ERASMO$_{NV}$ clustering approaches, demonstrating their superior empirical effectiveness across multiple datasets. The results, based on SS, DBI, and CHI, consistently revealed that our proposed models help to create well-defined, compact, and dense clusters, outperforming all strategies. Despite their high computational demands and dependency on data quality, we found the ERASMO models suited for practical clustering tools. Future work might focus on optimizing computational efficiency and enhancing robustness for diverse and noisy datasets.

Acknowledgements. This project was supported by the brazilian Ministry of Science, Technology and Innovations, with resources from Law nº 8,248, of October 23, 1991, within the scope of PPI-SOFTEX, coordinated by Softex and published Arquitetura Cognitiva (Phase 3), DOU 01245.003479/2024 -10.

References

1. Almazrouei, E., et al.: Falcon-40b: an open large language model with state-of-the-art performance (2023)
2. Borisov, V., Seßler, K., Leemann, T., Pawelczyk, M., Kasneci, G.: Language models are realistic tabular data generators. arXiv preprint arXiv:2210.06280 (2022)
3. Brooks, N.: Women's e-commerce clothing reviews (2017). https://www.kaggle.com/datasets/nicapotato/womens-ecommerce-clothing-reviews
4. Dataset, Y.: Yelp dataset (2014). http://www.yelp.com/dataset_challenge
5. Kaggle, PetFinder.my: Petfinder.my adoption prediction (2019). https://www.kaggle.com/c/petfinder-adoption-prediction/data/
6. Keraghel, I., et al.: Beyond words: a comparative analysis of llm embeddings for effective clustering. In: Int. Symposium on Intelligent Data Analysis, pp. 205–216. Springer, Heidelberg (2024)
7. Lewis, P., Perez, A., Petroni, W.t., Rocktäschel, T., et al.: Retrieval-augmented generation for knowledge-intensive nlp tasks. In: NIPS (2020)
8. Liu, Y., He, X., Zhong, T., et al.: understanding llms: a comprehensive overview from training to inference. arXiv preprint arXiv:2401.02038 (2024)
9. Min, B., Ross, D., others.: Recent advances in natural language processing via large pre-trained language models: a survey. ACM Comput. Surv. 1–40 (2023)
10. Moro, S. Rita, P., Cortez, P.: Bank marketing. UCI Machine Learning Repository (2012). https://doi.org/10.24432/C5K306
11. Olist: Brazilian e-commerce public dataset by olist (2023). https://doi.org/10.34740/KAGGLE/DSV/195341. acessado em: 03 de outubro de 2023
12. Ourabah, O., et al.: Large scale data using k-means. Mesopotamian J. Big Data **2023**, 36–45 (2023)
13. Pedregosa, F., Varoquaux, M., Prettenhofer, V., et al.: Scikit-learn: machine learning in python. J. Mach. Learn. Res. **12**, 2825–2830 (2011)
14. Petukhova, A., Matos-Carvalho, J.P., Fachada, N.: Text clustering with llm embeddings. arXiv preprint arXiv:2403.15112 (2024)
15. Pitafi, S., Anwar, T., Sharif, Z.: A taxonomy of machine learning clustering algorithms, challenges, and future realms. Appl. Sci. **13**(6), 3529 (2023)
16. Radford, A., Wu, J., Child, R., Luan, D., Amodei, D., Sutskever, I., et al.: Language models are unsupervised multitask learners. OpenAI blog **1**(8), 9 (2019)
17. Song, K., et al.: Mpnet: masked and permuted pre-training for language understanding. Adv. Neural Inf. Process. Syst. **33**, 16857–16867 (2020)

18. Tipirneni, S., Adkathimar, N., Hiranandani, V.N., Yuan, C., Reddy, C.K.: Context-aware clustering using large language models. arXiv (2024)
19. Tissera, M., Asanka, P., et al.: Enhancing customer segmentation using llms and deterministic, independent-of-corpus embeddings. In: 2024 4th (ICARC) (2024)
20. Touvron, H., Martin, L., Stone, K., et al.: Llama 2: open foundation and fine-tuned chat models. arXiv preprint arXiv:2307.09288 (2023)
21. Viswanathan, V., Gashteovski, K., Lawrence, C., Wu, T., Neubig, G.: Large language models enable few-shot clustering. arXiv preprint arXiv:2307.00524 (2023)
22. Warner, J., Sexauer, J., et al.: JDWarner/scikit-fuzzy: Scikit-Fuzzy version 0.4.2. Zenodo (2019). https://doi.org/10.5281/zenodo.3541386
23. Zhang, Y., Wang, Z., Shang, J.: Clusterllm: large language models as a guide for text clustering. arXiv preprint arXiv:2305.14871 (2023)
24. Zhu, P., Lang, Q., Liu, X.: Word embedding of dimensionality reduction for document clustering. In: 2023 35th Chinese Control and Decision Conference (CCDC), pp. 4371–4376. IEEE (2023)

Euclidean Alignment for Transfer Learning in Multi-band Common Spatial Pattern

Marcelo M. Amorim[✉][ID], Leonardo Prata[ID], João Stephan Maurício[ID], Alex Borges[ID], Heder Bernardino[ID], and Gabriel de Souza[ID]

Federal University of Juiz de Fora (UFJF), Juiz de Fora, MG, Brazil
marcelodmeloamorim@gmail.com

Abstract. Stroke is a major cause of death and disability worldwide. Motor-Imagery based Brain-Computer Interface (MI-BCI) models offer a post-stroke rehabilitation option. Existing studies for MI-BCI use Transfer Learning techniques like Euclidean Alignment (EA) but lose important brain information due to bandpass filtering. This study introduces new BCI architecture with multi-band temporal filters and EA. The methods considered here are Filter Bank (FB), Empirical Mode Decomposition (EMD), and Continuous Wavelet Transform (CWT). Results show performance improvements, especially with EA being applied before Filter Bank. These models offer promise for post-stroke rehabilitation, particularly when using EA before the multi-band filter.

Keywords: Electroencephalogram · Motor Imagery · Brain-Machine Interface · Medical applications

1 Introduction

Stroke is a leading cause of disability and mortality worldwide, demands urgent attention due to its deep impact on individuals and healthcare systems [18]. Over the years, significant strides have been made in stroke treatment, revolutionizing patient care and outcomes. One of the most recent methods of treatment is the utilization of Brain-Computer Interface models.

Brain-Computer Interface (BCI) is a system that allows for direct communication between the brain and an electronic device without peripheral muscles [3]. It translates the electrical signal of the brain in a task to a computational system. Invasive and non-invasive acquisition methods can collect the brain's signal. The invasive ways have a better signal-to-noise ratio (SNR), being between 10 to 100 times better than non-invasive methods, but are more dangerous for the subject, since they carry risks such as infection, tissue damage, and implant rejection due to their surgical nature. The non-invasive equipment is safer than the invasive ones, but they have a more imprecise signal. Electroencephalography (EEG) is the most used equipment for BCI as it is not invasive, has a satisfactory temporal resolution in the range of milliseconds (same as the invasive ones), and is portable, light, and 10 times cheaper than the invasive ones [2].

The BCI models are sensitive to the subject who provided the brain's signals to train the model. Due to the non-stationary behavior of the EEG signals and the anatomic difference between people's brains, the model is usually subject-dependent. Transfer Learning (TL) techniques can decrease the signal difference between subjects and reduce that problem. This way, we can use data from a set of subjects (source domain) to train a model to another one (target domain). Riemannian Alignment (RA) and Euclidean Alignments (EA) are the most used Transfer Learning methods [7]. However, those models only use a single band in the temporal filtering step, losing some temporal patterns. In this case, the alignment transforms the signals of the source and the target subjects to the same space.

We propose different ways to use multi-band models with EA, and the evaluated temporal filtering was Filter Bank (FB), Empirical Mode Decomposition (EMD), and Continuous Wavelet Transform (CWT). Using the PhysionetMI dataset, a well-known dataset from literature, the results obtained pointed out that multi-band with EA improves the results of BCI classifiers in this scenario. This is an advantage for the BCI models dedicated to post-stroke motor rehabilitation, as it reduces the calibration time and makes the procedure straightforward to be used with any patient.

2 Related Work

The variations in EEG signals between individuals are remarkable, reflecting the distinct nuances of their brain activities [12]. This condition is intensified in individuals who have suffered some form of stroke. As a result, these differences significantly impact the training phase in BCI, requiring a personalized approach. Therefore, in most cases, it is crucial to train a distinct model for each person. To deal with this situation, transfer learning methods were proposed to use many subjects in a single training step. One of the most used methods is the Riemannian Alignment (RA), which is a method to align the covariance matrix of the signals from each subject during their resting state. However, Riemannian space makes the model more computationally expensive and limited since the classifiers must perform efficiently in that space. To outperform that problem, the Euclidean Alignment (EA) was proposed [7]. EA aligns the EEG trials for each subject in Euclidean space using the mean of the covariance matrix of the subject's trials as a reference matrix. Thereafter, the transformed reference matrix regarding each subject becomes the identity matrix. This technique reduces the difference between data of distinct subjects, similarly than RA, but without transferring the data to Riemannian space. Therefore, it is possible to use non-Riemannian classifiers, such as Linear Discriminant Analysis (LDA) and Support Vector Machine (SVM) [14,15]. Moreover, the computational cost is reduced by operating EA instead of RA. When comparing models, EA points out better results than RA when transfer learning is used for BCI applications [7].

After its release, different approaches were proposed to improve EA. For instance, the unification of EA and RA alignment methods using a hybrid (Euclidean and Riemannian) space [12]. In addition, a selection of the subjects with data similar to that of the target one can be considered. [12]. Another approach to improve EA is weighting the subject importance based on the distance between the source and target data [5]. Comparisons of transfer learning methods applied to deep learning classifiers pointed out that RA and EA present better results than normalization techniques [13,21,22]. Those articles do not use EA with multi-band techniques. In this work, we propose novel pipelines for EA using multi-bands and parallel application of EA after signal decomposition.

3 Brain-Computer Interfaces for Post-Stroke Motor Rehabilitation

Stroke is a medical emergency that leaves sequels, including cognitive and motor deficits and oropharynxes dysphagia [10]. The rehabilitation of post-stroke patients has great importance to the recovery of their cognitive and motor functions. Also, BCI for post-stroke rehabilitation is an alternative to induce neural plasticity. BCIs support this process by delivering targeted stimuli that create a closed loop for neural reorganization. Engaging directly with external devices activates brain regions linked to movement, strengthening pertinent neural pathways and fostering positive plasticity.

Reinforcement learning promotes BCI for post-stroke motor rehabilitation, since it involves adapting behavior based on the consequences of actions. Within BCIs, individuals can receive instant and personalized feedback corresponding to their rehabilitation endeavors. This feedback creates a positive reinforcement, fortifying the neural connections linked to targeted movement patterns. Furthermore, BCIs aim to close the motor planning loop, aiding in the reconnection of brain circuits responsible for motor control. By converting brain signals into actionable commands for external devices, BCIs support the rehabilitation of motor planning, promoting iterative practice and the gradual enhancement of motor skills.

Furthermore, studies on upper-limb rehabilitation using BCI indicate that the accuracy of the BCI system can improve motor function. Other studies also showed a relationship between the BCI accuracy and clinical gains [2,4,16]. This relationship is based on the hypothesis that higher BCI accuracy may improve confidence and motivation in patients, which may better promote plasticity [16]. Moreover, another reason for the success of BCI rehabilitation can be the correlation between engagement and higher levels of patient attention.

Figure 1 represents the BCI cycle for post-stroke motor rehabilitation. First, brain signals are collected through sensors, such as electroencephalogram (EEG). The collected signals are filtered to remove interference and increase the signal-to-noise ratio. Techniques such as temporal and spatial filtering can be applied to extract relevant information. Then, feature extraction and selection steps are performed, and the chosen features are used to classify the signal. With the signals already classified, the system executes commands based on the label. For post-stroke motor rehabilitation, the feedback is usually presented as an image on a screen and an electrical stimulus in the muscle. In the context of BCI systems for patient rehabilitation, it is relevant to mention that a smaller number of electrodes brings benefits in terms of ease of treatment and patient comfort [20]. Furthermore, data alignment is emphasized for synchronizing stimuli and their responses and ensuring more accurate temporal analysis. Data Alignment also proves to be effective in enhancing system efficiency and reducing interference and signal noise. By utilizing Data Alignment, we can reduce the number of trials needed to train the model, resulting in faster and less exhausting treatment.

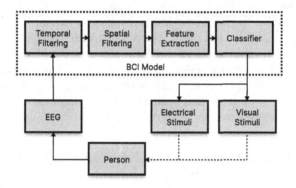

Fig. 1. BCI cycle for post-stroke motor rehabilitation.

4 Methods

This section presents the methods from the literature used as a baseline to develop the proposed BCI pipelines. We describe EA, the temporal filter methods used in the experiments, and the Common Spatial Pattern.

4.1 Euclidean Alignment

Euclidean Alignment (EA) was proposed as an alternative to Riemannian Alignment (RA) [7]. The advantages of EA are the reduction of the alignment time when compared to RA, and the maintenance of the data in the Euclidean space. EA uses a reference matrix, calculated through the EEG trials, to align the data from different subjects. A reference matrix $\bar{R}_s$ is the mean of the trial's covariance for the s-th subject, which is defined by

$$\bar{R}_s = \frac{1}{N_s} \sum_{n=1}^{N_s} X_s^n (X_s^n)^\mathrm{T} \qquad (1)$$

where N_s is the number of trials of subject s, X_s^n is the n-th trial of subject s. The matrix $\bar{R}_s$ reduces the difference between the subjects as

$$E_s^n = \bar{R}_s^{-\frac{1}{2}} \times X_s^n \qquad (2)$$

where E_s^n is the signal of subject s after alignment. The new reference matrix for all the transformed subjects is equal to the identity matrix. Therefore, the new space is the same for all the subjects.

4.2 Temporal Filtering

Temporal filtering is a technique used in data processing to extract or manipulate temporal information in data sets. Its purpose is to enhance signal quality by improving the signal-to-noise ratio. Temporal filtering removes or attenuates unwanted temporal components in the data, allowing the identification of more significant patterns or the reduction of noise. The choice of a temporal filtering technique relies on the particular context of the problem and the goals of the analysis. This section presents four temporal filters: Bandpass, Filter Bank (FB), Empirical Mode Decomposition (EMD), and Continuous Wavelet (CWT).

Bandpass and Filter Bank: Bandpass Filters are widely employed in various signal-processing applications, including wireless communication systems, audio and video processing, and EEG. Then, they are useful for isolating signals of specific frequencies and eliminating unwanted noise [1]. A bandpass filter allows for a specific range of frequencies to pass through the filter while attenuating frequencies outside that range.

A Filter Bank is a group of bandpass filters that divide an input signal into sub-bands. Figure 2 shows a diagram of the Analysis Filter Bank.

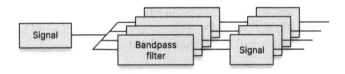

Fig. 2. Filter Bank.

Empirical Mode Decomposition: Empirical Mode Decomposition [9] is a method that decomposes non-stationary and non-linear data in different components to extract information that is not explicit in the original signal. The core of this method lies in the empirical identification of the oscillatory modes inherent in the data, distinguished by their characteristic time scales. After this identification, the data is decomposed accordingly.

The method uses the sifting process [9] to decompose the data into Intrinsic Mode Functions (IMFs) and a residual. Figure 3 presents a schematic representation of an EMD algorithm, with 3 IMFs and a residual as outputs. The sifting process of an x signal consists of the following steps:

- The local extrema of the signal are identified, and all the local maxima and minima are connected by a cubic spline line as the upper envelope. Their mean is denominated m_i, and the difference between the data and m_i is the first component h_i.
- If the component h_i is an IMF, this is separated from the remaining data and is designated as c_i. The result is a component r_n. Otherwise, Step (1) is repeated with h_i as the input signal.
- If r_n is a residual, the original dataset x is decomposed into n IMFs c_i and a residual r_n, which can be either the mean trend or a constant. Otherwise, Steps (1) and (2) are repeated.
- The EDM procedure can be represented as the following equation

$$x = \sum_{i=1}^{n} c_i + r_n \qquad (3)$$

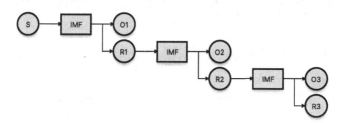

Fig. 3. Empirical Mode Decomposition with 3 levels.

Continuous Wavelet: It is an application of the wavelet transform that employs arbitrary scales and nearly arbitrary wavelets [19]. Wavelets are brief oscillations localized in time and frequency. In this transformation, the signal is decomposed into a collection of wavelets through the wavelet method, corresponding to a distinct frequency band. The Continuous Wavelet Transform (CWT) generates a time-frequency representation of a signal with best location found by the model in both the time and frequency dimensions.

The CWT generates a graphical representation called a spectrogram that shows the intensity of different frequencies along time. Each point in the spectrogram reflects the intensity of the associated frequency component at a specific time. CWT is especially useful for analyzing dynamic signals, where frequency characteristics can vary over time, and the choice of wavelet and scale parameters is critical in CWT, as it determines the sensitivity of the transform to various signal characteristics. CWT function of an input signal $x(t)$ can be represented as

$$W(a,b) = \frac{1}{\sqrt{a}} \int_{-\infty}^{\infty} x(t)\psi^*(\frac{t-b}{a})dt \quad (4)$$

where a controls the length of the wavelet, b controls the position of the wavelet along the time axis, and $\psi^*(\frac{t-b}{a})$ is the complex conjugate version of the wavelet $\psi(\frac{t-b}{a})$.

4.3 Common Spatial Pattern

Common Spatial Pattern (CSP) is a technique for spatial filtering to extract features from signals. CSP aims to find a set of spatial filters that maximize the difference between signals from two classes based on their covariance matrices. Its application is given by

$$Z_i = W^T \times E_i \quad (5)$$

where $Z_i \in \mathbb{R}^{S \times T_s}$ is the matrix obtained when we multiply the i-th trial $E_i \in \mathbb{R}^{S \times T_s}$ with the transformation matrix $W^T \in \mathbb{R}^{S \times S}$. S and T represent the number of electrodes and the size of the signal collected per electrode in each trial. The correlation matrices of each label are used to find the transformation matrix W, and it is defined by

$$\Sigma^{(c)} = \frac{1}{N_c} \sum_n^{N_c} E_n^{(c)} E_n^{(c)T} \quad (6)$$

where $\Sigma^{(c)} \in \mathbb{R}^{S \times S}$ is the correlation matrix between the electrodes, N_c is the number of training samples of class C, and $E_n^{(c)}$ is the n-th training sample of class C. Maximization is restricted so that the sum of the diagonal matrices formed by the eigenvalues is equal to the identity matrix, that is, $\Lambda^{(1)} + \Lambda^{(2)} = I$ This problem is equivalent to solving the generalized eigenvalue problem, defined as

$$\Sigma^{(1)}W = (\Sigma^{(1)} + \Sigma^{(2)})W\Lambda \quad (7)$$

Finally, the following equation is applied to Z to extract its features:

$$f_i = log\left(\frac{diag(Z_i Z_i^T)}{tr(Z_i Z_i^T)}\right) \quad (8)$$

where f_i represents the feature vector of Z_i, and $diag(\cdot)$ and $tr(\cdot)$ represent the main diagonal and trace of the matrix, respectively. The number of CSP filters is equal to the number of electrodes. However, only the first and last filters are used. As CSP maximizes the variance of class 1 in the first eigenvectors and class 2 in the last, a feature vector v_i is created where its elements are the $m/2$ first components and $m/2$ final components of f_i.

4.4 Proposed Multi Band Pipelines with Euclidean Alignment

A temporal filter improves the signal-to-noise ratio and, when applied parallel to the data, can create a new set of filtered data. In this work, we propose these filters with EA to evaluate their combinations. CSP is adopted as a base model, as it is a standard classification approach for MI [7]. We propose here two ways to use EA with multi-band filtering: using the EA before and after the multi-band filter. Also, one can observe that many multi-band filters can be used. This work focuses on the temporal filter models presented in Sect. 4.2. The pipelines described in Fig. 4 are in order: the baseline model with EA [7], the use of a temporal filter before the EA step, and the application of EA before the temporal filter. As shown in Fig. 4, we used LDA as the classifier [14,15].

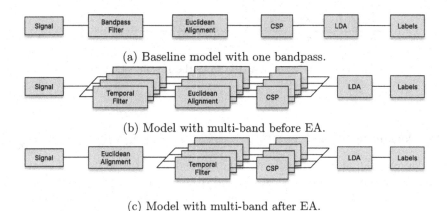

(a) Baseline model with one bandpass.

(b) Model with multi-band before EA.

(c) Model with multi-band after EA.

Fig. 4. Literature and proposed architectures using Euclidean Alignment.

5 Computational Experiments

We performed the experiments to evaluate the proposals in Sect. 4. A 5-fold stratified cross-validation was performed to evaluate the proposed approaches. The non-parametric Kruskall-Wallis statistical test was used as a multi-group test followed by Dunn's test as post-hoc with the Bonferroni correction. The codes implemented for this work are publicly available[1].

5.1 Datasets

The PhysionetMI dataset [6] was used to perform the experiments. That dataset has the recorded data of 109 subjects with 14 sessions per subject. The acquisition was performed using an EEG with a sample rate of 160 Hz and 64 monopolar

---
[1] https://github.com/ghdesouza/bci.

electrodes based on the 10–10 electrode system [17]. For the first two sessions, a baseline with eyes open and closed was recorded without any task to execute. For other sessions, there are two classes per session, which is sessions: (i) Left-fist and right-fist movement: 3, 7, 11; (ii) Left-fist and right-fist motor imagery: 4, 8, 12; (iii) Both-fists and both-feet movement: 5, 9, 13; and (iv) Both-fists and both-feet motor imagery: 6, 10, 14. Each session with a task has 15 trials, and each trial has 8 s. The cue for the task is presented for 1 s, followed by 3 s of the task execution. Moreover, there is a 4-second break between each trial. The steps of each trial are shown in Fig. 5. Due to the binary classification nature of the models presented here, we only used Left-fist and right-fist motor imagery in our experiments.

Fig. 5. Timing scheme of a trial from the PhysionetMI dataset.

5.2 Parameters Used

The computational experiments were performed with two subsets of electrode positions: (i) 3 electrodes: C3, Cz, C4; and (ii) 8 electrodes: FC3, C3, CP3, FCz, CPz, FC4, C4, CP4. We reduced the number of electrodes in those 2 cases, as it is proven that a smaller number of electrodes makes the treatment more comfortable for the patient, faster to prepare the treatment, and less expensive [11,20]. In the future, we will be able to use this electrode parameter to acquire data from more individuals, without losing model accuracy. The positions of the electrodes above the brain motor area are: (i) above the primary motor area when using 3 electrodes, and (ii) the supplementary motor and premotor areas in addition to those in (i) when using 8 electrodes [8].

Furthermore, we created two new classes: left-fist and right-fist, by combining the movement and motor imagery of each one. Only those 2 new classes were used as it is the most common approach for post-stroke rehabilitation BCIs [20]. All acquired data was down-sampled to 128 Hz using a cubic spline. When analyzing the dataset, the EEG trials of subjects 88, 94, and 101 were unavailable, resulting in these subjects being removed from our computational experiments. Considering this, the experiments were performed using data from the remaining 106 subjects. The data window for training started 0.5 s after the cue and lasted for 2 s. The same window bounds were used for the testing phase.

We created 11 models to evaluate our proposal. The first models serve as baselines, namely: CSP, EA-CSP, and $(\cdot)$-CSP, where $(\cdot) \in \{\text{FB, EMD, CWT}\}$. After that, the models with EA applied after the multi-band have the form $(\cdot)$-EA-CSP. Finally, we have the models where EA is applied before the multi-band in the form EA-(·)-CSP. We performed preliminary experiments, and after that, we set the hyperparameters of the models as: The bandpass was implemented using the Fast Fourier Transform (FFT), between 4 and 40 Hz. For CSP, we used $m = 4$. The filter bank used nine bandpass filters: (4, 8), (8, 12), (12, 16), (16, 20), (20, 24), (24, 28), (28, 32), (32, 36), (36, 40). EMD used the smallest amount of IMF less 1, with nine being the maximum limit of this value, and CWT was implemented with five levels. For each target subject, the source data for EA had all other subjects except the target, and the solver for LDA was SVD.

5.3 Results Using 3 Electrodes

The experiments were performed with the settings presented in Session 5.2. Table 1 presents the results for the three-electrode case, where the accuracy for EA-CSP and MB-CSP was worse than EA-MB-CSP and MB-EA-CSP for all MB methods. Filter Bank was the best temporal filter method regardless of the use of EA. EA-FB-CSP outperformed the remaining approaches. The statistical analysis presented a significant difference p-value < 0.001 for all methods when compared to EA-FB-CSP. Furthermore, FB-EA-CSP was the second-best model, with a statistical difference concerning the remaining models. These results pointed out that Filter Bank is the best temporal filtering evaluated here to be used with EA, regardless the order of its application in the BCI pipeline. It is also possible to note that EA before multi-band reached better results in all tested cases.

Table 1. Accuracy for the scenario with 3 electrodes.

	Max	Min	Median	Average ± std
CSP	0.7222	0.3556	0.5444	0.5384 ± 0.0756
FB-CSP	0.6667	0.3556	0.5333	0.5262 ± 0.0668
EMD-CSP	0.6778	0.3444	0.5000	0.4968 ± 0.0664
CWT-CSP	0.7000	0.3556	0.5111	0.5057 ± 0.0654
EA-CSP	0.7000	0.4667	0.5556	0.5577 ± 0.0542
FB-EA-CSP	0.8333	0.5333	0.7111	0.7109 ± 0.0552
EMD-EA-CSP	0.7556	0.4889	0.6333	0.6308 ± 0.0530
CWT-EA-CSP	0.7333	0.4556	0.5889	0.5934 ± 0.0509
EA-FB-CSP	**0.9444**	**0.7000**	**0.8000**	**0.7998 ± 0.0464**
EA-EMD-CSP	0.7889	0.5667	0.6889	0.6884 ± 0.0465
EA-CWT-CSP	0.8111	0.6000	0.6889	0.6906 ± 0.0488

5.4 Results Using 8 Electrodes

For the 8 electrodes case, and their results are presented in Table 2. The results are similar to those observed in the 3 electrodes case. EA-FB-CSP outperformed the remaining approaches. The use of EA before multi-band performed better than that applying EA after it, and Filter Bank is the temporal filter model with the best result. Such as in the 3-electrode case, the p-value for all methods compared to EA-FB-CSP was lower than 0.001. When comparing both electrode settings, 3 electrodes reached better results than 8 electrodes for all cases. That difference between 3 and 8 electrodes was statistically significant for all models. In the same way as the 3-electrode case, the EA-MB-CSP pipeline was more robust than the other two pipeline models. The results pointed out that EA as the first step of the BCI pipeline makes the model more robust between subjects.

Table 2. Accuracy for the scenario with 8 electrodes.

	Max	Min	Median	Average ± std
CSP	0.6667	0.3222	0.5111	0.5055 ± 0.0740
FB-CSP	0.6667	0.3778	0.5111	0.5122 ± 0.0587
EMD-CSP	0.6444	0.3556	0.4889	0.4899 ± 0.0685
CWT-CSP	0.6444	0.3333	0.5000	0.5042 ± 0.0591
EA-CSP	0.6889	0.4667	0.5111	0.5226 ± 0.0370
FB-EA-CSP	0.7778	0.5333	0.6778	0.6706 ± 0.0510
EMD-EA-CSP	0.7222	0.4889	0.5833	0.5862 ± 0.0523
CWT-EA-CSP	0.6778	0.4444	0.5556	0.5623 ± 0.0468
EA-FB-CSP	**0.8333**	**0.6000**	**0.7333**	**0.7268 ± 0.0453**
EA-EMD-CSP	0.7556	0.5333	0.6469	0.6501 ± 0.0462
EA-CWT-CSP	0.8111	0.5556	0.6667	0.6660 ± 0.0514

5.5 Performance Profile Analysis

The performance profiles (PPs) were used to conduct an overall analysis concerning the accuracy value [20], where each pair (subject, electrode case) is a problem. PPs present how many problems were solved with less than τ times the best value for each problem. Figure 6 contains PPs regarding the results of the 11 methods evaluated in this work. The following conclusions can be highlighted: (i) EA-FB-CSP was the model with the best result for most subjects, since it had the largest value for $\tau = 1$ (about 90% of the 212 problems). (ii) EA-FB-CSP obtained the best overall performance in the average case, as this approach achieved the biggest area under the PP curve; and (iii) EA-FB-CSP is the most robust method, with at more than 10% worse than the best case. Other results in PPs are: (i) FB-EA-CSP was the second method for the average result but the 4-th most robust method; (ii) the three most robust cases used EA before the multi-band; and (iii) EA-CSP was better and more robust than all cases which did not use EA.

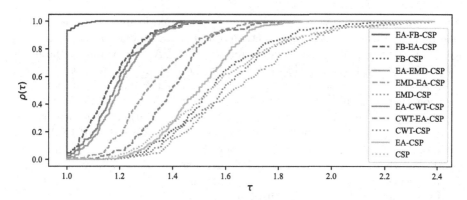

Fig. 6. Performance profiles of the results. Each problem is a combination of subject and number of electrodes.

6 Discussion

All models in the 3-electrode case obtained better results than their counterparts in the 8-electrode case. One can observe that the amount of electrode raises the dimensionality for LDA with no relevant information, leading to worse results. LDA performs better when the number of trials per area in the search space is high enough to fill it. When the dimension increases, the search space gets bigger and less dense. That explains that models in the 3-electrode case have better results than the 8-electrode case.

The standard BCI pipeline is formed by the following steps: Temporal Filtering, Spatial Filtering, Feature Extraction, Feature Selection, and Classifier. EA can be represented as a spatial filter due to its linear transformation on the input signal. Therefore, better results are expected when EA is used after the temporal filtering, such as in the BCI pipeline. However, the results found in the experiments demonstrate the opposite, leading to questions regarding the BCI stage's order. Spatial filters before the temporal step can result in new configurations for the BCI pipeline. Filter Bank was the best temporal filter tested here. This result is expected, as it takes the signal in frequency sub-bands, being able to select separately μ (8–13) and β (13–30) rhythms, which is crucial for motor imagery activities [23]. The other temporal filtering techniques applied in this work do not use the signal spectrum as effectively, which explains their results in comparison with Filter Bank.

7 Conclusion

BCI-based MI models can be used for post-stroke rehabilitation. There are some difficulties despite that, such as long calibration periods during the training phase and low resolution of the signal obtained. We proposed Euclidean Alignment (EA) with multi-band temporal filters to reduce the impact of these two conditions. The models were created using Filter Bank, Empirical Mode Decomposition, and Continuous Wavelet as temporal filter techniques. Each model uses one of these temporal filters, either before or after EA. We evaluated the models in 2 cases: with 3 and 8 electrodes in the PhysionetMI dataset. The models used in this work demonstrate how EA and multi-band temporal filters can be used together to acquire more robust models. They also demonstrate that changing the steps' order can improve the quality of the solutions. The best model in the computational experiments performed here is EA-FB-CSP-LDA, with an improvement of approximately 44% for three electrodes and 43% for eight electrodes when compared to EA-CSP-LDA, reaching an accuracy of 80% and 73.33%, respectively. EA before the temporal filtering presented results better than those obtained by models with EA after the temporal filtering.

New experiments using Feature Selection or Subject Selection can be performed to improve those results. Our results also pointed out that Spatial Filtering before Temporal Filtering has prominent potential for new BCI pipelines.

Acknowledgements. The authors thank the support provided by CAPES, CNPq, FAPEMIG, UFJF, and OpenBCI.

References

1. Ang, K.K., Chin, Z.Y., Wang, C., Guan, C., Zhang, H.: Filter bank common spatial pattern algorithm on bci competition iv datasets 2a and 2b. Front. Neurosci. (2012). https://doi.org/10.3389/fnins.2012.00039
2. Ang, K.K., Guan, C.: Eeg-based strategies to detect motor imagery for control and rehabilitation. IEEE Trans. Neural Syst. Rehabil. Eng. (2017). https://doi.org/10.1109/TNSRE.2016.2646763
3. Bashashati, A., Fatourechi, M., Ward, R.K., Birch, G.E.: A survey of signal processing algorithms in brain-computer interfaces based on electrical brain signals. J. Neural Eng. (2007). https://doi.org/10.1088/1741-2560/4/2/R03
4. Biasiucci, A., et al.: Brain-actuated functional electrical stimulation elicits lasting arm motor recovery after stroke. Nat. Commun. (2018). https://doi.org/10.1038/s41467-018-04673-z
5. Gao, Y., Li, M., Peng, Y., Fang, F., Zhang, Y.: Double stage transfer learning for brain-computer interfaces. IEEE Trans. Neural Syst. Rehabil. Eng. (2023). https://doi.org/10.1109/TNSRE.2023.3241301
6. Goldberger, A.L., et al.: Physiobank, physiotoolkit, and physionet: components of a new research resource for complex physiologic signals. Circulation (2000). https://doi.org/10.1161/01.cir.101.23.e215
7. He, H., Wu, D.: Transfer learning for brain-computer interfaces: a euclidean space data alignment approach. IEEE Trans. Biomed. Eng. (2020). https://doi.org/10.1109/TBME.2019.2913914
8. Homan, R.W., Herman, J., Purdy, P.: Cerebral location of international 10-20 system electrode placement. Electroencephalogr. Clin. Neurophysiol. (1987). https://doi.org/10.1016/0013-4694(87)90206-9
9. Huang, N.E., et al.: The empirical mode decomposition and the hilbert spectrum for nonlinear and non-stationary time series analysis. Proc. Math. Phys. Eng. Sci. (1998). https://doi.org/10.1098/rspa.1998.0193

10. Kelly-Hayes, M., Beiser, A., Kase, C.S., D'Agostino, R.B., Scaramucci, A., Wolf, P.A.: The influence of gender and age on disability following ischemic stroke: the framingham study. J. Stroke Cerebrovasc. Dis. (2003). https://doi.org/10.1016/S1052-3057(03)00042-9
11. Leeb, R., Lee, F., Keinrath, C., Scherer, R., Pfurtscheller, G., Bischof, H.: Brain-computer communication: motivation, aim, and impact of exploring a virtual apartment. IEEE Trans. Neural Syst. Rehabil. Eng. (2007). https://doi.org/10.1109/TNSRE.2007.906956
12. Li, Y., Wei, Q., Chen, Y., Zhou, X.: Transfer learning based on hybrid riemannian and euclidean space data alignment and subject selection in brain-computer interfaces. IEEE Access (2021). https://doi.org/10.1109/ACCESS.2020.3048683
13. Liang, Y., Ma, Y.: Calibrating eeg features in motor imagery classification tasks with a small amount of current data using multisource fusion transfer learning. Biomed. Signal Process. Control (2020). https://doi.org/10.1016/j.bspc.2020.102101
14. Lotte, F., et al.: A review of classification algorithms for eeg-based brain-computer interfaces: a 10 year update. J. Neural Eng. (2018). https://doi.org/10.1088/1741-2552/aab2f2
15. Lotte, F., Congedo, M., Lécuyer, A., Lamarche, F., Arnaldi, B.: A review of classification algorithms for eeg-based brain-computer interfaces. J. Neural Eng. (2007). https://doi.org/10.1088/1741-2560/4/2/R01
16. Mane, R., Chouhan, T., Guan, C.: Bci for stroke rehabilitation: motor and beyond. J. Neural Eng. (2020). https://doi.org/10.1088/1741-2552/aba162
17. Nuwer, M.R.: 10–10 electrode system for eeg recording. Clin. Neurophysiol. (2018). https://doi.org/10.1016/j.clinph.2018.01.065
18. Pacheco-Barrios, K., et al.: Burden of stroke and population-attributable fractions of risk factors in Latin America and the Caribbean. J. Am. Heart Assoc. (2022)
19. Samar, V.J., Bopardikar, A., Rao, R., Swartz, K.: Wavelet analysis of neuroelectric waveforms: a conceptual tutorial. Brain Lang. (1999). https://doi.org/10.1006/brln.1998.2024
20. de Souza, G.H., Bernardino, H.S., Vieira, A.B., Barbosa, H.J.C.: Differential evolution based spatial filter optimization for brain-computer interface. In: Proceedings of the Genetic and Evolutionary Computation Conference (2019). https://doi.org/10.1145/3321707.3321791
21. Wang, X., Yang, R., Huang, M.: An unsupervised deep-transfer-learning-based motor imagery eeg classification scheme for brain-computer interface. Sensors (2022). https://doi.org/10.3390/s22062241
22. Xu, L., Xu, M., Ma, Z., Wang, K., Jung, T.P., Ming, D.: Enhancing transfer performance across datasets for brain-computer interfaces using a combination of alignment strategies and adaptive batch normalization. J. Neural Eng. (2021). https://doi.org/10.1088/1741-2552/ac1ed2
23. Yu, H., Ba, S., Guo, Y., Guo, L., Xu, G.: Effects of motor imagery tasks on brain functional networks based on eeg mu/beta rhythm. Brain Sci. (2022). https://doi.org/10.3390/brainsci12020194

Evaluating CNN-Based Classification Models Combined with the Smoothed Pseudo Wigner-Ville Distribution to Identify Low Probability of Interception Radar Signals

Edgard B. Alves[1]([✉]), Jorge A. Alves[2], and Ronaldo R. Goldschmidt[1]

[1] Instituto Militar de Engenharia, Rio de Janeiro, RJ 22.290-270, Brazil
e6garbraz@icloud.com
[2] Escola Naval, Rio de Janeiro, RJ 20.010-060, Brazil

Abstract. Fundamental in defense, Radar Electronic Warfare (REW) requires adaptation to current threats. Automatic recognition algorithms for intrapulse modulations (ATR) of Low Probability of Interception (LPI) radar signals are essential in REW. Existing LPI signal ATR methods combine the Choi-Williams Distribution (CWD) pre-processing technique with Convolutional Neural Networks (CNN). This work proposes two new ATR combinations, based on SqueezeNet and GoogLeNet CNN. Both used the Smoothed Pseudo-Wigner-Ville distribution (SPWVD) pre-processing technique as an alternative to CWD. Replacing CWD by SPWVD was based on the hypothesis that the latter usually provides higher resolutions than the former. The proposed ATR overcame the SOTA ATR, achieving a 99.06% accuracy, under noisy environments and providing evidence to the hypothesis raised. Experiments involved two datasets with 13 types of modulations and 806,000 samples each.

Keywords: CNN · Automatic LPI Radar Signal Recognition · SPWVD

1 Introduction

Currently, Radar Electronic Warfare (REW)[1] has assumed a fundamental role in defense of countries around the world. Increasingly, REW Support Measures (ESM) have become essential to improve detection, identification, and protection processes against enemy equipment such as missiles, for instance. In the context of missile detection, efforts have been made to incorporate Electronic Intelligence systems (ELINT) into ESM equipment aimed at identifying Low Probability of Interception (LPI) radar signals [5]. Due to the use of robust

---

[1] Set of actions that aim to ensure the use of the electromagnetic spectrum by friendly forces and prevent, reduce or prevent its use by enemy forces [8].

© The Author(s), under exclusive license to Springer Nature Switzerland AG 2025
A. Paes and F. A. N. Verri (Eds.): BRACIS 2024, LNAI 15072, pp. 444–459, 2025.
https://doi.org/10.1007/978-3-031-79029-4_31

automatic recognition algorithms of intrapulse modulations (ATR - Automatic Target Recognition) of LPI radar signals, ELINT systems have good performance, even in environments with low signal-to-noise ratio (SNR) [6]. For text simplification purposes, in this article, the acronym ATR will be used as the expression *ATR model*.

In recent years, several ATR systems for Low Probability of Intercept (LPI) radar signals have been developed using different classification algorithms, such as classical Artificial Neural Networks (ANN) [9], Decision Trees (DT) [10], and Convolutional Neural Networks (CNN) [6], among others [5]. To apply these ATR systems, the signals need to be pre-processed using Time-Frequency Analysis (TFA) techniques[2]. The main TFA techniques employed in LPI radar signal ATR include the Choi-Williams Distribution (CWD) [6], Short-Time Fourier Transform (STFT) [13], and variations of the Wigner-Ville Distribution (WVD) [14,15], such as the Smoothed Pseudo-Wigner-Ville (SPWVD). The SPWVD is one of the most effective TFA techniques for estimating various temporal and spectral parameters of LPI radar signals, especially in noisy environments [7]. Notably, literature works that have combined the TFA-CWD[3] technique with CNN have achieved the best performances in identifying LPI radar signals [3,5,6].

When analyzing these studies, it becomes evident that although they all employed different Time-Frequency Analysis (TFA) techniques for signal pre-processing, there was no initiative to optimize the CNN's hyperparameters according to the pre-processing technique. Efforts to optimize hyperparameters in Machine Learning applications are justified because, in general, classification model performance varies based on the data and the pre-processing applied to it [2]. Nevertheless, all the mentioned studies relied on the *default* parameterization of the CNN implementations.

In light of the above mentioned, this study raises the following hypothesis: *Using a CNN with hyperparameters optimized accordingly to the TFA technique used in the pre-processing of LPI radar signals can lead to better performance when combining TFA+CNN*[4]. *Moreover, when comparing TFA+CNN with SPWVD as the TFA technique to cases where CWD is used, the former technique can lead to superior results.*

To gather pieces of evidence that support the hypothesis raised above, this study aimed to analyze the results produced by the following combinations: $SPWVD+SqueezeNet_{opt}$; $CWD+SqueezeNet_{opt}$; $SPWVD+GoogLeNet_{opt}$; and $CWD+GoogLeNet_{opt}$. All these combinations were tested using 13 types of intrapulse modulations in LPI radar signals. The following combinations,

---

[2] A TFA technique consists of converting signals from the time domain to the time-frequency domain [11].

[3] Also, to simplify the discourse, throughout the article, the expression *TFA-X technique* will be adopted to denote the expression *TFA X technique*, where X is a TFA technique.

[4] In this work, the TFA+CNN notation will be used to denote the application of the TFA technique in data pre-processing, followed by the application of the indicated CNN as the classifier.

$SPWVD+SqueezeNet_{opt}$, $CWD+SqueezeNet_{opt}$, $SPWVD+GoogLeNet_{opt}$ and $CWD+GoogLeNet_{opt}$, refer to versions of the CNN SqueezeNet and CNN GoogLeNet with optimized hyperparameters, trained after applying TFA-SPWVD and TFA-CWD to the data, respectively. The 13 tested modulations were generated by considering a range of random values for their respective parameters, resulting in two Time-Frequency Image (TFI) databases: one derived from the application of TFA-SPWVD (SPWVD-TFI) and the other from the application of TFA-CWD (CWD-TFI). Each of these databases contains 403,000 TFI. Remarkably, the $SPWVD+GoogLeNet_{opt}$ combination achieved an average classification accuracy of 99.06% at 0dB SNR, surpassing both previous related works and the results obtained by the $CWD+GoogLeNet_{opt}$ combination, thus confirming the considered hypothesis.

The present text is structured into five sections. Section 2 introduces the Time-Frequency Analysis (TFA) techniques used to generate the Time-Frequency Images (TFI) employed in the experiments. Section 3 discusses state-of-the-art studies in LPI radar signal identification. Section 4 outlines the experimental methodology, including TFI creation and the structure of classifiers based on SqueezeNet and GoogLeNet. Section 5 describes the obtained results and compares them with those from related works. Finally, Sect. 6 presents the research reflections, highlighting the main contributions and suggesting avenues for future investigations.

2 Fundamental Theory

Various pre-processing techniques for Low Probability of Intercept (LPI) radar signals have been developed. Among these, we can highlight TFA methods such as the WVD, its variations, and the CWD. These techniques play a crucial role in enhancing ATR performance.

The WVD allows optimal time-frequency concentration compared to other TFA methods [11]. However, it introduces Cross-Terms (CT)[5]. To mitigate the influence of CT, the Pseudo-Wigner-Ville Distribution (PWVD) was introduced, incorporating a sliding analysis window along the frequency axis [11]. Subsequently, as an enhancement of the PWVD, the CWD was developed to reduce the CT influence both in frequency and time domains. The CWD is defined based on the Fourier transform $X(\omega)$ of $x(t)$, as indicated in Eq. 1. Here, t represents the time variable, ω is the angular frequency variable, $*$ denotes complex conjugation, and $\sigma = 1$ serves as a scale factor, critical for CT suppression by smoothing the CWD distribution. The kernel ϕ acts as a low-pass filter for processing the two-dimensional Fourier Transform into a Cohen class ambiguity function [11]. Typically, an exponential kernel $\phi(\xi,\tau) = e^{-\xi^2\tau^2/\sigma}$ is adopted in this distribution [1].

---

[5] CT are spurious terms that arise in some time-frequency domain transformations.

$$CWD_x(t,\omega) = \frac{1}{2\pi} \int_{\xi=-\infty}^{\infty} e^{-j\xi t} \int_{\mu=-\infty}^{\infty} \sqrt{\frac{\sigma}{4\pi\xi^2}} e^{\frac{(\mu-\omega)^2}{4\xi^2/\sigma}} x(\mu + \frac{\xi}{2}) x^*(\mu - \frac{\xi}{2}) d\mu d\xi \tag{1}$$

Another approach proposed to mitigate the interference caused by CT in the WVD is the creation of the SPWVD. The SPWVD introduces smoothing windows both in the time and frequency domains. Mathematically, the SPWVD is defined by Eq. 2 [7]. In this equation, h(τ) and g(ν) represent the time and frequency window functions, respectively. The signal x(t) corresponds to the analytic signal of r(t), and $*$ denotes complex conjugation. Specifically, the analytic signal x(t) follows the Equation $x(t) = r(t) + jH[r(t)]$, where $H[r(t)]$ represents the Hilbert transform of the real signal r(t).

$$SPWVD_x(t,f) = \int\int x(t-\nu+\tau/2)x^*(t-\nu-\tau/2)h(\tau)g(\nu)e^{-j2\pi f\tau} d\nu d\tau \tag{2}$$

The visual effect of interference resulting from CT in the time-frequency domain can be observed in the examples shown in Fig. 1. These examples depict the results of applying different TFA techniques to an LPI radar signal. The resulting TFI from the WVD Fig. 1a and the PWVD Fig. 1b exhibit pronounced CT values (visible as highlighted vertical bars). In contrast, the TFI produced by the CWD Fig. 1c and the SPWVD Fig. 1d show reduced CT effects. Notably, although the TFI appears similar, the image in Fig. 1d seems to have a slightly higher resolution than that in Fig. 1c, suggesting a potential advantage of using SPWVD over CWD. More details about each TFA can be obtained in [11].

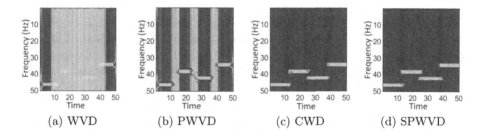

Fig. 1. LPI Costas-4 Signal Pre-processed by Different TFA

3 Related Work

Currently, the state-of-the-art studies in the field of LPI radar signal classification, specifically regarding modulation type, include those by [3,5,6]. Table 1 presents main characteristics of these works compared to the present proposal.

Notably, among these studies, only [3] explored different TFA techniques and compared results across classification models. However, unlike the current proposal, the three LPINet CNN evaluated in [3] experiments were generated using *default* hyperparameter values, regardless of the TFA technique used for signal pre-processing. Additionally, our proposal uniquely assessed and compared the effects of employing TFA-SPWVD and TFA-CWD on the same signals. This strategic choice was based on the expectation that even minor resolution differences, as illustrated in Figures 1d and 1c, could lead to improved classification results with TFA-SPWVD compared to TFA-CWD.

Table 1. Main Characteristics of Related Works and This Proposal

Works	TFA Techniques	CNN Type	Hiper. Otimiz.	Range of SNR (dB) from	to	Number of LPI signals	Total CNN Parameters
Kong [5]	CWD	LWRT	No	−20	10	12	2 097 512
Huynh-The [3]	CWD,STFT,WVD	LPINet	No	−20	10	13	288 768
Liu [6]	CWD	CV-LPINet	No	−18	10	12	9784
This Work's Proposal	CWD, SPWVD	SqueezeNet	Yes	−20	10	13	729 100
	CWD, SPWVD	GoogLeNet	Yes	−20	10	13	5 900 000

4 Proposed Methodology

To validate the hypothesis proposed in this study, we employed the methodology depicted in Fig. 2. This methodology comprises the following steps: Creation of LPI Radar Signal Instances Base, generated by adding Additive White Gaussian Noise (AWGN) and simulated channel loss to the noise-free LPI radar signal; Signal Pre-processing, generating the SPWVD-TFI and CWD-TFI bases by pre-processing the signals using SPWVD and CWD TFA techniques, respectively; Model training, validating and testing, where the classification models were trained, validated and tested based on the LPI modulation type using the pre-processed signals; and Evaluation of Results, where the performance of our proposed approach is evaluated. Notably, our method considered both TFA-SPWVD and TFA-CWD on the same signals, aiming to leverage their distinct characteristics for improved classification results. The next paragraphs detail each of these steps.

To create the Base of LPI Radar Signal Instances, we modeled the receiver of a radar system. We considered that the complex sample of an intercepted LPI radar signal is perturbed by Additive White Gaussian Noise (AWGN) and channel loss, as indicated by the equation: $y(k) = x(k) \circledast h(k) + n(k)$. In this equation: $x(k)$ represents the signal generated during the Signal Generation step, which is noise-free; $h(k)$ corresponds to the channel interference resulting from the Channel Loss Generation step; $n(k)$ characterizes the noise introduced during the AWGN Generation step; k denotes the sample index for each T_s (sampling period), considering a sampling frequency f_s.

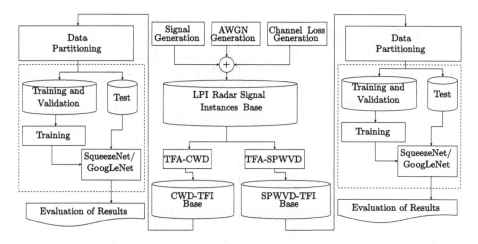

Fig. 2. Flowchart of the Methodology Adopted in the Experiments

It's important to emphasize that this study aimed to employ a data creation mechanism identical to that used in related works. To achieve this, we utilized the source code for generating LPI radar signals provided by the authors of [6] after corresponding via email. Additionally, we created the database using the same intrapulse modulations and parameter ranges specified in Table 1. Consequently, we generated 13 different types of LPI radar signal modulations, including linear frequency modulation (LFM), unmodulated (rectangular) signals, Costas modulation (Frequency Shift Key - FSK), binary Barker coding, five polyphase codes (Frank, P1, P2, P3, and P4), and four polytemporal codes (T1, T2, T3, and T4). Similar to [3], we introduced noise during the AWGN Generation step, varying the SNR from −20 dB to +10 dB with a 1.0 dB increment. In the same way, channel loss interference was modeled, using Rayleigh fading during the Channel Loss Generation step. The different signal instances constituting the LPI Signal Base were created by randomly varying the specific parameters for each intrapulse modulation, following the specifications outlined in [3] and detailed in Fig. 3.

As specified in [3], to generate the CWD-TFI and SPWVD-TFI bases, we initially pre-processed each LPI radar signal using the SPWVD and CWD techniques. During this step, we applied Kaiser filters with a size of 63 samples and a shape factor of 0.5 to smooth the time and frequency windows. Subsequently, we obtained the corresponding TFI, capturing images with 256 grayscale levels and dimensions of (50 × 50) pixels using bicubic interpolation for resizing.

Table 2. TFI Databases Generated in This Work

TFI Types	SNR Levels (−20 dB to 10 dB with the increase of 1 dB)	Intrapulse Modulations Types	LPI Signals Instances	Total of TFI Bases
CWD-TFI	31	13	1.000	403.000
SPWVD-TFI	31	13	1.000	403.000

Notations			**LPI waveforms**	
Param.	Description	Types	Param.	Range of value
U	uniform distribution	All	f_c	$U(f_s/6, f_s/5)$
f_s	sampling frequency	LFM	N	$[512, 1024]$
B	bandwidth		B	$U(f_s/20, f_s/15)$
N	number of samples	Rect	N	$[512, 1024]$
f_c	center frequency		N	$[512, 1024]$
FH	frequency hop	Costas	FH	$\{3, 4, 5, 6\}$
f_m	fundamental frequency		f_m	$U(f_s/32, f_s/25)$
L_c	code length	Barker	L_c	$\{7, 11, 13\}$
c_{pp}	cycles per phase code		c_{pp}	$[2, 5]$
M	number of frequency steps	Frank	c_{pp}	$[3, 5]$
n_s	number of sub-codes		M	$\{6, 7, 8\}$
n_g	number of segments	P1, P2	c_{pp}	$[3, 5]$
n_p	number of phase states		M	$\{6, 8\}$
ψ	path delay	P3, P4	c_{pp}	$[3, 5]$
G	average path gain		n_s	$\{36, 64\}$
f_{Dmax}	maximum Doppler shift	T1, T2	n_p	2
			n_g	$\{4, 5, 6\}$
Channel configuration			N	$[512, 1024]$
Model	Rayleigh fading	T3, T4	n_p	2
ψ	$U(1, 1000)$ ns		n_g	$\{4, 5, 6\}$
G	$U(-20, 0)$ dB		N	$[512, 1024]$
f_{Dmax}	$U(10, 1000)$ Hz		B	$U(f_s/20, f_s/15)$

Fig. 3. Parameter Ranges Used in Generating LPI Signal Instances [3]

Figure 4 illustrates examples of TFI obtained from TFA-SPWVD with an SNR of +10 dB.

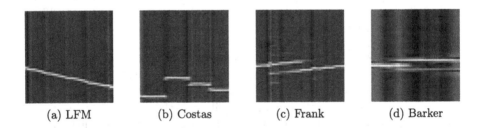

(a) LFM   (b) Costas   (c) Frank   (d) Barker

Fig. 4. TFI examples from the SPWVD-TFI base

As indicated in Table 2, two databases were generated, each containing 403.000 TFI (31 SNR levels * 13 LPI signals * 1000 LPI Signal Instances).

During the Data Partitioning step, each TFI base was sorted into five disjoint sets to facilitate cross-validation. In each cross-validation round, the data was stratified into training, validation, and test datasets in a 70-15-15 ratio, following a process and proportion similar to that used by [3].

The ATR proposed in this work was evaluated using classifiers based on two CNN models: SqueezeNet and GoogLeNet. Each of these CNN models was selected for distinct reasons.

SqueezeNet[6] was chosen primarily because it delivers strong performance comparable to AlexNet in image classification tasks, yet with 50 times fewer parameters and a model size 510 times smaller. This compactness makes SqueezeNet suitable for memory-constrained devices [4]. Additionally, as indicated in Table 1, SqueezeNet falls within a medium-sized network category compared to other CNN used in related works. Its topological architecture features unique *Fire Modules*, comprising *squeeze* layers with 1 × 1 convolution filters and *expand* layers with 1 × 1 and 3 × 3 convolution filters. You can visualize this architecture in Fig. 5. In this work, we have chosen version 1.1 of the SqueezeNet network because it requires 2.4 times fewer computations than version 1.0 while maintaining the same accuracy[7].

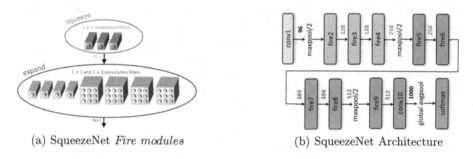

(a) SqueezeNet *Fire modules* (b) SqueezeNet Architecture

Fig. 5. SqueezeNet CNN [4]

GoogLeNet[8] was chosen primarily for its performance, which approaches the SOTA in classification [12]. This network architecture consists of 27 deep layers, formed by stacking 9 *inception modules* linearly, as illustrated in Fig. 6. These modules utilize 1 × 1, 3 × 3, and 5 × 5 convolution filters, along with maximum pooling layers. This structure enables image identification and categorization, even when limited information is available [12].

The fine-tuning of the SqueezeNet and the GoogLeNet was made through a MATLAB app known as Deep Designer[9]. The optimization process of the hyperparameter values of the SqueezeNet was made for the CWD-TFI ($CWD+SqueezeNet_{opt}$) and for the SPWVD-TFI ($SPWVD+SqueezeNet_{opt}$). In the same way, the optimization process of the hyperparameter values of the GoogLeNet was made for the CWD-TFI ($CWD+GoogLeNet_{opt}$) and for the

---

[6] https://github.com/forresti/SqueezeNet.
[7] https://github.com/forresti/SqueezeNet/tree/master/SqueezeNet_v1.1.
[8] https://github.com/BVLC/caffe/tree/master/models/bvlc_googlenet.
[9] https://www.mathworks.com/help/deeplearning/ref/deepnetworkdesigner-app.html.

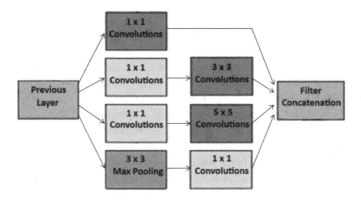

Fig. 6. *Inception module* [12]

SPWVD-TFI ($SPWVD+GoogLeNet_{opt}$). These optimization processes were possible due to the use of the MATLAB app known as Experiment Manager[10] using Bayesian searches. These searches employed stochastic gradient descent with momentum (SGDm) as the optimizer. The optimal hyperparameter values and their corresponding search ranges during training are summarized in Table 3. Additionally, a minibatch size of 128 was used for all models.

Table 3. Search Ranges and Optimal Values Found for SqueezeNet and GoogLeNet hyperparameters during Cross-Validation process

Selected Parameter	Range		Optimal Values Found			
			$SqueezeNet_{opt}$		$GoogLeNet_{opt}$	
	form	to	CWD	SPWVD	CWD	SPWVD
Initial Learn Rate	0.1	0.0005	0.01	0.0081	0.0092	0.01
Momentum	0.98	0.8	0.8631	0.8494	0.9421	0.9124
L2 Regularization	0.01	10^{-10}	$9.3111*10^{-7}$	0.0001	0.0001	0.0005
Learn Rate Drop Period	10 epochs	1 epoch	10 epochs	10 epochs	10 epochs	10 epochs
Learn Rate Drop Factor	0.2	0.01	0.01	0.01	0.01	0.01

Finally, the Evaluation of Results step was made by comparing the performance of the $SPWVD+SqueezeNet_{opt}$, $CWD+SqueezeNet_{opt}$, $SPWVD+GoogLeNet_{opt}$, and $CWD+GoogLeNet_{opt}$ combinations with each other and with the results reported by state-of-the-art studies.

All experiments were conducted on a hardware platform with an Intel® Core™ i5-12500H 12th Gen CPU running at 2.50 GHz, 16 GB of RAM, and a single NVIDIA GeForce RTX 3070 Ti GPU with 8 GB of memo

---

[10] https://www.mathworks.com/help/matlab/ref/experimentmanager-app.html.

5 Results

To obtain evidence supporting the validity of the hypothesis raised in this study, three analyses were conducted based on the evaluation of the obtained results.

The first analysis was conducted in two stages. In the first stage, the results of the $CWD+SqueezeNet_{opt}$ and $SPWVD+SqueezeNet_{opt}$ combinations obtained during the cross-validation process were compared. In the second stage, the results of the $CWD+GoogLeNet_{opt}$ and $SPWVD+GoogLeNet_{opt}$ combinations, also obtained during cross-validation, were compared. Both stages present precision values for each tested combination considering the 13 LPI signal types and varying SNR between −20 dB and 10 dB. Notably, regardless of the combination used, precision approaches 100% for higher SNR values (above 0 dB) across all LPI signal types.

The results of the first analysis using SqueezeNet can be visualized in Fig. 7. Comparing Fig. 7a and Fig. 7b, it can be noticed that, at −20 dB, the Barker signal achieved 97.4% precision with $CWD+SqueezeNet_{opt}$ and 93.2% precision with $SPWVD+SqueezeNet_{opt}$. Conversely, at the same SNR, the T2 signal achieved 31.2% precision with $CWD+SqueezeNet_{opt}$ and 55.5% with $SPWVD+SqueezeNet_{opt}$. Another important observation is the favorable results for signals T1 to T4 when using $SPWVD+SqueezeNet_{opt}$, especially at negative SNR values, compared to results obtained with $CWD+SqueezeNet_{opt}$. However, at −15 dB SNR, $CWD+SqueezeNet_{opt}$ outperformed $SPWVD+SqueezeNet_{opt}$ for the Costas and LFM signals.

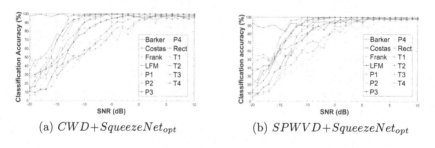

(a) $CWD+SqueezeNet_{opt}$ (b) $SPWVD+SqueezeNet_{opt}$

Fig. 7. Comparison of classification precision tested by SqueezeNet$_{opt}$

The results of the first analysis using the GoogLeNet are presented in Fig. 8. Comparing Fig. 8a and Fig. 8b, it can be noticed that, at −20 dB, the Barker signal achieved 97.1% precision with $CWD+GoogLeNet_{opt}$ and 95.2% precision with $SPWVD+GoogLeNet_{opt}$. Similarly, at the same SNR, the T1 signal achieved 66.7% precision with the $SPWVD+GoogLeNet_{opt}$ against 27.6% with the $CWD+GoogLeNet_{opt}$. Another notable result is the good performance of the LFM signal, using the $SPWVD+GoogLeNet_{opt}$, and the Costas signal, with the $CWD+GoogLeNet_{opt}$.

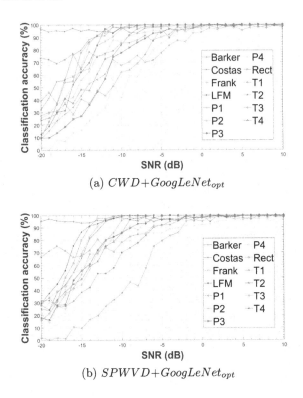

(a) $CWD+GoogLeNet_{opt}$

(b) $SPWVD+GoogLeNet_{opt}$

Fig. 8. Comparison of classification precision tested by GoogLeNet$_{opt}$

In summary, based on the results obtained in the first analysis, it is clear that both combinations yielded comparable outcomes. Specifically, each combination performed better for certain LPI radar signal types and worse for others. Consequently, there is no single combination that outperforms the other across all 13 LPI radar signal types and the entire SNR range.

The second analysis aimed to compare the average accuracy results obtained by considering all 13 LPI radar signal types together, across the 5 rounds of cross-validation adopted, and the four types of ATR combinations. Accordingly, the average accuracy value was calculated for each combination across the entire SNR range, as illustrated in Fig. 9. It can be noticed in Fig. 9 that the combinations using the CNN GoogLeNet$_{opt}$ achieved higher accuracy.

Another conclusion drawn from the results presented in Fig. 9 is that networks composed of the SPWVD-TFI datasets, specifically $SPWVD+SqueezeNet_{opt}$ and $SPWVD+GoogLeNet_{opt}$, exhibited superior average accuracy across the entire SNR range when compared to combinations using the CWD-TFI datasets ($CWD+SqueezeNet_{opt}$ and $CWD+GoogLeNet_{opt}$). Additionally, it is noteworthy that for high SNR values, average accuracy approaches 100%, regardless of the combination used. For instance, at 0 dB, $SPWVD+SqueezeNet_{opt}$ achieved an accuracy of 97.83%, while $SPWVD+GoogLeNet_{opt}$

reached 99.06%. Another noteworthy aspect from the curves in Fig. 9 that reflects the superior performance of $SPWVD+SqueezeNet_{opt}$ combination compared to $CWD+SqueezeNet_{opt}$ occurred at an SNR of -15 dB. At this point, $SPWVD+SqueezeNet_{opt}$ outperforms $CWD+SqueezeNet_{opt}$ by 4.4 p.p. Similarly, the optimal performance point for the $SPWVD+GoogLeNet_{opt}$ combination, comparing with $CWD+GoogLeNet_{opt}$, occurs at an SNR of -16 dB, with a margin of 2.7 p.p above $CWD+GoogLeNet_{opt}$. Ultimately, the SPWVD-TFI based combinations consistently achieve more prominent accuracy results when compared to the CWD-TFI based combinations, mainly at negative SNR values.

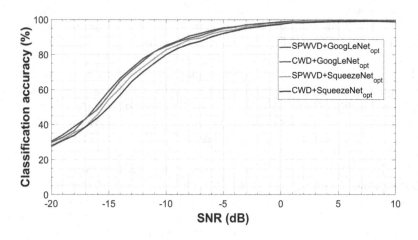

Fig. 9. Average accuracy obtained with the four proposed ATR combinations

Table 4 provides the mean, minimum, and maximum accuracy values and the standard deviation obtained by each ATR combination. It also shows the p-values and the results of the Wilcoxon Signed Rank Test applied to the ATR combination pairs $(CWD+SqueezeNet, SPWVD+SqueezeNet)$ and $(CWD+GoogLeNet, SPWVD+GoogLeNet)$ with significance level $\alpha = 0.05$ and null hypothesis H_0 stating that the means are statistically identical for each pair of ATR combinations. The Wilcoxon test was run in MATLAB with the Statistics and Machine Learning Toolbox.

Table 4. Accuracy Values and Results of Wilcoxon Signed Rank Test

ATR Combination	Accuracy				Wilcoxon Signed Rank Test	
	Mean Value	Standard Deviation	Min. Value	Max. Value	p Value	H_0 Hypothesis
CWD+SqueezeNet	79.91	0.34	79.57	80.25	4.9e–06	Rejected
SPWVD+SqueezeNet	81.13	0.23	80.9	81,36		
CWD+GoogLeNet	82.67	0.13	82.54	82.8	0.0073	Rejected
SPWVD+GoogLeNet	83.21	0.17	83.04	83.38		

The results obtained, shown by Table 4, suggest a slight superiority of SPWVD-TFI over CWD-TFI when combined with an optimized CNN for classification. On average, the worst result of $SPWVD+SqueezeNet_{opt}$ 80.9% outperformed the best result of $CWD+SqueezeNet_{opt}$ 80.25%. Similarly, the worst result of $SPWVD+GoogLeNet_{opt}$ 83.04% surpassed the best result of $CWD+GoogLeNet_{opt}$ 82,8%. Additionally, its important to note that the standard deviation of mean accuracy for $SPWVD+SqueezeNet_{opt}$ 0.22 was significantly lower than for $CWD+SqueezeNet_{opt}$ 0.34, indicating more consistent accuracy distributions for the SPWVD-TFI-based combinations, while both $CWD+GoogLeNet_{opt}$ and $SPWVD+GoogLeNet_{opt}$ standard deviation of average accuracy's have been similar and rather small values, 0.13 and 0.17 respectively. It is also important to highlight that the Wilcoxon test rejected the *null* hypothesis in both ATR combination pairs, indicating statistical evidence that SPWVD can lead to more accurate classification models than CWD.

In Fig. 10, we present the confusion matrices, at −8 dB SNR, for the two combinations that achieved the highest accuracy, considering their validation rounds, the $SPWVD+GoogLeNet_{opt}$ and $CWD+GoogLeNet_{opt}$. In both confusion matrices, certain modulation types are strongly confused with others. For instance, P1 signals are often misclassified as P4, while T1 signals are confused with T2. These misclassifications occur regardless of the TFA technique employed (i.e. CWD or SPWVD). However, a noticeable advantage lies in the classification capability of $SPWVD+GoogLeNet_{opt}$ over $CWD+GoogLeNet_{opt}$. This conclusion is based on two observations: The higher accuracy achieved by $SPWVD+GoogLeNet_{opt}$ 90.00% compared to $CWD+GoogLeNet_{opt}$ 89.38%; and the better precision provided by $SPWVD+GoogLeNet_{opt}$, which outperformed $CWD+GoogLeNet_{opt}$ when classifying 5 LPI radar signal types (Barker, LFM, P3, P4, and T1), and losing only at classifying 3 LPI signal types (P1, T2, and T3). These findings underscore the effectiveness of SPWVD-TFI in enhancing radar waveform recognition, particularly in low SNR environments.

The results observed in the first and second analyses provide evidence supporting the validity of the hypothesis proposed in this study. Specifically, the use of the SPWVD TFA technique, combined with SqueezeNet and GoogLeNet CNN trained with optimal hyperparameter values, appears to be a promising option for identifying the modulations of LPI radar signals.

Finally, the third analysis was conducted to ensure the promising performance of the ATR combinations proposed in this article compared with the related studies listed in Table 1. This comparison was feasible because all these studies evaluated their combinations using cross-validation, and the LPI radar signal instance base used in our study was generated following the same procedures and execution conditions as those adopted by the related works in their experiments.

The graph in Fig. 11 indicates the mean accuracy values obtained by our study and related works. From this figure, it is possible to note that the proposed combinations outperform all combinations from related studies across the entire tested SNR range. Notably, accuracy remains consistently high for SNR values

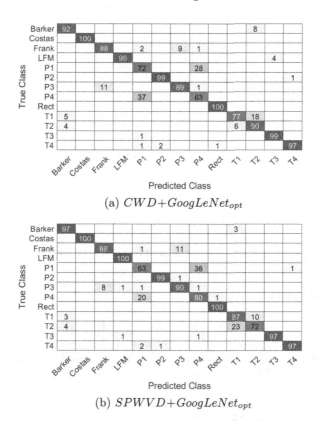

Fig. 10. Confusion Matrices

above $0 dB$. Considering the SNR range from –6 dB to 10 dB, the proposed combinations achieve an average accuracy of 97.41% with $SPWVD+SqueezeNet_{opt}$ and 98.05% with $SPWVD+GoogLeNet_{opt}$, while the combination proposed by [3] achieves 96.33%. Even at the lower end of the SNR spectrum (e.g., – 18 dB), our approach achieves an accuracy of 39%, surpassing the 25% accuracy reported by [3]. This 14 p.p difference represents a 56% improvement, demonstrating robust performance even in noisy conditions. The gain achieved by the present work, especially under low SNR values, gives the ELINT, incorporated into an ESM system, a better capacity for an early reaction against threats that employ guidance technology based on LPI radars, such as, missiles. It should be noted that the radar signals transmitted by a missile equipped with an LPI radar have very low power and are immersed in noise. Therefore, the highest detection range of an ESM will be limited to the lowest SNR value at which that ESM is capable of correctly detecting an LPI signal.

Finally, given the results of the three above analyses, it is possible to observe the effectiveness of our proposed ATR combinations across different SNR levels.

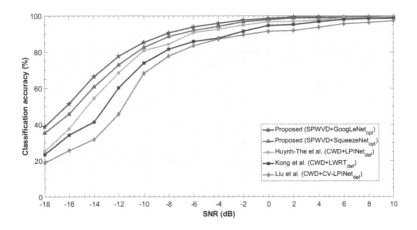

Fig. 11. Mean accuracy values of the proposed ATR and related works

6 Conclusions

In the context of automatic identification of LPI radar signals, this article presents the following contributions: (i) providing experimental evidence that employing the SPWVD TFA technique as a preprocessing step can lead to better classification model performance compared to models built using the CWD TFA technique, current state-of-the-art solution for this problem; (ii) proposing two new ATR combinations, where the pre-processing was based on SPWVD and the classifiers were based on CNN SqueezeNet and GoogLeNet. These ATR combinations outperform existing literature approaches across a wide SNR range from −18 dB to +10 dB, considering 13 distinct intrapulse modulations; (iii) providing two large image databases, each containing 403,000 samples, to be used in further experiments in this field.

As part of future work, in our quest for further evidence to validate the hypothesis proposed in this article and to enhance the state-of-the-art in LPI radar signal modulation detection, we intend to evaluate the performance of new ATR combinations and different signal types. For example, in the future, we intend to explore the use of STFT as the pre-processing TFA technique combined with different CNN models and with different signal input datasets, such as with acoustic signals.

Disclosure of Interests. The authors have no competing interests to declare that are relevant to the content of this article.

References

1. Choi, H.I., Williams, W.J.: Improved time-frequency representation of multicomponent signals using exponential kernels. IEEE Trans. Acoust. Speech Signal Process. **37**(6), 862–871 (1989)

2. Faceli, K., Lorena, A.C., Gama, J., Almeida, T.A.D., Carvalho, A.C.P.D.L.F.d.: Inteligência artificial: uma abordagem de aprendizado de máquina, vol. 1, p. 100. IEEE (2021)
3. Huynh-The, T., et al.: Accurate lpi radar waveform recognition with cwd-tfa for deep convolutional network. IEEE Wirel. Commun. Lett. **10**(8), 1638–1642 (2021)
4. Iandola, F.N., et al.: Squeezenet: alexnet-level accuracy with 50x fewer parameters and 0.5 mb model size. arXiv preprint arXiv:1602.07360 (2016)
5. Kong, S.H., et al.: Automatic lpi radar waveform recognition using cnn. IEEE Access **6**, 4207–4219 (2018)
6. Liu, Z., et al.: A method for lpi radar signals recognition based on complex convolutional neural network. Int. J. Numer. Model. Electron. Netw. Dev. Fields **37**(1), e3155 (2024)
7. Ma, N., Wang, J.: Dynamic threshold for spwvd parameter estimation based on otsu algorithm. J. Syst. Eng. Electron. **24**(6), 919–924 (2013)
8. MD: Política de ge de defesa. Portaria 333/MD, art. 4 (2004)
9. Milczarek, H., et al.: Automatic classification of frequency-modulated radar waveforms under multipath conditions. IEEE Sens. J. (2023)
10. Niranjan, R., Rama Rao, C., Singh, A.: Fpga based identification of frequency and phase modulated signals by time domain digital techniques for elint systems. Defence Sci. J. **71**(1) (2021)
11. Pace, P.E.: Detecting and classifying low probability of intercept radar. Artech house (2009)
12. Szegedy, C., et al.: Going deeper with convolutions. In: Proceedings of the IEEE Conference on Computer Vision and Pattern Recognition, pp. 1–9 (2015)
13. Walenczykowska, M., Kawalec, A., Krenc, K.: An application of analytic wavelet transform and convolutional neural network for radar intrapulse modulation recognition. Sensors **23**(4), 1986 (2023)
14. Wan, C., Si, W., Deng, Z.: Research on modulation recognition method of multicomponent radar signals based on deep convolution neural network. IET Radar Sonar Navigat. (2023)
15. Willetts, B., Ritchie, M., Griffiths, H.: Optimal time-frequency distribution selection for lpi radar pulse classification. In: 2020 IEEE International Radar Conference (RADAR), pp. 327–332 (2020)

Evaluating Large Language Models for Tax Law Reasoning

João Paulo Cavalcante Presa[✉][iD], Celso Gonçalves Camilo Junior[iD], and Sávio Salvarino Teles de Oliveira[iD]

Federal University of Goias (UFG), Goiânia, Brazil
joaopaulop@discente.ufg.br, {celsocamilo,savioteles}@ufg.br

Abstract. The ability to reason over laws is essential for legal professionals, facilitating interpreting and applying legal principles to complex real-world situations. Tax laws are crucial for funding government functions and shaping economic behavior, yet their interpretation poses challenges due to their complexity, constant evolution, and susceptibility to differing viewpoints. Large Language Models (LLMs) show considerable potential in supporting this reasoning process by processing extensive legal texts and generating relevant information. This study evaluates the performance of LLMs in legal reasoning within the domain of tax law for legal entities, utilizing a dataset of real-world questions and expert answers in Brazilian Portuguese. We employed quantitative metrics (BLEU, ROUGE) and qualitative assessment using a solid LLM to ensure factual accuracy and relevance. A novel dataset was curated, comprising genuine questions from legal entities in tax law, answered by legal experts with corresponding legal texts. The evaluation includes both open-source and proprietary LLMs, providing a assessment of their effectiveness in legal reasoning tasks. The strong correlation between robust LLM evaluator metric and Bert Score F1 suggests these metrics effectively capture semantic aspects pertinent to human-perceived quality.

Keywords: Legal Reasoning · Large Language Models (LLMs) · Legal Question Answering · Tax Law

1 Introduction

The ability to reason over laws is essential for legal professionals, enabling them to interpret and apply legal principles to complex real-world situations. Legal questions often lack straightforward answers, requiring thorough analysis, comprehensive research, and synthesis of multiple sources to develop well-founded arguments or solutions. Tax law, in particular, is crucial because it influences how governments fund public services and impacts economic activity by shaping investment decisions and individual spending. However, interpreting tax law presents significant challenges for Natural Language Processing (NLP) due to the inherent complexity and ambiguity of legal language, constant updates, amendments, and the need to contextualize regulations within specific jurisdictions.

Large Language Models (LLMs) show significant potential in enhancing the legal reasoning process [23]. These models can process extensive legal texts, including statutes, case law, and legal opinions, to extract relevant information and address the complexities of tax law. By leveraging advanced generation techniques, LLMs can answer legal questions using specific legal datasets, such as court cases and legal precedents, thus providing comprehensive and relevant information to legal professionals [29]. However, there is a gap in understanding how LLMs reason over legal texts, as existing question-answering tasks typically contain answers directly extractable from the provided texts, whereas legal reasoning often requires deeper comprehension and application of legal principles to nuanced scenarios [2].

To address this gap, we developed a novel dataset comprising real questions posed by legal entities in the domain of tax law, answered by legal experts with supporting legal texts (gold passages). This dataset allows us to assess the legal reasoning abilities of LLMs, focusing on their capacity to understand complex legal questions, use relevant law articles, and generate accurate and coherent responses. The evaluation compares LLM-generated answers to expert responses using metrics such as ROUGE [19], BLEU [27], and semantic similarity [40], alongside assessments by a strong LLM [41], contributing to a understanding of LLMs' legal reasoning capabilities.

This research evaluates both open-source and proprietary LLMs in scenarios requiring comprehensive understanding and application of the law, distinct from the extractive approaches used in datasets like SQuAD [28] and TriviaQA [14]. Our dataset requires LLMs to comprehend and apply the law to generate appropriate answers, often involving complex vocabulary and contexts not directly mirrored in the texts [2].

This paper presents two significant contributions to the field of legal NLP, particularly within the challenging domain of tax law. First, it introduces a novel dataset consisting of real-world tax law questions, expert-crafted answers, and supporting legal texts, moving beyond extractive question-answering tasks and requiring models to demonstrate legal reasoning abilities. Using this dataset, we evaluate how well LLMs understand complex tax law questions and generate accurate, well-supported answers, providing a better understanding of current LLM capabilities and limitations in handling legal reasoning tasks.

2 Related Works

While there are numerous works utilizing Large Language Models (LLMs) in the legal domain [5,22,23,25,39], our interest lies in those that apply LLMs for question and answer (Q&A) tasks. These works can be classified into three main categories: those using retrieval-augmented generation, those evaluating LLMs based on prior knowledge, and those performing fine-tuning and testing the models on Q&A tasks in the legal domain. Below, we discuss the key works found in each of these categories.

2.1 Tax Law Applications of LLMs

Evaluating Q&A LLMs with Retrieval-Augmented Generation. The LLeQA [21] dataset includes 1,868 legal questions annotated by experts, containing answers and legal references. This work applies the Retrieval-Augmented Generation (RAG) technique, retrieving statutory articles from an extensive corpus of Belgian legislation. The model's effectiveness is evaluated using the METEOR metric, demonstrating the feasibility of integrating information retrieval with LLMs to enhance the accuracy of legal responses. ChatLaw [5] addresses the creation of a large-scale language model for the legal domain, specifically in the Chinese context. This work combines vector database retrieval methods with keyword-based retrieval to increase the accuracy of responses. Integrating these techniques enables the model to provide more precise and contextually relevant answers.

Evaluating Q&A Legal Reasoning of LLMs. LAiW [6] proposes a benchmark for evaluating the capabilities of LLMs in the Chinese legal context. The aim of this work is to test how well models can handle specific legal tasks. The results show that some legal-specific LLMs perform better than their general counterparts, although there remains a significant gap compared to GPT-4 [26]. LawBench [8] offers a comprehensive assessment of LLM capabilities in legal tasks, including Q&A. This work extensively tested 51 popular LLMs, including 20 multilingual, 22 focused on Chinese, and 9 specific to law. The conclusion is that while fine-tuning LLMs on specific legal texts brings some improvements, the models still need to be usable and reliable for complex legal tasks.

Fine-Tuning and Evaluating Q&A Large Language Models. FedJudge [37] uses Federated Learning (FL) to overcome data privacy challenges. This framework optimizes federated legal LLMs, allowing the models to be trained locally on clients, with their parameters aggregated and distributed on a central server. FedJudge is evaluated on Q&A tasks using metrics such as ROUGE, BLEU, and BertScore to compare the quality of generated answers. This work demonstrates that the model provides more precise and relevant answers in different legal contexts. DISC-LawLLM [38] employs large language models trained on supervised datasets in the legal domain and incorporates a retrieval module to access and utilize external legal knowledge. This system assesses objective and subjective perspectives using DISC-Law-Eval, a benchmark that includes legal question answering. Additionally, subjective evaluation is carried out using the GPT-3.5 model as a judge.

3 Metodology

This section outlines the methodology utilized in our study, with a particular emphasis on the model selection process. We also detail the data collection process, the creation of a relevant corpus, and the experimental setups of the selected models, including the specific prompts and parameters used. Furthermore, we detail the evaluation approach, discussing both the metrics employed and the strategy for subjective evaluation.

3.1 Dataset Collection

Our dataset consists of a series of tax law questions related to legal entities. The questions were selected from a collection that is annually updated by the General Coordination of Taxation (Cosit) [9] of the Brazilian Federal Revenue Service. The dataset includes over a thousand question-answer pairs, with most answers being supported by a relevant normative or legal basis. The granularity of the references in the answers is as detailed as possible, citing the specific articles of law or other regulations used to formulate the responses. The questions represent real taxpayer doubts, and experts in the Brazilian tax field craft the answers. Below, we will discuss how the dataset was created.

Selection of Questions. We extracted a subsample from the comprehensive set of questions and answers provided by Cosit. In this selection process, we focused on questions that included responses with legal references rather than the entire regulation. Although the majority of responses included legal references, they were often elaborated by experts in a way that extended beyond the scope of the question or included excessive details such as tables and numerous examples. This complexity made them unsuitable for use in contexts like Retrieval-Augmented Generation (RAG). We excluded these overly detailed responses to ensure a fair evaluation with the LLMs. The initial outcomes of this selection process are depicted in the first three columns of Table 1.

Collection of Regulations (Gold Passages). After selecting the questions and their corresponding legal references, laws, and articles, we gathered each regulatory document referenced by the experts in their responses to the questions posed by legal entities. Although this task was time-intensive, it was essential for assessing the reasoning capabilities of LLMs in relation to legal texts. Upon completion of this stage, the dataset comprised the question, answer, reference to the regulation, and the regulation itself (gold passages). Table 1 presents the final dataset.

Legislation Corpus. In this stage, we collected over 30 documents, which included laws, instructions, decrees, and opinions. Each document contains up to thousands of articles comprising multiple provisions. These documents represent a fraction of the Brazilian tax legislation and include the regulations that underpin the experts' responses in the dataset. It is important to note that these regulations are constantly being amended, and many provisions have been revoked. All revoked provisions were excluded up to the dataset creation date to

Table 1. Questions and Answers with Legal References and Gold Passages

Question	Answer	Reference	Gold Passages
Which legal entities are exempt from presenting the ECF?	The following are exempt from presenting the ECF: I - those...	IN RFB No. 2004, 2021, art. 1, 1.	Art. 1 The Fiscal Accounting Bookkeeping (ECF) shall...
What are the tax effects incase the ECF is corrected?	When the correction of the ECF shows a higher tax due...	IN RFB No. 2055, 2021, art. 148.	Art. 148. The credit related to tax administered...
Does the exemption from IRPJ depend on prior recognition?	No. The benefit of the IRPJ exemption does not depend on...	RIR/2018, art. 192.	Art. 192. The exemptions referred to in this Section...
Under what circumstances is an individual considered equivalent to a legal entity?	For income tax purposes, individuals are considered...	RIR/2018, art. 162, 1, items I to III.	Art. 162. Individual enterprises are considered...
Are co-owners in property ownership subject to income tax?	Condominiums in property ownership are not subject to...	RIR/2018, art. 167.	Art. 167. Condominiums in property ownership shall...

ensure a high-quality corpus. Additionally, any questions that had their regulatory basis revoked were eliminated during the question selection phase. Figure 1 shows the corpus documents.

3.2 Experimental Setup

In this study, we conducted a comprehensive evaluation of large language models (LLMs) in terms of their ability to reason about laws, specifically focusing on corporate taxation for legal entities. We evaluated the LLMs using the datasets created in this paper, which were created from real-world questions and answers about the taxation of legal entities, which were provided by subject-matter experts.

We selected over 20 LLMs for evaluation, encompassing both proprietary and open-source models. The chosen models include notable examples such as Mistral AI, Llama, Gemma, Qwen, various community fine-tuned versions of these models, and a proprietary model. Each model possesses unique characteristics and capabilities, providing a diverse range of perspectives for our assessment.

In order to maintain consistency in our evaluations, we standardized the temperature parameter at 0.1 for all chosen models. This low-temperature setting was chosen to reduce randomness in the output, thereby encouraging more deterministic responses. Additionally, we did not impose a maximum token limit, allowing the models to generate responses without any constraints on their length.

A specific prompt (see Prompt Question Answer in Appendix A) was crafted to guide the models in reasoning about the law and generating appropriate

ADI SRF nº 5, de 2001	Lei nº 6.766, de 1979
ADN Cosit nº 4, de 1996	Lei nº 9.249, de 1995
Decreto-Lei nº 1.381, de 1974	Lei nº 9.316, de 1996
Decreto-Lei nº 1.510, de 1976	Lei nº 9.430, de 1996
Decreto-Lei nº 1.598, de 1977	Lei nº 9.532, de 1997
IN DPRF 21, de 1992	Lei nº 9.718, de 1998
IN RFB nº 1.252, de 2012	Lei nº 11.051, de 2004
IN RFB nº 1.520, de 2014	PN CST nº 1, de 1983
IN RFB nº 1.700, de 2017	PN CST nº 2, de 1983
IN RFB nº 2.004, de 2021	PN CST nº 4, de 1981
IN RFB nº 2.055, de 2021	PN CST nº 58, de 1977
IN SRF nº 213, de 2002	PN CST nº 72, de 1975
IN SRF nº 51, de 1978	PN CST nº 146, de 1975
IN nº 122, de 1989	Portaria MF nº 356, de 1988
Lei nº 6.404, de 1976	RIR 2018

Fig. 1. Documents from the legislative corpus

responses. The prompt explicitly instructs the models to reason through the legal context provided and formulate an answer. If a model is unable to generate a satisfactory response, it is instructed to state that it does not know the answer.

The legal information needed to answer each question, like an article from a law or a legal document, is included in the prompt. This information is the same one used by the experts to create the reference answers, ensuring a fair basis for comparison. By using standardized prompts and incorporating relevant legal provisions, it ensures that the models have access to the same information as human experts. This enables a thorough evaluation of their reasoning capabilities.

It is important to note that the questions and the reference answers are presented in Brazilian Portuguese. This aspect of the study tests the models' reasoning abilities and evaluates their proficiency in generating accurate and contextually appropriate responses in the Portuguese language. Given that many LLMs are primarily trained on English-language datasets, assessing their performance on Brazilian Portuguese legal texts for understanding the applicability and limitations of these models in non-English-speaking jurisdictions.

Although the dataset used in this experiment contains a corpus suitable for Retrieval-Augmented Generation (RAG), our evaluation focused solely on the tasks of generation and reasoning. This decision was inspired by other prominent datasets, such as SQuAD 2.0 and HotpotQA, which also provide the expected passages alongside the ground truth answers, allowing for a direct assessment of the model's generation capabilities without the retrieval step. By concentrating on these aspects, we aimed to isolate and thoroughly evaluate the LLMs' ability to generate accurate and reasoned responses based solely on the provided legal context.

3.3 Evaluation Metrics

Our study evaluated large language models (LLMs) using a comprehensive approach integrating quantitative and qualitative methods. For the quantitative assessment, we employed the BLEU (Bilingual Evaluation Understudy) and ROUGE (Recall-Oriented Understudy for Gisting Evaluation) metrics [19,27]. In the field of natural language processing, these metrics are crucial for assessing the quality of text generation by comparing the models' responses to a predefined set of reference answers. Specifically, in the domain of questions and answers related to corporate taxation, these metrics provide a quantitative measure of how closely the generated responses align with the ideal answers.

Despite their widespread use, metrics such as BLEU, ROUGE [37], and METEOR [21], which are widely used for evaluating language models, they primarily provide a quantitative perspective and may not fully capture the accuracy of responses in question-answering scenarios [20]. This limitation arises because these metrics do not adequately assess the factual accuracy or the relevance of the generated responses, which are critical in determining whether the questions were answered correctly.

In order to address this gap, we adopted a more nuanced qualitative approach, utilizing the capabilities of a powerful language model as a surrogate for human judgment. Specifically, we employed GPT-4 to evaluate the performance of other models. This approach is premised on the notion that a robust LLM, such as GPT-4, can effectively emulate human judgment in evaluating responses [7,10,17,20,31,35,41] to open-ended questions, thereby providing a closer approximation to human evaluative criteria.

For the qualitative evaluation, we used a carefully designed prompt to assess the factual accuracy of the models' responses. The accuracy of each model was then calculated based on this assessment. The specific prompt used for this evaluation can be found in Prompt Evaluation in the Appendix A for further details.

4 Results

In this section, we will present the results of the experiments described in the Experimental Setup section. We used the metrics outlined in the Metrics section, both of which are situated in the Methodology (Sect. 3). We selected the main language models operating in Portuguese to evaluate how well they can reason and answer questions in the context of tax law, with the aim of identifying potential improvements (Table 2).

4.1 Model Performance Analysis

The latest versions of the Llama, Qwen, and Mistral families exhibit significant advancements compared to their predecessors. These models incorporate several architectural enhancements, including SwiGLU activation [30] and Grouped Query Attention (GQA) [4]. Both the Qwen2-72B-Instruct [1] and Llama-3-70b-chat-hf [3] models benefited from these improvements, particularly the modifications to the tokenizer and the inclusion of GQA, leading to notable performance gains. As a result, the Qwen2-72B-Instruct [1] model achieved the highest accuracy. Similar results have been observed in other LLM evaluation benchmarks [1], highlighting the superior performance of models incorporating these techniques.

The performance analysis of the models revealed that model size significantly impacts the results, but this impact is not always straightforward. Larger models, such as Qwen2-72B-Instruct [1] and Mixtral-8x22B-Instruct-v0.1 [13], achieved superior performance, exhibiting the highest ROUGE-L, BLEU, Bert Score F1, and GPT-4 evaluated accuracy metrics. However, we observed that smaller models, such as Mistral-7B-Instruct-v0.3 [12] and OpenHermes-2p5-Mistral-7B [32], outperformed some larger models in specific metrics. For instance, Mistral-7B-Instruct-v0.3 attained a Bert Score F1 of 0.71, surpassing several larger models, and OpenHermes-2p5-Mistral-7B demonstrated remarkable performance with accuracy comparable to significantly larger models. These findings suggest that

Table 2. Model Performance Metrics

Model	ROUGE-L	BLEU	Bert Score F1	Acc. GPT-4
Mistral-7B-Instruct-v0.2	0.35	0.20	0.67	0.54
Mistral-7B-Instruct-v0.3	0.40	0.26	0.71	0.55
Mixtral-8x7B-Instruct-v0.1	0.38	0.24	0.70	0.53
Mixtral-8x22B-Instruct-v0.1	**0.44**	**0.30**	**0.73**	0.59
Llama-2-70b-chat-hf	0.38	0.19	0.69	0.49
Llama-2-13b-chat-hf	0.37	0.20	0.68	0.43
Llama-2-7b-chat-hf	0.32	0.14	0.65	0.34
Llama-3-70b-chat-hf	0.34	0.16	0.65	0.60
Llama-3-8b-chat-hf	0.35	0.15	0.65	0.54
Qwen1.5-110B-Chat	0.39	0.21	0.71	0.60
Qwen1.5-72B-Chat	0.41	0.24	0.71	0.62
Qwen1.5-14B-Chat	0.34	0.16	0.68	0.48
Qwen2-72B-Instruct	0.43	0.29	**0.73**	**0.64**
gemma-7b-it	0.40	0.22	0.70	0.45
Yi-34B-Chat	0.38	0.26	0.70	0.52
gpt-3.5-turbo	0.38	0.15	0.69	0.56
Platypus2-70B-instruct	0.41	0.29	0.70	0.57
vicuna-13b-v1.5	0.41	0.27	0.71	0.50
vicuna-7b-v1.5	0.37	0.23	0.69	0.39
openchat-3.5-1210	0.42	0.28	0.72	0.55
WizardLM-13B-V1.2	0.36	0.25	0.68	0.49
SOLAR-10.7B-Instruct-v1.0	0.36	0.23	0.70	0.51
OpenHermes-2p5-Mistral-7B	0.41	0.25	0.71	0.55

while larger models generally deliver better results due to their ability to capture more complex information, well-trained and fine-tuned smaller models can offer competitive performance in specific contexts. This trend indicates that the training quality and the model's suitability to the particular dataset are crucial factors that can mitigate the size disparity among models.

Although the volume of Portuguese data used in training these models has yet to be verified, the architectural and training improvements suggest the enhanced performance of the LLMs in Q&A tasks in corporate tax law. In the Mistral family, the Mixtral-8x22B-Instruct-v0.1 [24] model stood out with the highest scores in ROUGE-L, BLEU, and Bert Score F1, indicating the potential of the mixture of experts architecture [13] for legal texts in Portuguese.

The analysis of fine-tuned open-source models reveals significant improvements over the base models. The openchat-3.5-1210 [34] and OpenHermes-2p5-Mistral-7B [32], both derived from the Mistral-7B-v0.1 [12], showed notable increases in accuracy after fine-tuning. Similarly, the vicuna-13b-v1.5 and vicuna-7b-v1.5 models [41], fine-tuned from Llama 2 [33], also demonstrated advances in response accuracy. Furthermore, models such as WizardLM-13B-V1.2 [36], SOLAR-10.7B-Instruct-v1.0 [15,16], and Platypus2-70B-instruct [18], derived from Llama 2, improved the results of their base models. Notably, these fine-tuning processes were conducted on diverse datasets, not on the experimental dataset itself, yet still led to enhanced metrics within the experimental dataset. These improvements suggest that fine-tuning can effectively enhance the capabilities of Q&A and legal text generation tasks when applied to specific datasets.

4.2 Evaluation Metrics Analysis

The traditional metrics like BLEU and ROUGE may not fully capture the nuances needed for accurate question answering in Q&A tasks. The Bert Score F1 metric is extensively recognized for its alignment with human evaluation due to its capacity to capture deep semantic similarities between texts [40], surpassing the lexical matching capabilities of traditional metrics like ROUGE-L and BLEU [40].

While this study does not aim to prove that LLM as a judge for evaluation is aligned with human evaluation, recent studies have been exploring this alignment [20,31,35,41]. Our research evaluates the quality of responses generated by LLMs in legal domain Q&A tasks. The strong correlation between the LLM (GPT-4) Accuracy Evaluation and Bert Score F1, as evidenced by the Pearson (0.657) and Kendall (0.491) correlations (see Table 3), suggests that both metrics capture semantic aspects relevant to human-perceived quality. The results are in line with studies [40] recommending using Pearson and Kendall correlations to evaluate metric quality.

Furthermore, the Bland-Altman analysis, which is particularly suitable for comparing measurement methods [11], confirms that Bert Score F1 and LLM (GPT-4) Accuracy Evaluation are more concordant, as shown by the lower variation in point dispersion and the narrower width of the limits of agreement (see

Table 3. Correlation Matrices for Evaluation Metrics

Metrics	Pearson	Kendall
Accuracy (GPT4) ↔ Bert Score F1	0.658	0.491
Accuracy (GPT4) ↔ ROUGE-L	0.539	0.393
Accuracy (GPT4) ↔ BLEU	0.373	0.289
Bert Score F1 ↔ ROUGE-L	0.908	0.838
Bert Score F1 ↔ BLEU	0.781	0.666
ROUGE-L ↔ BLEU	0.821	0.670

Fig. 2). In contrast, ROUGE-L and BLEU demonstrated higher mean differences and wider limits of agreement. Bert Score F1 exhibited a mean difference close to zero and narrower limits of agreement, indicating better concordance with LLM Accuracy Evaluation measurements.

These findings imply that LLM (GPT-4) Accuracy Evaluation, similar to Bert Score F1, could be a valuable and representative metric for assessing the actual performance of language models. While ROUGE-L and BLEU show

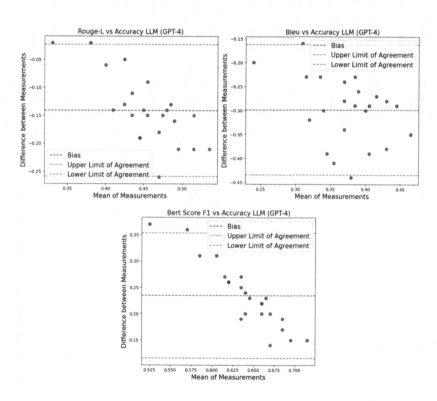

Fig. 2. Bland-Altman Plots for Metrics vs LLM (GPT-4) Accuracy Evaluation

higher correlations with Bert Score F1, the stronger correlation and concordance of LLM Accuracy Evaluation with Bert Score F1 indicate its potential alignment with human evaluation. This supports the development of evaluation metrics that more accurately reflect human-perceived quality, aligning with the direction of current research investigating the potential of LLMs used to align with human evaluation [20,31,35,41].

5 Conclusion

This study underscores the importance of tax law in society and the potential of language models to assist in its understanding and application. We developed a novel dataset of real-world tax law questions and expert answers in Brazilian Portuguese and conducted a rigorous evaluation of various language models. While our findings suggest that these models show promise in comprehending and reasoning about complex legal texts, further research is necessary to fully demonstrate their effectiveness in legal reasoning across a broader range of scenarios and tasks.

Our evaluation showed that advancements in model architecture have a noticeable impact on performance, and fine-tuning open-source models, even when done on diverse datasets rather than those specific to the legal domain, can still improve their ability to generate relevant and accurate responses. This suggests that continuous improvements and adaptations are valuable in enhancing the capabilities of language models in legal tasks.

For assessing model performance, we used Bert Score F1, known for its strong correlation with human evaluations in tasks involving descriptive and structural understanding, and a newer metric, LLM Accuracy Evaluation. While Bert Score F1 is already established as an effective measure aligned with human judgment, especially in descriptive tasks, our results showed that LLM Accuracy Evaluation also demonstrated strong correlation with Bert Score F1 through Pearson and Kendall correlations. The Bland-Altman analysis further confirmed that the LLM metric aligns closely with Bert Score, suggesting its potential as a reliable alternative in evaluations. However, it is important to note that while these findings are encouraging, the use of these metrics for reasoning-based tasks, such as those in this study, still requires further validation. The LLM metric is a promising tool, but more research is needed to establish its effectiveness fully, particularly in capturing the nuances of legal reasoning.

Limitations and Future Work. A limitation of our study is that while it focused on evaluating the generation and reasoning capabilities of LLMs, it did not require the models to identify specific legal provisions as part of their responses. Our dataset includes a comprehensive corpus containing the necessary laws, enabling the application of Retrieval-Augmented Generation (RAG) techniques. This allows models to retrieve relevant legal provisions and incorporate them into their answers, which is essential for a more complete statutory reasoning process. By providing the relevant legal articles, which do not directly mirror the answers to the questions, our study did assess a portion of the statutory reasoning by testing the models' ability to apply the law to generate accurate responses. Future work could leverage this corpus to explore the integration of RAG, aiming to enhance the models' ability to not only generate correct answers but also to identify and cite the appropriate legal provisions, thereby achieving a more robust and comprehensive statutory reasoning.

Dataset and Code Availability. The dataset used in this study, as well as the code for reproducing the experiments and analyses, are publicly available. The dataset can be accessed at the following link: https://github.com/joaopaulopresa/dataset. The code can be found here: https://github.com/joaopaulopresa/code.

A Appendix: List of Prompts Used

This appendix provides a list of the prompts utilized throughout the experiments detailed in this study. The inclusion of these prompts ensures the transparency and reproducibility of the experimental procedures, allowing other researchers to replicate the study and verify the findings.

Prompt Question Answer:

```
Use the following pieces of legal information from laws to answer the user's question.
If the answer is not clear in context, try to figure out by interpreting the information.
If you don't know the answer, just say that you don't know, don't try to make up an answer.
Context: {context}
Question: {question}
Do not quote the "contextual information" provided in the answer, do not say "according to the
    information" or anything like that, use the information only to answer the question.
Only return the helpful answer below and nothing else.
Answer the question in Portuguese.
Helpful answer:
```

Prompt Evaluation:

```
Evaluate the AI-generated response based on the following criteria:
1. Verify if the AI Response is contained within the Expert Response, meaning there are no
    contradictions. Ignore different terms or small additional or missing information.
2. The Expert Response may contain more information than requested in the question. If the information
    in the Expert Response is not necessary to answer the question, do not use it to evaluate the
    AI Response.
3. If the AI Response contains more information than the Expert Response, it should not be considered
    for evaluation as long as the information is correct.
4. Check if the response answers the question. Ensure the response provides the information requested
    in the question and is sufficient. For example, if the question can be answered with a simple "
    No," that is acceptable.
Include reasoning that justifies the Evaluation. If the criteria are met, return 'CORRECT.' If any of
    the criteria are not met, return 'WRONG.'
The Evaluation should be a JSON object with keys 'result' and 'reasoning.'
Examples:
1.
### Question:
Are the earnings from technical consulting services provided by a legal entity domiciled in Brazil to
    its parent company abroad subject to transfer pricing legislation?
### Expert Response:
Firstly, it is necessary to distinguish whether the provision of services in Brazil involved
    technology transfer. If technology transfer is proven with the consent of the National Institute
    of Industrial Property (INPI), the transaction will not be subject to transfer pricing rules as
    established by art. 55 of IN RFB No. 1,312, of 2012. In this case, the deduction of such
    expenses is subject to the limits established by arts. 362 to 365 of RIR/2018. If there is no
    technology transfer, these services become subject to transfer pricing rules.
### AI Response:
Yes, they are subject to transfer pricing unless there is technology transfer with INPI consent.
### Evaluation:
```

```
{{
    "reasoning": "The AI response aligns with the expert's response, correctly addressing the question
        without contradictions, although it is shorter.",
    "result": "CORRECT"
}}
2.
### Question:
What should be considered as 'accrued considerations'?
### Expert Response:
For the purposes of art. 175 of Normative Instruction RFB No. 1,700, of 2017, accrued considerations
    are considered due considerations.
### AI Response:
Due considerations.
### Evaluation:
{{
    "reasoning": "The AI response covers the main points mentioned by the expert without presenting
        contradictions, though it is less detailed.",
    "result": "CORRECT"
}}
3.
### Question:
Is there a deadline for offsetting rural activity tax losses?
### Expert Response:
There is no deadline for offsetting rural activity tax losses.
### AI Response:
The deadline is 7 days from the date of the loss, which can be extended up to 30 days for offsetting
    rural activity tax losses.
### Evaluation:
{{
    "reasoning": "The AI response is incorrect because it mentions '7 days' and 'up to 30 days,' which
        contradicts the expert's response.",
    "result": "WRONG"
}}
4.
### Question:
Can the negative CSLL calculation base be offset against results determined in subsequent periods?
### Expert Response:
Yes. The CSLL calculation base, when negative, can be offset up to 30% of the results determined in
    subsequent periods, adjusted by the additions and exclusions provided for by law.
### AI Response:
No. The CSLL calculation base can be offset against results determined in subsequent periods.
### Evaluation:
{{
    "reasoning": "The AI response contradicts the expert's response, providing an opposite answer
        regarding the possibility of CSLL offset.",
    "result": "WRONG"
}}
Now think step by step and make this Evaluation:
### Question:
{questao}
### Expert Response:
{resposta_especialista}
### AI Response:
{resposta_ia}
### Evaluation:
```

References

1. Qwen2 blog (2024). https://qwenlm.github.io/blog/qwen2/. Accessed 08 June 2024
2. Abdallah, A., Piryani, B., Jatowt, A.: Exploring the state of the art in legal qa systems. J. Big Data **10**(1), 127 (2023)
3. AI@Meta: Llama 3 model card (2024). https://github.com/meta-llama/llama3/tree/main
4. Ainslie, J., Lee-Thorp, J., de Jong, M., Zemlyanskiy, Y., Lebron, F., Sanghai, S.: Gqa: training generalized multi-query transformer models from multi-head checkpoints. In: Proceedings of the 2023 Conference on Empirical Methods in Natural Language Processing, pp. 4895–4901 (2023)
5. Cui, J., Li, Z., Yan, Y., Chen, B., Yuan, L.: Chatlaw: open-source legal large language model with integrated external knowledge bases. arXiv preprint arXiv:2306.16092 (2023)
6. Dai, Y., et al.: Laiw: a Chinese legal large language models benchmark (a technical report). arXiv preprint arXiv:2310.05620 (2023)
7. Du, Y., Wei, F., Zhang, H.: Anytool: self-reflective, hierarchical agents for large-scale api calls. arXiv preprint arXiv:2402.04253 (2024)

8. Fei, Z., et al.: Lawbench: benchmarking legal knowledge of large language models. arXiv preprint arXiv:2309.16289 (2023)
9. General Coordination of Taxation (Cosit): Questions and answers for legal entities (2023). https://www.gov.br/receitafederal/pt-br/assuntos/orientacao-tributaria/declaracoes-e-demonstrativos/ecf/perguntas-e-respostas-pj-2023.pdf. Accessed 11 Nov 2023
10. Hackl, V., Müller, A.E., Granitzer, M., Sailer, M.: Is gpt-4 a reliable rater? evaluating consistency in gpt-4's text ratings. In: Frontiers in Education. vol. 8, p. 1272229. Frontiers Media SA (2023)
11. Haghayegh, S., Kang, H.A., Khoshnevis, S., Smolensky, M.H., Diller, K.R.: A comprehensive guideline for bland-altman and intra class correlation calculations to properly compare two methods of measurement and interpret findings. Physiol. Meas. **41**(5), 055012 (2020)
12. Jiang, A.Q., et al.: Mistral 7b (2023). arXiv preprint arXiv:2310.06825
13. Jiang, A.Q., et al.: Mixtral of experts (2024). arXiv preprint arXiv:2401.04088
14. Joshi, M., Choi, E., Weld, D.S., Zettlemoyer, L.: Triviaqa: a large scale distantly supervised challenge dataset for reading comprehension. In: Proceedings of the 55th Annual Meeting of the Association for Computational Linguistics, vol. 1: Long Papers, pp. 1601–1611 (2017)
15. Kim, D., et al.: sdpo: don't use your data all at once (2024)
16. Kim, D., et al.: Solar 10.7b: scaling large language models with simple yet effective depth up-scaling (2023)
17. Koutcheme, C., Dainese, N., Sarsa, S., Hellas, A., Leinonen, J., Denny, P.: Open source language models can provide feedback: evaluating llms' ability to help students using gpt-4-as-a-judge. arXiv preprint arXiv:2405.05253 (2024)
18. Lee, A.N., Hunter, C.J., Ruiz, N.: Platypus: quick, cheap, and powerful refinement of llms (2023)
19. Lin, C.Y.: Rouge: a package for automatic evaluation of summaries. In: Text Summarization Branches Out, pp. 74–81 (2004)
20. Liu, Y., Iter, D., Xu, Y., Wang, S., Xu, R., Zhu, C.: G-eval: Nlg evaluation using gpt-4 with better human alignment. In: The 2023 Conference on Empirical Methods in Natural Language Processing (2023)
21. Louis, A., van Dijck, G., Spanakis, G.: Interpretable long-form legal question answering with retrieval-augmented large language models. arXiv preprint arXiv:2309.17050 (2023)
22. Ma, S., Chen, C., Chu, Q., Mao, J.: Leveraging large language models for relevance judgments in legal case retrieval. arXiv preprint arXiv:2403.18405 (2024)
23. Martin, L., Whitehouse, N., Yiu, S., Catterson, L., Perera, R.: Better call gpt, comparing large language models against lawyers. arXiv preprint arXiv:2401.16212 (2024)
24. Mistral.ai: Introducing the mixtral-8x22b-instruct-v0.1 model (2024). https://mistral.ai/news/mixtral-8x22b/. Accessed 15 May 2024
25. Niklaus, J., et al.: Flawn-t5: an empirical examination of effective instruction-tuning data mixtures for legal reasoning. arXiv preprint arXiv:2404.02127 (2024)
26. OpenAI: Gpt-4 technical report (2023). https://cdn.openai.com/papers/gpt-4.pdf. Accessed 10 Jan 2024
27. Papineni, K., Roukos, S., Ward, T., Zhu, W.J.: Bleu: a method for automatic evaluation of machine translation. In: Proceedings of the 40th Annual Meeting of the Association for Computational Linguistics, pp. 311–318 (2002)

28. Rajpurkar, P., Zhang, J., Lopyrev, K., Liang, P.: Squad: 100,000+ questions for machine comprehension of text. In: Proceedings of the 2016 Conference on Empirical Methods in Natural Language Processing, pp. 2383–2392 (2016)
29. Lai, J., Gan, W., Wu, J., Qi, Z., Philip, S.Y.: Large language models in law: a survey (2023). https://arxiv.org/abs/2312.03718
30. Shazeer, N.: Glu variants improve transformer. arXiv preprint arXiv:2002.05202 (2020)
31. Sottana, A., Liang, B., Zou, K., Yuan, Z.: Evaluation metrics in the era of gpt-4: reliably evaluating large language models on sequence to sequence tasks. In: The 2023 Conference on Empirical Methods in Natural Language Processing (2023)
32. Team, I.: Openhermes-2-5-mistral-7b (2024). https://github.com/inferless/OpenHermes-2-5-Mistral-7B. Accessed 15 May 2024
33. Touvron, H., A.: Llama 2: open foundation and fine-tuned chat models (2023)
34. Wang, G., Cheng, S., Zhan, X., Li, X., Song, S., Liu, Y.: Openchat: advancing open-source language models with mixed-quality data. In: The Twelfth International Conference on Learning Representations (2023)
35. Wei, F., Chen, X., Luo, L.: Rethinking generative large language model evaluation for semantic comprehension. arXiv e-prints pp. arXiv–2403 (2024)
36. Xu, C., et al.: Wizardlm: empowering large language models to follow complex instructions. arXiv preprint arXiv:2304.12244 (2023)
37. Yue, L., et al.: Fedjudge: federated legal large language model. arXiv preprint arXiv:2309.08173 (2023)
38. Yue, S., et al.: Disc-lawllm: fine-tuning large language models for intelligent legal services. arXiv preprint arXiv:2309.11325 (2023)
39. Zhang, R., et al.: Evaluation ethics of llms in legal domain. arXiv preprint arXiv:2403.11152 (2024)
40. Zhang, T., Kishore, V., Wu, F., Weinberger, K.Q., Artzi, Y.: Bertscore: evaluating text generation with bert (2020)
41. Zheng, L., et al.: Judging llm-as-a-judge with mt-bench and chatbot arena. In: Thirty-Seventh Conference on Neural Information Processing Systems Datasets and Benchmarks Track (2023)

Explaining Biomarker Response to Anticoagulant Therapy in Atrial Fibrillation: A Study of Warfarin and Rivaroxaban with Machine Learning Models

Adriano Veloso[1,2], Gianlucca Zuin[1,2(✉)], and Luan Sena[3]

[1] Computer Science Department, Universidade Federal de Minas Gerais, Belo Horizonte, Brazil
{adrianov,gzuin}@dcc.ufmg.br
[2] Kunumi, Belo Horizonte, Brazil
{gianlucca,adriano}@kunumi.com
[3] Electrical Engineering Department, Universidade Federal de Minas Gerais, Belo Horizonte, Brazil
luanborges@ufmg.br

Abstract. Atrial fibrillation (AF) is a common arrhythmia that originates in the heart's upper chambers and can lead to serious complications like strokes and systemic embolism due to atrial thrombi. To mitigate these risks, anticoagulants such as warfarin and rivaroxaban are frequently prescribed. This study examines how known biomarkers behave under treatment with either warfarin or rivaroxaban. We developed high-performance models (AUROC ¿ 0.94) to distinguish patient subpopulations based on their treatment. Additionally, we built models (AUROC ¿ 0.97) to differentiate individuals with AF from a control group without the condition. Using synthetic data generation for training augmentation and explainable machine learning techniques, we analyzed biomarker behavior, uncovering distinct patterns based on whether patients received warfarin or rivaroxaban. Our approach provides valuable insights into the critical factors influencing biomarker variations across different treatments, enhancing our understanding of their roles in AF management.

Keywords: Atrial Fibrillation · Explainable Machine Learning · Medical Data Analysis

1 Introduction

Atrial fibrillation (AF) is the most prevalent cardiac arrhythmia, impacting millions of people worldwide [31]. Its prevalence escalates with age, affecting approximately 10% of individuals by the age of 80 [12]. Additionally, underlying cardiac conditions, such as coronary artery disease, significantly heighten the risk of developing AF [12]. AF is characterized by erratic electrical impulses in the atria, leading to an irregular and rapid heart rhythm that disrupts blood flow, causing stasis and increasing the risk of thromboembolic events such as ischemic stroke and systemic embolism [16]. This condition is associated with a fourfold increase in mortality risk, as evidenced by a population-based study in Korea [19].

The primary goal in managing AF is to reduce the risk of thromboembolism and stroke, primarily through anticoagulation therapy [3,17,20]. Warfarin, a traditional anticoagulant, has been extensively used for this purpose but poses challenges due to its food and drug interactions and the need for frequent monitoring [26]. Maintaining therapeutic levels of warfarin requires regular international normalized ratio (INR) checks and frequent dose adjustments to mitigate bleeding risks. Recently, direct oral anticoagulants (DOACs), such as rivaroxaban, have emerged as more favorable alternatives due to their improved pharmacological profiles [1,26].

Machine learning has become a valuable tool in healthcare [9], aiding in the development of robust risk prediction algorithms and enhancing our understanding of disease-related features to improve patient outcomes [33]. These techniques are particularly effective in uncovering patterns and associations in large datasets that may be challenging to discern manually [28]. However, the application of machine learning in healthcare demands a focus on explainability due to the high-stakes nature of medical decision-making.

Explainability involves articulating the internal mechanisms and outputs of an algorithm in a way that is comprehensible to humans. For healthcare professionals to trust and effectively utilize these tools in clinical settings, they must understand the rationale behind a model's predictions. For example, when a machine learning model forecasts a high risk of stroke in an AF patient, clinicians need to grasp the factors influencing this prediction to make well-informed treatment decisions [14,21]. Trust in these models is built through transparency and interpretability, aligning model predictions with clinical knowledge and patient data [6]. Similarly, clear and transparent communication of these predictions fosters patient adherence to treatment plans [10].

In this study, we built a dataset comprising biomarkers from 195 individuals, including 109 in the control group (without AF) and 86 patients with AF. Among the AF patients, 47 were using warfarin and 39 were using rivaroxaban. The biomarkers used comprise patient data from characterization, blood count, lipid profile, coagulation, inflammatory and cardiac diseases of the individuals. We built high-performance machine learning models using this data (i) to identify biomarkers that differentiate individuals with AF from those without, and (ii) to compare biomarker responses between patients on different anticoagulant therapies (warfarin × rivaroxaban). Additionally, we conducted ablation studies to better understand the significance of two major biomarker groups (coagulation and inflammatory) given their known strong correlation with AF [11,23]. In summary, we utilize synthetic data generation [32], machine learning algorithms [18], and explainable machine learning [22] to create a framework that provides insights into AF pathophysiology and the effects of anticoagulant therapy.

2 Related Work

Machine learning is changing healthcare by enabling the analysis of vast amounts of patient data to uncover patterns and insights that improve diagnosis, treatment, and patient outcomes [2,29,34]. In [25], authors used convolutional neural networks to predict the evolution of Alzheimer's disease, and explainable machine learning to explain patterns within this evolution. In [5], authors employed Raman Spectroscopy to identify traces of proteins and developed machine learning models to find which proteins are more related to melanoma. In [4], authors developed a machine learning model for predicting the risk of death in COVID-19 patients. These models employ blood count data and present high prognostic performance. Authors in [7] developed a new ensemble algorithm to predict the evolution of pain relief in patients suffering from chronic pain.

Specific patterns in patients with AF were already studied using typical procedures in the biomedical literature. In [11], authors showed that platelet and coagulation activation profiles in patients with AF were reduced by warfarin or rivaroxaban use and that patients with AF using rivaroxaban were less hypocoagulable than patients using warfarin. This means that endogen thrombin potential was less reduced in patients using rivaroxaban compared to those taking warfarin. Similarly, authors in [23] showed that patients with AF (both warfarin and rivaroxaban groups) presented increased levels of inflammatory cytokines in comparison with controls. The use of rivaroxaban was associated with decreased levels of inflammatory cytokines in comparison with warfarin. On the other hand, patients with AF using rivaroxaban presented increased levels of the chemokines (MCP-1 in comparison with warfarin users; MIG and IP-10 in comparison with controls). Finally, authors in [8] showed that Hp levels and Hp1-Hp2 polymorphism are not associated with AF, and authors in [24] provided a basis for the study of inflammatory markers that had not yet been well addressed in AF, especially IP-10, besides supporting evidence about molecules that had previously been associated with the disease.

We employ the aforementioned findings to select the biomarkers to be included in our models. To the best of our knowledge, our study is the first to use machine learning models to provide a deep understanding of AF in terms of its known biomarkers and given the possible therapies.

3 Materials and Methods

Let D be a dataset comprising biomarkers from patients, including a control group without AF and patients diagnosed with AF. Specifically, D consists of individuals in the control group and AF patients who are either on warfarin or rivaroxaban. Further, let $X = \{x_1, x_2, \ldots, x_n\}$ be a feature vector containing the biomarker values for each individual. Missing values in numerical features are imputed using the mean of the feature within the corresponding class, and categorical features are imputed using the mode of the feature within the same class. Formally, for a numerical feature x_i in class C_k, the imputation is given by:

$$x_i^{(j)} = \frac{1}{|C_k|} \sum_{x_i \in C_k} x_i \tag{1}$$

For a categorical feature x_i in class C_k, the imputation is:

$$x_i^{(j)} = \text{mode}(x_i \in C_k) \tag{2}$$

The task is formulated as a binary classification problem, in which we are concerned with two distinct contexts: classifying individuals based on the presence or absence of AF, and classifying AF patients based on their medication (warfaring × rivaroxaban). The goal is to learn a function $f(X; \phi) \to \{0, 1\}$, where X represents the biomarker features and ϕ represents the learning parameters of the model. We perform ablation experiments to evaluate and better understand the impact of different biomarker subsets on classification performance. Thus, the evaluated models are trained on three different sets of variables:

- All biomarkers are used in our models.
- All biomarkers, excluding coagulation biomarkers.
- All biomarkers, excluding inflammatory biomarkers.

This ablation approach allows us to understand how different groups of biomarkers contribute to the classification task and identify which features are more informative for distinguishing between the patient groups.

3.1 Data

The dataset for our analysis comprises biomarkers from 195 individuals, including 109 in the control group (without AF) and 86 patients with AF. Among the patients, 47 were using warfarin and 39 were using rivaroxaban. The biomarkers used comprise patient data from characterization, blood count, lipid profile, coagulation, inflammatory and cardiac diseases of the individuals and can be found in Table 1.

Prior to further analysis, missing feature values for patients were imputed based on their respective classes (control, rivaroxaban, warfarin). Our data was divided into two distinct datasets for separate analysis. The first subset D_1 segregates individuals based on the presence or absence of atrial fibrillation:

$$y_i = \begin{cases} 1 & \text{if AF is present} \\ 0 & \text{if AF is absent} \end{cases} \tag{3}$$

The second subset, D_2, includes only AF patients and further classifies them based on their medication:

Table 1. Biomarkers used to build our machine learning models.

Group	Variable
Characterization	HAS (Systemic Arterial Hypertension)
	DM (Diabetes Mellitus)
	Dyslipidemia (hypercholesterolemia)
Blood Count	PLT (Platelets, THSD/CU MM)
Lipid Profile	TRIG (Triglycerides, mg/dL)
	Total Cholesterol (mg/dL)
	HDLc (High-density Lipoprotein, mg/dL)
	LDLc (mg/dL)
Coagulation	F1+2 (Prothrombin Fragment 1+2)
	TAFI (Thrombin Activatable Fibrinolysis Inhibitor)
	t-PA (Tissue-Type Plasminogen Activator)
	PAI-1 (Plasminogen Activator Inhibitor-1)
	MPP/ul
	MPEs/ul
	Lagtime_(min)
	ETP_(nMmin)
	Peak_(nM)
	ttPeak_(min)
Inflammatory	IP-10 (Interferon-gamma-induced Protein, pg/mL)
	MCP-1 (Monocyte Chemoattractant Protein, pg/mL)
	MIG (Monokine Induced by Interferon-gamma, pg/mL)
	RANTES (regulated on Activation, Normal T Cells Expressed and Secreted, pg/mL)
	IL-8 (Interleukin, pg/mL)
	IFN (Interferon, pg/mL)
	TNF (Tumor Necrosis Factor, pg/mL)
	IL-6 (Interleukin, pg/mL)
	IL-4 (Interleukin, pg/mL)
	IL-2 (Interleukin, pg/mL)
	IL-10 (Interleukin, pg/mL)
	TGFβ (Human Transforming Growth Factor-beta, pg/mL)
	Haptoglobin (Hp)
	Genotype Hp
Cardiac Diseases	GDF-15 (Growth Differentiation Factor)
	sICAM-1 (Soluble Intercellular Adhesion Molecule)

$$y_i = \begin{cases} 1 & \text{if the patient is on warfarin} \\ 0 & \text{if the patient is on rivaroxaban} \end{cases} \qquad (4)$$

As previously mentioned, for numerical features imputation was performed using the mean value of the feature within the class in either D_1 or D_2. For categorical features, the mode of the values within the class was used. By imputing missing values based on class-specific statistics we obtain results more closely related with the underlying data distribution, resulting in more realistic and reliable values.

Synthetic Data and Training Augmentation: To improve the visualization and analysis of explanatory factors, we generate synthetic data and create an expanded dataset D' using two techniques: Gaussian Copula [27,30] and Tabular Variational Autoencoder (TVAE) [32]. The Gaussian Copula method models the dependency structure between variables by transforming the original data into a space where the variables follow a multivariate normal distribution. This involves computing the empirical cumulative distribution function for each variable and transforming them using their inverse of the standard normal to produce data that follows a standard normal distribution. We then proceed to estimate the covariance matrix of these transformed variables to capture their dependencies. New samples are generated from a multivariate normal distribution and are transformed back to the original space using the inverse of the initial transformations. Formally, let $X = (X_1, X_2, \ldots, X_n)$ represent the original data and $U = (U_1, U_2, \ldots, U_n)$ be the transformed variables that follow a standard normal distribution. The Gaussian Copula C is defined as:

$$C(u_1, u_2, \ldots, u_n) = \Phi_\Sigma(\Phi^{-1}(u_1), \Phi^{-1}(u_2), \ldots, \Phi^{-1}(u_n)) \quad (5)$$

in which Φ represents the CDF of the standard normal distribution, Φ^{-1} its inverse, and Φ_Σ is the CDF of the multivariate normal distribution with the covariance matrix.

In contrast, TVAE uses a variational autoencoder to learn the underlying data distribution and generate synthetic samples. It consists of an encoder and a decoder structure, in which the encoder $q_\phi(Z|X)$ maps the input data X to a latent representation Z using a neural network characterized by a mean μ and variance σ^2. Formally:

$$q_\phi(Z|X) = \mathcal{N}(Z; \mu(X), \sigma^2(X)) \quad (6)$$

The decoder $p_\theta(X|Z)$ then maps the latent representation Z back to the original data space, attempting to reconstruct the original data $\hat{X}$:

$$p_\theta(X|Z) = \mathcal{N}(X; \hat{\mu}(Z), \hat{\sigma}^2(Z)) \quad (7)$$

We create synthetic data points by sampling from the latent space and feeding it to the decoder. For each technique, synthetic samples equivalent to 33% of the total number of individuals in each subset are generated, resulting in the expanded dataset D'. While synthetic data generation can raise concerns about the representativeness of the data distribution and the potential for introducing bias in real-world applications [15], validating the model on real datasets helps mitigate these issues and ensures its reliability in real-world scenarios.

3.2 Model Training

For each comparison, the model is trained on three different sets of variables: one including all biomarkers, one excluding coagulation markers, and one excluding inflammatory markers. The target variable corresponds to the group that the individual belongs to. Our models were built using an implementation of the LightGBM algorithm [18]. It works by generating a model made up of hundreds of simple decision trees, these trees are integrated into a unified model through boosting [13].

Explainability: As a key point in any health domain application, we employ explainability tools to better understand the decision-making process of the model. Specifically, we employ SHAP [22]. We represent how model $f(X; \phi)$ explains a phenomenon as a d-dimensional vector $E(f(X)) = e_1, e_2, ..., e_d$ showing which features are contributing to the model's prediction. Specifically, e_i takes a value that corresponds to the influence that the respective feature x_i had on the model decision. This influence is quantified by calculating SHAP values, which provide a unified measure of feature importance across different instances.

SHAP values are calculated using Shapley values from cooperative game theory, ensuring that each feature's contribution to the prediction is fairly attributed. For each prediction, the SHAP values are computed by considering all possible combinations of feature inputs, thus capturing the complex interactions between features. In particular, SHAP contains three desirable properties:

- Additivity: the explanations are truthfully explaining the model and each evaluated instance. This means that the sum of the SHAP values for all features equals the model output for that instance, ensuring that the attributions are consistent with the model's actual predictions.
- Consistency: if a model changes such that some feature's contribution increases or stays the same regardless of the other features, that feature's attribution should not decrease. This property ensures that SHAP values remain stable and interpretable even if the underlying model changes.
- Missingness: A missing feature gets an attribution of zero. That is, a feature not present in a model does not impact its output or explanation.

4 Experimental Results

Model performance was evaluated by the area under the receiver operating characteristic curve (AUROC). It is used to show how much the model is capable of distinguishing between competing classes (i.e., patient groups). In our experiments we employed stratified 5-fold cross-validation by dividing the dataset into five parts and, at each iteration, we use four folds for training and the remaining one as the testing set for the model. No hyperparameter tuning was performed. Following, we discuss the results of each study.

4.1 Atrial Fibrillation and Control Group

Table 2 shows the model performance in differentiating AF patients from the control group, with all models achieving high AUROC values. Literature suggests that physiological patterns in AF patients are markedly different from those in the control group, which the trained models were able to learn. Given the high performance of our models, our main interest lies in understanding these distinctive patterns. We used SHAP values for this analysis.

Table 2. Model performance on atrial fibrillation and control group classification.

	ROC AUC Average ± SD
Full Dataset	0.999 ± 0.001
Without Coagulation Biomarkers	0.989 ± 0.017
Without Inflammation Biomarkers	0.981 ± 0.023

When using SHAP to find the most significant features using all biomarkers, we found a great impact of IL-4, Peak, MPEs/ul, IL-2, ETP, F1+2, LDLc, IFN, TRIG, IL-10, TGFβ, and HDLc in model prediction (Fig. 1 (Top)). If we exclude the coagulation biomarkers the most important features are IL-4, IL-10, IFN, Total Cholesterol, IL-6, sICAM-1, LDLc, MCP-1, IL-2, IL-8, GDF-15, and HDLc (Fig. 1 (Middle)). Furthermore, without inflammation biomarkers the most impacting features are Peak, TAFI, F1+2, Lagtime, TRIG, MPEs/ul, PAI-1, Total Cholesterol, sICAM-1, GDF-15, MPP/ul and ETP (Fig. 1 (Bottom)).

These findings suggest that inflammation, lipid metabolism, and coagulation processes are all critical in the model's predictions. Without coagulation biomarkers, cytokines and lipid measures remain important, indicating that inflammation and lipid metabolism are still significant. Additional markers like sICAM-1 and MCP-1 become more prominent, highlighting the role of inflammation and immune response. Without inflammation biomarkers, coagulation and lipid measures (such as TAFI, F1+2, Lagtime, PAI-1) are significant. This shift underscores the importance of blood clotting mechanisms and lipid metabolism.

4.2 Warfarin × Rivaroxaban

Table 3 shows the model performance in differentiating AF patients on warfarin from AF patients on rivaroxaban. Again, all models achieved very high AUROC values, and given the high performance of our models, our main interest lies in understanding the distinctive patterns in both groups. We used SHAP values for this analysis.

When using SHAP to find the most significant features using all biomarkers, we found a great influence of IL-2, F1+2, ETP, GDF-15, sICAM-1, RANTES, TNF, t-PA, IL-4, TGFβ, PLT, and IL-8 in model prediction (Fig. 2 (Top)). When

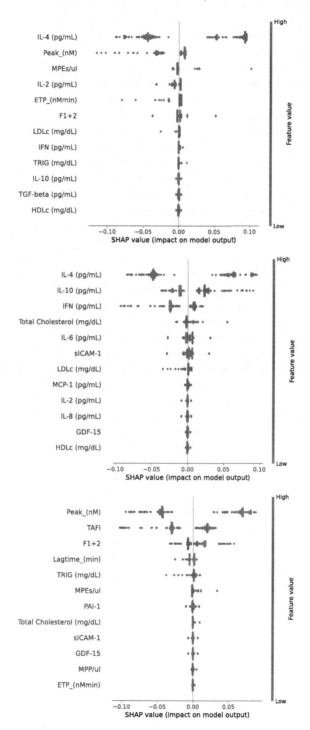

Fig. 1. (Top) All features. (Middle) All, but coagulation biomarkers. (Bottom) All, but inflammatory biomarkers

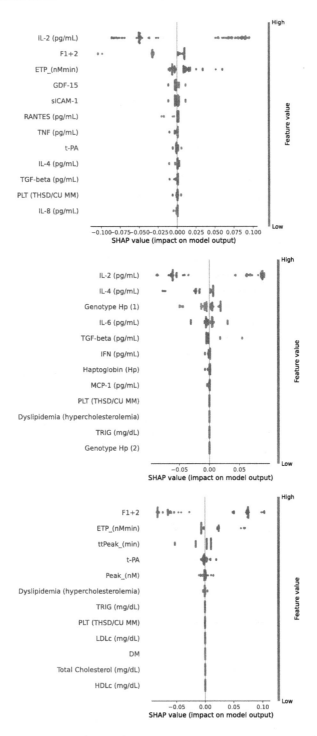

Fig. 2. (Top) All features. (Middle) All, but coagulation biomarkers. (Bottom) All, but inflammatory biomarkers

Table 3. Model performance on rivaroxaban and warfarin group classification.

	ROC AUC Average ± SD
Full Dataset	0.986 ± 0.027
Without Coagulation Biomarkers	0.945 ± 0.029
Without Inflammation Biomarkers	0.971 ± 0.036

excluding the coagulation biomarkers the most impacting features became IL-2, IL-4, Genotype Hp (1), IL-6, TGFβ, IFN, Haptoglobin (Hp), MCP-1, PLT, Dyslipidemia (hypercholesterolemia), TRIG and HAS (Fig. 2 (Middle)). Furthermore, without inflammation biomarkers the most impating features are F1+2, ETP, ttPeak, t-PA, Peak, Dyslipidemia (hypercholesterolemia), TRIG, PLT, Total Cholesterol, HDLc, LDLc, and TAFI (Fig. 2 (Bottom)).

5 Conclusion

This study underscores the potential of machine learning models in providing a deeper understanding of atrial fibrillation (AF). By leveraging ML models, we successfully identified and analyzed key biomarkers that differentiate AF patients from healthy individuals and elucidate the effects of different anticoagulant therapies. Our high-performance models demonstrate robust predictive capabilities, offering significant insights into the distinct biomarker patterns associated with warfarin and rivaroxaban treatments. The application of explainable machine learning methods enabled us to uncover critical factors influencing biomarker behavior, providing a deeper understanding of the pathophysiological processes in AF and the impacts of specific treatments (we can clearly see patients using rivaroxaban as having lower inflammation markers in contrast to those using warfarin). These insights contribute to a more nuanced view of how anticoagulant therapy affects biomarker profiles, which can inform clinical decision-making and optimize treatment strategies. Finally, synthetic data generation proved invaluable in augmenting our dataset, enhancing model robustness, and ensuring reliable predictions despite potential limitations in real-world data availability. This approach highlights the efficacy of combining data augmentation techniques with machine learning to overcome data scarcity issues common in clinical research.

Disclosure of Interests. The authors have no competing interests to declare that are relevant to the content of this article.

References

1. Alberts, M., Chen, Y.W., Lin, J.H., Kogan, E., Twyman, K., Milentijevic, D.: Risks of stroke and mortality in atrial fibrillation patients treated with rivaroxaban and warfarin. Stroke **51**(2), 549–555 (2020)

2. Amador, T., Saturnino, S., Veloso, A., Ziviani, N.: Early identification of ICU patients at risk of complications: regularization based on robustness and stability of explanations. Artif. Intell. Med. **128**, 102283 (2022)
3. Amin, A., Houmsse, A., Ishola, A., Tyler, J., Houmsse, M.: The current approach of atrial fibrillation management. Avicenna J. Med. **6**(01), 8–16 (2016)
4. Araújo, D.C., Veloso, A.A., Borges, K.B.G., das Graças Carvalho, M.: Prognosing the risk of COVID-19 death through a machine learning-based routine blood panel: a retrospective study in brazil. Int. J. Med. Inf. **165**, 104835 (2022)
5. Araújo, D.C., et al.: Finding reduced raman spectroscopy fingerprint of skin samples for melanoma diagnosis through machine learning. Artif. Intell. Med. **120**, 102161 (2021)
6. Bayer, S., Gimpel, H., Markgraf, M.: The role of domain expertise in trusting and following explainable ai decision support systems. J. Decis. Syst. **32**(1), 110–138 (2022)
7. Costa, A.B.D., Moreira, L., Andrade, D.C.D., Veloso, A., Ziviani, N.: Predicting the evolution of pain relief: Ensemble learning by diversifying model explanations. ACM Trans. Comput. Heal. **2**(4), 36:1–36:28 (2021)
8. Costa, L.B.X., et al.: Evaluation of serum haptoglobin levels and hp1-hp2 polymorphism in the haptoglobin gene in patients with atrial fibrillation. Molec. Biol. Rep. **49**, 7359–7365 (2022)
9. Deo, R.C.: Machine learning in medicine. Circulation **132**(20), 1920–1930 (2015)
10. DiMatteo, M.R., et al.: Physicians' characteristics influence patients' adherence to medical treatment: results from the medical outcomes study. Health Psychol. **12**(2), 93 (1993)
11. Duarte, R.C.F., et al.: Thrombin generation and other hemostatic parameters in patients with atrial fbrillation in use of warfarin or rivaroxaban. J. Thrombosis Thrombolysis **51**, 47–57 (2021)
12. Fang, M.C., Chen, J., Rich, M.W.: Atrial fibrillation in the elderly. Am. J. Med. **120**(6), 481–487 (2007)
13. Freund, Y., Schapire, R.E.: A decision-theoretic generalization of on-line learning and an application to boosting. J. Comput. Syst. Sci. **55**(1), 119–139 (1997)
14. Gerlings, J., Jensen, M.S., Shollo, A.: Explainable ai, but explainable to whom? an exploratory case study of xai in healthcare. In: Handbook of Artificial Intelligence in Healthcare: Practicalities and Prospects, vol. 2 pp. 169–198 (2022)
15. Giuffrè, M., Shung, D.L.: Harnessing the power of synthetic data in healthcare: innovation, application, and privacy. npj Dig. Med. **6**(1), 186 (2023)
16. Hindricks, G., et al.: ESC Scientific Document Group: 2020 ESC guidelines for the diagnosis and management of atrial fibrillation developed in collaboration with the European association for Cardio-Thoracic surgery (EACTS): the task force for the diagnosis and management of atrial fibrillation of the European society of cardiology (ESC) developed with the special contribution of the European heart rhythm association (EHRA) of the ESC. Eur. Heart J. **42**(5), 373–498 (2021)
17. Joglar, J.A., et al.: 2023 acc/aha/accp/hrs guideline for the diagnosis and management of atrial fibrillation: a report of the American college of cardiology/American heart association joint committee on clinical practice guidelines. Circulation **149**(1), e1–e156 (2024)
18. Ke, G., et al.: Lightgbm: a highly efficient gradient boosting decision tree. In: Guyon, I., Luxburg, U.V., Bengio, S., Wallach, H., Fergus, R., Vishwanathan, S., Garnett, R. (eds.) Advances in Neural Information Processing Systems, vol. 30. Curran Associates, Inc. (2017)

19. Lee, E., et al.: Mortality and causes of death in patients with atrial fibrillation: a nationwide population-based study. PLoS ONE **13**(12), e0209687 (2018)
20. Lip, G.Y., Tse, H.F.: Management of atrial fibrillation. Lancet **370**(9587), 604–618 (2007)
21. Lötsch, J., Kringel, D., Ultsch, A.: Explainable artificial intelligence (xai) in biomedicine: making ai decisions trustworthy for physicians and patients. BioMed. Inf. **2**(1), 1–17 (2021)
22. Lundberg, S.M., Lee, S.I.: A unified approach to interpreting model predictions. In: Guyon, I., Luxburg, U.V., Bengio, S., Wallach, H., Fergus, R., Vishwanathan, S., Garnett, R. (eds.) Advances in Neural Information Processing Systems, vol. 30. Curran Associates, Inc. (2017)
23. Martins, G.L., et al.: Comparison of inflammatory mediators in patients with atrial fibrillation using warfarin or rivaroxaban. Front. Cardiovasc. Med. **7**, 47–57 (2020)
24. Martins, G.L., et al.: Evaluation of new potential inflammatory markers in patients with nonvalvular atrial fibrillation. Int. J. Molec. Sci. **24**, 3326 (2023)
25. Nigri, E., Ziviani, N., Cappabianco, F.A.M., Antunes, A., Veloso, A.: Explainable deep cnns for mri-based diagnosis of alzheimer's disease. In: 2020 International Joint Conference on Neural Networks, IJCNN 2020, pp. 1–8. IEEE (2020)
26. Patel, M.R., et al.: Rivaroxaban versus warfarin in nonvalvular atrial fibrillation. New Engl. J. Med. **365**(10), 883–891 (2011)
27. Patki, N., Wedge, R., Veeramachaneni, K.: The synthetic data vault. In: 2016 IEEE International Conference on Data Science and Advanced Analytics (DSAA), pp. 399–410 (2016)
28. Phillips, P.J., et al.: Four principles of explainable artificial intelligence (2021)
29. Silcox, C., et al.: The potential for artificial intelligence to transform healthcare: perspectives from international health leaders. npj Dig. Med. **7**(1) (2024)
30. Sun, Y., Cuesta-Infante, A., Veeramachaneni, K.: Learning vine copula models for synthetic data generation (2018)
31. Wyndham, C.R.: Atrial fibrillation: the most common arrhythmia. Tex. Heart Inst. J. **27**(3), 257 (2000)
32. Xu, L., Skoularidou, M., Cuesta-Infante, A., Veeramachaneni, K.: Modeling tabular data using conditional gan. In: Wallach, H., Larochelle, H., Beygelzimer, A., d'Alché-Buc, F., Fox, E., Garnett, R. (eds.) Advances in Neural Information Processing Systems, vol. 32. Curran Associates, Inc. (2019)
33. Zuin, G., et al.: Prediction of sars-cov-2-positivity from million-scale complete blood counts using machine learning. Commun. Med. **2**(1), 72 (2022)
34. Zuin, G., et al.: A modified louvain approach for medical community detection using geographic data. In: 2023 IEEE 36th International Symposium on Computer-Based Medical Systems (CBMS), pp. 143–148. IEEE (2023)

Author Index

A

Abreu, Pedro Henriques 200
Almeida, Diogo M. 17
Almeida, Renato 185
Althoff, Paulo Eduardo 156
Aluísio, Sandra M. 33
Alves, Edgard B. 444
Alves, Jorge A. 444
Amorim, Marcelo M. 430
Angonese, Silvio Fernando 383
Antonelo, Eric A. 80
Araújo, Flávio H. D. 399

B

Balreira, Dennis Giovani 127, 215
Barbosa, Luciano 169
Bernardino, Heder 430
Boll, Antônio Oss 127, 247
Boll, Heloísa Oss 247
Bollis, Edson 185
Borges, Alex 430
Borges, Rodrigo N. 399
Braz, Camila Santana 368

C

Cação, Flávio 185
Cameron, Helena Freire 215
Camilo Junior, Celso Gonçalves 460
Candido Junior, Arnaldo 33, 340
Canuto, Anne Magaly de Paula 352
Carvalho, Levi Cordeiro 281
Claro, Maíla L. 399
Cruz, Michael 169
Cunha, Daniel C. 17

D

da Rosa, Augusto Seben 340
da Silva, Tiago 3

da Silva Junior, Waldir S. 64
de Mattos Neto, Paulo S. G. 17
de Oliveira, Douglas Amorim 231
de Paiva, Anselmo Cardoso 260, 324
de Paulo Faleiros, Thiago 156
de Souza, Gabriel 430
Delgado, Karina Valdivia 96, 231
dos Reis, Julio Cesar 414
dos Santos Silva, Fillipe 414
dos Santos, Eulanda M. 64

F

Farias, Giovani P. 48
França, Hellen Guterres 324
Freire, Valdinei 96
Freitas, Carla Maria Dal Sasso 127, 215

G

Galante, Renata 383
Gatto, Bernardo B. 64
Giusti, Rafael 296
Goldschmidt, Ronaldo R. 444

K

Kakimoto, Gabriel Kenzo 414
Koerich, Alessandro L. 64

L

Lauretto, Marcelo de Souza 231
Leal, Sidney E. 33
Lima, Rodrigo 33
Lira, Thiago 185
Lorena, Ana Carolina 141, 200
Luz, Eduardo 111

M

Machado, Vinicius P. 399
Magalhães, Marcio 185

Malossi, Rodrigo Mor 247
Mangussi, Arthur Dantas 200
Matos, Helder 270
Maurício, João Stephan 430
Mesquita, Diego 3
Mollinetti, Marco A. F. 64
Moreira, Benjamin Grando 310
Mota, Marcelle 270

N
Negrão, Arthur 111
Neto, Carlos Mendes dos Santos 324
Nunes, Rafael Oleques 127, 215

O
Olival, Fernanda 215
Oliveira, Andre 185
Oliveira, Cristiano S. 80
Oliveira, Frederico S. 340
Oliveira, Otavio 185
Oliveira, Saulo A. F. 281
Oliveira, Sávio Salvarino Teles de 460
Oliveira, Vitor H. T. 80

P
Panisson, Alison R. 48
Pappa, Gisele Lobo 368
Pedrosa, Rodrigo 111
Pellicer, Lucas 185
Pereira, Ricardo Cardoso 200
Pessoa, Alexandre César Pinto 324
Poloni, Katia 185
Prata, Leonardo 430
Presa, João Paulo Cavalcante 460
Puttlitz, Letícia Maria 127, 247

Q
Quintanilha, Darlan Bruno Pontes 324

R
Reis, Marcelo S. 414
Reis, Willy Arthur Silva 96
Rivolli, Adriano 141
Rocha, Thiago Alves 281

S
Santos, Araken de Medeiros 352
Santos, Eulanda Miranda dos 296
Santos, Joaquim 215
Santos, Reginaldo 270
Schulz, Hans Herbert 310
Sena, Luan 475
Silva, Anderson Lopes 324
Silva, Aristófanes Corrêa 260, 324
Silva, Guilherme 111
Silva, Jesaías Carvalho Pereira 352
Silva, Lucas Almeida da 296
Silva, Maxwell Pires 260
Silva, Pedro 111
Silva, Romuere R. V. 399
Soares, Anderson da Silva 340
Sousa, Leonardo P. 399
Souza, Cinthia 185
Spritzer, Andre 127, 215

T
Tavares, Anderson Rocha 127
Teixeira, Matheus Cândido 368
Toledo, Rafael S. 80

U
Ueda, Patricia S. M. 141

V
Valejo, Alan Demétrius Baria 156
Valentini, João 185
Veloso, Adriano 475
Veras, Rodrigo M. S. 399
Viana, Joaquim 270
Vieira, Renata 215
Virgilli, Rafaello 340

W
Wangenheim, Aldo von 80

Z
Zuin, Gianlucca 475

Printed in the United States
by Baker & Taylor Publisher Services